T0256159

5G Technology

5G Technology

3GPP Evolution to 5G-Advanced

Second Edition

Edited by

Harri Holma and Antti Toskala
Nokia
Finland

Takehiro Nakamura
NTT DOCOMO
Japan

Registered Offices
John Wiley & Sons, Inc., 111 River Street, Hoboken, NJ 07030, USA
John Wiley & Sons Ltd, The Atrium, Southern Gate, Chichester, West Sussex, PO19 8SQ, UK

For details of our global editorial offices, customer services, and more information about Wiley products visit us at www.wiley.com.

Wiley also publishes its books in a variety of electronic formats and by print-on-demand. Some content that appears in standard print versions of this book may not be available in other formats.

Library of Congress Cataloging-in-Publication Data applied for:
Hardback ISBN: 9781119816034

Cover Design: Wiley
Cover Image: TK

Set in 9.5/12.5pt STIXTwoText by Straive, Chennai, India
Printed and bound by CPI Group (UK) Ltd, Croydon, CR0 4YY

C9781119816034_060224

Contents

About the Editors

Harri Holma
Bell Labs Fellow, Nokia Technology Office, Espoo, Finland

Harri Holma joined Nokia Research Center in 1994 and received his M.Sc. from Helsinki University of Technology in 1995. He has been located both in Finland and in the United States during that time. Harri Holma is currently working as a fellow and advisor in the Technology Leadership Office in Nokia, with a special interest in radio technologies and mobile networks.

He completed a Doctor of Technology degree from Helsinki University of Technology (now Aalto University) in 2003. Dr. Holma has edited the books "WCDMA for UMTS," "HSDPA/HSUPA for UMTS," "LTE for UMTS," "Voice over LTE," "LTE Advanced," "HSPA+ Evolution," "LTE Small Cell Optimization," and "5G Technology," and has also contributed to a number of other books in the radio communication area.

Antti Toskala
Bell Labs Fellow, Nokia Standards, Espoo, Finland

Antti Toskala (M.Sc.) joined the Nokia Research Center in 1994, where he undertook WCDMA system studies as a research engineer and later as a senior research engineer and CDMA specialist.

He chaired the UMTS physical layer expert group in ETSI SMG2 in 1998, and from 1999 until 2003, he worked in 3GPP as chairman of the TSG RAN WG1. From 2003 to 2005, he worked as a senior standardization manager with system technologies at Nokia Networks and contributed to product development as the HSDPA chief architect.

From 2005 onwards, he worked with Nokia Networks as a senior standardization manager, focusing on HSPA and LTE standardization, and later as head of radio standardization with Nokia Siemens Networks, focusing on LTE and LTE-Advanced work in 3GPP.

He has coauthored eight books in 3G, 4G, and 5G, with further editions from many of them. As part of the 2010 LTE World Summit LTE Awards, he received the "Award for Individual Contribution for LTE Development," recognizing his contribution to both LTE standardization and LTE knowledge spreading in the industry.

He was nominated as a Nokia fellow in 2015, and a Bell Lab fellow in 2016. Currently, he is with Nokia Standards, in Espoo, Finland, heading Nokia 3GPP RAN standardization, with technical focus on 5G and 5G-Advanced towards 6G.

Takehiro Nakamura

Mr. Takehiro Nakamura joined NTT Laboratories in 1990. He is now chief standardization officer at NTT DOCOMO, Inc.

Mr. Nakamura has been engaged in R&D and the standardization activities for advanced radio and network technologies of W-CDMA, HSPA, LTE/LTE-Advanced, 5G, and 6G, and has been involved in strengthening inter-industry collaboration.

He has been contributing to standardization activities in ARIB, ITU, and 3GPP since 1997, including as vice-chair and chair of 3GPP TSG-RAN from 2005 to 2013.

Currently, he plays important role in promoting and accelerating 5G and 6G in Japan and globally as the acting chairman of strategy and planning committee, the leader of millimeter wave promotion ad hoc of 5G mobile communications promotion forum (5GMF), the leader of cellular system task group of ITS info-communications forum, the leader of white paper subcommittee in beyond 5G promotion consortium in Japan, and the board member of 5G-ACIA.

List of Contributors

Atarashi Hiroyuki
NTT Docomo
Japan

Brunel Dominique
Skyworks
France

Chen YihShen
MediaTek
Taiwan

Enescu Mihai
Nokia
Finland

Guey Jiann-Ching
MediaTek
Taiwan

Henttonen Tero
Nokia
Finland

Holma Harri
Nokia
Finland

Honkala Mikko
Nokia Bell Labs
Finland

Huttunen Janne
Nokia Bell Labs
Finland

Hwang Chienhwa
MediaTek
Taiwan

Imai Tetsuro
NTT Docomo
Japan

Iwamura Mikio
NTT Docomo
Japan

Jayasinghe Keeth
Nokia
Finland

Kalyanasundaram Suresh
Nokia
India

Kela Petteri
Nokia
Finland

Kishiyama Yoshihisa
NTT Docomo
Japan

Korpi Dani
Nokia Bell Labs
Finland

Koskela Jarkko
Nokia
Finland

Lauridsen Mads
Nokia
Denmark

Liao PeiKai
MediaTek
Taiwan

Lunttila Timo
Nokia
Finland

Luostari Riku
Nokia
New Zealand

Metsälä Esa
Nokia
Finland

Nagata Satoshi
NTT Docomo
Japan

Nakamura Takehiro
NTT Docomo
Japan

Noël Laurent
Skyworks
Canada

Poikselkä Miikka
Nokia
Finland

Ranta-aho Karri
Nokia
Finland

Ratasuk Rapeepat
Nokia
USA

Reunanen Jussi
Nokia
Thailand

Salmelin Juha
Nokia
Finland

Schober Karol
Nordic Semiconductors
Finland

Sébire Benoist
Nokia
Japan

Sébire Guillaume
MediaTek
Finland

Thepchatri Puripong
Nokia
Thailand

Toskala Antti
Nokia
Finland

Uitto Tommi
Nokia
Finland

Vasenkari Petri
Nokia
Finland

Venkatesan Venkat
Nokia Bell Labs
USA

Wu Weide
MediaTek
Taiwan

Yang Weidong
MediaTek
USA

Foreword

The last 25 years have witnessed an impressive evolution in global mobile communications from voice-centric 1G/2G services to 3G data services starting in 2001, 4G mobile broadband starting in 2010, and 5G starting in 2020. In Japan, NTT DOCOMO launched the world's first mobile internet service "i-mode" in February 1999 on the PDC-Packet(2G) system, and the service spread widely in Japan in the 3G era. Smartphones were introduced on the 3G system and have spread in the 4G era. The number of mobile subscribers and devices have increased rapidly and data usage has exploded. Data speed has grown from 2.4 kbps in 1993 to beyond 4 Gbps in 2022; that is a 2,000,000-fold increase in less than 30 years. I have been lucky to be part of this evolution in NTT DOCOMO, to have seen firsthand such great achievements and the great challenges the industry has gone through to manage the fast changes. We are now witnessing the beginning of the next major step in the evolution – that is 5G. We believe that 5G is one of the main pillars to provide value and excitement to customers and opportunities for value co-creation with partners in the 2020s. 5G also brings a number of new challenges for the industry to fulfill all these promises.

NTT DOCOMO has been one of the leading companies in developing advanced technologies and services for new-generation mobile communication systems including 5G. NTT DOCOMO launched pre-commercial 5G services in September 2019 and fully commercial 5G services in March 2020. We've been conducting aggressive activities to create new services towards the 5G era, such as the DOCOMO 5G Open Partner Program, and we believe that many new enterprise services can be created together with a variety of industry partners and that the market size of enterprise services can be expanded drastically utilizing 5G characteristics.

This book provides a combination of 5G specification explanations, performance evaluations, device design aspects, practical deployment considerations, and field experiences, which will help to roll out and tune 5G networks. The book has been written by more than 30 experts from multiple companies around the world, including network infrastructure and chipset vendors and mobile operators.

My colleague Takehiro Nakamura has been a key contributor inside NTT DOCOMO and also globally in 5G standardization, development, and early testing. He has nearly 30 years of experience in the field. Harri Holma and Antti Toskala from Nokia bring worldwide knowledge of 5G technology and its details. Harri and Antti have 30 years of experience in the industry starting from the early days of 3G research.

I believe that you will find this book enjoyable and useful in helping you to enhance your understanding of the potential of 5G technology. I hope that we can witness together a successful evolution of 5G technology towards 2030.

Takaaki Sato
Executive Vice President, Chief Technology Officer
Executive General Manager of R&D Innovation Division
NTT DOCOMO, INC.

Preface

Long term evolution (LTE) networks were launched commercially in 2009, and the technology turned out to be hugely successful in boosting mobile broadband capabilities. Global mobile data traffic has grown by a factor of 10 during the last six years. LTE has enabled a large number of new applications in smartphones and has brought high-speed internet access to hundreds of millions of people who never had internet access earlier. 5G targets are set far beyond LTE in terms of technical capabilities and potential use cases. 5G is designed to provide ultra-reliable, low-latency communication which opens completely new application areas for enterprise communication, like remote control, or for consumer communication, like e-sports and cloud gaming. 5G will also boost mobile broadband performance to data rates beyond 10 Gbps. These impressive targets require new solutions for the 5G mobile networks including new spectrum options, new antenna structures, new physical layers, protocol designs, and new network architectures. A deep understanding of the underlying 5G technology allows us to take full benefit of new capabilities. This book describes details of 5G specifications and practical deployment aspects. The second edition of the book added more than 100 pages about 5G-Advanced which enhances 5G technology with new capabilities towards 2030. We hope you enjoy reading the book!

The contents of the book are summarized as follows: Chapters 1 and 2 provide an introduction to 5G targets and standardization organizations. Chapter 3 presents 5G spectrum options, and Chapter 4 network architecture options. The main new technology components in 5G are shortly presented in Chapter 5 and in more detail in the following chapters. Physical layer is described in Chapter 6 and radio protocols in Chapter 7. 5G network performance is also defined by the deployment aspects like site density and transport network. These topics are discussed in Chapters 8 and 9. The overall 5G performance aspects, including data rates, coverage, and latency, are presented in Chapter 10, followed by examples of practical field measurements in Chapter 11. Device design aspects are discussed in Chapter 12 for the radio frequency (RF) part, and in Chapter 13 for the baseband modem part. Internet of Things optimization is described in Chapter 14. The latest updates in 4G LTE evolution are presented in Chapter 15. The second edition added Chapters 16–21: Chapter 16 illustrates an overview of 5G-Advanced, Chapter 17 focuses on the radio enhancements in 5G evolution, and Chapter 18 on industrial Internet of Things evolution. 5G-Advanced also enables new use cases that are described in Chapter 19, open radio access network

(RAN) and cloud RAN aspects are considered in Chapter 20, and the utilization of artificial intelligence is discussed in Chapter 21.

January 18, 2024

Harri Holma and Antti Toskala
Espoo, Finland

Acknowledgment

The editors would like to acknowledge the hard work of the contributors from Nokia Bell Labs, Nokia, NTT Docomo, Mediatek, and Skyworks: Hiroyuki Atarashi, Dominique Brunel, YihShen Chen, Mihai Enescu, Jiann-Ching Guey, Tero Henttonen, Mikko Honkala, Janne Huttunen, Chienhwa Hwang, Mikio Iwamura, Suresh Kalyanasundaram, Petteri Kela, Yoshihisa Kishiyama, Dani Korpi, Jarkko Koskela, Mads Lauridsen, PeiKai Liao, Timo Lunttila, Riku Luostari, Esa Metsälä, Satoshi Nagata, Laurent Noël, Miikka Poikselkä, Karri Ranta-aho, Rapeepat Ratasuk, Jussi Reunanen, Juha Salmelin, Benoist Sébire, Guillaume Sébire, Satoshi Suyama, Puripong Thepchatri, Tommi Uitto, Petri Vasenkari, Venkat Venkatesan, Weide Wu, and Weidong Yang.

We also would like to thank the following colleagues for their valuable comments and contributions: Brian Cho, Frank Frederiksen, Hiroki Harada, Rauli Järvelä, Ahlem Klass, Matti Laitila, Danieala Laselva, Vesa Lehtinen, Henrik Liljeström, Pertti Lukander, Pekka Marjelund, Peter Merz, Amit Mukhopadhyay, Klaus Pedersen, Johanna Pekonen, Rauno Ruismäki, Juha Sipilä, Tami Stegmaier, Samuli Turtiainen, and Chunli Wu.

We appreciate the fast and smooth editing process provided by Wiley publishers and especially Sherlin Benjamin, Nandhini Karuppiah, Jeevaghan Devapal, and Sandra Grayson.

We are grateful to our families, as well as the families of all the authors, for their patience during the late-night writing and weekend editing sessions.

The editors and authors welcome any comments and suggestions for improvements or changes that could be implemented in forthcoming editions of this book. The feedback is welcome to editors' email addresses harri.holma@nokia.com, antti.toskala@nokia.com, and nakamurata@nttdocomo.com.

1

Introduction

Harri Holma[1], Antti Toskala[1], Takehiro Nakamura[2], and Tommi Uitto[1]

[1] Nokia, Finland
[2] NTT DOCOMO, Japan

CHAPTER MENU

1.1 Introduction

5G radio represents a major step in mobile network capabilities. So far, mobile networks have mainly provided connectivity for smartphones, tablets, and laptops for consumers. 5G will take traditional mobile broadband to the extreme in terms of data rates, capacity, and availability. Additionally, 5G will enable new services including industrial Internet of Things (IoT) connectivity and critical communication. 5G targets are set very high, with data rates up to 20 Gbps and capacity increases up to 1000 times, and also provides a flexible platform for device connectivity, ultra-low latency, and high reliability. A number of new use cases and applications can be run on top of 5G mobile networks. It is expected that 5G will fundamentally impact the whole society by improving efficiency, productivity, and safety. 4G networks were designed and developed more than ten years ago, mainly by telecom operators and vendors for the smartphone use case. There is a lot more interest in 5G networks by other parties, including different

5G Technology: 3GPP Evolution to 5G-Advanced, Second Edition.
Edited by Harri Holma, Antti Toskala, and Takehiro Nakamura.
© 2024 John Wiley & Sons Ltd. Published 2024 by John Wiley & Sons Ltd.

industries and communities, to understand 5G capabilities and to take full benefit of 5G networks. 4G was about connecting people. 5G is about connecting everything.

5G has the ingredients to have a much more profound impact on society and enterprises compared to earlier mobile technology generations, relatively speaking. First, 2G, 3G, and 4G were predominantly about people connectivity – enabling persons to call one another, or access the internet from virtually anywhere, anytime. 5G, with its capabilities for Ultra Reliable Low Latency Communication (URLLC) connectivity, has been designed from the outset for high-performance IoT. 5G will enable operators to help their corporate customers to automate their business processes. It is worth noting that productivity improvements in physical business processes, such as manufacturing, construction, and logistics, have been lagging in service industries that have been able to digitalize and automate their processes during the last decades. Hence, industry verticals have shown great interest in 5G and plan to build their own private dedicated wireless network or use operator spectrum and network with slicing technology.

Second, we are seeing an interesting coinciding of the three inflection points – 5G as a new radio standard, the proliferation of cloud concepts in wireless networks, and the rapidly increasing use of artificial intelligence (AI) and machine learning (ML). Hyper-successes often happen in business when several major inflection points coincide. An example of earlier success was the 2G Global System for Mobile Communications (GSM); the first ever global standard for mobile communication, deregulation of the operator field, and electronics component price reduction making mobile telephone accessible to the masses. The major inflection points around 5G technology are illustrated in Figure 1.1.

3G and 4G brought enhanced capability for data connectivity. There were some attractive technological improvements happening at the same time, including touch-screen devices, tablet form factors, and new business models with application stores. These inflection points boosted the success of 4G technology globally. IoT capability was added later on top of 4G, while IoT optimization was inbuilt into 5G from the beginning, providing better performance and economics. URLLC will cross the chain for robust low latency communication enabling, for example, wireless robots. Further massive Machine Type Communication (mMTC) makes it more economical than previous generations to connect a very large number of objects wirelessly to the network, than with previous generations.

5G system design and deployment need to be different from the earlier mobile network generations because of the new requirements. 4G solutions are not good enough to deliver the true 5G promises. This book describes 5G specifications, technologies, network architectures, and 5G deployment and optimization aspects as well as some practical aspects related to implementing 5G devices.

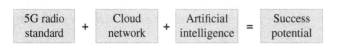

Figure 1.1 Major inflection points leading to potential success.

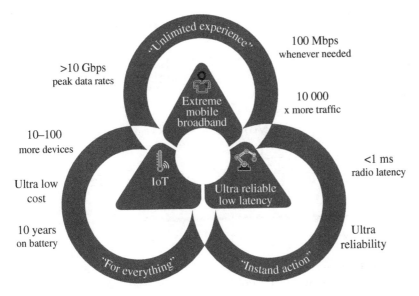

Figure 1.2 Main 5G targets.

1.2 5G Targets

5G targets are illustrated in Figure 1.1. The three main cornerstones are extreme mobile broadband, massive IoT communications, and Ultra Reliable Low Latency Communication (URLLC). Extreme mobile broadband focuses on higher data rates beyond 10 Gbps, more consistent data rates with a minimum capability of 100 Mbps everywhere, and 10.000× more traffic capacity. Massive IoT aims at optimizing networks and devices for connectivity with billions of low-cost devices with long battery lifetimes. The third corner targets to provide very low latency below 1 millisecond (ms) with ultra-high 99.9999% reliability (Figure 1.2).

1.3 5G Technology Components

The targets of 5G networks are beyond the capabilities of existing mobile networks. A number of new technologies are needed to fulfill all those targets. The main new technology components are shown in Figure 1.3.

1. *New spectrum*: 5G is the first mobile radio technology that is designed to operate on any frequency band between 400 MHz and 90 GHz. The low bands are needed for coverage and the high bands are for high data rates and capacity. The initial 5G deployments use Time Division Duplex (TDD) between 2.5 and 5.0 GHz, Frequency Division Duplex (FDD) below 2.5 GHz, and TDD at millimeter waves at 24–39 GHz.
2. *Massive Multiple Input Multiple Output (MIMO) beamforming* can increase spectral efficiency and network coverage substantially. Beamforming becomes more practical at higher frequencies because the antenna size is relative to the wavelength and the antenna size gets smaller at higher frequencies. In practice, massive MIMO can be

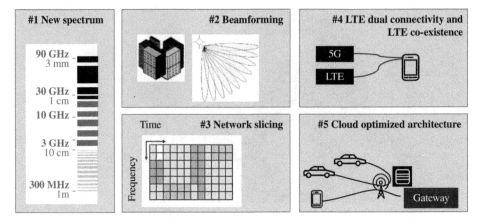

Figure 1.3 Key 5G technology components.

utilized at frequencies above 1 GHz in the base stations, and millimeter waves even in the devices. Massive MIMO will be part of 5G specifications and deployments from day 1.

3. *Network slicing*: Physical and protocol layers in 5G need flexible design in order to support different use cases, and different frequency bands, to maximize the energy and spectral efficiency. Network slicing will create virtual network segments for the different services within the same 5G network. This slicing capability allows operators to support different use cases and enterprise customers without having to build dedicated networks.

4. *Dual connectivity and LTE co-existence*: 5G can be deployed as a standalone (SA) system but, more typically, 5G will be deployed together with LTE in the early phase. 5G devices can have simultaneous radio connections to 5G and to LTE. Dual connectivity can make the introduction of 5G simpler, increase the user data rate, and improve reliability. 5G is also designed for LTE co-existence which makes spectrum sharing feasible and spectrum refarming simpler.

5. *Support for cloud implementation and edge computing*: The current architecture in LTE networks is fully distributed in the radio and fully centralized in the core network. The low latency requires bringing the content close to the radio which leads to local break out and edge computing. The scalability requires bringing the cloud benefits to the radio networks with edge cloud architecture. 5G radio and core are specified for native cloud implementation, including new interfaces inside the radio network.

1.4 5G Spectrum

5G radio is designed for flexible utilization of all available spectrum options from 400 MHz to 90 GHz including licensed, shared, and unlicensed, FDD and TDD duplexing, narrowband and wideband allocations. The three main spectrum options are illustrated in Figure 1.4. Millimeter wave spectrum above 20 GHz can provide wide bandwidths up to 1–2 GHz, which brings very high data rates up to 5–20 Gbps for

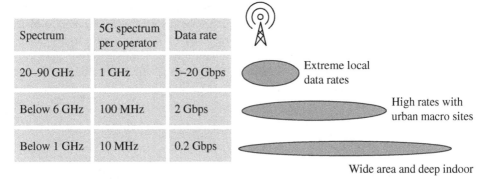

Spectrum	5G spectrum per operator	Data rate
20–90 GHz	1 GHz	5–20 Gbps
Below 6 GHz	100 MHz	2 Gbps
Below 1 GHz	10 MHz	0.2 Gbps

Extreme local data rates

High rates with urban macro sites

Wide area and deep indoor

Figure 1.4 5G can utilize all spectrum options.

extreme mobile broadband capacity. Millimeter waves are mainly suited for local usage like mass events, outdoor and indoor hotspots, and fixed wireless use cases. Millimeter waves can also be used for offloading traffic from low bands in busy hotspot areas. One use case for millimeter waves is providing very high capacity to public transport systems like trains or trams.

Mid-band spectrum at 2.5–5.0 GHz will be used for 5G coverage and capacity in urban areas by reusing existing base station sites. The spectrum around 3.5 GHz is attractive for 5G because it is available almost globally, and the amount of bandwidth can be up to 100 MHz or more per operator at that frequency. The peak data rate is 2 Gbps with 100 MHz bandwidth and 4×4MIMO. 5G coverage at 3.5 GHz can be similar to LTE1800 coverage by using massive MIMO beamforming.

Low bands, below 3 GHz, for FDD, are needed for wide area rural coverage, for low latency and high reliability, and deep indoor penetration. Extensive coverage is important for new use cases like IoT and critical communication. The low band could be 700 MHz, which is made available in many countries at the same time as 5G. Another option is 900 MHz which is mostly occupied by 2G and 3G today, or 600 MHz in the USA. Any other FDD bands can also be refarmed to 5G. LTE and 5G can be deployed on the same band using a dynamic spectrum-sharing solution which makes the refarming a smooth process.

Global 5G spectrum options are shown in Figure 1.5. Mid-band spectrum, between 2.5 and 5.0 GHz, can be found in most countries, as well as millimeter wave spectrum

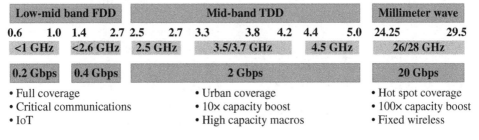

Figure 1.5 5G spectrum options globally.

in a growing number of countries. The first 3GPP phase for 5G provides support of up to 52.6 GHz frequency range while higher frequency bands up to 71 GHz were added in later releases. At low bands, 5G can utilize new 600 or 700 MHz allocations or do a refarming of the existing bands. 5G can combine multiple bands together to boost the performance beyond what is achieved with a single band only. The solution can be carrier aggregation or dual connectivity.

1.5 5G Capabilities

5G radio can bring major benefits in terms of network performance and efficiency compared to LTE radio, see the summary in Figure 1.6. We expect substantially higher data rates, clearly lower cost per bit, higher spectral efficiency, higher network energy efficiency, improved IoT device power efficiency, and lower latency. The values are based on the following assumptions:

– Peak rate of 10 Gbps assumes 1 GHz bandwidth with 2×2MIMO and 256QAM modulation with 10 bps/Hz.
– Cost per byte considered network capital expenditures (capex) and operating expenditures (opex). The calculation assumes busy hour average throughput of 1 Gbps for 3 sectors macro base station, busy hour share 7% of daily traffic, depreciation period of 5 years, 20% of base stations carry 50% of traffic, base station capex 40 000 EUR, and opex equals capex. The cost per byte in 5G can be 10× lower than in the reference case of LTE because of higher spectral efficiency and wider bandwidth, which simply provide a lot more capacity per carrier in 5G.
– The transmission time in 5G is 0.125 ms, and even less with mini-slot, which enables a round trip time of 1–2 ms, together with the 5G architecture enabling local service provisioning. Practical LTE round trip time is 10–15 ms.
– Average power consumption for IoT devices is assumed to be 300 mW during active time, and 0.02 mW during deep sleep. One transmission per minute and each transmission lasting 0.1 s. That leads to energy consumption of less than 10 µWh per transmission.

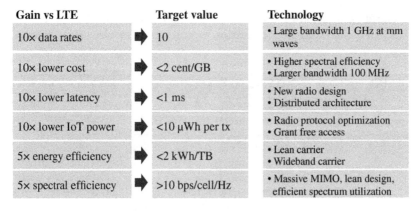

Gain vs LTE		Target value	Technology
10× data rates	➡	10	• Large bandwidth 1 GHz at mm waves
10× lower cost	➡	<2 cent/GB	• Higher spectral efficiency • Larger bandwidth 100 MHz
10× lower latency	➡	<1 ms	• New radio design • Distributed architecture
10× lower IoT power	➡	<10 µWh per tx	• Radio protocol optimization • Grant free access
5× energy efficiency	➡	<2 kWh/TB	• Lean carrier • Wideband carrier
5× spectral efficiency	➡	>10 bps/cell/Hz	• Massive MIMO, lean design, efficient spectrum utilization

Figure 1.6 Summary of 5G technology capabilities.

- Energy efficiency assumes 3 sector 100 MHz macro base station busy hour average throughput of 1 Gbps, busy hour share 7%, base station average power consumption 200 W, and 20% of base stations carry 50% of traffic. The efficiency improvement of 5× compared to LTE is obtained with the power-saving techniques at low load, and with wideband carriers up to 100 MHz.
- Spectral efficiency of 10 bps/Hz/cell assumes the use of massive MIMO beamforming and four antenna devices. The typical LTE downlink efficiency is 1.5–2.0 bps/Hz/cell in the live networks with 2×2MIMO.

1.6 5G Capacity Boost

The first use case for 5G is more capacity and a higher data rate for mobile broadband. 5G brings more capacity on top of LTE networks, with wider bandwidth and massive MIMO and other technologies improving spectral efficiency. The maximum bandwidth in LTE is 20 MHz. Typical spectral efficiency in LTE networks is 1.5–2.0 bps/Hz. That is the average efficiency during busy hours leading to a cell throughput of 40 Mbps. 5G has bandwidth up to 100 MHz at frequencies of 2.5–5.0 GHz. The spectral efficiency of 4–10 bps/Hz leads to a cell throughput of up to 800 Mbps. That corresponds to 20 times higher cell capacity in 5G compared to a typical LTE case. 5G brings a major capacity boost on top of LTE networks by using more spectrum, and by providing higher spectral efficiency (Figure 1.7).

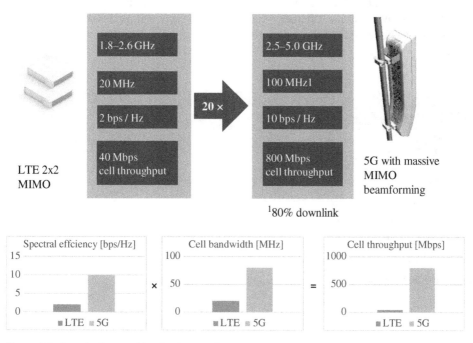

Figure 1.7 Capacity boost with 5G at 2.5–5.0 GHz.

1.7 5G Standardization and Schedule

3GPP started work towards 5G specifications with a workshop in 2015. 5G study items were kicked off in 2016, followed by work items in 2017. The first set of 5G specifications was completed from the content point of view in Release 15 in December 2017, and Abstract Syntax Notation One (ASN.1) backward compatibility was established in September 2018. A few important corrections were still needed from the December 2018 version. December 2018 version allows Non-standalone (NSA) 5G deployment together with LTE. The first full radio and core Standalone (SA) release was completed in June 2018, with ASN.1 in December 2018, with the major corrections finalized for March 2019. March 2019 allowed SA 5G deployment.

The first commercial 5G networks were launched in Korea and in the United States of America in April 2019, based on Release 15 specifications using the NSA version. The first SA network was launched in August 2020. 3GPP continued the specification work with Release 15 late drop for other architecture options in the first half of 2019. Release 16 was completed during the second half of 2020. The high-level 3GPP schedule is illustrated in Figure 1.8. The focus of the work in Release 15 was clearly on the mobile broadband aspects, with more focus on other use cases in Release 16 and beyond, as can be seen from Chapter 16.

The first 5G networks were launched during the first half of 2019 and more than 130 operators launched commercial 5G running during 2019 and 2020. A number of 5G devices were made available in 2019, followed by iPhone 5G launch in October 2021. If we compare the early phases of 5G and LTE commercialization, we can note that 5G took off substantially faster than LTE in terms of network launches and in terms of device availability. If we just look at the 3GPP specifications, 5G started 9.5 years after LTE, but the practical difference in the device penetration is less than eight years. The benchmarking of LTE and 5G commercialization schedules is illustrated in Figure 1.9.

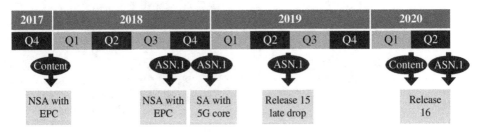

NSA = Non-standalone
SA = Standalone
EPC = Evolved packet core

Figure 1.8 5G timing in 3GPP.

5G commercialization was much faster than LTE

Figure 1.9 5G timing in early commercial deployments.

1.8 5G Use Cases

5G networks enable a number of new use cases for consumers, enterprises, homes, and public domains. Figure 1.10 illustrates example use cases and the corresponding requirements. Video experience can be enhanced with a 5G higher data rate to support 360-degree viewing, especially in mass events where 5G provides substantially higher capacity. Virtual reality and augmented reality, with gaming and other use cases, can enhance user experience. Cloud gaming and esports experience can be improved by lower latency and higher data rates. One use case can be in-vehicle entertainment, allowing one to enjoy high data rates and high-quality video while moving.

The 5G use cases are far beyond consumer applications. Many industrial applications are enabled by high data rates, low latency, and extreme reliability, for example, industry automation with low latency robot control, remote control of machinery, or traffic control, including support for autonomous driving. 5G also brings new capabilities to

Figure 1.10 5G use case areas.

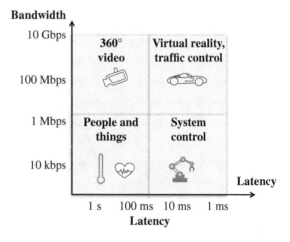

public safety networks, agriculture automation, health care monitoring, and fixed wireless access to homes and small enterprises.

1.9 Evolution Path from LTE to 5G

Part of 5G capabilities can be provided on top of 4G networks with 4.5G, 4.5G Pro, and 4.9G enhancements, also known as LTE-Advanced Pro. 4.5G was commercially available in 2016, supporting 600 Mbps peak rates. 4.5G also brought the first set of IoT optimization and public safety capabilities into LTE networks. 4.5G Pro supports 1 Gbps with carrier aggregation and 4×4 MIMO. LTE evolution can co-exist on the same frequency with legacy LTE devices, which makes the LTE evolution a smooth step towards 5G. LTE evolution can complement 5G, especially in the early phase when 5G coverage is still limited. Figure 1.11 shows high-level evolution path from 4G LTE to 5G.

1.10 5G-Advanced

3GPP continued enhancing 5G capabilities after the successful launch of 5G technology for mobile broadband. 3GPP Releases 16 and 17 completed URLLC features and enabled a wider ecosystem where 5G radio is applied to new use cases beyond mobile broadband. The next major step in 5G evolution is 5G-Advanced, which refers to 5G evolution defined in 3GPP Releases 18 and beyond. 5G-Advanced brings many enhancements to 5G radio, boosting its capabilities towards the 6G era beyond 2030. The first part of 5G-Advanced specifications with Release 18 will be completed in 3GPP during the first half of 2024, and is expected to be available commercially during 2025, followed by Release 19, planned to be ready at the end of 2025, with likely market availability in 2027 timeframe.

5G-Advanced targets utilizing 5G mobile networks for a number of vertical use cases beyond mobile broadband and ultra-reliable communications. Metaverse future

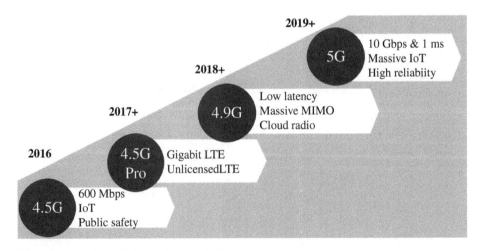

Figure 1.11 4G LTE evolution path towards 5G.

5G mobile broadband
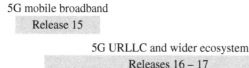

5G URLLC and wider ecosystem

Figure 1.12 5G-Advanced schedule in 3GPP.

requires eXtended Reality (XR) support, which is one of the key 5G-Advanced capabilities. Other focus areas include uplink improvements, mobility enhancements, IoT connectivity, super accurate positioning and timing, AI, and ML utilization, and improved network energy efficiency. Figure 1.12 illustrates the 5G-Advanced schedule on top of the first 5G Release and Release 16 and 17, with more details of the main topics included in the specifications. The expected benefits and new vertical use cases are covered from Chapters 16–19. Work will continue beyond Release 19, with Release 20 adding further topics to 5G-Advanced as well as starting studies towards 6G, as discussed in Chapter 16.

1.11 Summary

5G is not only about new radio or new architecture or new core, but about a number of new use cases. There is a lot of interest in 5G networks by other parties, including different industries and cities, to understand 5G capabilities and push the 5G availability. 5G is about connecting everything in the future which will make 5G much more integrated in our society to keep it running, compared to earlier generations focusing on mobile broadband use cases. The very high targets of 5G networks need new technologies and new deployment models to make it happen in the live networks. It is expected that 5G, combined with cloud and artificial intelligence, can fundamentally impact the whole society in terms of improving efficiency, productivity, and safety. 5G will first come with greatly improved mobile broadband capabilities operating together with LTE, and then move to provide wider selection of new use cases addressing new market segments with SA 5G operation. Read the following chapters to understand more about how 5G works, what 5G can do to answer your needs, and how much 5G will evolve next with 5G ecosystem expansion and 5G-Advanced.

2

5G Targets and Standardization

Hiroyuki Atarashi[1], Mikio Iwamura[1], Satoshi Nagata[1], Takehiro Nakamura[1], and Antti Toskala[2]

[1] *NTT DOCOMO, Japan*
[2] *Nokia, Finland*

CHAPTER MENU

2.1 Introduction

Standards are defined in the 3rd generation partnership project (3GPP), with the actual contributions coming from members of the regional standards bodies that participate in 3GPP activity. More than 600 companies are members of 3GPP via their regional standards organizations. Before standards work begins, however, the requirements and use cases are defined. This occurs on the one hand in forums like ITU-R (International Telecommunication Union Radio sector) and on the other hand in industry forums, especially in the Next Generation Mobile Network (NGMN) Alliance. Pre-standards research forums are also involved, for example, the projects part of the 5G Infrastructure Public Private Partnership (5G-PPP) and 5G Mobile Communications Promotion Forum (5GMF), which aim to collect research inputs both from industry and academia, before actual standards work begins. There are other forums that take interest in specific areas of 5G development for the next phase of 5G development, as discussed in Chapter 15. This chapter covers the key activities leading to 5G specification finalization in the International Telecommunication Union (ITU), NGMN, and 3GPP.

2.2 ITU

ITU has been contributing to international standardization of mobile communication systems through the standard known as IMT (International Mobile

5G Technology: 3GPP Evolution to 5G-Advanced, Second Edition.
Edited by Harri Holma, Antti Toskala, and Takehiro Nakamura.
© 2024 John Wiley & Sons Ltd. Published 2024 by John Wiley & Sons Ltd.

Table 2.1 Previous international standardization for mobile communication systems by ITU.

Name	Example of radio interface technologies.
IMT-2000	WCDMA/HSPA/LTE, cdma2000, TD-SCDMA, WiMAX.
IMT-Advanced	LTE-Advanced, Wireless-MAN-Advanced

Telecommunications). This standardization has been achieved through development of ITU-R recommendations on radio interface technologies for the terrestrial components of IMT-2000 and IMT-Advanced, and these technologies have been used for 3rd and 4th generation mobile communication systems (see Table 2.1).

In early 2012, by taking into account 5G research activities around the world, ITU initiated studies on the development of "IMT for 2020 and beyond," called "IMT-2020." This section explains the vision of IMT-2020 and the development of IMT-2020 radio interface technologies by ITU in relation to the international standardization of 5G.

2.2.1 IMT Vision for 2020 and Beyond

In September 2015, ITU published Recommendation ITU-R M.2083 [1], "IMT Vision – Framework and Overall Objectives of the Future Development of IMT for 2020 and beyond." This recommendation was developed on the basis of a number of inputs by ITU members, taking into account the 5G research activities by industries, research organizations, and academia around the world at that time. The pre-standard research activities included 5G-PPP projects like METIS 1 and 2, or 5G_NORMA in Europe, and activities in Japan within 5GMF.

One of the key messages described in the Recommendation is identification of the following three major usage scenarios in IMT-2020:

- *Enhanced mobile broadband (eMBB)*: Supporting eMBB usage scenarios with improved performance and an increasingly seamless user experience compared to existing mobile broadband applications.
- *Ultra-reliable and low latency communications (URLLC)*: Supporting stringent requirements for capabilities such as throughput, latency, and availability for those applications, such as wireless control of industrial manufacturing, or production processes, remote medical surgery, distribution automation in a smart grid, and transportation safety.
- *Massive machine-type communications (mMTC)*: Supporting a very large number of connected devices, typically transmitting a relatively low volume of non-delay-sensitive data.

Recommendation ITU-R M.2083 also identifies a number of key capabilities of IMT-2020, such as peak data rate, user experience data rate, spectrum efficiency, mobility, latency, connection density, network energy efficiency, and area traffic capacity.

It should be noted that while all these key capabilities may be important, the relevance of certain key capabilities may be significantly different depending on the usage scenarios presented above (see Table 2.2).

Table 2.2 Enhancement of key capabilities from IMT-Advanced to IMT-2020.

Key capabilities	IMT-Advanced	IMT-2020
Peak data rate	1 Gbps	20 Gbps
User experience data rate	10 Mbps	100 Mbps
Spectrum efficiency	1×	3×
Mobility	350 km/h	500 km/h
Latency	10 ms	1 ms
Connection density	10^5 devices/km^2	10^6 devices/km^2
Network energy efficiency	1×	100×
Area traffic capacity	0.1 Mbps/m^2	10 Mbps/m^2

2.2.2 Standardization of IMT-2020 Radio Interface Technologies

In order to embody the "IMT vision" presented above, ITU started its process to develop a new ITU-R Recommendation for the IMT-2020 radio interface technologies and to achieve international standardization of 5G. The timeline and process for the development of IMT-2020 radio interface technologies are summarized in the table below [2] (see Table 2.3).

In this process, ITU invites its members and other organizations to submit their proposals on candidate radio interface technologies for IMT-2020. The submitted proposals will be evaluated by ITU members and other external evaluation groups to ascertain whether the proposed radio interface technologies can satisfy the minimum requirements defined by ITU.

For this IMT-2020 development process, ITU has published three reports; Reports ITU-R M.2410, 2411, and 2412 [3–5]. These reports provide the minimum requirements for candidate IMT-2020 radio interface technologies and their evaluation methodologies. The minimum requirements defined by ITU consist of the "service," "spectrum," and "technical performance" requirements, respectively, which are summarized in the table below (see Table 2.4).

Table 2.3 Timeline for standardization of IMT-2020 radio interface technologies.

Timeline	Process
2016 to the first half of 2017	ITU defines minimum requirements for IMT-2020 radio interface technologies and calls for proposals from ITU membership and external organizations.
Second half of 2017 to the first half of 2019	Proponents submit their proposals on candidate IMT-2020 radio interface technologies to ITU.
Second half of 2018 to the first half of 2020	External evaluation groups evaluate the submitted proposals and whether the proposals satisfy the ITU minimum requirements.
Second half of 2019 to the second half of 2020	ITU agrees on IMT-2020 radio interface technologies to be included in an ITU-R Recommendation.

Table 2.4 IMT-2020 minimum requirements defined by ITU.

Item	Requirements		
Service requirements			
Support for a wide range of services	Support a range of services across different usage scenarios (eMBB, URLLC, and mMTC)		
Spectrum requirements			
Frequency bands identified for IMT	Able to utilize at least one frequency band identified for IMT in the ITU Radio Regulations?		
Higher-frequency range/band(s)	Able to utilize the higher-frequency range/band(s) above 24.25 GHz?		
Technical performance requirements			
Peak data rate	Downlink – 20 Gbps; Uplink – 10 Gbps.		
Peak spectral efficiency	Downlink – 30 bps/Hz; Uplink – 15 bps/Hz.		
User experience data rate	In the dense urban eMBB test environment, downlink – 100 Mbps, and uplink – 50 Mbps.		
5th percentile user spectral efficiency	**Test environment**	**Downlink (bps/Hz)**	**Uplink (bps/Hz)**
	Indoor Hotspot – eMBB	0.3	0.21
	Dense Urban – eMBB	0.225	0.15
	Rural – eMBB	0.12	0.045
Average spectral efficiency	**Test environment**	**Downlink (bit/s/Hz/TRxP)**	**Uplink (bit/s/Hz/TRxP)**
	Indoor Hotspot – eMBB	9	6.75
	Dense Urban – eMBB	7.8	5.4
	Rural – eMBB	3.3	1.6
Area traffic capacity	In the indoor hotspot eMBB test environment, downlink – 10 Mbps/m^2		
User plane latency	4 ms for eMBB, 1 ms for URLLC		
Control plane latency	20 ms		
Connection density	In the mMTC usage scenario, 1 000 000 devices per km^2		
Energy efficiency	In the eMBB usage scenario, the capability to support a high sleep ratio and long sleep duration		
Reliability	In the URLLC usage scenario, $1-10^{-5}$ success probability of transmitting a layer 2 PDU (protocol data unit) of 32 bytes within 1 ms in channel quality of coverage edge for the urban macro-URLLC test environment, assuming small application data (e.g. 20 bytes application data + protocol overhead).		
Mobility	**Test environment**	**Normalized traffic channel link data rate (bps/Hz)**	**Mobility (km/h)**
	Indoor Hotspot – eMBB	1.5	10
	Dense Urban – eMBB	1.12	30
	Rural – eMBB	0.8	120
		0.45	500
Mobility interruption time	In the eMBB and URLLC usage scenarios, 0 ms		
Bandwidth	At least 100 MHz		
	Supports bandwidth up to 1 GHz for operation in higher-frequency bands (e.g. above 6 GHz) and scalable bandwidth.		

Table 2.5 Test environments for the evaluation of technical performance requirements.

Usage scenario	Test environment	Description
eMBB	Indoor hotspot-eMBB	An indoor isolated environment at offices and/or in shopping malls based on stationary and pedestrian users with very high user density.
	Dense urban-eMBB	An urban environment with high user density and traffic loads focusing on pedestrian and vehicular users.
	Rural-eMBB	A rural environment with larger and continuous wide area coverage, supporting pedestrian, vehicular, and high-speed vehicular users.
URLLC	Urban macro-URLLC	An urban macro-environment targeting ultra-reliable and low-latency communications.
mMTC	Urban macro-mMTC	An urban macro-environment targeting continuous coverage focusing on a high number of connected machine-type devices.

As shown in the above table, some of the requirements are associated with the following five test environments, taking into account the three major usage scenarios (eMBB, URLLC, and mMTC) envisaged in IMT-2020 (see Table 2.5).

3GPP had submitted its proposal to the ITU's IMT-2020 development process, which was explained above. The proposal is "5G" developed by 3GPP, i.e. Release 15 and beyond, in accordance with (1) a set of radio interface technologies (SRITs) with new radio (NR) and E-UTRA (Evolved universal terrestrial radio access)/long term evolution (LTE), and (2) a radio interface technology (RIT) with NR only. These 3GPP technologies are expected to be included in the ITU-R Recommendation on the radio interface technologies for IMT-2020 to be available in early 2021.

2.3 NGMN

The NGMN Alliance is an industry organization of leading worldwide telecom operators, vendors, and research institutes. Founded in 2006 by international telecom operators, its objective is to ensure that the functionality and performance of NGMN infrastructure, service platforms, and devices will meet the requirements of operators and, ultimately, will satisfy end-user demand and expectations.

Driven by the profound interest of operators to make 5G truly global, free of fragmentation, and open to innovations, NGMN established the 5G initiative in 2014 to put together their vision for 5G. Led by a core team of operators, the NGMN 5G initiative, despite accommodating considerations for different market contexts and business priorities, successfully put together use cases, requirements, and architecture design principles for 5G in an "NGMN 5G White Paper" [6]. Endorsed by 21 operators, i.e. AT&T, Bell, British Telecom, China Mobile, Deutsche Telekom, KPN, Korea Telecom, NTT DOCOMO, Orange, Singtel, SK Telecom, TELE2, Telecom Italia, Telefonica, Telekom

Austria Group, TeliaSonera, Telstra, Telus, Turkcell, VompelCom, and Vodafone, the white paper was published on March 2015.

According to NGMN:

5G is an end-to-end ecosystem to enable a fully mobile and connected society. It empowers value creation towards customers and partners, through existing and emerging use cases, delivered with consistent experience, and enabled by sustainable business models.

As such, the envelope of network capabilities in terms of throughput, lower latency, and higher connection density needs to be greatly expanded. To cope with a wide range of use cases and business models, a high degree of flexibility and scalability needs to be embedded in design. Business orientation and economic incentives with a foundational shift in cost, energy, and operational efficiency should make 5G feasible and sustainable.

2.3.1 NGMN 5G Use Cases

In addition to supporting the evolution of the established prominent mobile broadband use cases, NGMN identified the need for 5G to support countless emerging use cases with a high variety of performance attributes and support for a wide range of devices. Studies on various representative use cases resulted in the following families of use cases:

- *Broadband access in dense areas*: Improve service availability in densely populated areas, including pervasive video, smart office, operator cloud services, and HD video/photo sharing at stadiums/open-air gatherings.
- *Broadband access everywhere*: Provide consistent user experience in terms of throughput, and digital inclusion of people from all geographies, to realize 50+ Mbps everywhere and ultra-low-cost networks.
- *Higher user mobility*: Support the growing demand for mobile services, such as entertainment, internet access, and remote computing, in vehicles, trains, and aircrafts, including moving hot spot scenarios such as walking/cycling demos.
- *Massive Internet of Things (IoT)*: Support massive number of devices (e.g. sensors, actuators, and cameras) from low-cost/long-range/low power to broadband machine type communications (MTC), including smart wearables (clothes), sensor networks, and mobile video surveillance.
- *Extreme real-time communications*: Support services that require extreme real-time interaction, including safety support for autonomous driving vehicles, remote computing, and tactile internet.
- *Lifeline communication*: Realize high levels of availability and resilience to support enhanced public safety and emergency services, even in cases of traffic surge, as well as efficient network recovery in case of natural disasters.
- *Ultra-reliable communications*: Support various applications of industries that require very high reliability, such as remote operation and control of machines, collaborative robots, remote surgery, and life-critical services.
- *Broadcast-like services*: Support efficient distribution of information from one source to many destinations with possible feedback for interactive services, varying in range from local to regional and national levels.

In addition to the conventional business models, NGMN highlights the need for 5G to support various types of customers and partnerships, such as to mobilize vertical industries and support industry processes. Partnerships will be established on multiple layers ranging from sharing infrastructure to exposing network capabilities or integrating partner services into the overall system.

2.3.2 NGMN 5G Requirements

The 5G use cases demand very diverse and sometimes extreme requirements. A single solution may not satisfy all the requirements at the same time at a reasonable cost. Several use cases may be active concurrently at the same time, requiring a high degree of flexibility and scalability of the 5G network. NGMN sets the following requirements:

- Excellent and consistent user experience.
  - Consistent user experience across time and service footprint in a highly heterogeneous environment, depending on the use case.
  - A much higher user data rate, required to be available in at least 95% of locations (including the cell edge), for at least 95% of the time:
    o *Dense urban*: 300 Mbps DL
    o *Smart office*: 1 Gbps DL
    o Multi-Mbps data rates everywhere – including in stadiums, airplanes, and areas currently not connected to the internet.
  - *Much lower latency*: Less than 1 ms end-to-end latency for certain car-to-car and industry automation communication needs.
  - Seamless service experience to moving users (up to 500 km/h) and also static/nomadic users/devices.
- *System performance*: Significantly expanded network capabilities to cope with the variety and variability of use cases.
  - Connection/traffic density:
    o *Users in a crowd*: Several tens of Mbps for tens of thousands of users in crowded areas.
    o *Smart office*: 15 Tb/s/km^2 traffic density for smart offices.
    o *Massive sensor deployments*: Up to several hundred thousand simultaneous connections per square kilometer.
  - Significantly enhanced spectral efficiency (average and cell edge, across bands) to keep the number of sites reasonable.
  - Enhanced resource and signaling efficiency to minimize resource and energy consumption.
- Smart devices with growing capabilities.
  - High degree of programmability and configurability of any device by the network (over the air testing, OTA).
  - Flexible and dynamic device capability handling.
  - Devices to support multiple bands simultaneously and multiple modes (Frequency Division Duplex [FDD], Time Division Duplex [TDD], mixed) for true global roaming.
  - *Significantly increased battery life*: At least 3 days for smartphones, up to 15 years for low-cost MTC device.

- *Enhanced services*: Value creation toward customers and partners through capabilities enhancing today's overall service delivery.
  - Seamless and always-best-experience connection without user intervention, across existing, new, and non-3GPP radio access technologies (RATs).
  - Unnoticeable mobility across existing, new, and non-3GPP RATs.
  - Network-based positioning with accuracy from 10 to <1 m outdoors and <1 m indoors, in real-time.
  - Strengthened security for services and networks in highly heterogeneous environments, working also when the user is roaming.
  - Protection of users' trusted information.
  - Ultra-high reliability rate of ≥99.999% for specific use cases.
- *New business models*: Expansion of current opportunities and creating opportunities for new business models within the 5G ecosystem.
  - *Partner service provider, XaaS asset provider*: Configure and manage services, for example, via OpenAPI – exposing network capabilities in a flexible, configurable, and programmable manner.
  - *Connectivity provider*: Connectivity is delivered using only necessary network functions – provisioning and configuration on demand, and in a programmable manner.
  - *Network sharing model*: Enabling various sharing schemes to maximize the overall synergies of sharing agreements, and to allow for flexibly/rapidly changing models and relationships.
- *Highly efficient network operation*: High efficiency in cost, energy, innovation, deployment, and operations and maintenance (O&M) while minimizing the total cost of ownership.
  - Half of total network energy consumption for 1000× traffic growth.
  - Significant reduction of O&M complexity and cost.
  - Ultra-low-cost for very low average revenue per user (ARPU) areas and/or MTC services.
  - Flexible and fast introduction of new services and technologies.
  - *Ease of deployment*: Plug and play, self-configuration/heal.
  - *Flexibility and scalability*: Openness and multi-vendor capability at all levels, modular provisioning, functional split of core/RAN network domains/elements, decoupling hardware and software.
  - Fixed-mobile convergence, for seamless user experience and unified subscriber management.

2.3.3 NGMN 5G Architecture Design Principles

The 5G use cases demand very diverse, and sometimes extreme requirements, with each use case requiring its own specific set of performance, scalability, and availability. NGMN anticipated that the relatively monolithic architecture of previous-generation networks could not efficiently address the necessary flexibility and scalability, in supporting a considerably wider range of business needs. To accommodate more agile and efficient provisioning of new services, the 5G architecture should include modular network functions that could be deployed and scaled on demand. To enable concurrent active services sharing the same operator network, NGMN envisions an architecture

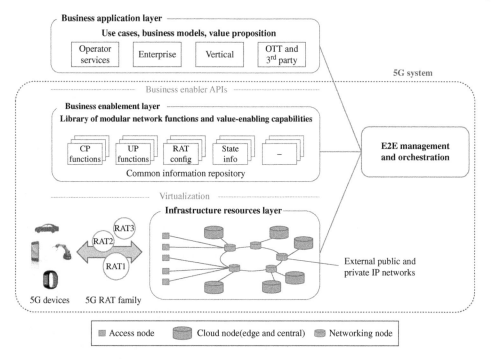

Figure 2.1 NGMN 5G architecture.

that splits the overall system into three distinctive layers, namely, the business application layer, the business enablement layer, and the infrastructure resources layer, building on network function virtualization and software-defined networks as shown in Figure 2.1. The business enablement layer comprises various C-plane and U-plane functions, as well as various RAT configurations and state information, forming a common information repository. The business enabler APIs are exposed to the various applications, including operator services, enterprise, and vertical services, as well as over-the-top (OTT) services. Those business enabler capabilities are called upon by various applications for tailoring to individual applications in a flexible and scalable manner. This capability to tailor to individual application needs using a common infrastructure, often referred to as "network slicing," is laid out as the foundational architecture for 5G in the NGMN White Paper.

2.3.4 Spectrum, Intellectual Property Rights (IPR), and Further Recommendations by NGMN

The NGMN White Paper also gives recommendations on spectrum and IPR policies for 5G. With regard to spectrum, NGMN highlights the following:

- Because of the different requirements, operators must rely on a larger spectrum portfolio:
  - Possibility to refarm their existing mobile spectrum holdings for 5G and to have access to additional spectrum that may be identified at the ITU WRC-15.

- Access to low-frequency spectrum (below 6 GHz), especially below 1 GHz, absolutely essential for an economical delivery of mobile services in rural and indoor
environments; and access to higher-frequency bands above 6 GHz suitable for very
high data rates and shorter-range connectivity, with potentially large bandwidths
of 500–1000 MHz/operator.
- Integrating 5G under the umbrella of IMT within ITU is vital for global spectrum
identification and harmonization.
- Backhaul requirements for 5G may include wireless solutions and need spectrum.

With regard to IPR, NGMN highlights the following:

- Improve 5G Standard Essential Patent (SEP) declarations to improve transparency
and limit abusive patent declarations related to 5G standards, while still encouraging
early declarations.
- Establish independent 5G SEP assessments to ensure quality declarations, transparency, and effectiveness.
- Explore and establish patent pool licensing for 5G to determine appropriate licensing
terms and conditions (including royalties) within the 5G patent pool framework to
meet the overall NGMN business objectives.

After the publication of the NGMN 5G White Paper, NGMN went on to look into the
details of the end-to-end architecture [7], where network slicing concepts and implications were further elaborated. The work has further led to the definition of service-based
architecture [8]. In addition to developing a deeper understanding of the vertical industry implications of 5G, NGMN hosted workshops in Europe, the United States, and Asia,
devising dialogues with actual vertical industry organizations and companies. Iterative
discussions with the actual vertical industries have led to the publication of a comprehensive set of vertical use cases and requirements, as detailed in [9]. Extreme reliability
and low latency requirements and their feasibility were further studied in a subsequent
project [10]. NGMN also studied the business perspective of 5G in their deliverable
created by the Business Principles Group [11], in which the mutual implications of the
business and technology contexts are elaborated. In addition, NGMN established the
testing and trials initiative to devise a strategy to guarantee the efficiency and success of
different trial activities pursued by different parties [12].

2.4 3GPP Schedule and Phasing

3GPP has been contributing to the production of technical reports and specifications
that cover telecommunications network technologies, including radio access, the core
transport network, and service capabilities. In 3GPP, a radio access network (RAN) 5G
workshop was held in September 2015 during the course of which the requirements and
potential technologies of 5G were discussed. In this workshop, there was an emerging
consensus that there will be a new, non-backward-compatible radio access technology as part of 5G, supported by the need for LTE-Advanced evolution in parallel. A
consensus emerged that there should be two phases for the 5G normative work considering various usage scenarios, requirements, and urgent commercial demands. The first
phase addresses a more urgent subset of the commercial needs, and the second phase

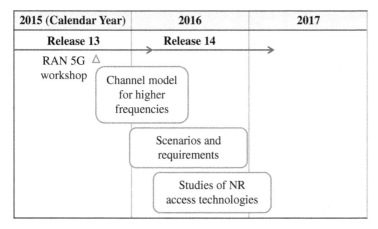

Figure 2.2 3GPP schedule related to NR access technology studies.

addresses all the identified usage scenarios and requirements. At this time, it seemed widely agreed that, while the normative work can be phased to initially specify support for only a subset of the identified usage scenarios, and requirements, the design of the NR should be forward compatible, so that it can optimally support the remaining usage scenarios and requirements that will be added in a later phase.

After this workshop, in Releases 13 and 14, channel modeling for higher frequencies was studied to take care of the spectrum above 6 GHz, and usage scenarios and requirements were also studied. Technical reports on higher-frequency channel modeling, usage scenarios, and requirements were summarized in Refs. [13, 14], respectively. In Release 14, from 2016 to the beginning of 2017, studies of NR access technologies were conducted to identify and evaluate the candidate technologies, as shown in Figure 2.2. Technical reports of these studies were summarized in multiple reports such as [15, 16].

In Release 15, from the beginning of 2017 to 2019, the initial NR specification work was conducted to produce the initial 5G specifications called phase 1 as shown in Figure 2.3. The target objectives of the initial NR specification were set to specify the functionalities for eMBB and URLLC considering a more urgent subset of the commercial needs. At the same time, new specification work for LTE-Advanced was also conducted to extend functionality and improve performance.

In phase 1, both non-standalone and standalone operations of NR were specified. The non-standalone operation was specified in the early timing of Release 15. Standalone-specific functionalities were specified in the later timing of Release 15 to keep functional commonalities between non-standalone and standalone operations as much as possible.

In addition, 3GPP allows late drop of Release 15 specifications to take care of additional migration network architectures such as NR-NR dual connectivity.

The Release 16 specification work started in mid-2018 and was completed towards the end of 2020, as progress was impacted by COVID-19 which forced us to run meetings in electronic format. Chapters 16 and 17 address new functionalities being added to enhance Release 15 functionalities and to enable new use cases like industrial IoT covered in Chapter 18.

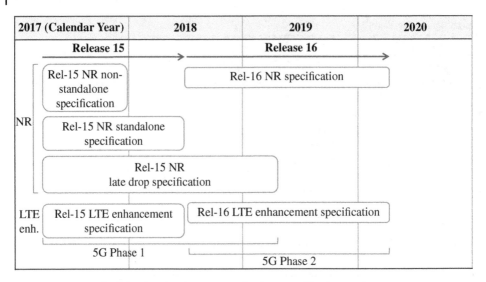

Figure 2.3 3GPP schedule related to NR access technology specification work.

Release 17 standardization work started in early 2020 rather slowly due to the pandemic and it took until mid-2022 before ASN.1 could be frozen. 3GPP had concluded the Release 17 scope in December 2019, just before physical 3GPP meetings had to be put on hold. In 3GPP, it turned out that starting work, especially with new topics, was challenging with the lack of physical meetings when having to resort to having only electronic meetings.

The 3GPP Release 15 specifications took some time to mature after the first specifications were published in December 2017. The NR non-standalone specifications of the first version of ASN.1 were completed in March 2018, and it experienced non-backward-compatible (NBC) changes until September 2018 as indicated in Figure 2.4. For the December 2018 version, the protocol remained backward compatible (and has been maintained backward compatible in 2019 as well). The December 2018 version required some corrections to reflect some important items like finalization of the user equipment (UE) capabilities in terms of which features the UEs are required to support and which they are not. This requires the functionality from December 2018 to be taken into account in the commercial implementations. December 2018 is thus the required baseline for commercial non-standalone NR implementations for ensuring interoperability in an environment with UEs from multiple vendors.

For standalone NR, the March 2019 version was still experiencing several critical changes (even if the ASN.1 remained backward compatible), and thus the March 2019 version is the first version that has a suitable baseline for commercial standalone 5G implementations. The standalone ASN.1 has remained backward compatible since March 2019.

The late drop started formal backward compatibility from June 2019 onward, with stability getting benefit from the corrections done for the non-standalone version already earlier. The late drop covers especially synchronous 5G–5G dual connectivity as well as additional dual connectivity architecture options between LTE and 5G, as described in Chapter 5.

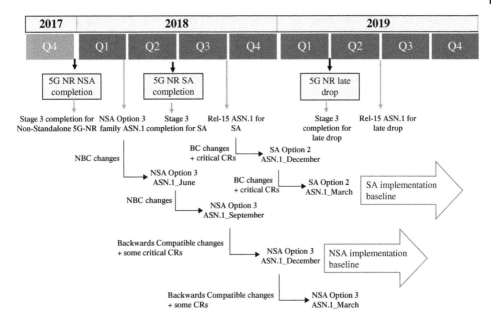

Figure 2.4 Release 15 NR specification stability.

2.5 Evolution Towards 5G-Advanced and 6G

The work in 3GPP had progressed further with Release 18 having started the 5G-Advanced work, with the first 5G-Advanced Release 18 specifications created at the end of 2023 and ASN.1 freeze targeted in June 2024. Release 19 work starts in early 2024, with the specification milestone for ASN.1 freeze set for the end of 2025, as discussed further in Chapter 16, and the overall timeline from Release 15 to 19 is shown in Figure 2.5. Technology studies for 6G are expected to be start from Release 20 onwards, as discussed in Chapter 16, to meet IMT-2030 requirements and timelines.

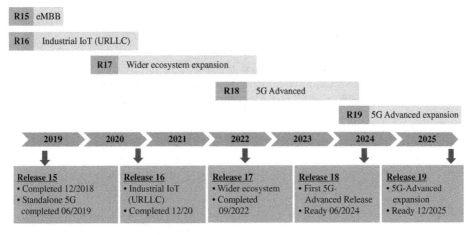

Figure 2.5 5G evolution beyond Release 15.

The official schedules for Release 20 and beyond have not yet been concluded on 3GPP, but are expected to be concluded in March 2024 with the expected timing disussed in Chapter 16.

References

1 Recommendation ITU-R M.2083-0 (2015). IMT Vision – Framework and overall objectives of the future development of IMT for 2020 and beyond. September 2015.
2 ITU-R Working Party 5D web-page (2019). IMT-2020 submission and evaluation process. https://www.itu.int/en/ITU-R/study-groups/rsg5/rwp5d/imt-2020/Pages/submission-eval.aspx.
3 Report ITU-R M.2410-0 (2017). Minimum requirements related to technical performance for IMT-2020 radio interface(s). November 2017.
4 Report ITU-R M.2411-0 (2017). Requirements, evaluation criteria and submission templates for the development of IMT-2020. November 2017.
5 Report ITU-R M.2412-0 (2017). Guidelines for evaluation of radio interface technologies for IMT-2020. November 2017.
6 NGMN Alliance (2015). NGMN 5G White Paper. March 2015.
7 NGMN Alliance (2018). 5G End-to-End Architecture Framework v2.0. February 2018.
8 NGMN Alliance (2018). Service-Based Architecture in 5G. January 2018.
9 NGMN Alliance (2016). Perspectives on Vertical Industries and Implications for 5G. September 2016.
10 NGMN Alliance (2018). 5G Extreme Requirements: E2E Considerations. August 2018.
11 NGMN Alliance (2016). 5G Prospects – Key Capabilities to Unlock Digital Opportunities. July 2016.
12 NGMN Alliance (2019). Definition of the Testing Framework for the NGMN 5G Pre-Commercial Network Trials (Version 2). January 2019.
13 3GPP Technical Report TR38.900 (2016). Study on channel model for frequency spectrum above 6 GHz. Version 14.0.0 June 2016.
14 3GPP Technical Report TR38.913 (2016). Study on scenarios and requirements for next generation access technologies. Version 14.0.0 September 2016.
15 3GPP Technical Report TR38.802 (2017). Study on new radio access technology Physical layer aspects. Version 14.0.0 March 2017.
16 3GPP Technical Report TR 38.801 (2017). Study on new radio access technology: Radio access architecture and interfaces. Version 14.0.0 March 2017.

3

Technology Components

Harri Holma

Nokia, Finland

CHAPTER MENU

3.1 Introduction

5G targets are set very high, not only in terms of data rates up to 20 Gbps and capacity increase by 1000 times but also in terms of providing a flexible platform for the new services like massive Internet of Things (IoT) and critical communication. The high targets of 5G networks require a number of new technology components. This chapter presents an overview of the main new technologies in 5G compared to Long Term Evolution (LTE). A summary is shown in Table 3.1. The chapter is divided into spectrum utilization, beamforming, physical layer and protocols, network slicing, dual connectivity, and new architecture with radio cloud and edge computing. Each topic is explained in the corresponding section in this chapter.

3.2 Spectrum Utilization

3.2.1 Frequency Bands

5G radio is designed for flexible spectrum utilization to take advantage of all available spectrum options from 400 MHz to 90 GHz including licensed, shared access and

5G Technology: 3GPP Evolution to 5G-Advanced, Second Edition.
Edited by Harri Holma, Antti Toskala, and Takehiro Nakamura.
© 2024 John Wiley & Sons Ltd. Published 2024 by John Wiley & Sons Ltd.

Table 3.1 Main new technology areas in 5G.

Spectrum utilization	• Spectrum above 6 GHz • Bandwidth 100–400 MHz • Spectrum occupancy up to 98% • Flexible control channel locations in frequency domain
Beamforming	• User-specific demodulation reference signals • Common channel beamforming • Optimized MIMO feedback signaling
Physical layer and protocols	• Low latency: mini-slot, self-contained subframe, pipeline processing • Lean carrier with flexible reference signals • LDPC channel coding • OFDM uplink waveform
Network slicing	• Service-specific end-to-end network slices • Flow-based QoS support • Dynamic Quality of Experience
Dual connectivity	• Simultaneous radio connection to 5G and LTE radios • Usage of LTE protocols and core network • Higher data rate with 5G and LTE aggregation
Radio cloud and edge computing	• Interface between remote radio unit and centralized baseband • Edge computing support for local content and local breakout

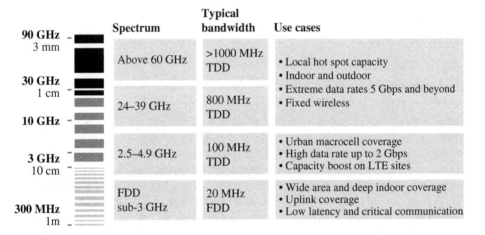

Figure 3.1 5G can utilize all spectrum options.

unlicensed, Frequency Division Duplex (FDD) and Time Division Duplex (TDD) bands, and narrowband and wideband allocations. The three main spectrum options are illustrated in Figure 3.1. Millimeter wave spectrum above 20 GHz can provide bandwidth even above 1 GHz, which allows a very high data rate up to 20 Gbps and extreme mobile broadband capacity. Millimeter waves are mainly suited for local usage like mass events, outdoor and indoor hotspots, and fixed wireless use case.

Spectrum at 2.5–5.0 GHz will be used for 5G coverage and capacity in urban areas by reusing the existing base station sites. The spectrum around 3.5 GHz is attractive for 5G because it is available globally, and the amount of spectrum is high. The bandwidth is typically up to 100 MHz per operator at that frequency. 5G coverage at 3.5 GHz can be comparable to LTE1800 coverage if massive Multiple Input Multiple Output (MIMO) beamforming is used.

Low FDD bands are needed for wide area rural coverage, ultra-high reliability, and deep indoor penetration. Extensive coverage is important for the new uses cases like IoT and critical communication.

For more details about spectrum options, see Chapter 4.

3.2.2 Bandwidth Options

5G radio is designed to support a large number of bandwidth options. Narrowband carrier is needed at low bands, and wideband carrier is beneficial at high bands. The maximum bandwidth in LTE is 20 MHz, while 5G supports up to 100 MHz for sub-6 GHz cases and up to 400 MHz for millimeter waves. The maximum bandwidths are shown in Table 3.2. Larger bandwidths can be supported with carrier aggregation, but wideband carrier is more efficient for the wideband spectrum allocations; see Figure 3.2. If 100 MHz spectrum is available, it can be utilized by LTE with carrier aggregation: five carriers, each 20 MHz. The carrier aggregation solution needs common channels on every carrier, load balancing is needed between carriers, and unnecessary guard bands are needed. It is more efficient to use a single 100 MHz carrier than five 20 MHz carriers. The maximum bandwidth of 20 MHz in LTE was a good choice because the spectrum allocation per operator was typically 20 MHz or less in those frequencies where LTE is deployed. There is, however, more spectrum available in the new frequencies where 5G is deployed, like the 3.5 GHz band and millimeter waves, which justifies the larger bandwidth in 5G.

3.2.3 Spectrum Occupancy

LTE has been designed to include a 10% guard band to avoid interference between adjacent carriers. 20 MHz LTE has 18 MHz transmission bandwidth, while 2 MHz is reserved for guard bands. The practical deployments show that a 10% guard band is excessive. Also, the RF requirements can be made tighter in 5G compared to LTE. Therefore, the spectrum utilization can be increased from 90% in LTE to up to 98% 5G. Figure 3.3 shows the 5G spectrum utilization compared to the channel spacing for different carrier bandwidths and subcarrier spacings. The deployment case with

Table 3.2 Maximum carrier bandwidth with LTE and 5G.

LTE	20 MHz
5G sub-6 GHz	100 MHz
5G millimeter waves	400 MHz

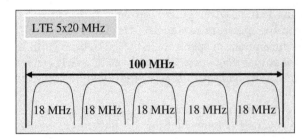

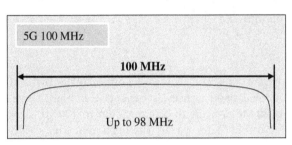

Figure 3.2 5G wideband carrier is efficient for wide spectrum allocations.

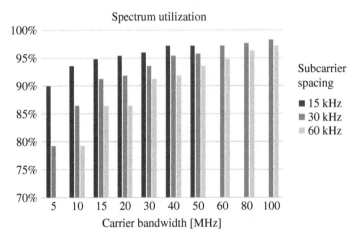

Figure 3.3 5G spectrum utilization.

100 MHz and 30 kHz has a spectrum utilization of 98%. The case with 10 MHz and 15 kHz has a spectrum utilization of 94%.

3.2.4 Control Channel Flexibility

5G is designed to support flexible bandwidth allocation. There may be cases where a part of the band cannot be utilized, or a part of the band is interfered, or the amount of spectrum does not match the predefined 3GPP bandwidths. It is important that the system include flexibility that allows utilization of the band according to the specific requirements. The LTE control channel has some limitations: LTE downlink Physical Downlink Control Channel (PDCCH) is allocated over the full band, and LTE uplink Physical Uplink Control Channel (PUCCH) is allocated at the band edges. There is no flexibility in LTE control channel allocations if we need, for example, to squeeze down the carrier

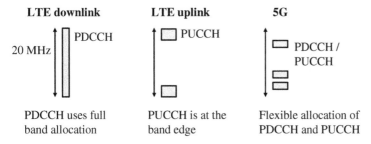

Figure 3.4 Control channel allocations in frequency domain.

narrower from one side. The 5G control channel includes more flexibility, which allows allocation of control channels in suitable locations in the frequency domain. The control channel allocations are illustrated in Figure 3.4. For more details, see Chapter 6.

3.2.5 Dynamic Spectrum Sharing

Spectrum refarming from 2G to 3G and to LTE was a slow and relatively complex process. Legacy technology must be eliminated from a part of the spectrum, which must then be allocated for the new technology. It was not possible to share the spectrum between different technologies. Spectrum refarming from LTE to 5G will be simpler with Dynamic Spectrum Sharing (DSS). 5G and LTE can occupy the same spectrum band from the control channel point of view, and the resources are allocated dynamically between the two technologies depending on the instantaneous device distribution and capacity needs. DSS is described in more detail in Chapters 8 and 10.

3.3 Beamforming

Beamforming is an attractive solution for boosting mobile network performance. Beamforming can provide higher spectral efficiency, which adds a lot more capacity to the existing base station sites. Beamforming can also improve link performance and provide an extended coverage area. Beamforming has been known in academic studies for many years, but beamforming was not supported by the first LTE specifications in 3GPP Release 8. Beamforming was included in the 5G specifications already in the first 3GPP Release 15. Beamforming benefits are obtained in practice with massive MIMO antennas. The target is to make 5G radio design fully optimized for massive MIMO beamforming. The underlying principle of beamforming is illustrated in Figure 3.5. The traditional solution transmits data over the whole cell area while beamforming sends the data with narrow beams to the users. The same resources can be reused for multiple users within a sector, interference can be minimized, and the cell capacity can be increased.

Massive MIMO is the extension of traditional MIMO technology to antenna arrays having large number of controllable transmitters. 3GPP defines massive MIMO as more than eight transmitters. Beams can be formed in number of different ways to deliver either a fixed grid of beams or user equipment (UE)-specific beamforming. If the antenna has two transceiver (TRX) branches, it can send with two parallel streams

Non-beamforming
transmission

Beamforming with
example 8 beams

Figure 3.5 Beamforming enhances radio capacity and coverage.

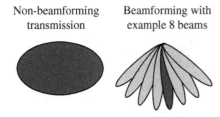

to one UE. If the antenna has four TRXs, it can send four streams to one UE having four antennas, or dual-stream to two UEs simultaneously with Multi-user Multiple Input Multiple Output (MU-MIMO). If the antenna has 64 TRXs, it can send data to multiple UEs in parallel. The number of TRXs is an important design factor in massive MIMO antennas. The greater the number of TRXs used, the more beams can be generated, which gives more capacity. But incorporating more TRXs also makes the antenna larger and increases the costs. Another important antenna design factor is the number of antenna elements, which can be greater than the number of TRXs. Figure 3.6 illustrates an example antenna with 192 elements: 12 vertical, 8 horizontal, and 2 different polarizations. The number of antenna elements defines the antenna gain and the coverage. The use of more antenna elements makes the antenna larger and increases the antenna gain. The spacing of the antenna elements depends on the frequency: the physical size of the antenna is larger at lower frequencies.

- The number of antenna elements defines the antenna gain and antenna size. The size is also heavily dependent on the frequency. The antenna becomes smaller at high bands.
- The number of TRXs can be equal to or lower than the number of antenna elements, and it defines the capacity gain.
- The number of MIMO streams can be equal to or lower than the number of TRXs. The number of MIMO streams defines the peak data rate capability, and it is mainly dependent on the baseband processing capability.

When the number of antenna elements is larger than the number of TRXs, the additional elements are typically added as more rows. A typical MIMO antenna could have 192 antenna elements, 64 transmitters, and support up to 16 MIMO streams. There are three rows per TRX in this antenna.

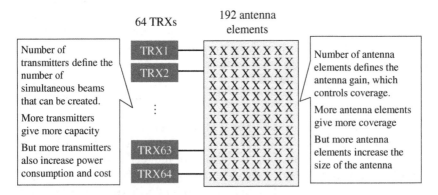

Figure 3.6 Massive MIMO principles with more transmitters and more antenna elements.

Figure 3.7 Common channel beamforming in 5G with beam sweeping.

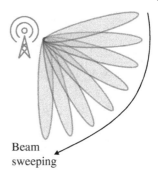

Beam
sweeping

There are a number of reasons why beamforming can yield more gains in 5G than in LTE:

- Beamforming is supported for 5G common channels with beam sweeping; see Figure 3.7. Beam sweeping refers to the operation where the synchronization signal and broadcast channel are transmitted in different beams in the time domain. Common channel beamforming is not supported in LTE.
- User-specific reference signals in 5G support user-specific beamforming. LTE must use cell-specific reference signals (CRSs) that cannot be used for beamforming.
- There are no legacy device limitations in 5G since beamforming is included in 5G from the first specifications. LTE beamforming must be based on uplink Sounding Reference Signal (SRS) measurements since legacy devices do not support beamforming feedback.
- More transmission branches are supported in 5G. 5G supports initially feedback for 64TX, while LTE supports 4TX in Release 8, 8TX in Release 10, 16TX in Release 13, and 32TX in Release 14.

Massive MIMO active antennas combine an antenna and a large number of small RF units into a single package. The traditional solution has been to separate the passive antenna and the RF unit. An active antenna enables practical beamforming implementation when the phasing of the small power amplifiers can be controlled with digital processing. The typical number of RF units can be 16, 32, 64, or 128 inside the active antenna. The active antenna also enables a simpler practical installation since there are no cables between the antenna and the radio frequency (RF). The power efficiency can also be enhanced since there are no losses in RF cables and connectors. A Nokia active antenna with 64 TRXs is shown in Figure 3.8. See Chapter 6 for more details about mMIMO signaling and Chapter 10 for mMIMO antenna structures and performance.

3.4 Flexible Physical Layer and Protocols

5G radio design requires enhanced flexibility in the physical layer and in the protocols. The main new solution areas are highlighted in this section. The details can be found in Chapters 6 and 7.

3.4.1 Flexible Numerology

5G radio must have the flexibility to support all the different spectrum options. The solution is flexible numerology as shown in Table 3.3. 5G is designed to support several

Figure 3.8 Massive MIMO antenna for 3.5 GHz band with 64 TRXs.

Table 3.3 5G numerology is designed for all spectrum options.

Subcarrier spacing	15 kHz	30 kHz	60 kHz	120 kHz	240 kHz
Symbol duration	66.7 μs	33.3 μs	16.7 μs	8.33 μs	4.17 μs
Nominal cyclic prefix	4.7 μs	2.3 μs	1.2 μs	0.59 μs	0.29 μs
Maximum bandwidth	50 MHz	100 MHz	100 MHz	400 MHz	400 MHz
Slot length	1.0 ms	0.5 ms	0.25 ms	0.125 ms	0.125 ms

subcarrier spacing and scheduling intervals depending on the bandwidth and on the latency requirements. Subcarrier spacings of 15–240 kHz are defined in Release 15. In later releases, higher subcarrier spacings can be accommodated. With higher subcarrier spacings, more symbols can be accommodated in a subframe, resulting in a lower acquisition time. The narrow spacings are used with narrow 5G bandwidths and are better for the extreme coverage. If we consider typical 5G deployment at the 3.5 GHz band, the bandwidth could be 40–100 MHz, the subcarrier spacing 30 kHz, and slot length 0.5 ms. The corresponding numbers in LTE are 20 MHz bandwidth, 15 kHz subcarrier spacing, and slot length 1 ms. 5G subcarrier spacing is designed to be 2^N multiples of 15 kHz. If very low latency is required, then the so-called mini-slot can be used where the transmission time is shorter than one slot. It is also possible to combine multiple slots together.

Most spectrum allocations globally below 2.5 GHz are limited to 20–40 MHz per operator per band. Therefore, the LTE maximum bandwidth is defined as 20 MHz. Wider spectrum allocations are generally available at 2.5 and 3.5 GHz or higher frequencies. 5G radio is designed to support wider bandwidths like 100 MHz or even 400 MHz at higher frequencies.

3.4.2 Short Transmission Time and Mini-slot

LTE Release 8 has the minimum transmission time of 1 ms, which leads to a minimum round trip time of 10–15 ms in practice. The transmission time is 1 ms in both directions,

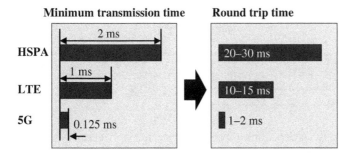

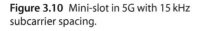

Figure 3.9 Minimum transmission time and corresponding round trip time.

Figure 3.10 Mini-slot in 5G with 15 kHz subcarrier spacing.

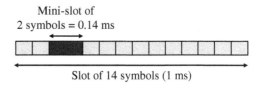

the buffering time is 0.5 ms, and the decoding time is a multiple of 1 ms. If we want to make the round trip time shorter, we need to define a shorter transmission time. The minimum transmission time in High Speed Packet Access (HSPA) is 2 ms, in LTE 1 ms, and in 5G 0.125 ms, which enables a 1–2 ms round trip time in 5G. A shorter frame size also requires different structures for the control channels and faster processing times in UEs and base stations. Figure 3.9 shows transmission times and the corresponding round trip times.

The 5G subframe is 1 ms long and has 1, 2, 4, or more slots depending on the subcarrier spacing. The slot length equals 14 symbols. The typical scheduling interval is one slot. In addition it is possible to schedule data using a mini-slot with a length of typically 2, 4, or 7 symbols. The purpose of the mini-slot is to enable low latency communication for all subcarrier spacings. Mini-slot brings in the flexibility that allows a low latency service to use a very short transmission time while other services can use a longer transmission time. The mini-slot concept is illustrated in Figure 3.10 with a 1 ms slot and a 0.14 ms mini-slot of two symbols.

3.4.3 Self-Contained Subframe

The self-contained subframe in 5G includes both downlink and uplink control parts. It enables low latency operation with acknowledgment/negative acknowledgment (ACK/NACK) feedback in the same TDD subframe. A single subframe contains everything related to the data transmission: downlink grant, downlink data, and uplink ACK. The modularity of the self-contained subframe is handy when introducing new services later in 5G networks. Since 5G with the self-contained subframe supports blank sub-frames, it enables forward compatibility with future services on the same carrier.

The self-contained subframe is divided into the downlink control part, data transmission part, and uplink control part. The data part can be utilized for downlink or for uplink in the case of dynamic TDD operation. Since the control channels are at the

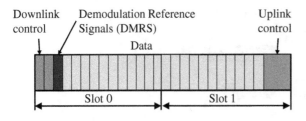

Figure 3.11 Self-contained subframe in 5G.

edges of the subframe, there is no downlink–uplink interference between the control channels.

The self-contained subframe also brings other benefits including unlicensed band operation and beamforming optimization. Fast switching between the downlink and uplink helps in the case of the reciprocal channel to optimize the performance with massive MIMO antennas (see Figure 3.11).

3.4.4 Asynchronous HARQ

5G supports asynchronous Hybrid Automatic Repeat Request (HARQ) retransmission. Asynchronous HARQ allows retransmission to occur with flexible delay after the first transmission, which gives more freedom for the packet scheduler optimization. LTE uses synchronous HARQ in the uplink where the retransmission must occur with an 8 ms delay after the first transmission. The need for the retransmission is signaled in the downlink by an NACK message in LTE. 5G does not use explicit ACK/NACK signaling, but the need for the retransmission is indicated with the new uplink resource allocation with the same process number. The uplink retransmission schemes are shown in Figure 3.12

LTE synchronous uplink retransmission

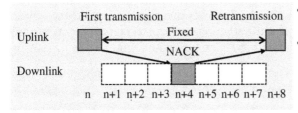

- Fixed retransmission timing in uplink with 8 ms delay
- ACK/NACK signaling

5G asynchronous uplink retransmission

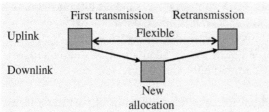

- Flexible retransmission timing
- No ACK/NACK signaling
- Retransmission indicated with new allocation with same process number

Figure 3.12 Uplink retransmission schemes in LTE and in 5G.

3.4.5   Lean Carrier

The LTE base station transmits CRSs four times per millisecond. Reference signals are needed by the UE for cell search, mobility measurements, and channel estimation and decoding. LTE reference signals must be transmitted all the time even if there are no connected UEs in the cell. 5G system design is different: there are no CRSs at all. 5G has user-specific reference signals instead, where the reference signals are transmitted together with data. If no user data transmission occurs, there is no reference signal transmission either. There are a number of benefits in flexible reference signal transmission:

- Lower base station power consumption. The 5G base station can utilize power saving mode during low load cases because there is no need to keep up with frequent reference signal transmission. The ultimate target is "zero users–zero power consumption."
- Less interference from reference signals, which minimizes inter-cell interference and improves network capacity. LTE cell reference signals make 10% of the total base station power during maximum loading, but the share of reference signals is substantially higher – even 50% of the interference – during the low load.
- More efficient beamforming. CRSs are useless for user-specific beamforming, which requires user-specific reference signals. Since beamforming is an essential part of 5G systems, the reference signal structure needs to be designed accordingly.

Figure 3.13 illustrates the transmission of LTE reference signals: 20 times in a 5 ms period, that is, four times every millisecond. 5G does not have any CRS. Only synchronization and broadcast channels are transmitted with a typical frequency of every 20 ms. The relative share of 5G common channels becomes even lower when the subcarrier

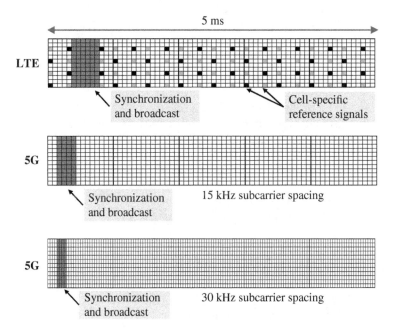

Figure 3.13 Lean carrier in 5G.

spacing increases. Both 15 and 30 kHz cases are shown. Synchronization and broadcast channels use four symbols, i.e. 1.4% of all symbols in 20 ms with 15 kHz and 0.7% of all symbols with 30 kHz.

3.4.6 Adaptive Reference Signals

Reference signals are required for the channel estimation and demodulation of user data. The reference signal transmission frequency is adaptive in 5G. The benefit is that the transmission can be optimized according to the environment and the expected mobile speed. The Doppler frequency Δf can be calculated from the mobile speed v, carrier frequency f, and speed of light c:

$$\Delta f = \frac{v}{c} f.$$

If we want to support 500 km/h with 3.5 GHz, the Doppler frequency is 1600 Hz. The phase of the received signal changes considerably 1600 times per second, and the required reference signal frequency needs to be considerably higher. This means the reference signal transmission must occur multiple times every millisecond. On the other hand, if the mobile speed is lower, like 3 km/h, the Doppler frequency is just 10 Hz, and the channel remains similar for more than 10 ms. LTE radio is designed for a Doppler frequency of up to 750 Hz with a fixed frequency of the reference symbols. The adaptive reference signal frequency is illustrated in Figure 3.14. The frequency of the reference signals also depends on the Doppler spread. When there is a dominant propagation path as in the 3GPP High Speed Train (HST) models, the residual Doppler spread is small after frequency offset correction. The greatest need for a large number of reference signals is for channels that have a U-shaped Doppler spread and an average frequency offset of zero. We also note that the UE synchronizes to the downlink signal including the Doppler spread. Therefore, the experienced Doppler shift can be up to two times higher in the base station receiver than that shown here.

3.4.7 Adaptive UE Specific Bandwidth

5G allows the bandwidth to be configured for each UE differently, which is also known as bandwidth part. Figure 3.15 illustrates an example with a 100 MHz carrier when UE1 uses instantaneously full 100 MHz, UE2 uses 20 MHz, and UE3 10 MHz. The UE with

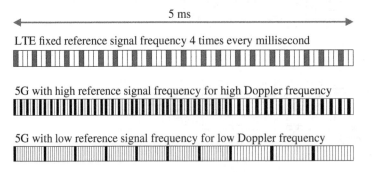

Figure 3.14 Adaptive reference signal frequency in 5G.

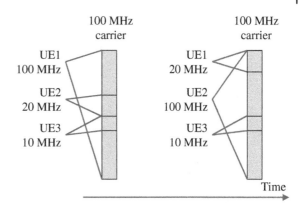

Figure 3.15 Adaptive UE specific bandwidth part.

100 MHz could support high-data-rate mobile broadband, while smaller bandwidths can be used for less demanding applications. Bandwidth can be adapted dynamically per UE depending on the data rate requirements. The next moment, the UE1 bandwidth is changed to 20 MHz and UE2 to 100 MHz. Adaptation brings benefits in term of device power consumption from the RF and from the baseband point of view. The UE is required to receive and transmit only within its own bandwidth part. Note that there are no changes to the UE categories in Release 15: all UEs must support full 100 MHz bandwidth even if UEs are allowed to instantaneously use less bandwidth. For more details about device design aspects, see Chapter 13.

3.4.8 Distributed MIMO

Distributed MIMO refers to the downlink transmission from two or more cells toward the UE, or to the uplink reception from the UE to two or more cells. Distributed Multiple Input Multiple Output (dMIMO) is also known as Coordinated Multipoint Transmission (CoMP) and Multi-Transmission Reception Point (TRP). The target of distributed MIMO is to improve the cell edge performance by increasing signal power toward the UE and by minimizing inter-cell interference, and to improve connection reliability. Distributed MIMO can be considered as an extension of massive MIMO where the transmitters are not within a single antenna but are distributed to different sites. The concept is illustrated in Figure 3.16 where three radio units transmit data simultaneously to a single UE. The transmission is controlled by a centralized unit with low latency connections to each radio unit. Distributed MIMO has major potential to improve user data rates, especially in the cell edge conditions.

CoMP technology has been utilized successfully in uplink in the LTE, especially in busy mass events where the uplink capacity is the bottleneck. The field results indicate a three times higher uplink capacity. CoMP gains in downlink have been difficult to obtain in LTE because UE support was missing and because CoMP in FDD requires fast feedback signaling. 5G has better potential to take advantage of CoMP in downlink also compared to LTE.

3.4.9 Waveforms

A number of different waveforms have been used in mobile technologies from Time Division Multiple Access (TDMA) and Frequency Division Multiple Access (FDMA)

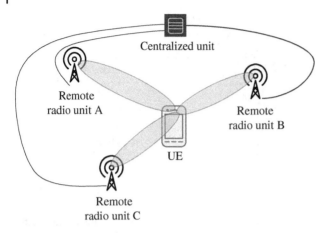

Figure 3.16 Distributed MIMO transmission.

in 2G Global System for Mobile Communications (GSM) to Code-Division Multiple Access (CDMA) in 3G Wideband Code-Division Multiple Access (WCDMA), and to Orthogonal Frequency Division Multiple Access (OFDMA) in 4G LTE downlink. The uplink solution in LTE is Single Carrier Frequency Division Multiple Access (SC-FDMA), which was selected to minimize UE RF requirements. The SC-FDMA option is included also in 5G for maximum coverage. The main waveform in 5G uplink is OFDMA, which is different from the LTE uplink. The motivation is that OFDMA can be shown to have clearly better performance than SC-FDMA, especially with multiantenna transmission and higher-order modulation. Multiantenna uplink transmission was not implemented in LTE, while it is expected to be part of 5G deployments. The drive for a higher data rate in 5G explains why OFDMA is preferred in the 5G uplink. Table 3.4 illustrates the multiple access solutions in different mobile generations. Table 3.5 shows the link performance with OFDMA and SC-FDMA with 4×4 MIMO and with Quadrature Phase Shift Keying (QPSK), 16-QAM, and 64-QAM. There is a 1.6–3.9 dB performance benefit for OFDMA compared to SC-FDMA.

Table 3.4 Waveforms and multiple access solutions from 2G to 5G.

	2G GSM	3G WCDMA	4G LTE	5G
Downlink	TDMA	CDMA	OFDMA	OFDMA
Uplink	TDMA	CDMA	SC-FDMA	OFDMA/SC-FDMA

Table 3.5 Uplink SNR requirement with 4×4 MIMO.

	QPSK (dB)	16-QAM (dB)	64-QAM (dB)
OFDMA	6.2	14.2	19.9
SC-FDMA	7.8	16.8	23.8

3.4.10 Channel Coding

Channel coding is needed in mobile systems to provide more reliable connection for data and control in the presence of fading and interference. The channel coding in 3G and in LTE was Turbo coding for data channels, while 5G uses a different solution: Low Density Parity Check (LDPC) for data channels and Polar coding for control channels. LDPC, Turbo, and Polar have similar link performance for data-intensive applications, while LDPC has a clear advantage over Turbo and over Polar for implementation complexity. These implementation aspects are important when the peak data rate in 5G increases even beyond 10 Gbps. Table 3.6 shows the benchmarking of coding solutions in terms of area efficiency and energy efficiency. LDPC shows the highest efficiency in term of Gbps per silicon area, and LDPC also shows highest energy efficiency in terms of bits per nanojoule. LDPC has most efficient implementations due to parallelized architecture and flexibility of code design, and can satisfy new radio access requirements. Polar coding was proposed in 3GPP for 5G, but Polar coding is less efficient than LDPC for high data rates. As a compromise, Polar coding was selected to be used for the control channels.

3.4.11 Pipeline Processing and Front-Loaded Reference Signals

The 5G physical layer includes a concept called pipeline processing that aims to minimize both decoding latency and UE power consumption. The LTE subframe has the control part followed by the reference signals and data. The LTE UE must receive the complete subframe before decoding can be started. 5G has the Demodulation Reference Signal (DMRS) in the beginning of the subframe followed by data symbols that can be decoded one by one. This approach is called front-loaded DMRS, and the target is to minimize decoding latency. Pipeline processing enables faster and continuous decoding. Pipeline processing also enables a faster UE sleep mode if no data are transmitted to that UE. The concept of pipeline processing is illustrated in Figures 3.17 and 3.18.

3.4.12 Connected Inactive State

5G requires new structures for the Radio Resource Control (RRC) protocols in order to minimize setup time and signaling. LTE RRC is designed with two RRC states: idle and connected. The LTE UE sits normally in the idle state to minimize UE power consumption. When there is a need to transmit any data, the UE is first transferred to the RRC-connected state, data transmission is completed, and the UE returns to the idle after state the inactivity timer expires. The LTE approach is not optimal for the frequent transmission of low data volumes, which is typical from smartphones or IoT devices. 5G

Table 3.6 Channel coding complexity benchmarking Adapted from [1].

	LDPC	Turbo	Polar
Silicon area efficiency	>10 Gbps/mm^2	<5 Gbps/mm^2	<5 Gbps/mm^2
Energy efficiency	>100 bit/nJ	<1 bit/nJ	<100 bit/nJ

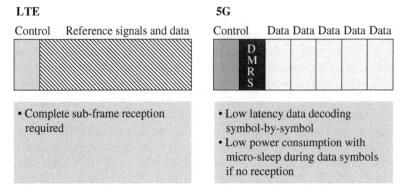

Figure 3.17 Pipeline processing in 5G.

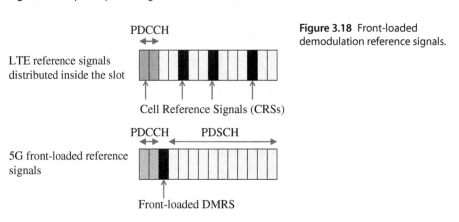

Figure 3.18 Front-loaded demodulation reference signals.

specifications include the RRC-connected inactive state, where the target is to maintain the RRC connection for a long time and minimize the RRC setup signaling. The UE can send data even after long inactivity without any RRC setup signaling, which also minimizes latency. Minimized signaling is beneficial for the UE, radio network, and core network. The high-level RRC structure is illustrated in Figure 3.19. We note that the Connected Inactive State has similarities with the 3G Cell_PCH (Paging Channel) state,

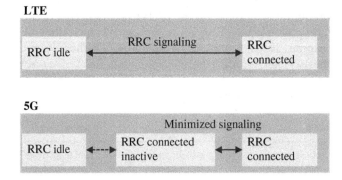

Figure 3.19 Always-on RRC protocol in 5G.

where the RRC connection was maintained for long time while minimizing UE power consumption.

3.4.13 Grant-Free Access

Any uplink data transmission in LTE requires a capacity request from the UE to the base station followed by a packet scheduling decision and resource allocation. That is true even when the RRC connection exists. This allocation cycle causes an additional delay and signaling. 5G design includes grant-free access where the UE is allowed to send some data without any scheduling. Grant-free access is known also as contention-based access. Grant-free access minimizes latency and related signaling. Grant-free access is similar to data transmission on 3G Random Access Channel (RACH) (see Figure 3.20).

3.4.14 Cell Radius of 300 km

5G radio is designed to support multiple different deployment scenarios. Typical cell sizes are small, from a few hundred meters to a few kilometers. There may also be cases where very large cell sizes are required, for example, when providing coverage over open areas like the sea or when providing coverage to aircraft or when using balloons or drones for providing coverage to extreme locations. The very large cell size must be taken into account in the physical layer design because of the propagation delay. 5G supports up to 300 km cell range, which corresponds to a 1 ms one-way propagation delay as shown in Figure 3.21. LTE was designed up to 100 km. The long propagation delay is considered in the reception of the RACH preamble and in the range of the timing advance. Figure 3.22 illustrates the random access case where the uplink signal arrives

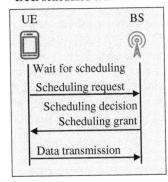
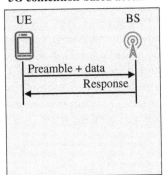

Figure 3.20 Grant-free access in 5G.

Figure 3.21 Propagation delay and 300 km cell range.

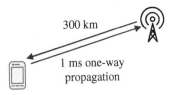

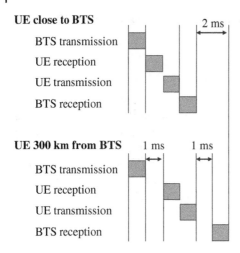

UE close to BTS

BTS transmission

UE reception

UE transmission

BTS reception

2 ms

UE 300 km from BTS   1 ms   1 ms

BTS transmission

UE reception

UE transmission

BTS reception

Figure 3.22 Transmission and reception timing with long propagation delay.

2 ms later from the far-way UE. The UE is not aware of the propagation delay during the preamble transmission. Therefore, the propagation delay must be considered in the uplink reception window.

The uplink reception in 5G is synchronous, which requires that the transmissions from distant UEs must start earlier than those from close UEs in order to simultaneous uplink reception in the base station. The uplink transmission timing is adjusted with a timing advance, and the range of the signaling is large enough to accommodate a 300 km cell range.

3.5 Network Slicing

5G networks are designed to support very diverse and extreme requirements for latency, throughput, capacity, and availability. Network slicing offers a solution to meet the requirements of all use cases in a common network infrastructure. The concept of network slicing is illustrated in Figure 3.23. The same network infrastructure can support, for example, smartphones, tablets, virtual reality connections, personal health devices, critical remote control, or automotive connectivity. LTE supports Quality of Service (QoS) differentiation, but we need more in 5G. Network slicing is described in more detail in Chapter 5.

3.6 Dual Connectivity with LTE

5G is the first radio solution which is closely integrated with the legacy radio network for a smooth rollout and seamless experience. The solution is called dual connectivity, where a 5G UE can have simultaneous connections to a 5G radio network and an LTE radio network. The first version of 5G in 3GPP supports the non-standalone (NSA) architecture with dual connectivity. Both 5G base stations (gNodeB) and LTE base stations (eNodeB) are connected to Evolved Packet Core (EPC). The control plane goes

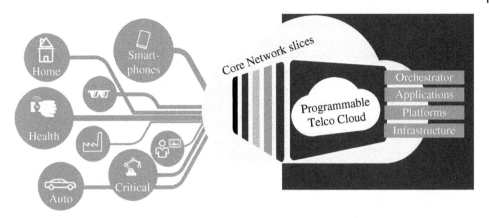

Figure 3.23 Network slicing concept.

via LTE. This architecture is called Option 3x. Usage of the existing EPC and LTE mobility procedures make 5G introduction fast with Option 3x.

5G can also be deployed as a standalone (SA) solution without LTE using architecture Option 2 and 5G core network (5G-CN). The SA solution is simpler than NSA since there is no need for interworking between 5G and LTE. SA has the benefit that it can provide lower latency than the NSA solution since there are no LTE protocols included which may cause additional latency. The 5G core network enables new end-to-end services.

It is also possible later to have an NSA architecture with both 5G and LTE nodes connected to the new 5G core network. The control plane can go via LTE or via 5G in Option 7 and Option 4.

5G architecture options are illustrated in Figure 3.24. See Chapter 5 for more details about architecture options.

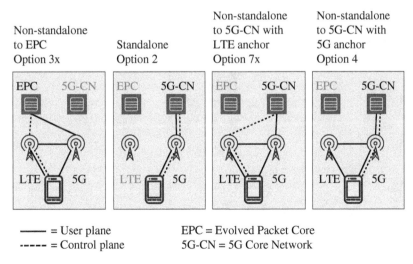

Figure 3.24 Main 5G architecture options.

3.7 Radio Cloud and Edge Computing

Radio network architecture has typically been distributed, where all the radio processing is done close to the antenna in the base station. The core network architecture has been highly centralized with only a small number of core sites. Future architecture will be different: radio processing will becomes more centralized for better scalability, and core processing will become more distributed for lower latency. This evolution will bring edge cloud servers to the mobile networks. These server locations can then host both radio and core network functionalities. The architecture evolution is illustrated in Figure 3.25.

The 5G specifications are designed to support radio cloud by defining a new interface within the radio network for splitting the functionality between a distributed RF site and the centralized edge cloud site. Radio cloud implementation enables network scalability, for example, when adding a large number of IoT-connected devices. Figure 3.26 illustrates the main functionality split options and the related transport requirements. The rightmost figure shows the distributed option where all radio functions are close to the antenna. That option is used in most LTE networks and is also typical in 5G networks. The leftmost figure shows another option that has been used in LTE networks. It is baseband hoteling, where all the baseband processing is located in a centralized place. The Common Public Radio Interface (CPRI) interface can be used between the baseband and RF. CPRI requires very low latency, and the CPRI data rate requirement increases as a function of the bandwidth and the number of antennas. If we have a 100 MHz 5G carrier with 64TRX massive MIMO with 16 streams, the CPRI data rate requirement is close to 1 Tbps without CPRI compression, which makes it difficult to use the CPRI interface in 5G. Therefore, two other functionality split options with a more relaxed transport requirement are considered. Low-layer and higher-layer split options have the delay critical functions located close to the RF and the less delay critical functions at the edge cloud site. That allows the transport requirements to be minimized. If the transport network can provide low latency (<1 ms), then just part of Layer 1 (Fast Fourier Transform [FFT] and channel coding) is located close to the

Typical network architecture

Future architecture

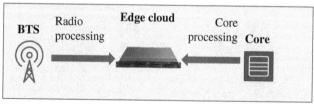

Figure 3.25 Network architecture evolution.

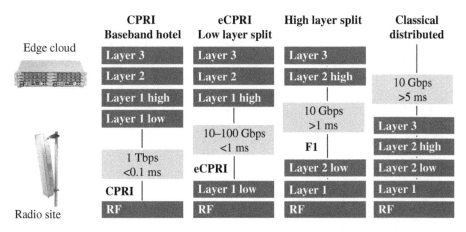

Figure 3.26 Functionality split options in 5G radio network and approximate transport requirements for 3-sector base station with 100 MHz and massive MIMO antenna.

RF while the other part of Layer 1 and the other layers are in the edge cloud. The interface in that case is called enhanced Common Public Radio Interface (eCPRI). If the transport latency is higher, then more functionalities are located at the antenna site, and only part of Layer 2 and Layer 3 are in the edge cloud. The interface in that case is called F1.

3.8 Summary

5G networks are expected to deliver substantially higher capabilities, higher data rates and a number of new services from low latency and high reliability to low-cost IoT modules. New technology components are needed to fulfill those targets. This chapter summarized the main new technologies: new spectrum options with wideband carrier, beamforming optimized radio design, flexible physical and protocol layers, network slicing, dual connectivity, and cloud optimized network architecture. 5G implies a massive change in the network design and optimization. The following chapters provide more details about 5G design and specifications.

Reference

1 Nokia, Alcatel-Lucent Shanghai Bell 2016. Performance and complexity of Turbo, LDPC and Polar codes. 3GPP TSG-RAN WG1 #84bis, R1-162897. Busan, Korea, 11th – 15th April. https://www.3gpp.org/ftp/tsg_ran/WG1_RL1/TSGR1_84b/Docs/R1-162897.zip

4

Spectrum

Harri Holma[1] and Takehiro Nakamura[2]

[1] *Nokia, Finland*
[2] *NTT DOCOMO, Japan*

CHAPTER MENU

4.1   Introduction

Spectrum is the key asset for mobile operators. The available spectrum is a major factor in defining how much capacity and how wide a coverage the mobile network can provide. This chapter discusses the different spectrum options for 5G, their characteristics, and the expected spectrum for the early phase of 5G deployments. 5G is the first ever mobile radio system that is designed to utilize any spectrum between approximately 400 MHz and 90 GHz. 5G is also designed to be deployed in licensed, shared, and unlicensed spectrum bands. 5G can utilize Frequency Division Duplex (FDD) technology for paired spectrum and Time Division Duplex (TDD) technology for unpaired spectrum. When the same 5G technology is deployed on different bands, it will be efficient to fully exploit the capacity and coverage properties of the bands optimally together.

The motivations for 5G deployments on different spectrum bands are summarized in Figure 4.1. The high bands at millimeter wave have lot of spectrum and enable high capacity and data rates. The low bands have great propagation and provide wide area coverage. Millimeter wave generally refer to the frequencies where the wavelength is in millimeters, that is, frequencies between 30 and 300 GHz. In practice, 5G frequencies at 24–30 GHz also come under millimeter wave. The amount of spectrum per operator

5G Technology: 3GPP Evolution to 5G-Advanced, Second Edition.
Edited by Harri Holma, Antti Toskala, and Takehiro Nakamura.

Band	5G use case and motivation	Spectrum per operator
Above 52 GHz	• Very high data rate >10 Gbps • Licensed and unlicensed frequencies	>1000 MHz
24–39 GHz	• High data rate >5 Gbps • Hot spot capacity and fixed wireless	800 MHz
5 GHz unlicensed	• Local solution • No spectrum license needed	Up to 500 MHz shared
3.3–5.0 GHz and 2.6 GHz	• 2 Gbps with100 MHz and 4x4MIMO • Urban capacity with massive MIMO	100 MHz
1.5–2.6 GHz	• 5G brings massive MIMO capability • Gradual refarming from LTE to 5G	Combined 2 × 50 MHz
Sub 1 GHz	• Wide area and deep indoor coverage • Low latency with FDD	Combined 2 × 25 MHz

Figure 4.1 Motivation for 5G deployment on different spectrum bands.

at 24–39 GHz is typically up to 800 MHz, which enables user data rates of 5 Gbps. Higher frequencies above 50 GHz offer more spectrum, which enables user data rates above 10 Gbps. The challenge with high spectrum is the short propagation. Millimeter wave signals attenuate fast, and the cell range is limited to a few hundred meters.

The spectrum at 3.3–5.0 GHz is an attractive combination of high data rate and extensive coverage. The spectrum per operator is typically 100 MHz. The combination of 100 MHz and 4×4 Massive Multiple Input Multiple Output (MIMO) gives a peak data rate of 2 Gbps. The coverage area of 3.3–5.0 GHz frequencies can be close to the 2 GHz band when combined with high-gain base station antennas and beamforming techniques. That spectrum uses TDD technology. Another mainstream TDD spectrum for 5G is 2.5 GHz.

The spectrum at 1.5–2.6 GHz is widely used by Long Term Evolution (LTE) networks, especially for urban area coverage and capacity. The maximum spectrum per band is typically 2 × 20 MHz per operator. That spectrum is mostly FDD, while unpaired blocks at 2.3 and 2.6 GHz also use TDD technology. 5G deployment at these frequencies can be implemented by refarming from LTE to 5G. One motivation for refarming to 5G is that 5G is designed for massive MIMO.

The spectrum below 1 GHz is essential for providing wide area coverage and deep indoor penetration. Therefore, low bands can be utilized for critical communications with ultra-high reliability in wide area networks, including rural cases too. FDD technology has also a latency benefit compared to TDD since FDD supports simultaneous transmission and reception. The total spectrum in sub-1 GHz bands is typically 2 × 20–30 MHz on two or three different spectrum blocks.

Spectrum licensing can use a number of different models: exclusive licensing, shared licensing, or unlicensed spectrum; see Figure 4.2. 2G and 3G radios are only deployed on licensed spectrum, and LTE is also mostly deployed on licensed spectrum. The starting point in 5G is that it can utilize spectrum with any licensing model. Most frequencies

Figure 4.2 5G will support different spectrum licensing methods.

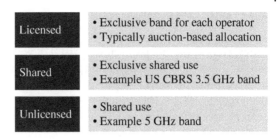

Licensed	• Exclusive band for each operator • Typically auction-based allocation
Shared	• Exclusive shared use • Example US CBRS 3.5 GHz band
Unlicensed	• Shared use • Example 5 GHz band

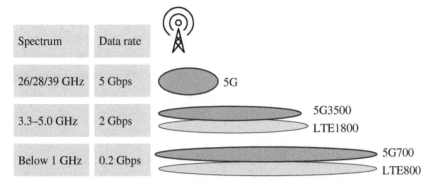

Spectrum	Data rate	
26/28/39 GHz	5 Gbps	5G
3.3–5.0 GHz	2 Gbps	5G3500 / LTE1800
Below 1 GHz	0.2 Gbps	5G700 / LTE800

Figure 4.3 Example of early phase 5G spectrum usage and coverage.

below 5 GHz have been allocated with exclusive licensing. Shared licensing is used today, for example, at 3.5 GHz Citizen Broadband Radio Service (CBRS) band in the United States. 5G will also be designed to operate in unlicensed spectrum, for example, 5 GHz band. Unlicensed 5G allows local 5G networks to be provided without any spectrum license.

Figure 4.3 illustrates typical spectrum usage in the early phase of 5G rollout. The mainstream spectrum below 6 GHz globally will be 3.5 GHz, covering up to 400 MHz from 3.4 to 3.8 GHz with possible later extension up to 4.2 GHz. In addition, China and Japan will use a part of 4.4–5.0 GHz. The spectrum around 3.5 GHz is attractive for 5G because it is available globally and the amount of spectrum is relatively high. Operators are able to get 100 MHz continuous spectrum in many countries which, is a perfect match with 5G device capabilities. The lower bands tend to have a maximum of 20 MHz continuous spectrum per band. The target of 3.5 GHz is to provide similar coverage as lower frequencies by using a higher-gain beamforming antenna to compensate for the higher path loss. Therefore, 5G at 3.5 GHz can utilize existing base station sites.

5G will also need to have low bands below 1 GHz in order provide deep indoor penetration and large coverage areas. Extensive coverage is important for the new use cases like Internet of Things (IoT) and critical communication. The low band could be 700 MHz, which is available in many countries at the same time as 5G. Another option is 900 MHz, which is mostly occupied by 2G and 3G today, or 850 MHz. The spectrum allocation for legacy technologies could be minimized, or the legacy networks could be closed down. One option for 5G low band in the United States is also 600 MHz. It is also possible to share spectrum between LTE and 5G with dynamic spectrum sharing.

Millimeter wave at 24, 28, and 39 GHz is used to provide extremely high local data rates for smartphone users and for fixed wireless users.

4.2   Millimeter Wave Spectrum Above 20 GHz

Millimeter wave spectrum above 20 GHz will be essential in 5G for providing the required increase in capacity and data rates for the future requirements. Regulatory bodies around the world are currently working toward opening up new spectrum bands from 6 to 100 GHz. These higher frequencies provide much more bandwidth than the current spectrum below 6 GHz. If only a fraction of that 94 GHz of spectrum were made available to the mobile industry, it could easily be 20–30 GHz of new spectrum, which is significantly higher than all the spectrum available below 6 GHz. In practice, up to 800 MHz bandwidth per operator can be provided in the early phase below 40 GHz and up to 2 GHz above 50 GHz frequencies later. All these spectrum blocks use TDD technology. The following spectrum blocks will be available in millimeter wave and are illustrated in Figure 4.4:

- *24–29 GHz band.* The upper part of this band, known as the 28 GHz band, covers 26.5–29.5 GHz, and the lower part, known as the 26 GHz band, covers 24.25–27.5 GHz.
- *37–43.5 GHz*, known as the 40 GHz band.
- *57–64 GHz band*, also known as the V-band or 60 GHz band.
- *71–76 GHz and 81–86 GHz bands*, also known as the E-band or 70 and 80 GHz band.
- *92–95 GHz band*, also known as the W-band or 90 GHz band.

The first millimeter wave spectrum for 5G will be at the 28 GHz band in the United States, which is called Band n261 in 3rd Generation Partnership Project (3GPP) at 27.5–28.35 GHz. A few other countries, including Korea and Japan, use band n257, covering 26.5–29.5 MHz. Korea allocated 800 MHz for all three operators at 26.5–28.9 GHz during 2018. Europe and China will use band n258, covering 24.25–27.5 GHz (Figure 4.5).

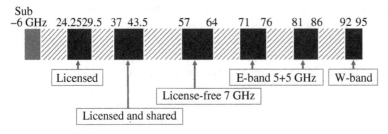

Figure 4.4 Main spectrum blocks above 20 GHz bands.

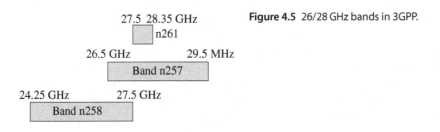

Figure 4.5 26/28 GHz bands in 3GPP.

The 60 GHz band is currently provisioned for unlicensed operation. This band has a large amount of oxygen absorption, but it does not significantly impact communications in the small cells. The 60 GHz unlicensed band could be one option for local 5G networks without spectrum license.

70 and 80 GHz bands operate under a lightly licensed paradigm, and this band can be aggregated up to a total of 2×5 GHz. Rain attenuation can be severe with longer distances, but will not be a problem for small distances such as less than 100–200 m. Any system operating in the 70 and 80 GHz bands must coexist with fixed satellite service, automotive radar (77–81 GHz), and radio astronomy. The 70 and 90 GHz bands are commonly used for microwave backhauling. No 5G usage is planned for these bands in the short term.

The 90 GHz band is provisioned for unlicensed operation but only for indoor applications. This band can also be used for outdoor point-to-point license-light operation.

The spectrum characteristics below 6 GHz bands are well known based on the existing LTE deployments, while there are new learnings required for millimeter wave. Table 4.1 illustrates the main differences between low bands and millimeter wave. The low-band deployments are relatively narrowband, and the total spectrum in downlink in general has a maximum of 100–200 MHz, while the millimeter wave cases can have up to 1–2 GHz of spectrum. The cell sizes at low bands can be large: more than 1 km in urban areas, and more than 10 km in rural areas. The cell size at millimeter wave is typically 100 m, or a few hundreds of meters in the open area. Low bands tend to be interference and bandwidth limited, and solutions are needed for inter-cell interference minimization. Millimeter wave cases are typically noise limited, and solutions are needed for cell range maximization. The millimeter wave can take better advantage of beamforming with a large number of antenna elements because the size of the antenna becomes smaller at high frequencies.

Signal attenuates faster as a function of distance at higher frequencies. That is true in the line-of-sight (LOS) case and even more true in the non-line-of-sight (NLOS) case. Additionally, the penetration loss through walls typically increases as a function of

Table 4.1 New spectrum and new challenges.

	Below 6 GHz	Above 20 GHz
Bandwidth	Limited spectrum. Typically <200 MHz per operator	Large spectrum resources. Up to 800 MHz per operator
Interference	Inter-cell interference main limitation	Coverage (noise) main limitation
Cell size	Large >1 km	Small <1 km
Antenna elements	Less than 256 at 3–5 GHz Less than 128 at <2 GHz	Even more than 1024
Base station maximum power	Up to 200 W per band	Less than 10 W
Main challenge	Inter-cell interference control for maximizing capacity with limited spectrum	Link performance improvement with beamforming antennas for maximizing coverage area

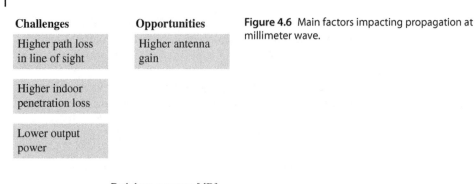

Figure 4.6 Main factors impacting propagation at millimeter wave.

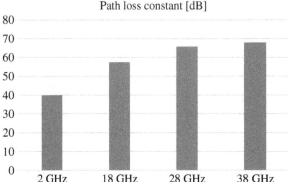

Figure 4.7 Measurement-based path loss constant for line-of-sight conditions Adapted from [1].

the frequency. We also need to consider that the base station output power is generally lower at millimeter wave than at low bands. When all these factors are considered, the cell range at millimeter wave tends to be very small. We illustrate the impact of high frequency on propagation in this section. On the other hand, the antenna gain can be increased at millimeter wave because the wavelength becomes smaller, which allows a higher-gain antenna in a smaller form factor. These main factors impacting the cell range are listed in Figure 4.6.

Figure 4.7 illustrates path loss measurements in the LOS case with different frequencies from 2 to 38 GHz. There is a constant difference between the frequencies regardless of the distance. The path loss is approximately 26 dB higher at 28 GHz vs. 2 GHz, and 2 dB more at 38 GHz. These values clearly illustrate that the path loss at millimeter wave is considerably higher than at low band, and it will impact the cell range.

When there are obstacles between the transmitter and receiver, the signal attenuation increases further, and the impact increases generally as a function of the frequency. Figure 4.8 illustrates the outdoor to indoor penetration loss at 28 GHz for residential family houses. The measurements compare the path loss when the receiver is outside by the window and indoors 1.5 m from the window. The median measured loss is 9 dB in the case of a plain glass window. The loss increases to 15 dB with low-emissivity windows and to 17 dB with foil-backed wall insulation. The building material was based on wood in these measurements. The penetration losses would be even higher if the device were deeper inside the building and if the building material were concrete. Propagation deeper into the building will increase the loss due to internal walls, furniture, and so on.

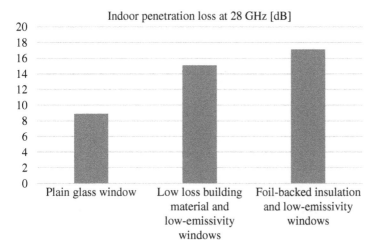

Indoor penetration loss at 28 GHz [dB]

Figure 4.8 Measured outdoor to indoor penetration loss at 28 GHz in wooden building Adapted from [2].

This additional loss appears to be rather weakly frequency dependent but rather strongly dependent on the composition of the interior of the building. The observed losses were 0.2–2 dB/m over the 2–60 GHz range. The total indoor penetration loss could increase to 30–40 dB in the case of concrete building material.

Frequencies around 60 GHz experience some absorption by oxygen. The absorption, however, has only a marginal impact on the cell range since the attenuation is below 10 dB/km. Assuming a typical cell range of clearly less than 1 km shows that the worst-case oxygen absorption is just a few decibels. Since the presence of oxygen is fairly consistent at ground level, its effect on 60 GHz radio propagation can be easily modeled. Weather-related attenuation is even more insignificant at those bands. Even heavy rainfall of 25 mm per hour causes just 5 dB/km attenuation and makes only a very small percentage contribution to the aggregate attenuation in the 60 GHz oxygen absorption region. More details about millimeter wave measurements can be found in Chapter 13.

4.3 Mid-Band Spectrum at 3.3–5.0 GHz and at 2.6 GHz

Spectrum at 3.3–5.0 GHz, also known as the C-band, is the mainstream spectrum for 5G and provides a nice combination of high bandwidth, wide area coverage, and global availability. This spectrum is emerging as a core band for 5G. Figure 4.9 illustrates 3GPP frequency variants in 3.3–4.2 GHz and the typical global allocations. Many countries have – and will – allocate up to 400 MHz of spectrum at 3.4–3.8 GHz, while other countries are able to allocate only a part of that spectrum. Those allocations fall into the 3GPP band n78 covering 3.3–3.8 GHz. More spectrum is expected to be available later at 3.8–4.2 GHz, which is covered by the 3GPP band n77. Japan already allocated spectrum at 3.8–4.1 GHz in 2019.

Some regulators reserve a part of the 3.5 GHz spectrum for local and regional use for industrial networks. The target is that this part of the spectrum can be utilized locally

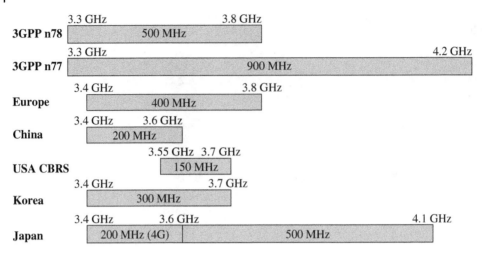

Figure 4.9 Global spectrum allocation at the 3.3–4.2 GHz band.

in the factories or other similar cases to provide private networks for ultra-reliable low-latency communication. For example, in Germany, 300 MHz is reserved for nationwide licenses, while 100 MHz is allocated for local cases.

3.5 GHz spectrum auctions have been completed in a number of countries so far. Figure 4.10 illustrates the spectrum allocations in a few countries. In many cases, operators are able to access 100 MHz, or nearly 100 MHz, continuous spectrum, which is a good starting point for 5G. In some cases, the allocation is less than 100 MHz per

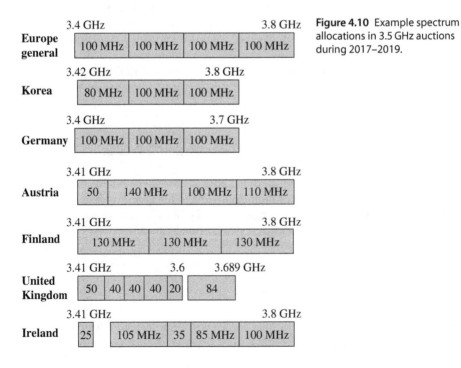

Figure 4.10 Example spectrum allocations in 3.5 GHz auctions during 2017–2019.

Figure 4.11 2.6 GHz band allocation.

2.496 GHz	Band n41 TDD	2.69 GHz
	194 MHz	

Band n7 Uplink	Band n38 TDD	Band n7 Downlink
70 MHz	50 MHz	70 MHz

operator, which allows operation of the 5G network, but the data rates and the capacity will be lower.

Another important mid-band spectrum for 5G is 2.6 GHz TDD, is referred to as band n41 in 3GPP. The total spectrum is 190 MHz, or even 194 MHz from 2496 to 2690 MHz, which enables wideband 5G usage. Time Division LTE (TD-LTE) has been deployed on that band in most countries. Therefore, 5G and TD-LTE need to coexist in practical networks. Some countries use a combination of FDD and TDD allocations with bands 7 and 38 (Figure 4.11).

The spectrum price is highly market dependent: more operators, less available spectrum, and high demand for more spectrum generally increase the spectrum price in the auction. A few spectrum auction results are shown in Figure 4.12. The price is shown as Euros per MHz per person, with the total spectrum price being divided by the amount of spectrum and by the country's population. The auction in Italy resulted in relatively high prices of 0.36 EUR/MHz/person. The auctioned spectrum was 200 MHz, and the total price was more than 4.3 billion Euros with the Italian population of 60 M. The spectrum price in Finland was 10 times lower than in Italy. Other auction prices, for example, Spain, the United Kingdom, Korea, and Ireland, fall between those extremes.

The spectrum prices are highly dependent on the band. The general trend is that low bands have a higher price per MHz than high bands. The first reason is that the low bands have better propagation and coverage than high bands. The second reason is that the low bands have less spectrum. Figure 4.13 illustrates the spectrum prices in

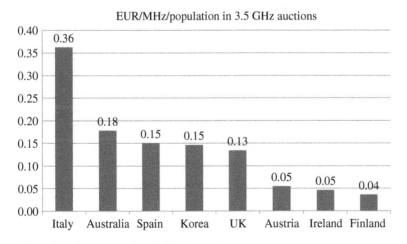

Figure 4.12 Spectrum prices in 3.5 GHz auctions.

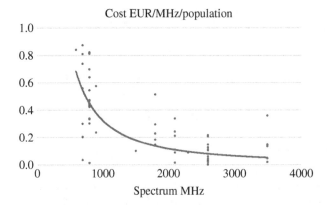

Figure 4.13 Spectrum prices with different frequencies and trend line.

several auctions. Each dot represents one spectrum auction in one country. The variance between different auctions is large. The trend line is added to show the general differences between the bands. The trend is that the sub-1 GHz band is approximately 10 times more expensive per MHz than high bands at 2.6 and 3.5 GHz.

4.4 Low-Band Spectrum Below 3 GHz

The low bands, especially sub-1 GHz bands, are important for providing extensive 5G coverage. The first-phase low bands are 600 MHz in the United States and 700 MHz in other markets. 700 MHz spectrum in 3GPP is band n28, which covers 2×45 MHz for Asia Pacific and 2×30 MHz in Europe and other markets where 800 MHz (band 20) is used. 600 MHz spectrum in 3GPP is band n71. These two bands are illustrated in Figure 4.14. Both bands are already used by LTE, which leads to coexistence of LTE and 5G technologies.

All existing LTE bands will eventually be refarmed to 5G. If there are no new frequencies available for 5G at low bands, then refarming will be needed already in the early phase of the 5G rollout. Dynamic spectrum sharing can be utilized for flexible refarming. The main LTE bands in the current networks are illustrated in Figure 4.15: the left side shows typical bands in Europe, and the right side shows typical bands in Americas. The frequencies used in the Asia Pacific region are a mix of these two cases.

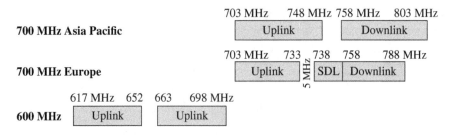

Figure 4.14 Mainstream 5G low bands at 700 MHz (band n28) and 600 MHz (band n71). SDL = Supplemental Downlink.

Europe

2600 MHz	Band 7+38 / 41
2300 MHz	Band 40
2100 MHz	Band 1
1800 MHz	Band 3
1500 MHz	Band 32 / 50 / 51
900 MHz	Band 8
800 MHz	Band 20
700 MHz	Band 28

Americas

2600 MHz	Band 41
2300 MHz	Band 30
2100 MHz	Band 4 / 66
1900 MHz	Band 2
850 MHz	Band 5 / 26
700 MHz	Band 12–14, 17
600 MHz	Band 71

Figure 4.15 Main LTE frequency bands.

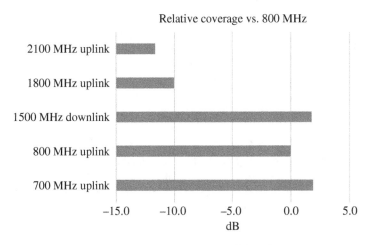

Figure 4.16 Coverage of 1500 MHz supplemental downlink.

One potential new frequency for 5G could be the 1500 MHz supplemental downlink band, which has a total bandwidth of 90 MHz. This band is an attractive solution for providing more downlink capacity with wide area coverage. Figure 4.16 illustrates the relative coverage. The 1500 MHz downlink has similar or better coverage as the 700 and 800 MHz uplinks. Therefore, 1500 MHz can be efficiently used for capacity boosting in those cells where large coverage and deep indoor penetration are required.

4.5 Unlicensed Band

The unlicensed 5 GHz band offers a lot of spectrum, as shown in Figure 4.17, even over 500 MHz, which is substantially more than any operator has in terms of licensed spectrum. 3GPP has specified band 46 covering 775 MHz at 5150–5925 MHz. The use of unlicensed spectrum is governed by some regulatory requirements, such as being able

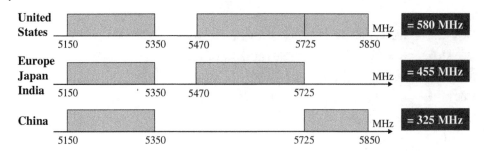

Figure 4.17 5 GHz unlicensed band.

to detect if a radar system is using the band (Dynamic Frequency Selection, DFS) or being able to coexist with other users of the band. The latter is often referred to as Clear Channel Assessment (CCA) or Listen-Before-Talk (LBT) and means that it is not always possible to transmit immediately if the intended channel is occupied. The maximum allowed transmission power also varies depending on the region and on the part of the 5 GHz band. Typically, some parts of the band are restricted to indoor use, with a maximum transmit power of 250 mW or less. Other parts allow higher transmission powers, typically around 1 W, and consequently are also suitable for outdoor deployments. In some cases, such as the US 5.725–5.850 GHz, there are no specific requirements other than the transmission power limitation.

2G and 3G technologies can only operate on licensed frequencies. The same was true also for LTE in 3GPP Releases 8–12, while Release 13 added Licensed Assisted Access (LAA), which allows utilization of the 5 GHz unlicensed band together with some of the lower licensed frequencies with carrier aggregation. The unlicensed band can be utilized with different options, as shown in Figure 4.18. The first option is to use the unlicensed band for boosting downlink capacity and data rates on top of the licensed band. This solution is called LAA, where the primary cell is on the licensed band and secondary cells are on the unlicensed band, providing higher capacity and higher data rates. LAA rollout

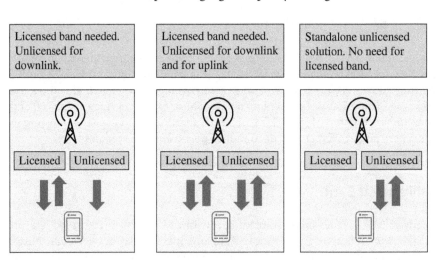

Figure 4.18 Options for utilizing unlicensed band in 5G.

started from the United States in 2017 and has been available in devices since 2018. LAA includes solutions to comply with the emission and coexistence rules of the 5 GHz band. The second option is to boost uplink data rates also by using unlicensed bands while still having primary cell on the licensed bands. The third option is a standalone unlicensed band version where radio coverage can be provided without any spectrum license. The standalone unlicensed version can provide local connectivity, for example, in factories or offices or campus areas. A standalone unlicensed version of LTE was never defined in 3GPP but the technical work was done by the MulteFire Alliance, and the solution is called MulteFire. The mandatory changes are included in the MulteFire specifications so that the existing Wi-Fi networks and MulteFire networks can coexist in the same band and in the same area.

Unlicensed band support for 5G will be defined in Release 16. The target is to include all these deployment options. 5G has a few benefits compared to LTE when operating in the unlicensed bands. LTE bandwidth per cell is limited to 20 MHz, which leads to the need for intra-band carrier aggregation at 5 GHz bands. The 5G bandwidth enables wider carrier bandwidth, which is more practical for taking advantage of the 5 GHz band. Another 5G benefit is the flexibility of the control channel allocations.

Multiple networks and technologies can coexist in the unlicensed bands with CCA and LBT rules. Coexistence simulations with Wi-Fi and 5G-Unlicensed (5G-U) sharing the same band are illustrated in Figure 4.19. The Wi-Fi system here is 802.11ac. The 20 MHz case is studied for both Wi-Fi and 5G. The results are shown for a medium load. A total four cases are studied: The Wi-Fi and 5G mean throughput is studied when the interfering technology is another Wi-Fi and another 5G. The results show that the Wi-Fi throughput is marginally higher if the interfering system is changed from Wi-Fi to 5G. That means 5G can nicely coexist with Wi-Fi and is even less harmful to Wi-Fi than another Wi-Fi system. The results also show that 5G provides 3–4 times higher throughput than Wi-Fi.

In order to understand the utilization of the 5 GHz unlicensed band, we conducted scanner measurements in a major shopping mall in Finland. A large number of Wi-Fi access points are visible in the band 5150–5350 MHz and quite a few also in the band

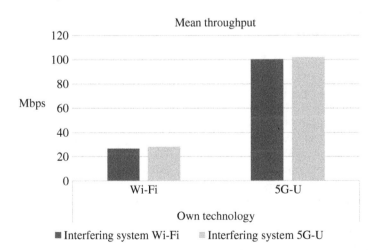

Figure 4.19 Mean throughput with 5G unlicensed and with Wi-Fi sharing the spectrum Adapted from [3].

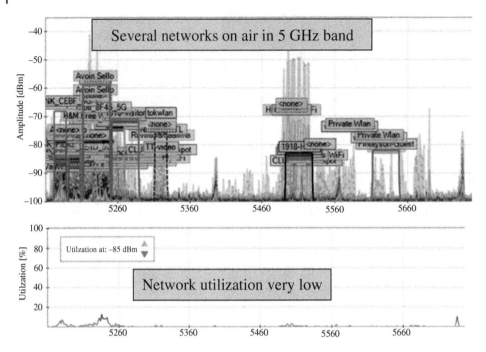

Figure 4.20 5 GHz band scanner measurements in a large shopping mall.

5470–5725 MHz. The access points in the frequency domain are shown in the upper part of Figure 4.20. It appears as if the spectrum is heavily occupied by the Wi-Fi signal. But it is not enough to check only the number of access points; we also need to analyze the utilization of the networks, which is shown in the lower part of Figure 4.20. The utilization threshold in these measurements was −85 dBm. The results show that the utilization is very low. There are many access points in that shopping mall, but there is still a lot of room for additional traffic because of the low utilization. The conclusion is that the unlicensed band 5G network could be used in this specific area to boost the capacity of licensed band networks.

4.6 Shared Band

Shared licensing is used today, for example, at the 3.5 GHz CBRS band in the United States. The shared access is commonly known as Licensed Shared Access (LSA). It allows a limited number of licensees to use the frequency band with one or more incumbent users already present. CBRS is controlled by a three-level spectrum authorization framework to combine a number of commercial uses on a shared basis with incumbent federal and non-federal users of the band. Access and operations will be managed by a dynamic spectrum access system. The three tiers are Incumbent Access, Priority Access, and General Authorized Access. Incumbent Access users include authorized federal and Fixed Satellite Service users currently operating in the 3.5 GHz band, including radar services. These users will be protected from harmful interference from Priority Access and General Authorized Access users. The Priority Access tier consists of Priority Access Licenses (PALs) that will be assigned using competitive bidding within the 3550–3650 MHz portion of the band. Each PAL is defined as a

Table 4.2 3GPP frequency variants [5, 6].

Operating band	Uplink	Downlink	Duplex
n1	1920–1980 MHz	2110–2170 MHz	FDD
n2	1850–1910 MHz	1930–1990 MHz	FDD
n3	1710–1785 MHz	1805–1880 MHz	FDD
n5	824–849 MHz	869–894 MHz	FDD
n7	2500–2570 MHz	2620–2690 MHz	FDD
n8	880–915 MHz	925–960 MHz	FDD
n12	699–716 MHz	729–746 MHz	FDD
n14	788–798 MHz	758–768 MHz	FDD
n18	815–830 MHz	860–875 MHz	FDD
n20	832–862 MHz	791–821 MHz	FDD
n25	1850–1915 MHz	1930–1995 MHz	FDD
n28	703–748 MHz	758–803 MHz	FDD
n29	N/A	717–728 MHz	SDL
n30	2305–2315 MHz	2350–2360 MHz	FDD
n34	2010–2025 MHz	2010–2025 MHz	TDD
n38	2570–2620 MHz	2570–2620 MHz	TDD
n39	1880–1920 MHz	1880–1920 MHz	TDD
n40	2300–2400 MHz	2300–2400 MHz	TDD
n41	2496–2690 MHz	2496–2690 MHz	TDD
n48	3550–3700 MHz	3550–3700 MHz	TDD
n50	1432–1517 MHz	1432–1517 MHz	TDD
n51	1427–1432 MHz	1427–1432 MHz	TDD
n66	1710–1780 MHz	2110–2200 MHz	FDD
n70	1695–1710 MHz	1995–2020 MHz	FDD
n71	663–698 MHz	617–652 MHz	FDD
n74	1427–1470 MHz	1475–1518 MHz	FDD
n75	N/A	1432–1517 MHz	SDL
n76	N/A	1427–1432 MHz	SDL
n77	3300–4200 MHz	3300–4200 MHz	TDD
n78	3300–3800 MHz	3300–3800 MHz	TDD
n79	4400–5000 MHz	4400–5000 MHz	TDD
n80	1710–1785 MHz	N/A	SUL
n81	880–915 MHz	N/A	SUL
n82	832–862 MHz	N/A	SUL
n83	703–748 MHz	N/A	SUL
n84	1920–1980 MHz	N/A	SUL
n86	1710–1780 MHz	N/A	SUL
n257	26 500–29 500 MHz	26 500–29 500 MHz	TDD
n258	24 250–27 500 MHz	24 250–27 500 MHz	TDD
n260	37 000–40 000 MHz	37 000–40 000 MHz	TDD
n261	27 500–28 350 MHz	27 500–28 350 MHz	TDD

non-renewable authorization to use a 10 MHz channel in a single census tract for three years. Up to seven total PALs may be assigned in any given census tract with up to four PALs going to any single applicant. Applicants may acquire up to two consecutive PAL terms in any given license area during the first auction. The General Authorized Access tier is licensed by rule to permit open, flexible access to the band for the widest possible group of potential users. General Authorized Access users are permitted to use any portion of the 3550–3700 MHz band not assigned to a higher-tier user and may also operate opportunistically on unused Priority Access channels. The process of assigning spectrum is automated, with several Spectrum Allocation Servers (SAS) coordinating the scheme nationwide. Each base station must report its position. On the basis of the propagation data, SAS estimates the impact of the new transmitter on other nearby small cells. Additional outdoor radio measurement receivers are also used to assess background levels. If the RF power density is less than −80 dBm, then the SAS authorizes spectrum use. CBRS rules are shown in detail in [4].

4.7 3GPP Frequency Variants

3GPP has defined frequency variants for all relevant bands where 5G likely will be deployed. The work for each frequency variant is initiated by mobile operators and completed by equipment vendors. Frequency variants are released independently: vendors can take a frequency variant from the Release 16 specifications but still use Release 15 functionality. More frequency variants are defined during the evolution of 3GPP specifications. Table 4.2 shows the list of frequency variants in September 2019.

4.8 Summary

5G is the first mobile radio technology that can utilize any spectrum from below 1 GHz up to 90 GHz, both FDD and TDD, and both licensed and unlicensed bands. The millimeter wave frequencies above 20 GHz provide extremely high local data rates with bandwidths up to 800 MHz initially. Mid-band spectrum at 2.5–5.0 GHz offers high capacity and data rates in urban areas with existing cell sites. Low-band spectrum is beneficial for providing wide and reliable coverage for low-latency critical communication. Unlicensed bands are useful for local networks and for boosting the capacity of the licensed bands. The target of 5G is to combine the benefits of different spectrum blocks together for optimized performance.

References

1 Sun, S., Thomas, T., Rapparport, T. et al. (2015). Path loss, shadow fading, and line-of-sight probability models for 5G urban macro-cellular scenarios. *IEEE Globecom Workshop*.
2 Du, J., Chizhik, D., Feick, R. et al. (2018). Suburban residential building penetration loss at 28 GHz for fixed wireless access. *IEEE Wireless Communication Letters* 7 (6): 890–893.

3 3GPP TSG RAN WG1 (2018). Evaluation Results for NR-U, R1–1812659.

4 Electronic Code of Federal Regulations (2017). Title 47: Telecommunication, PART 96—CITIZENS BROADBAND RADIO SERVICE.

5 3GPP Technical Specifications 38.101-1 (2019). User Equipment radio transmission and reception; Range 1, v. 15.5.0. March 2019.

6 3GPP Technical Specifications 38.101-2 (2019). User Equipment radio transmission and reception; Range 2, v. 15.5.0. March 2019.

5

5G Architecture

Antti Toskala and Miikka Poikselkä

Nokia, Finland

CHAPTER MENU

5.1 Introduction

The 5G network architecture was influenced by different factors. There was the preparation for cloud-based implementations, readiness to deal with larger data rates and lower latencies than previous generations, enabling new services, as well as the need to interwork with Long-Term Evolution (LTE), especially in the first phase. All these factors have impacted the 5G architecture. Besides the radio technology development, a new 5G core network has also been defined which allows new service elements in terms of local and global services as well as the new concept of flow-based quality of service and support for network slicing and many other features enabling a more efficient cloud-based implementation when compared to the LTE core (EPC, Evolved Packet Core). This chapter presents an overview of the different 5G architectures and then covers in more details the radio access network (RAN) architecture and interfaces as well as the 5G Core (5GC) network with key elements and functionalities. This chapter concludes with the introduction of network slicing support in Release 15.

5.2 5G Architecture Options

Originally there were eight architecture options on the table in 3GPP discussions [1] when all possible aggregation and core network combinations were considered, but as

5G Technology: 3GPP Evolution to 5G-Advanced, Second Edition.
Edited by Harri Holma, Antti Toskala, and Takehiro Nakamura.
© 2024 John Wiley & Sons Ltd. Published 2024 by John Wiley & Sons Ltd.

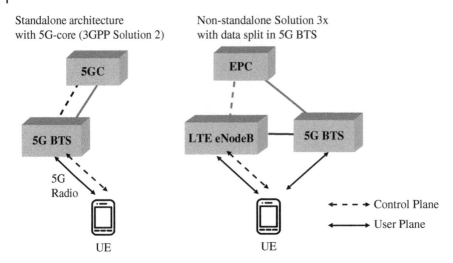

Standalone architecture with 5G-core (3GPP Solution 2)

Non-standalone Solution 3x with data split in 5G BTS

Figure 5.1 Main architecture options for initial 5G roll-outs.

the work progressed the following two main architecture options emerged, with slight difference in timing as discussed in Chapter 2. First in the finalization schedule was the approach with LTE as the connection anchor and using the existing LTE core, with 5G radio as the secondary cell (Architecture option 3). Next to be completed was the standalone 5G radio with the new 5G core (architecture option 2). The options illustrated in Figure 5.1 can be described as follows (in the order of finalization in 3GPP):

- Option 3 refers to the architecture option which uses LTE as the anchor for the connection as well as connecting for via the existing EPC. 5G is only on the user plane in the radio part, which is then used with dual connectivity with LTE. For routing the user plane data, there exist further alternatives, creating different variants from option 3, as described later. From the service point of view, only enhanced Mobile Broadband (MBB) is available.
- Option 2 refers to architecture with 5G radio connected to the 5G core network. This enables all new functionality brought by the 5G core, such as Service-Based Architecture (SBA) or network slicing as described later in this chapter. The Release 15 services offered include the enhanced MBB as well as Ultra Reliable Low Latency Communication (URLLC), but cover also basic services such as voice over 5G. There is no need to operate with LTE-5G dual connectivity in this architecture, and thus this is similar for roll-out as LTE was originally (though 5G-5G dual connectivity was enabled in a later phase). Option 2 enables interworking with LTE (handovers/reselection to/from LTE) as covered later in this chapter.

The data routing can be done in different ways depending where to split the user plane between LTE and 5G. The different alternatives (option 3 variants) are as follows:

- Option 3 with Data split in the LTE eNB. The user plane data received from the EPC is divided by the LTE eNB between LTE radio and the 5G gNB (data over the X2 interface). This allows a single service (with bearer split) to be transmitted from both the gNB and eNB and respectively received from both sides in the uplink.

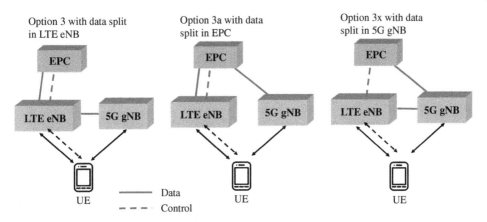

Figure 5.2 Different data split options for the option 3 family.

- Option 3A with data split in the EPC. This does not allow data rate aggregation for a single service, but a single service is only serviced either by an LTE eNB or a 5G gNB. In this case, the X2-interface is not used for user plane transmission. It can be anticipated that for some low-data-rate service like VoLTE, this could be used while mobile broadband would use one of the other solutions enabling data rate aggregation.
- Option 3X with data split in the 5G gNB. Data received from the EPC is divided by the 5G gNB between 5G radio and LTE eNB (data over the X2 interface).This is usually favored as it minimizes the investment on the LTE side since typically the LTE eNB has not been dimensioned originally to handle such high data rates, including the backhaul transmission part. Also, the latency is reduced if most of the packets go via 5G in any case (Figure 5.2).

There exist also further architecture options which were finalized in June 2019, addressing the additional dual connectivity options as follows. These are not foreseen to be supported by the first phase UEs to be used with 5G network openings as these are finalized only as part of "late drop" in Release 15:

- Option 4, with the 5G gNB being the connection anchor. LTE connection is the secondary on the connection with the control plane going via the 5G gNB toward the 5G core network. This is suitable for the cases when there is sufficient coverage available for the 5G connection, either building a dense enough network or then using a low enough frequency band for a 5G carrier (for at least one of the 5G carriers). This option avoids the need to connect LTE eNB for the 5G core network
- Option 7, with the LTE eNB being the connection anchor. Now the 5G connection is the secondary on the connection with the control plane going via the LTE eNB toward the 5G core network. This assumes that the LTE eNB is updated to allow connectivity toward the 5G core network. This is suitable for the cases when the LTE coverage is expected to be much better than the available 5G coverage, for example, when using 5G on mmW bands with a relatively small cell radius. Option 7X would then refer to the case when the user plane from the 5G core is routed via the 5G gNB (Figure 5.3).

From the architectural point of view, the late drop of Release 15 enables also the pure 5G–5G dual connectivity between two 5G gNBs. There was interest expressed in this

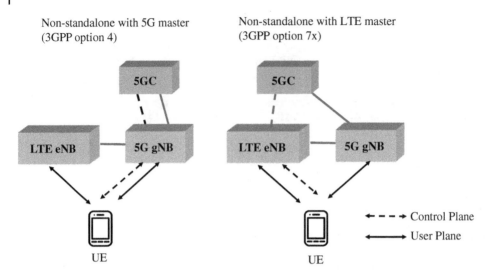

Figure 5.3 Further 3GPP architecture options part of Release 15 "Late drop."

both from the standpoint of using mmW bands together with 5G on lower bands, as well as of considering 5G with low band as the coverage band in dual connectivity with the 2.5 GHz or 3.5 GHz band as the capacity band. The 5G–5G dual connectivity is limited in Release 15 to the synchronized case only, and is to be extended for the asynchronous case part of 5G Phase 2 in Release 16, as covered in Chapter 15

5.3 5G Core Network Architecture

In recent years, many communication service providers (CSPs) have been virtualizing their EPC, changing the deployment from a series of dedicated physical appliances (specific blade servers for each function), to virtualized network functions hosted on general-purpose servers in the cloud. These virtualized core networks allow CSPs to expand into new services, such as cellular Internet of Things (IoT) or to provide enterprise-specific core services, or to refresh or expand legacy EPC networks. In practice, this virtualization of EPC does not work well in multivendor deployments as it was not built into the EPC specifications in the first place.

Therefore, support for Network Function Virtualization and Software Defined Networking was one of the first architecture requirements for the 5G core network. In addition, it was equally important to require capability for the 5G core network to be deployed in a cloud-native way. Applications designed for the cloud must also follow scaling design principles such as splitting the control plane and the user plane as well as separating compute and storage resources, capabilities which were not built into the EPC architecture.

Separation of the control and user planes provides the ability to utilize distributed edge cloud architectures for bandwidth-intensive and latency-critical applications requiring independent distribution of the user plane functions from that of the control plane functions. For example, for certain low-latency use cases it will be important to place user plane resources close to the access point where the user/device is attached, to reduce end-to-end latency and the transport network load. The 5G core network will further enable users to be connected simultaneously to the edge and central cloud and move between them.

The core network needs to support stateless network functions where the compute resource is decoupled from the storage resource. Compute and storage separation enables unlimited linear scalability and extreme resilience. In the 5G architecture, the so-called Unified Data Management (UDM) will host subscriber data, session data, policies, operational data, charging, and accounting data. It will also enable an open ecosystem for data exposure and analytics.

Another major difference compared to EPC is that multi-access support is built in right from the beginning, enabling efficient use of different fixed and mobile network assets to provide access-independent value-added services, maximized data rates, increased reliability, and improved user experience.

These new capabilities will further enable building of end-to-end network slicing, enabling public and private operators to address the needs of various verticals by dynamically adapting the network to their specific requirements. Network slices can be isolated from each other in terms of connectivity and resource allocation, creating "virtual private service networks." These can be configured in specific ways to achieve, for example, low latency, or high reliability, or both, and new network slices can be introduced easily via automation, thus enabling new business models that were not feasible with dedicated physical networks.

The resulting 5G Core Network Architecture (simplified) is depicted in Figure 5.4.

The point-to-point architecture that has been used up to 4G still applies to some interfaces in 5G. In the point-to-point model, different network functions are connected over standardized interfaces that allow for multivendor networks. This is well understood both conceptually and operationally; it has served mobile operators for decades. In the 5G system architecture, interfaces between the user equipment (UE) and the core (N1),

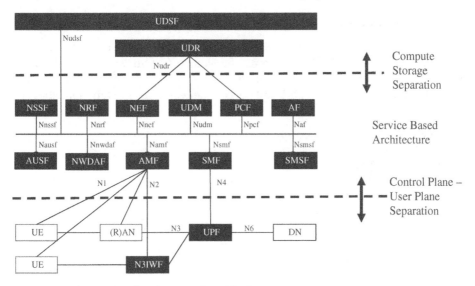

Figure 5.4 5G core network architecture (simplified).

the access network and the core (N2 and N3), and the control and user plane (N4) will still use a point-to-point architecture.

Now, however, that with the transition to a cloud infrastructure and the need for greater service agility, the point-to-point model is no longer the best option for all the core network interfaces. For CSPs that view 5G as an opportunity for transformative change both in terms of functionality and cost per bit, a service-based interface instead of point-to-point interfaces looks more attractive.

The challenge with the point-to-point architecture is that it contains many unique, or quasi-unique, interfaces between functional elements, each connected to multiple adjacent elements. This "tangle" of connections creates dependencies between functions and makes it difficult to change the deployed architecture. If a new function is introduced, or an existing function expanded or upgraded, the operators needs to reconfigure multiple adjacent functions and test the new configuration before going live. This raises the business cases thresholds to experiment with, and deploy, new services.

In effect, the end user service is tied to the network, and consequently the operator's addressable market is artificially limited. This is acceptable where the service set is well defined and relatively fixed (voice, broadband, etc.), but in the 5G era, where operators expect to offer a multitude of diverse services that must be able to adapt to fast-changing demand or industry-specific requirements, a more dynamic and agile architecture is needed.

The SBA decouples the end user service from the underlying network and platform infrastructure, and in doing so, enables both functional and service agility. By virtue of the SBA operating on a cloud-native foundation, it is much simpler for an operator to add, remove, or modify virtual network functions (VNFs) from a network processing path (functional agility) and create new service paths on demand (service agility).

3GPP has selected SBA for application in the 5G Core Control Plane (the middle part of Figure 5.4. Service-based interfaces are all based on HTTP and are all denoted with the legend Nxxx in Figure 5.4.

5.3.1 Access and Mobility Management Function

The Access and Mobility Management Function (AMF) is part of the control plane processing on the core side. The AMF receives the Non-Access Stratum (NAS) signaling from the UE via the access network and is also connected to the gNB for control plane signaling between the access network and the core network. The key functionalities are as follows:

- Terminates (Radio) Access Network Control Plane signaling (N2).
- Terminates NAS Signaling (N1).
- Provides Access Authentication and Access Authorization.
- Provides NAS ciphering and integrity protection.
- Enables registration management, connection management, reachability management, and mobility management.
- Provides transport and proxying for session management (SM) messages between the UE and the Session Management Function (SMF).
- Transports SMS messages between the UE and the Short Message Service Function (SMSF).
- Supports location services (e.g. routes messages between the UE and the Location Management Function (LMF) as well as between the RAN and the LMF).

5.3.2 Session Management Function

The SMF takes care of the session management, such as session establishment, modification, and release, including maintenance of the tunnel between the User Plane Function (UPF) and the access network node, for example, the gNB. Further functionality includes the following:

- Provides UE IP address allocation and management
- Selects and controls of the UPF
- Interfaces with policy functions
- Controls and co-ordinates charging data collection at the UPF
- Determines the Session and Service Continuity (SSC) mode of a protocol data unit (PDU) session

5.3.3 User Plane Function

The user plane processing is covered by the UPF. There can be more than one UPF in the network, for example, one UPF handling the local traffic with the other UPF handling all other traffic. Services with very low latency requirements (such as those related to URLLC, etc.) are well suited for local UPF handling as additional latency due to the long transmission distance can be avoided. The key functionalities of the UPF are the following:

- Perform all UPFs on the packets: forwarding, routing, marking, quality of service (QoS), inspection, and so on.
- Enforce policies received from the Policy Control Function (PCF) (via the SMF).
- Send traffic usage reporting to the SMF for charging.
- Act as the anchor point for intra-/inter-radio access technology (RAT) mobility.
- Connect to external networks (N6) or other UPFs.

5.3.4 Data Storage Architecture

The Unified Data Repository (UDR) is a common database for all kinds of standardized data structures including Subscription Data, Policy Data, Structured Data for exposure, and Application Data. Three different network functions can access the UDR to read, update, delete, and subscribe to notifications of data changes.

UDM hides application logic from the UDR. UDM takes care of, for example, generation of authentication credentials, identity handling (both user and subscription), information storage about serving entities like the AMF and SMF, and SMS delivery support. In addition to standardized structure data, the 5G architecture enables any network function to store unstructured in the Unstructured Data Storage Function (UDSF). This capability enables a network function to store all UE-related data in the UDSF, and thus the network function can become stateless. In Release 15, this is capability is defined for the AMF.

5.3.5 Policy Control Function

The PCF governs network behavior by applying a unified policy framework. This is enabled with the following primary capabilities:

- Creation of both SM and access management policy association
- Provision and deletion of policy and charging management decisions

- Provision and deletion of access and mobility management decisions
- Delivery of UE access selection and PDU session selection policies to the UE via the AMF
- Ability to utilize policy-control-related subscription information and application-specific information stored in the UDR
- Ability to obtain spending limit reports from the Charging Function
- Interactions with the NEF (Network Exposure Function), for example, to obtain Application-Function-influenced traffic steering authorization

5.3.6 Network Exposure Function

- Exposes 5G Core capabilities and events to external applications. Uses the UDR as its data source.
- Allows third parties to securely provision information to the 5G Core (e.g. Expected UE Behavior).
- Handles masking of sensitive network information toward external parties.

5.3.7 Network Repository Function

- Maintains the network function (NF) profile of available NF instances and their supported services, and provides NF discovery and selection
- Selection criteria can include location (latency), load, Data Network Name (application), access network type, slice, etc.
- Provides much more granular policies and dynamic capabilities when compared to Domain Name System (DNS) selection

5.3.8 Network Slice Selection

- Determines the candidate AMF(s) or AMF Set to be used to serve the UE
- Selects which slices the UE can connect
- Takes care of mapping slice specific identifiers

5.3.9 Non-3GPP Interworking Function

The Non-3GPP InterWorking Function (N3IWF) provides support to connect UEs via non-3GPP access (only untrusted Wi-Fi access in Release 15). This is enabled with the following primary capabilities:

- Support of IPsec tunnel establishment with the UE in order to authenticate and authorize it to access the 5G core
- Establishes IPsec Security Association to support PDU Session traffic
- Performs tasks related to user plane traffic such as relaying uplink and downlink user plane packets between UE and UPF and packet marking
- Handling of control plane traffic such as relaying NAS signaling between UE and AMF and passing signaling from SMF related to PDU session and QoS

5.3.10 Auxiliary 5G Core Functions

Table 5.1 presents other network functions as introduced in 3GPP Release 15.

Table 5.1 Auxiliary 5G core network functions.

Name of network function	Primary functionality
Short Message Service Function (SMSF)	Provides necessary interworking functions to enable Short Message service between 5G UE and legacy SMS functionality in the network
Authentication Server Function (AUSF)	Authenticates the UE and provides related keying material for it
Security Edge Protection Proxy (SEPP)	Provides Message filtering and policing on inter-PLM control plane interfaces and hides network topology from other vendors as it can act as a single point of contact
Network Data Analytics Function (NWDAF)	Provides slice specific network data analytics to PCF and NSSF
Location Management Function (LMF)	Obtains location measurements equally from UE and Radio Access Network

5.4 5G RAN Architecture

The 5G radio access architecture consists of two basic elements as follows:

- Central Unit (CU), which is intended to handle relatively higher layers above the physical layer. The physical realization of CU could be either dedicated hardware or a radio-cloud type of implementation. The CU is connected on one hand to the Distributed Unit (DU) and on the other hand to the 5G core network (Next Generation Core (NGC), as shown in Figure 5.5.
- The DU, which is located on the cell site together with the antennas and the RF unit. The DU is expected to handle especially time-critical processing which does not tolerate too much delay to enable processing away from the cell site.

In the considerations in 3GPP, the approach with RF only on the site (with antennas) has not been considered desirable due to the very high transmission requirements of sending IQ samples. Although such an approach was used in some markets with LTE, the larger number of antennas as well as the larger bandwidths considered make the required data rates so massive that they are not often seen in large-scale network deployments.

The following interfaces are defined as part of the 5G radio access architecture:

- The F1 interface connects the CU to one or more DUs. A single CU is expected to be able to handle multiple DUs.
- The Xn-interface connects the different CUs together. When considering the early deployment architecture with the LTE core (EPC), the interface between LTE eNB and 5G gNB is called the X2-interface.
- The E1 interface facilitates separation of the control and user plane processing of the CU for the CU-CP and CU-UP parts, as shown in Figure 5.2. The E1 interface is available from the June 2018 specification versions onward.
- The NG interface toward the 5G Core (5GC). It is divided into the user plane part (NG-U) and the control plane part (NG-C). In the overall reference architecture in 3GPP, NG-U is called the N3 interface, and NG-C is called the N2 interface.

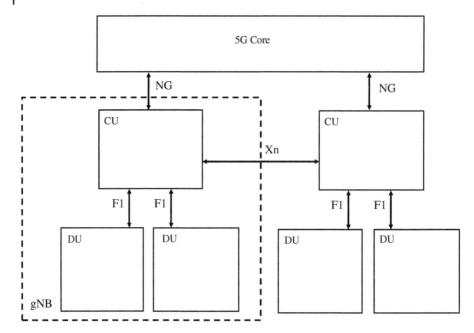

Figure 5.5 Overall 5G radio access network architecture.

The detailed functionalities and protocol stacks for these interfaces are covered in more detail in later sections of this chapter.

The functional split shown in Figure 5.6 is also called the higher layer functional split. This is the only functional split defined in 3GPP in Release 15. This functional split is suitable also for the cases when the backhaul has some latency limitations since the retransmission are controlled locally in the DU. This also allows the CU to be located relatively farther away than if the retransmissions were controlled on the DU side.

The aspects of the lower layer functional split have not been covered in 3GPP so far; the different alternatives were anyway studied during 3GPP Release 15. The approach was to aim for more functionality in the CU to facilitate more central handling of functionality and to have less functionality in the RF site together with antennas and RF units. Some L1 functionality was anyway seen to be necessary in addition to the RF and antennas in the site to avoid sending I/Q samples, as that would have resulted in massive requirements on the backhaul connectivity. This is illustrated in Figure 5.7, which shows that there is a factor of 40 difference in the data rates estimated when comparing sending bits after the Packet Data Convergence Protocol (PDCP) layer and when sending I/Q samples for the RF only in the site solution. Considering larger bandwidths than 100 MHz and larger number of antennas would result in an even bigger difference compared to RF only at the site solution. Moving MAC/RLC to the CU naturally results in somewhat higher data rates, as now retransmission handling has moved to the CU, and thus some packets end up being sent multiple times through the F1 interface in addition to the extra control signaling that becomes necessary.

A lower layer split with enhanced CPRI (eCPRI) shown in Figure 5.7 is being considered in the O-RAN Alliance [2] organization outside 3GPP, and the interface has also

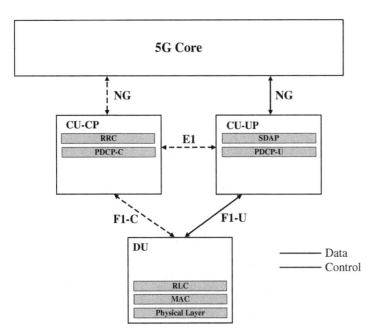

Figure 5.6 5G radio access network architecture with control and user plane separation for CU with the higher layer split.

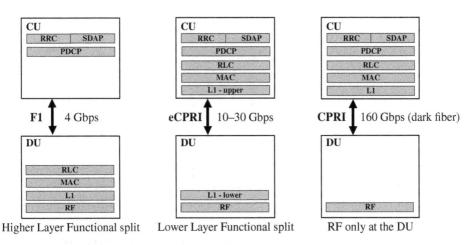

Figure 5.7 Impact on backhaul connectivity with different functional split options.

been defined in the Common Public Radio Interface (CPRI) under the name eCPRI, corresponding to the lower layer split options as studied in 3GPP [3]. The key difference with the eCPRI interface is that in unlike the earlier CPRI interface versions, eCPRI is packet based and thus facilitates options other than dark fiber, such as Ethernet. 3GPP will not address in Release 15 or 16 further such lower layer functional splits in later releases, but by Release 19 this has not been considered further. Thus, from that perspective, the F1 specification in Release 15 covers only the higher layer functional split. The term F2-interface referring to the eCPRI interface has been used outside 3GPP to refer to the

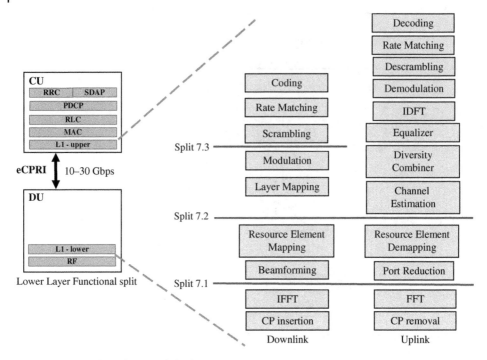

Figure 5.8 Lower layer functional slit alternatives.

interface between the DU (with RF and lower L1) and the CU (upper L1 and other higher layer functions), but the eCPRI/F2 interface is not covered by the 3GPP specifications.

In CPRI, multiple different options for the lower layer split functional split with eCPRI were defined, as well as studied also in 3GPP [4]. The options include the following alternatives:

- Option 7.1, with the split in the downlink direction between the beamforming and Inverse Fast Fourier Transform (FFT) block, and in the uplink between FFT and beamforming
- Option 7.2 with placement between the layer mapping and resource element mapping in the downlink and FFT and port reduction in the uplink direction
- Option 7.3 is for the downlink only, between scrambling and modulation functionality

There is an impact on the resulting data rates for the fronthaul interface as addressed in Chapter 9, but clearly the needed transmission capacity is always largely reduced compared to sending pure RF samples to the cell site. Between the different options, there is the trade-off always between how easy the network upgrades are compared to which operations are easier to be performed closer to the RF than further away. Work done in the O-RAN Alliance has taken option 7.2 as the basis for the work on the open fronthaul interface (Figure 5.8), as addressed more in Chapter 20.

5.4.1 NG-Interface

The NG interface connects the gNB (5G-RAN) to the 5G Core. It is based on the use of IP transport based on GTP-U and UDP. Similar to the S1-interface between LTE and

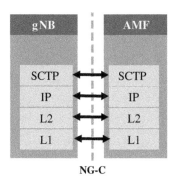
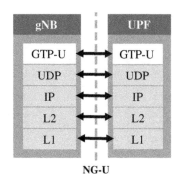

Figure 5.9 NG-interface control and user plane protocol stacks.

the EPC core, the NG interface is also defined to be an open multivendor interface. The NG-interface has been designed in an access-agnostic way, to facilitate convergence toward and use of the 5G core for other access technologies as well. This would simply require from the end devices the ability to use 3GPP-defined NAS signaling for connection setup/registration.

The NG-interface protocol stacks for the control and user planes are shown in Figure 5.9. The control plane part is based on the use of the Stream Control Transmission Protocol (SCTP) on top of an IP connection. While the user plane part transports the user data, the key functions of the control plane part of the NG-interface are as follows:

- NG interface management
- UE context and mobility management
- Transport of NAS messages
- Paging
- PDU session management

5.4.2 Xn-Interface

The Xn-interface connects gNBs to another gNB (or to eNB). The term Xn-interface is always used when operating together with the NG-core (5G-core). When the architecture option 3 is used with the 4G EPC core, then the interface is called the X2-interface. The Xn-interface protocol stack is shown in Figure 5.10. As with the X2-interface in LTE networks, the Xn-interface is defined to enable operation as a multivendor interface. Figure 5.9 shows the Xn-protocol split. From the dimensioning point of view, the Xn-interface differs from the LTE-based X2 interface as the user data is routed via Xn/X2 more continuously than in LTE. In LTE, it was used only for temporary packet forwarding. Dimensioning is addressed in Chapter 9.

For the user plane side, the key functionality is data forwarding and flow control, while on the control plane side of the Xn-interface key functionalities are as follows:

- Xn-interface management
- UE mobility management, including context transfer and RAN paging
- Dual connectivity

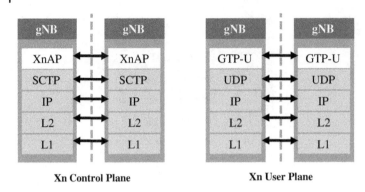

Xn Control Plane   **Xn User Plane**

Figure 5.10 Xn control and user plane protocols.

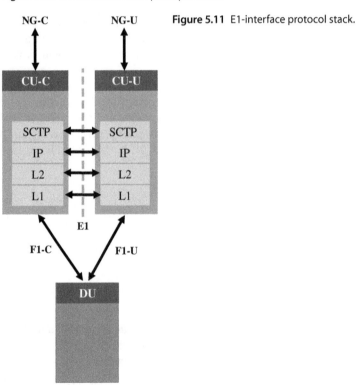

Figure 5.11 E1-interface protocol stack.

5.4.3 E1-Interface

The E1 interface connects the user plane and control plane parts of the CU functionality. The E1 interface was added in the June 2018 version of the specifications, together with the standalone 5G option and the 5G core. The E1 interface protocol stack is shown in Figure 5.11.

5.4.4 F1-Interface

The F1-interface connects the CU and DU. In Release 15, the supported functional split, as presented earlier in this chapter, is the higher layer functional split. The F1-interface protocol stack is shown in Figure 5.12.

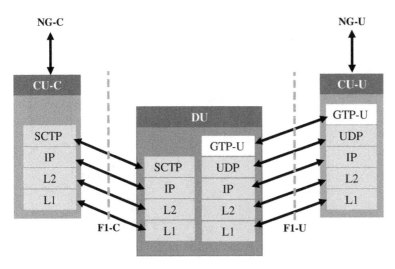

Figure 5.12 F1-interface protocol stacks.

5.5 Network Slicing

Network slicing in 5G is intended to address services with different kinds of requirements, while at the same time improving the efficiency of the network. Some services require low latencies and need to address local service content, some require high mobility, while other services are close to the current mobile broadband type characteristic without specific requirements on reliability or latency. The slicing principles in 5G allow a UE to belong to more than a single slice, while still maintaining only a single signaling connection with the network.

The UE is provided information on the available slices when registering for the network (via NAS signaling), and a single UE may be configured for up to eight difference slices. Once a PDU session is set up, the UE is then signaled the Network Slice Selection Assistance Information (NSSAI), assuming this has been provided earlier to the UE. On the basis of this information, the network will select the appropriate slice instance (and related resources), with the AMF coordinating the actions in the 5G core side. There is one AMF that is common for all the slices a single UE has. This AMF will obtain information on which slices are allowed for a given subscription by interfacing with the Network Slice Selection Function (NSSF). The 5G slicing approach allows a single UE to access both services belonging to a specific slice as well as services which will be handled as the "normal" traffic. Earlier generations were using user-specific priorities, which would mean the same handling for all services provided for a single UE.

On the 5G-RAN, not too much has been specified in terms of reserving the resources for the particular slice, but in general the scheduler is in a key position to ensure that each slice is provided with suitable handling and resources so that the intended QoS can be provided. There is no point, for example, reserving some dedicated piece of spectrum for a particular slice; rather, the scheduler should always be allowed to select the most suitable resource taking into account the situation for the whole band (or set of frequency bands) that is available at the gNB. The configuration of radio parameters may be different for each slice (but does not have to be). If a particular part of spectrum is

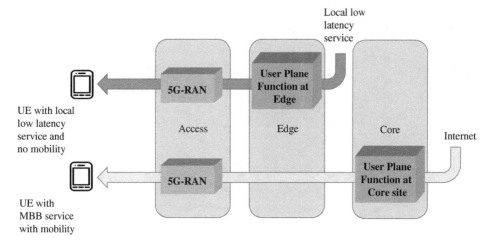

Figure 5.13 Example network slicing use case.

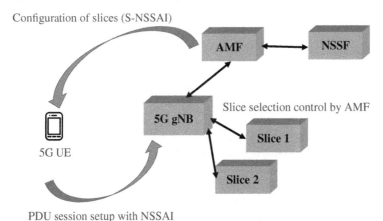

Figure 5.14 Setting up a slice for UE.

used for a specific slice, the tolerance for interference would decrease and the frequency domain scheduling benefits would be degraded.

The example in Figure 5.13 shows the different handling for two slices with different types of service requirements. The service requiring very low latency is handled with more local UPF resources compared to the "regular" mobile broadband service.

With slicing, it is possible to consider the customers needing different level of service. The NSSAI has one or more S-NSSAIs (Single NSSAI). The S-NSSAI identifies each slice configured for the UE. An example of signaling flow for setting up the slicing operation is shown in Figure 5.14 with NAS signaling being used to provide slide information for the UE

5.5.1 Interworking with LTE

The focus on the any work related to interworking has been on interworking with LTE. Release 15 contains the methods that enable interworking between LTE and 5G for both

Figure 5.15 Reselection from 5G to LTE.

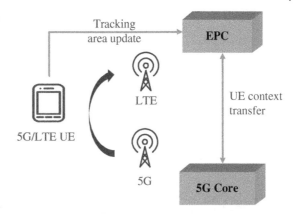

idle and active modes of connection. When losing the coverage of the 5G signal (when connected to a standalone 5G network), the following alternatives are available:

- In idle mode, the UE can reselect the LTE cell, in which case the UE mobility management (MM) and SM context will be transferred from the 5G core to the EPC. This has been defined both ways, so when 5G connectivity resumes, the UE can again reselect 5G, and now the context is transferred back from the EPC to the 5G core. Moving from 5G to 4G is shown in Figure 5.15.
- In the active mode, the handover has been defined in both ways between LTE and 5G, on the basis of the measurements done by the UE, as discussed in Chapters 6 and 7. For services like voice, the same IMS can be used, and there is no need to involve the CS domain in 5G and LTE interworking. An example of active mode handover is shown in Figure 5.16.

The active mode mobility is normally based on UE measurements, as discussed in Chapter 7. Following this, the network side will make a decision on the need for handover. In Figure 5.15, an example of 5G to LTE mobility is shown following the UE measurement report and gNB decisions that inter-system handover to LTE is needed.

3GPP specifications do not support in Release 15 5G interworking with any other radio technologies other than LTE. It is being addressed in Release 16 for voice service if there is still a need to add interworking with 3G to cover for possible deployments without existing LTE coverage in the area (and only 3G coverage). A Release 15 UE, when running out of 5G coverage and not finding an LTE network can naturally select 3G or even 2G, but there is no support for UE context transfer in such a case or active mode handover possibility directly from 5G to 2G/3G. Once the UE has moved to the LTE side, the existing interworking mechanisms on the LTE side are naturally available including 2G/3G related interworking if the UE supports 2G and 3G. There is no radio level interworking with Wi-Fi defined in Release 15. However the untrusted non-3GPP-related access technologies, such as Wi-Fi, may use the 5G core via the earlier discussed Non-3GPP InterWorking Function (N3IWF) if it is implemented in the network and the UE is capable of sending the 5G NAS signaling via Wi-Fi. There is no data aggregation possible with this kind of interworking. So far, 3GPP is not planning to introduce any of the radio level Wi-Fi interworking solutions (which have been specified for LTE) in the 5G side in Release 15 or 16 scope.

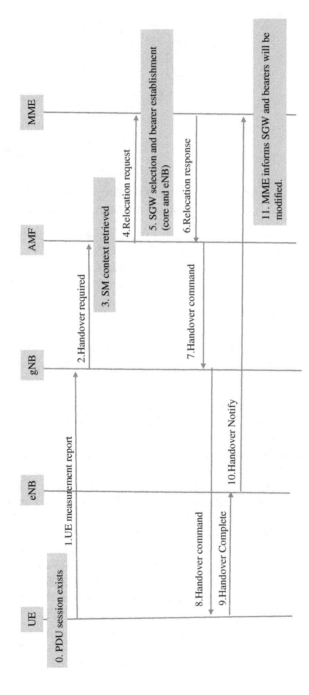

Figure 5.16 5G to LTE inter-system handover.

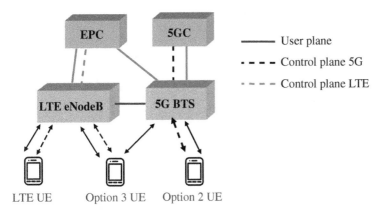

Figure 5.17 Example interworking between architecture options 2 and 3.

When one considers the practical networks with option 3 and 2 enabled, the architecture looks as depicted in Figure 5.17, In this case, the gNB is acting both as the secondary cell for first phase UEs which are not expected to support other architecture alternatives as well as the primary cell for UEs which support standalone 5G operation. It is foreseen that some UE types, especially ones intended for services like URLLC operation, may end up supporting only standalone 5G operation. Similarly, the LTE eNB will support both legacy LTE-only UEs as well as UEs which support dual connectivity with 5G.

5.6 Summary

The 5G radio operated with the 5G core enables many features beyond the capabilities of LTE, including the new flow-based QoS, support for local services, and network slicing. Further, the new core has been designed to enable support for the URLLC service, which allows the use of 5G to be expanded for industrial applications in addition to many other use cases. The overall architecture has been designed to enable cloud-native implementation, not only for the core but also for the radio functionality in the CU side.

5G radio and the 5G core provide different means for interworking with LTE, ranging from the connected mode and idle mode mobility, all the way to dual connectivity. The architectures becoming operational first, non-standalone 5G operation with EPC followed by standalone 5G with 5G core, can interwork well and provide a solution for legacy co-existence together with full end-to-end 5G.

The 5G core also enables connectivity from access networks other than 3GPP-based networks, as long as the devices support the 3GPP-based connection establishment signaling (3GPP NAS), as the 5G-core-related procedures have been made access independent.

References

1 3GPP Tdoc RP-161266 (2016).5G architecture options – full set, Deutsche Telecom. June 2016.
2 O-RAN Alliance. https://www.o-ran.org
3 Common Public Radio Interface. www.cpri.info
4 3GPP TR 38.801 (2017). Study on new radio access technology: Radio access architecture and interfaces. March 2017.

6

5G Physical Layer

Mihai Enescu, Keeth Jayasinghe, Karri Ranta-Aho, Karol Schober, and Antti Toskala

Nokia, Finland

CHAPTER MENU

6.1 Introduction

5G multiple access, and the physical layer in general, has differences from Long Term Evolution (LTE) as discussed in Chapter 1. In the downlink direction, the multiple access is based on Orthogonal Frequency Division Multiple Access (OFDMA), similar to LTE. In the uplink direction, 5G has adopted both OFDMA and Single Carrier Frequency Division Multiple Access (SC-FDMA), while LTE uses only SC-FDMA. SC-FDMA is often denoted also as DFT-Spread OFDMA (DFT-S-OFDMA). Further, the use of beamforming is now much more built in, as with LTE, and the range of frequency bands supported extends to much higher frequencies than LTE can achieve. The frame structures with 5G offer more flexibility to cover different cases, depending on the frequency

5G Technology: 3GPP Evolution to 5G-Advanced, Second Edition.
Edited by Harri Holma, Antti Toskala, and Takehiro Nakamura.

band to be used or whether one needs to optimize for high capacity or for low power consumption of the network and UEs. This chapter presents the 5G physical layer principles, including the mapping of the transport channel to the physical channel, 5G frame structure, and key physical layer procedures. Special emphasis is on the new features compared to LTE, such as full support of beamforming, larger bandwidths, and a more flexible frame structure. This chapter also introduces the 5G channel coding solutions, physical layer procedures, and measurements as specified in 3GPP. This chapter concludes with an overview of the physical layer aspects of the 5G UE capabilities.

6.2   5G Multiple Access Principle

The role of OFDMA remains strong with 5G, similar to what it was with LTE. In a sense, the role of OFDMA is even stronger as now it is used also in the uplink direction, in addition to the single carrier transmission. With 5G, the single carrier transmission is only intended to be used at the cell edge when there are limitations with the link budget, otherwise OFDMA is used. The motivation for the use of OFDMA comes from better performance with the multiple antenna transmission case (so much better performance that even the transmission power reduction due to the increased peak-to-average ratio (PAR) is acceptable). Traditionally, the use of OFDMA has caused about 1–2 dB loss in the available power for the uplink transmission compared to the SC-FDMA waveform. However, even if the resulting transmission power is less, the better link performance with multi-antenna multi-stream transmission makes OFDMA a better choice when the link budget has some margin.

The other aspect is interference management in Time Division Duplex (TDD) networks. This is clearly easier to handle if both uplink and downlink use the same multiple access solution. The principle of OFDMA is shown in Figure 6.1, and more information on OFDMA fundamentals is given in [1] and the references therein.

The uplink direction uses SC-FDMA (DFT-S-OFDMA) for the cases when the transmission power is limited and the use of uplink multi-stream transmission is not possible. OFDMA is intended to be used when the link budget allows the use of features like

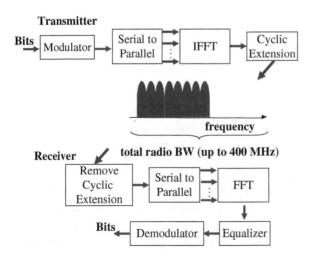

Figure 6.1 OFDM principle.

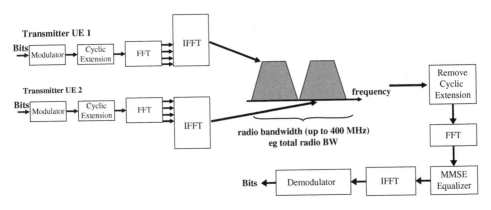

Figure 6.2 SC-FDMA principle.

higher-order modulation and uplink Massive Multiple Input Multiple Output (MIMO) transmission. 5G will fall back to SC-FDMA operation when there is not enough link quality for multi-stream operation. The SC-FDMA principle is illustrated in Figure 6.2. The same principle used in LTE is maintained, with only one symbol at a time sent using SC-FDMA transmission. The FFT/IFFT pair at the transmitter side allows the transmission to be placed accurately and without filtering complexity in the correct place within the carrier (following the grants given by the gNB).

Compared to LTE, the new 5G radio needed to be possible to be operated with:

- Higher frequency bands, with the Release 15 frequency range reaching up to 52.6 GHz and studies in Release 17 considering ranges up to 114 GHz
- Higher bandwidths with up to 100 MHz on lower frequency bands (below 7.125 GHz) and up to 400 MHz with above-24-GHz bands
- Shorter latency down to the sub-millisecond level

For these reasons, higher subcarrier spacing (resulting in shorter symbol duration) was employed, especially toward the higher frequency bands and where a larger bandwidth was possible. It was decided to specify a set of subcarrier spacings that represents a power-of-two relationship to the LTE subcarrier spacing. This approach was selected to facilitate efficient multi-technology support with platforms using both LTE and 5G. Compatible sampling rates make multimode implementation more cost efficient. The subcarrier spacings range from 15 kHz all the way up to 240 kHz (for data up to 120 kHz), with parameters as shown in Table 6.1. The smaller subcarrier spacings are intended to be used below 7.125 GHz bands (Frequency Range 1, FR1), while the higher subcarrier spacing are intended for above 24 GHz (FR2) due to the better tolerance for phase noise with higher frequency bands. The border of FR1 was lifted from 6 GHz to 7.125 GHz only as late as December 2018 to cover the extension of the unlicensed 5 GHz frequency bands to above 6 GHz range in some regions. The lowest part of the FR1 range is as low as 410 MHz.

The basic approach is to reduce the OFDM symbol duration in the time domain, which increases the subcarrier spacing. For example, increasing the symbol rate to twice the LTE symbol rate leads to the 30 kHz subcarrier spacing. If one retains the same number of symbols per sub-frame, then the duration of the transmission time interval (scheduling interval) would be 0.5 ms. The use of N as a multiplier, and staying with powers

Table 6.1 5G L1 subcarrier spacing values.

Subcarrier spacing (kHz)	15	30	60	120	240
Symbol duration (us)	66.7	33.3	16.6	8.33	4.17
Nominal CP (us)	4.7	2.41	1.205	0.60	0.30
Nominal max carrier BW (MHz)	49.5	99	198	396	—
Max FFT size	4096	4096	4096	4096	—
Min scheduling interval (symbols)	14	14	14	14	—
Min scheduling interval (slots)	1	1	1	1	—
Min scheduling interval (ms)	1.0	0.5	0.25	0.125	—

of 2, i.e. 2, 4, 8, and 16, allows an effective multimode base transceiver station (BTS) implementation with LTE. If the subcarrier spacing (15 kHz) and symbol duration used in LTE are unchanged, and only the number of subcarriers is increased, a very large FFT size in the receiver and transmitter side would be the result, which is not desirable either. Further, with higher frequencies, the phase noise would cause the subcarriers to be non-orthogonal, and thus the system performance would be degraded, or not possible at all with higher frequency bands. Also, a shorter symbol duration would not be achieved with the 15 kHz subcarrier spacing.

Another approach for enabling shorter latency is the introduction of the mini-slot, which means using allocations as low as 2, 4, or 7 symbols. Besides the latency, the use of the mini-slot will also make dealing with the hybrid beamforming easier, as is explained later in this chapter.

The use of higher bands also has impacts from the increased phase noise. The higher the carrier frequency, the more phase noise is generated, as shown in the example depicted in Figure 6.3. Thus, increasing the subcarrier spacing is also a must for the use of OFDM with higher frequencies. In particular, the use of higher-order modulation would be difficult otherwise as the system design would limit the maximum signal-to-noise ratio (SNR) that can be achieved. The CMOS oscillator shown in Figure 6.3, for example, will exhibit seriously degraded performance with 15 kHz subcarrier spacing when going to the 10–30 GHz frequency range. For this, 5G deployments with sub-6 GHz bands will use 15 or 30 kHz subcarrier spacing, while the bands above 24 GHz will use 120 kHz subcarrier spacing. The 240 kHz subcarrier spacing is only for L1 control information, as covered later in this chapter.

All this will naturally drive the use of higher subcarrier spacing with shorter symbol duration with 5G radio design. Due to the large range of spectrum to be covered with 5G, a single numerology cannot serve all the cases in an optimal way, and in some cases cannot even serve them at all; therefore, different parameterizations are needed to cover different operational scenarios. The use of 15 kHz spacing provides the largest cyclic prefix, which may be necessary in large cells in low frequency bands.

Another aspect of the subcarrier spacing selection to consider is the maximum bandwidth that can be used. As shown in Table 6.1, 15 kHz subcarrier spacing supports up to 50 MHz carrier bandwidths (as effectively some guard band left unused); in the same way, 30 kHz subcarrier spacing allows for bands with up to 100 MHz carrier bandwidth.

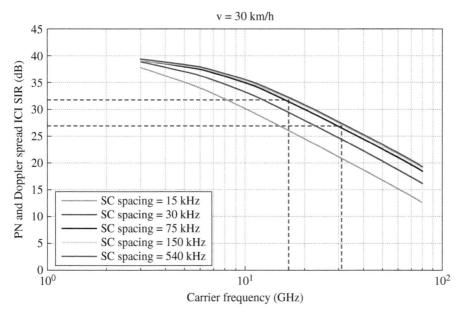

Figure 6.3 Impact of carrier frequency on phase noise.

The actual occupied bandwidth is now somewhat more than in LTE (though not in all cases), as shown in Table 6.2. The 5G numerology allows an implementation with 4k FFT size to keep the complexity to a reasonable level. The LTE channel occupancy was constant at 90%, while with 5G it varies depending on the bandwidth and subcarrier spacing in question. In some cases, the channel occupancy is approximately 95%, while in other cases it becomes worse than in LTE. As can be seen from Table 6.2, the use of 60 kHz subcarrier spacing with low bands would decrease the channel occupancy efficiency. The UE capabilities, as discussed in later in this chapter, leave support for 60 kHz subcarrier spacing as an optional feature and thus something not foreseen as being supported in the first-phase UE implementations, when there is no obvious need for that.

For the higher frequency bands, LTE, of course, is not relevant as a reference as LTE cannot be operated there. For the 26 and 28 GHz ranges, it is possible to operate up to 400 MHz bandwidth, while 200 MHz is the minimum requirement for Release 15 UEs. Thus, for example, an allocation of 800 MHz would need to be operated with 200 MHz carriers if it is desired to serve such UEs. The channel occupancy numbers for Frequency

Table 6.2 Channel occupancy of LTE compared to 5G for sub 6 GHz operation (in MHz) for different nominal channel bandwidths.

SCS	5 MHz	10 MHz	15 MHz	20 MHz	25 MHz	40 MHz	50 MHz	60 MHz	80 MHz	100 MHz
LTE 15 kHz	4.50	9.00	13.50	18.00	NA	NA	NA	NA	NA	NA
5G 15 kHz	4.50	9.36	14.22	19.08	23.94	38.88	48.60	NA	NA	NA
5G 30 kHz	3.96	8.64	13.68	18.36	23.40	38.16	47.88	58.32	78.12	98.28
5G 60 kHz		7.92	12.96	17.28	22.32	36.72	46.80	56.88	77.04	97.20

Table 6.3 Channel occupancy for frequencies above 24 GHz
(Release 15 Frequency range 2).

SCS	50 MHz	100 MHz	200 MHz	400 MHz
5G 60 kHz	47.520	95.040	190.080	N/A
5G 120 kHz	46.080	95.040	190.080	380.160

Range 2 (FR2) are shown in Table 6.3. As shown in Table 6.3, the channel occupancy is around 95% with the mandatory 200 MHz channel bandwidth with both 60 and 120 kHz subcarrier spacing. The practical phase noise performance is foreseen to limit the use of 60 kHz subcarrier spacing on higher frequencies.

6.3 Physical Channels and Signals

L1 channels are defined in Release 15 for downlink [2]:

- Physical Downlink Control Channel (PDCCH), which carriers only L1 Downlink Control Information (DCI) and uses only Quadrature Phase Shift Keying (QPSK) modulation. There are 1 or more Control Channel Elements (up to 16) used for PDCCH transmissions, which allows one to adjust the amount of physical layer resources used for each PDCCH transmission. This enables basically link adaptation for PDCCH transmission depending on the link conditions.
- Physical Downlink Shared Channel (PDSCH) carrying the user data, as well as control information for higher layers (MAC and above). The PDSCH provides the physical layer to transport the Downlink the Downlink Shared Channel (DL-SCH), which carries information from higher layers. Also, the Paging Channel (PCH) is carried on the PDSCH. The PDSCH can use either QPSK, 16-QAM, 64-QAM, or 256-QAM modulation. PDSCH allocation, and other parameters used for PDSCH transmission, are signaled on the PDCCH.
- Physical Broadcast Channel (PBCH), which carries the necessary system information to enable a UE to access the 5G network. As described in Chapter 7, the "System information on demand" concept is introduced in 5G: only the information essential for system access is actually broadcasted continuously (periodically), while other system information is provided only on the need basis. The transport channel Broadcast Channel (BCH) is mapped on the PBCH. The modulation on the PBCH is QPSK. The BCH together with the Primary Synchronization Signal (PSS) and Secondary Synchronization Signal (SSS) form the SS Block, which is the fundamental 5G structure with beamforming operation as well as for the cell search and system access in general. The SS Block is shown in Figure 6.4.

The PDCCH is place before the PDSCH as shown in Figure 6.5, which allows the PDCCH to carry information on the downlink scheduling for the PDSCH. Also, the uplink allocations are normally signaled on the PDCCH. The relative positions of the PDCCH and PDSCH are such that in the downlink the PDCCH is always just before the PDSCH. As described later in this chapter, the PDCCH can carry information other than the downlink or uplink allocations.

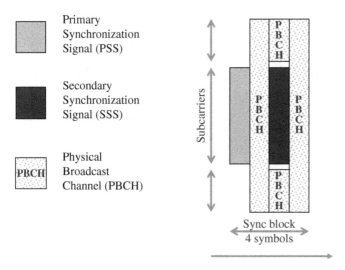

Figure 6.4 5G SS Block structure.

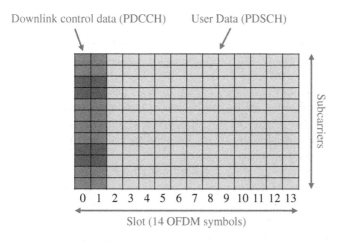

Figure 6.5 5G downlink control and data channels.

On the uplink side, the following channels are available:

- Physical Uplink Control Channel (PUCCH), which carriers the L1 control information (when there is no PUSCH allocation) including Hybrid Automatic Repeat Request (HARQ) feedback as well as Channel State Information (CSI). The modulation on the PUCCH is QPSK, or in some cases Pi/2-BPSK too. There are different types of PUCCH, which can be divided into be either short or long PUCCH, as shown in Figure 6.6.
- Physical Uplink Shared Channel (PUSCH), which carriers the user data as well as higher layer and L1 control information. The transport channel Uplink Shared Channel (UL-SCH) is mapped on the PUSCH. The PUSCH uses the same set of modulation options as the PDSCH (QPSK, 16-QAM, 64-QAM, and 256-QAM), but Pi/2-BPSK is also included as an optional modulation. The use of Pi/2-BPSK was studied initially

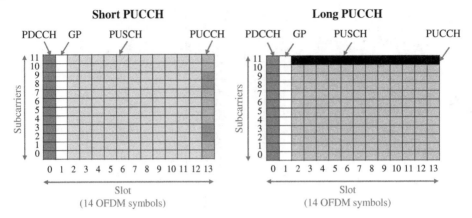

Figure 6.6 5G short and long PUCCH.

with LTE development during Release 8 studies [1], but was not included at the time in the specifications as LTE was able to reach comparable coverage with QPSK already.

- Physical Random Access Channel (PRACH), for enabling the random access procedure. The corresponding transport channel is the Random Access Channel (RACH).

Besides the physical channels, the following physical signals are defined in Release 15:

- Demodulation References Signals (DM-RSs), which are used to aid channel estimation. Note that as in LTE, there is no longer the concept of Common Reference Signals (CRS). DM-RS is needed naturally for both uplink and downlink directions. As described later, the amount of DM-RS can vary depending on the environment, that is, on whether there are high-velocity UEs or not.

- Phase-tracking Reference Signals (PT-RSs) are used with high frequency bands (if configured by the network). PT-RS is repeated periodically in the frequency domain while continuing a sequence of known data. It allows reduction of the common phase error on the receiver side (phase noise compensation). PT-RS is also available in both downlink and uplink directions.

- Channel State Information Reference Signals (CSI-RSs) (in the downlink) are also known as signals which allow a UE to make a CSI measurement report. The approach is similar to the LTE with known CSI-RS location in the time and frequency domains, making feedback necessary for frequency domain scheduling in the gNB.

- PSSs, together with the SSSs, form the physical cell ID. There are three different sequences for the PSS and 336 different sequences for the SSS, thus together they create a space of 1008 different physical cell IDs. The PSS and the SSS are transmitted together with the PBCH as shown in Figure 6.4. The so-called SS Block, also referred to as the sweeping sub-frame in connection with beamforming, has a duration of four symbols.

- The Sounding Reference Signals (SRSs) are used in the uplink direction to enable uplink frequency domain scheduling. SRS is especially useful in the Frequency Division Duplex (FDD) case when there is not channel reciprocity available, but it is also useful in the TDD case as allocation does not often cover the full band anyway.

6.4 Basic Structures for 5G Frame Structure

The frame structure in 5G differs from the LTE structure. The one major difference is larger flexibility in terms of symbol allocation between uplink and downlink. The symbols can be configured to be basically downlink, uplink, or flexible in the case of TDD operation. While in the ideal situation everything would be dynamic all the time, in real 5G networks there needs to be alignment on the uplink downlink split with the adjacent cells; also, in Release 15, the assumption is that operators in the same graphical area would use the aligned uplink downlink allocation when operating in the same TDD band. Figure 6.7 shows some examples of configurations when dealing with a 5G-only band. If LTE is used in the same TDD band, the alignment requires a more downlink-heavy configuration. 5G naturally also supports the FDD operation, which is needed with low frequency bands. In this case, uplink and downlink resources are always available.

Besides flexibility, another driver for differences has been the use of pipeline processing and the assumption that full carrier bandwidth should not be used for control signaling as the total carrier bandwidth might be too much in some cases for all UE types. Pipeline processing is enabled in 5G with the placement of the reference signals early in the frame, for example, in the second symbol (the first symbol is used to control information), which avoids having to receive the full slot before actual decoding can start. This, together with the efficient channel coding (see the later section on 5G channel coding with data transmission), allows fast processing to achieve low-latency operation (Figure 6.8).

The frame structure depends also on the subcarrier spacing. If the subcarrier spacing is 15 kHz, then the sub-frame duration is 1 ms, as is the minimum scheduling interval. If the subcarrier spacing becomes shorter, then the slot duration (and the minimum scheduling interval) becomes shorter, as shown in Figure 6.9. The typical subcarrier spacings are either 15 kHz (for bands below 1 GHz), 30 kHz for bands such as 3.5 GHz,

Figure 6.7 Example 5G TDD structures.

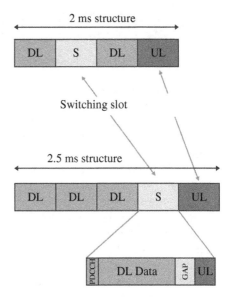

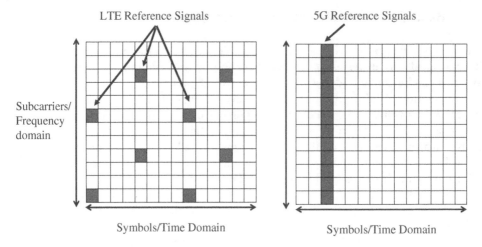

Figure 6.8 5G and LTE reference symbols in the frame structure.

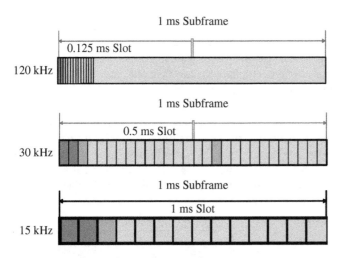

Figure 6.9 Typical frame durations expected in the first-phase 5G devices.

and 120 GHz subcarrier spacing with 26/28 GHz bands. 60 GHz subcarrier spacing is not supported by the first-phase 5G devices (an optional UE feature).

In the downlink direction, the first two symbols carry the PDCCH while the rest of the frame carries the PDSCH. After the PDCCH part follows the DM-RS. In the example shown in Figure 6.8, there is only one instance of DM-RS, the so-called front-loaded DM-RS. In an environment with high mobility, one would need to add additional DM-RS later in the slot (Figure 6.10).

Besides the allocation of the 14 symbols with the "regular" frame structure, there is also the approach called mini-slot. With the mini-slot structure, instead of 14 symbols, one can also allocate 2, 4, or 7 symbols. This allows shorter latencies to be achieved even with the 15 kHz subcarrier spacing, which otherwise would use the 1 ms allocation period. With the very high frequency bands, the use of the mini-slot allows also avoids having the transmission "locked" to one direction for the duration of 14 symbols when

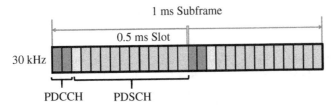

Figure 6.10 5G downlink with PDCCH and PDSCH in 5G sub-frame structure.

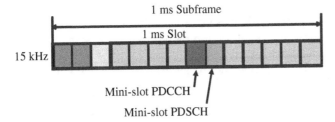

Figure 6.11 5G. Mini-slot example.

using hybrid beamforming. More details of the beamforming operation are explained later in this chapter. An example of the mini-slot structure is illustrated in Figure 6.11.

There is a further special frame structure called the self-contained sub-frame. The idea with the self-contained sub-frame is to minimize the latency by reserving space for the feedback signaling in the sub-frame itself. As shown in Figure 6.12, there are a few empty symbols after the downlink data transmission has finalized and then space for a small amount of uplink control information (UCI) transmission. This would enable (if the UE is fast enough in the decoding) faster feedback to be provided than in a regular TDD setup, where one has to wait for the next free available uplink resource. The first-phase networks and devices are not foreseen to support the self-contained sub-frame, as Release-15-based TDD deployments assume that the TDD network is synchronized between operators for the uplink and downlink split. If other operators in the same band would not be using this feature, the uplink/downlink interference likely would not enable this, except in deployments which are isolated or otherwise have a large enough minimum coupling loss between the base stations of different operators. Also, the use case for the self-contained sub-frame is rather the Industrial Internet of Things (IoT) type of operation.

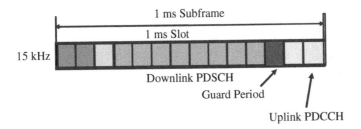

Figure 6.12 Self-contained sub-frame.

6.5   5G Channel Structures and Beamforming Basics

One of the key enablers for beamforming is the sweeping sub-frame. Each four-symbol downlink sweeping sub-frame consists of the PSS, SSS, and PBCH, as shown in earlier in Figure 6.4. This structure is often referred to as the SS Block. The SS Block is the basis for UE measurement taking from a cell, as well as gaining access to the network. Further, with beamforming, the beam identification is based on the detection and measurement of the SS Block. With the beam sweeping operation, the synch block is repeated several times, as shown in Figure 6.13. With the lower frequency bands (below 2.5 GHz), typically one can support at most 4 beams (with 15 kHz or 30 kHz subcarrier spacing), while with bands up to 6 GHz, up to 8 beams could be supported. With the mmW bands, the 120 kHz or 240 kHz subcarrier spacing enables up to 64 beams. As the symbol duration is shorter with the higher subcarrier spacing, the necessary synch block sequence duration remains reasonable with higher frequency bands even with a larger number of beams in use.

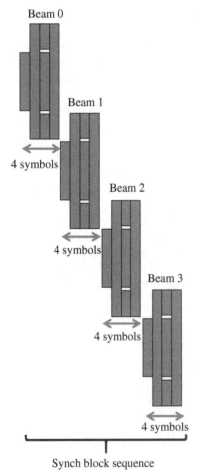

Figure 6.13 SS Block sequence enabling beam-related feedback.

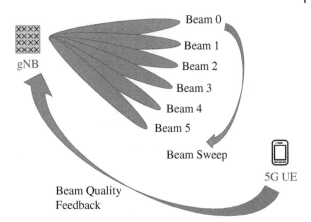

Figure 6.14 Beam quality feedback.

In the case of the beamforming operation, the SS Block is repeated for each beam during the sweeping sub-frame. The aim is to have the UE identify which beam was received the best and provide corresponding feedback for the gNB (Figure 6.14).

Important beamforming concepts have been introduced in 3GPP. One of the fundamental aspects is beam correspondence, which basically means the UE capability to send in the uplink to the beam direction corresponding to the one it has measured in the downlink direction to be the best one. This becomes even more important when a UE can use more than one set of antennas, as is likely to be the case with frequency bands above 24 GHz (3GPP Frequency Range 2). The 3GPP specifications define the concept of beam correspondence and define two different UE types for that:

- UEs than meet the correspondence without uplink beam sweeping.
- UEs that need uplink beam sweeping to meet the beam correspondence requirements. These UEs need to support the uplink beam management feature.

An example of uplink beam sweeping is shown in Figure 6.15.

Beamforming is addressed in more detail later in this chapter.

The expected order of MIMO transmissions will also vary as a function of the frequency bands. While going to higher frequency bands allows a UE to handle more antennas, the number of supported MIMO layers does not increase when going to the mmW range; on the contrary, as illustrated in Figure 6.16, with the mmW spectrum, UEs are only expected support at most two MIMO layers even if the overall number of antennas likely to be more. This is due to the reduced diversity available in typical 28-GHz-range deployment cases with a relatively small cell radius. Thus, 3GPP specifications also define the UE capabilities as follows:

- Bands below 2.3 GHz: UEs support two receiver antennas.
- Bands above 2.3 GHz: UEs support four receiver antennas.
- Above 24 GHz: Only two MIMO streams are required.

These definitions will impact the peak data rates achievable, as discussed later in this chapter. The UE with support for the mmW spectrum (24 GHz or higher) is, however, expected to have directivity in the transmission, allowing the UE to select different transmission directions from the set of antennas it has. Since antenna sizes with the mmW spectrum are very small, some UEs will have two or more antenna groups, and thus will have capability to change between the antenna groups.

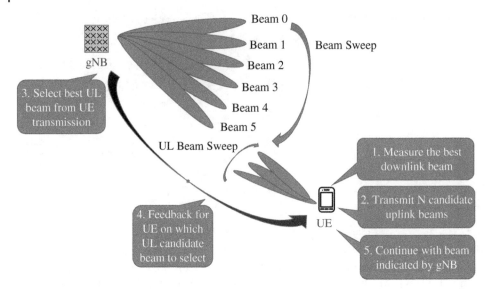

Figure 6.15 Uplink beam sweeping with beam correspondence operation.

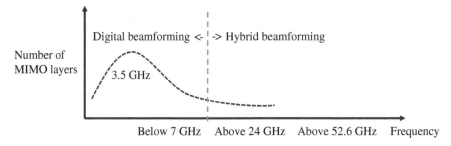

Figure 6.16 Number of MIMO layers feasible as a function of frequency.

The use of the mini-slot together with hybrid beamforming allows the transmission resources to be reserved for a shorter time for a particular direction, compared to the use of the full 14-symbol slot. This improves the resource usage efficiency, especially with smaller packet sizes and durations.

6.6 Random Access

Random access with 5G supports different PRACH formats and numerologies to cope with different deployments. The random access sequences themselves are based on the Zadoff-Chu sequences:

- Sequence length $L = 839$, with subcarrier spacings of 1.25 kHz and 5 kHz. This sequence can be used for very large cells (up to 100 km).
- Sequence length of $L = 139$, with subcarrier spacings of 15, 30, 60, and 120 kHz. These offer suitable solutions for different coverage situations.

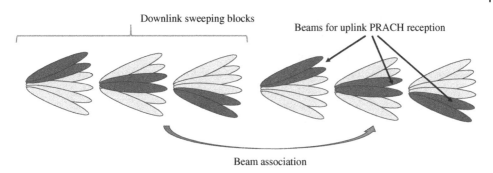

Downlink sweeping blocks

Beams for uplink PRACH reception

Beam association

Figure 6.17 Beam association for uplink PRACH transmission.

Several formats are supported for different coverage situations. Due to the cyclic prefix duration, a short sequence length is suited for cell radio on the order of 4.7 km. After this distance, some degradation appears; thus, for larger cells, it is better to use a longer sequence. With 30 kHz TDD deployment and a single uplink slot in the frame structure, only a shorter sequence is possible. Use of a longer sequence would require allocation of two consecutive slots in the uplink, something not supported with either of the TDD patterns shown earlier in Figure 6.7.

The beamforming operation has an impact on the RACH procedure. Once the UE has detected the cell from the strongest SS Block, it needs to use RACH resources that correspond to the best beam direction toward the UE. Thus, the UE needs to establish beam association, as shown in Figure 6.17. Otherwise, the gNB may not receive the UE correctly and will not use the right beam direction in the downlink for the PRACH response. Compared to LTE, the UE needs to select correct resources in time, frequency, and additionally in the spatial domain.

The UE will use a power ramping procedure if it does not receive the response for the PRACH preamble, and the UE will then transmit the preamble with a higher transmit power at the next suitable resource. The gNB will use the beam direction based on PRACH detection until further measurements are received from the UE for possible beam adjustment.

6.7 Downlink User Data Transmission

The user data in the downlink direction is carried on the PDSCH. As indicated in connection with the frame structure, the allocation period in the time domain can have different values depending on the subcarrier spacing, and the mini-slot structure can also be used. The PDCCH informs the device about the downlink allocation and other important parameters like modulation and the coding format being used, what resources are actually allocated as well as HARQ information (which HARQ process and whether the transmission is a new packet or not).

The modulation alternatives used in the downlink direction are QPSK, 16-QAM, 64-QAM, and 256-QAM. For the users in the cell edge area, one naturally uses either QPSK or 16-QAM, while users closer to the cell center may use also 64-QAM or even 256-QAM.

The channel encoding chain has the following functionality:

o Cyclic redundancy check (CRC) attachment, which enables detection of the erroneous decoding in the receiver side.
o Low density parity check (LDPC) encoding for low complexity and high performance of user data coding. See the later channel coding section for more details.
o Rate matching to fit for the available downlink resource.
o Code block concatenation (Figure 6.18).

After the encoding, the following operations take place:

– *Scrambling*. This is different from ciphering, as the aim is to ensure that sufficient variation exists in the data to ensure proper modulation detection (compared to the case of sending longer periods of a single constellation point only).
– The modulation mapper maps the bits for the modulation in use, from 2 up to 8 bits per symbol (from QPSK to 256-QAM).
– The layer mapper maps the code words to different layers, with up to 8 layers supported in 5G Release 15. This is equal to the maximum MIMO order supported, while typically the UE will use only two layers, with the exception of the bands around 2.5 to 3.5 GHz, where four layers are supported as the baseline.
– Antenna port mapping to map to the antenna ports in use. There may be more antenna ports available than layers used /supported by the UE.
– Resource block mapping covers mapping to virtual resource blocks and then further mapping to the physical resource blocks (PRBs) for actual transmission.
– OFDM signal generation creates the actual OFDM signal to be transmitted. The figure does not cover RF parts such as TX filtering, the power amplifier, or the antenna. Note that with the beamforming operation there are often more antennas than actually antenna ports in the baseband operation (Figure 6.19).

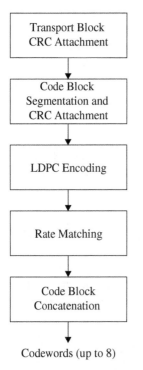

Figure 6.18 PDSCH channel encoding.

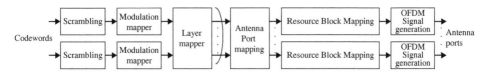

Figure 6.19 PDSCH transmitter chain after encoding.

6.8 Uplink User Data Transmission

The user data in the uplink direction is transmitted either with OFDM or with DFT-S-OFDMA (single carrier) transmission depending on the link budget situation. The PUSCH carries user data in the uplink direction.

The uplink allocation (uplink grant) is received on the PDCCH, which informs the UE how much resources are allocated for the uplink transmission as well as other transmission parameters to be used. The gNB scheduler will get feedback information not only on the channel condition (Channel Quality Indicator [CQI]) and power availability in the UE but also on the amount of data in the UE buffers. Power headroom and buffer status reports are covered in the MAC layer signaling as explained in Chapter 7. The PDCCH signaling will also determine if the UE is to use OFDM or DFT-s-OFDM transmission. The gNB can use also SRSs to determine which part of the uplink band is best suited for the uplink allocation.

The uplink modulation alternatives are QPSK, 16-QAM, 64-QAM, and 256-QAM, and in addition, $\pi/2$-BPSK is possible when using DFT-S-OFDM. The normal UE is able to reach maximum transmission power with QPSK, and thus $\pi/2$-BPSK support is optional for the UE side. $\pi/2$-BPSK is intended to be used with transform precoding only, as the key driver was to minimize the transmitted signal PAR. Use of $\pi/2$-BPSK was already considered for LTE [1] but it was not adopted at the time as LTE UE requirements require reaching maximum TX power with QPSK. This avoids reducing the uplink data rates when operating at the maximum power level.

The uplink encoding chain is similar to the downlink, but with the uplink allocation (grant) in place the UCI is multiplexed with uplink data, as shown in Figure 6.20. The same LDPC coding as in the downlink direction is used also in the uplink.

In the uplink direction, an important element is the use of non-contiguous uplink transmission. While on the capability level each UE supports the minimum bandwidth assigned for each frequency band, the use of a bandwidth part (BWP) will temporary limit the part of spectrum that a UE will operate in. If the uplink allocations would always need to be contiguous, this would limit the available data rate greatly if some narrow transmission of uplink control channel were present in between. The benefit of non-contiguous (or almost contiguous) transmission is shown in Figure 6.18, as it allows larger data rates in presence of UEs with BWPs configured. It is also expected that there will be new UE categories in the future which do not support full bandwidth (see Chapter 15), and operating with them will be much more efficient when almost contiguous transmission is supported in the UEs.

In order not to increase the waveform PAR variations, the feature is defined only for use with OFDM in the uplink. This is anyway the most likely case when the UE has enough power reserves for full-bandwidth transmission (Figure 6.21).

The uplink signal generation in Figure 6.19 shows an example with 2 TX antennas and support for two-stream MIMO operation. In the first phase, the UEs based on Release 15 are foreseen to support only 1 TX per technology and frequency band, especially in

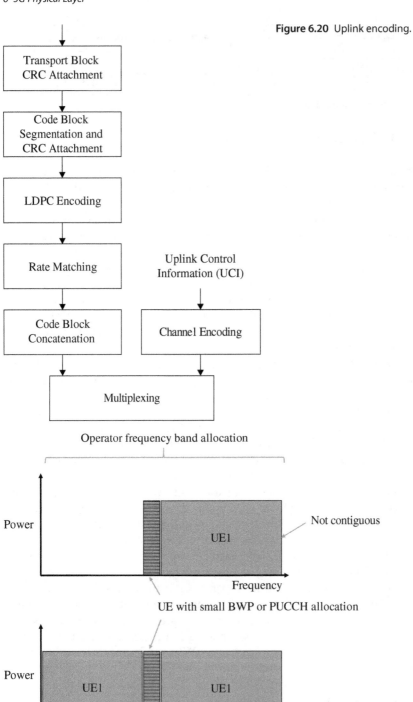

Figure 6.20 Uplink encoding.

Figure 6.21 Improvement for uplink allocation flexibility with almost contiguous uplink operation.

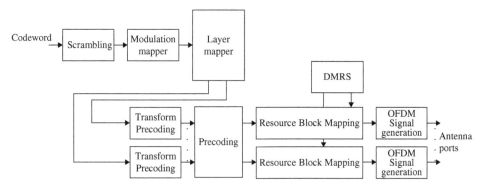

Figure 6.22 PUSCH transmitter chain after encoding.

connection with LTE–5G dual connectivity. In a later phase, this is foreseen to become more popular, especially when operating with standalone 5G when transmitters on both LTE and 5G will not need to be used simultaneously (Figure 6.22).

6.9 Uplink Signaling Transmission

UCI consists of

- HARQ-ACK feedback, used to provide the HARQ-ACK feedback of the downlink user data.
- Scheduling Request (SR) is used, as the name indicates, by the UE to request to be scheduled in the uplink due to data arriving at its transmit buffer.
- CSI is used to support downlink scheduling with a set of different subcomponents – CQI, Rank Indicator (RI), LI, precoding matrix indicator (PMI), SS/PBCH Block Resource indicator (SSBRI), CSI-RS resource indicator (CRI), and Reference Signal Received Power (RSRP) – as explained in more detail in the MIMO section.

The UCI payload utilizes different encoding techniques depending on the payload size. A CRC of 6 bits is used if the UCI is 12–19 bits, and a CRC of 11 bits is used if the UCI is 20 bits or larger. For UCI sizes of 11 bits or less, no CRC is used. The following UCI channel coding techniques are used:

- Simplex for 1 or 2 bits (HARQ-ACK or SR)
- Reed–Muller for 3–11 bits (More complex HARQ-ACK)
- Polar for >11 bits (CSI feedback or very large HARQ-ACK)

5G provides a number of different ways to transmit UCI. There are five different PUCCH formats (Table 6.4) to carry, and in addition, UCI can be multiplexed within the PUSCH as simultaneous PUCCH and PUSCH transmissions are not supported.

The PUCCH formats 0 and 1 with a small payload are designed for HARQ-ACK and SR transmission only; the short format (0) can be used if there is no issue with uplink coverage, and the long format (1) for coverage-limited scenarios. The short PUCCH can also be used in a TDD carrier's switching slot with ease, as shown in Figure 6.13. In addition, the short formats are more suitable for a gNB utilizing analog Rx beam

Table 6.4 Characteristics of the PUCCH formats.

PUCCH format	UCI size	Waveform	Modulation	Duration	Bandwidth	Multiplexing capability
0	1–2 bits	DFT-S-OFDM	Sequence selection with phase rotation	1–2 symbols	1 PRB	6 UEs
1				4–14 symbols		84 UEs
2	> 2 bits	OFDM	QPSK	1–2 symbols	1–16 PRBs	1 UE
3		DFT-S-OFDM	QPSK and Pi/2 BPSK	4–14 symbols	1–16 PRBs	1 UE
4					1 PRB	4 UEs

forming, as it enables time-multiplexing several users' feedback without adding much to the feedback delay.

The PUCCH formats 2–4 can be used for more complicated HARQ-ACK feedback, where several downlink slots' HARQ-ACKs are mapped to one uplink slot in a TDD configuration. The HARQ-ACK feedback size further increases if carrier-aggregation- or code-block-group-based HARQ-ACK is employed in the downlink. The formats 2–4 are also the formats to use for CSI feedback, again categorized as a short format (2) and long formats (3 and 4). The differentiation between short and long formats is again similar as with the small payload size formats, but there are two variants of the long format. Format 3 provides the maximum payload size allowed by the polar code, while format 4 is able to support a lower UCI payload size, but it allows for multiplexing up to four users on the same resource for better user capacity (Figure 6.23).

One restriction with the LTE PUCCH design was symmetric frequency hopping, which is quite suitable for a typical setup but somewhat suboptimal for cases where one band edge is experiencing more interference or needs to be left unused due to adjacent systems. New Radio (NR) PUCCH frequency hopping is supported for the long PUCCH formats (1, 3, and 4) is fully flexible, so that it is possible to place the first hop of the PUCCH freely in frequency, and the second hop again freely and independently of the first hop to another location in frequency. The number of symbols allocated for the long PUCCH is divided equally between the first and the second hop, and the PUCCH transmission remains contiguous in time.

Long PUCCH formats (1, 3, and 4) also support multi-slot PUCCH using repetition. This mechanism is designed for maximum coverage, although it costs more in terms of delay and overhead. When multi-slot PUCCH is used, inter-slot frequency hopping can be employed (Figure 6.24).

The UE is configured with PUCCH resource sets, each resource set containing several PUCCH resource configurations including the PUCCH format and its time/frequency location. When the UE receives a DCI on the PDCCH that triggers a corresponding

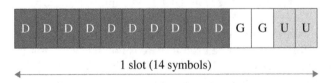

1 slot (14 symbols)

Figure 6.23 Switching slot with a two-symbol PUCCH after the switching gap.

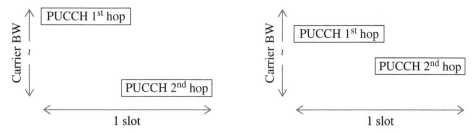

Figure 6.24 PUCCH intra-slot frequency hopping. Left: Edge-to-edge hopping (LTE). Right: Each hop can be freely placed in frequency, enabling more flexible frequency occupancy (NR).

PUCCH transmission (HARQ-ACK, aperiodic CSI report), the size of the triggered payload and well as the PUCCH resource indicator determine the PUCCH configuration to use for the PUCCH transmission in a slot, and the slot in which to transmit is determined by the slot timing indicated by the DCI. Periodic CSI reports and SRs are configured with the PUCCH resources, formats, and slot timings by the Radio Resource Control (RRC) layer.

Uplink beamforming applies to the PUCCH just as to the PUSCH. Each PUCCH resource set is configured with a set of candidate spatial relation signals by the source's RRC, and a MAC control element is used to pick one set out of the configured set to be active at a given time. This source signal is used to determine the uplink beam the UE transmits the PUCCH with. If the source signal is SRS, it is straight forward to transmit the PUCCH using the same beam used to transmit the indicated SRS. If the source signal is a Synchronization Signal Block (SSB) or CSI-RS, the UE needs to be able to transmit the PUCCH using the uplink beam that directionally corresponds with the downlink beam the active spatial relation source signal is received in.

Simultaneous PUSCH and PUCCH transmissions are not supported because the independent PUCCH and PUSCH setups lead to a transmission waveform that is not contiguous in frequency. This in turn would yield a very high peak-to-average power ratio to the transmission, and to maintain the peaks within the linear region of the power amplifier the average transmit power would need to be severely reduced, limiting the actual transmitted energy of an OFDM symbol and thus cutting coverage. Thus, simultaneous PUSCH and PUCCH transmissions are not supported, although obviously nothing prevents frequency multiplexing one user's PUSCH with another user's PUCCH. When the UE is scheduled to transmit uplink data on the PUSCH, it is possible to time-multiplex a short PUCCH in the same slot with a truncated PUSCH allocation, but this would limit the PUCCH usage to short PUCCH formats and force a large portion of the frequency resources to go unused on the PUCCH symbols. Similarly, one could time-multiplex PUSCH and long PUCCH formats on a slot level, but this would further reduce the uplink throughput a single UE can achieve as no data could be sent on the slots carrying UCI. Because of this, when there is a need to transmit UCI simultaneously with PUSCH, the UCI is multiplexed in the PUSCH. This principle has been used since the first LTE release (3GPP Release 8) as well.

When the UCI is multiplexed on the PUSCH, the PDCCH scheduling the PUSCH indicates which uplink resources allocated for the PUSCH are to be used for UCI instead of the PUSCH. This is possible as the HARQ-ACK and CSI feedback are both deterministic in nature. The code rate of the UCI is determined by the UCI payload size and the amount of resources the scheduling allocates for it. In addition, UCI-type specific power control offsets can be used to boost the PUSCH transmit power when it is carrying UCI to ensure the desired reliability for a given UCI message, simultaneously compensating for the lost PUSCH bits as well. The UCI and the PUSCH payloads are still processed independently, and the PUSCH frees the resource elements for UCI delivery either by puncturing (one or two bits of the UCI payload) or rate matching (more than two bits of UCI payload).

6.10 Downlink Signaling Transmission

The heart of the NR system is downlink control signaling, designed to address a variety of scenarios covering, for example, enhanced mobile broadband (eMBB), ultra-reliable low-latency communications (URLLC), and massive machine-type communications (mMTC). In addition, 5G must support an extremely wide frequency range from below 1 GHz to above 50 GHz, which includes the frequencies that have not been utilized for cellular networks to date, for example, millimeter waves.

Physical control resources, over which the PDCCH is transmitted, are configured using COntrol REsource SETs (CORESET). In the frequency domain, a CORESET other than zero is configured on a BWP in full chunks of 6 RBs on a 6RB grid as shown in Figure 6.22. The 6RB grid originates in Point A, which determines also the start of the cell's common RB (CRB) grid. Consequently, the gNB may configure CORESET to be discontinuous in frequency, to, for example, accommodate a narrow-band system within the serving cell or to improve frequency diversity. On the other hand, CORESET zero used in initial access is contiguous, does not need to be 6RB grid aligned, and is configured using 4 bits in the Master Information Block (MIB) with a limited set of parameters (Figure 6.25).

An additional property of CORESET is its time-domain length of 1 to 3 OFDM symbols. The physical resources of a configured CORESET are separated into Resource

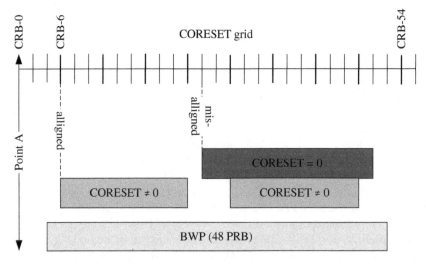

Figure 6.25 CORESET configuration in frequency domain.

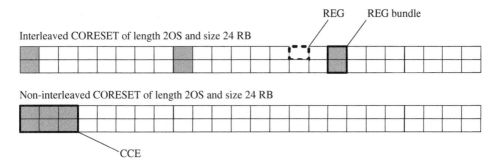

Figure 6.26 CORESET structure.

Element Groups (REGs), each containing 9 Resource Elements (REs) to carry the payload and 3 REs of Demodulation Reference Signals (DMRSs) in an RB-symbol cluster. These REGs are further clustered into REG bundles of 2, 3, or 6 REGs, for which the precoder is constant. The REG bundles may be further interleaved and mapped to the logical control channel element (CCE) consisting of 6 REGs. Figure 6.26 illustrates the units of a CORESET. Whether REG bundles are interleaved or not as well as the REG bundle size is configurable for a CORESET, addressing the operator's needs for frequency diversity and channel estimation quality in its deployment. Up to 3 CORESETs can be configured per BWP. However, the minimum requirement for a UE is to support CORESET zero and one additional configured CORESET.

A CCE carries exactly 108 bits, that is, 54 REs modulated by QPSK. To improve reliability and/or coverage, the PDCCH payload can be transmitted over 1, 2, 4, 8, and 16 CCEs, the so-called aggregation levels. Before the payload is mapped to one or more CCEs, it is channel-coded with polar code and concatenated with CRC. To reduce parsing and false alarms, NR increased CRC from 16 to 24 bits compared to LTE.

The time-domain location of the first symbol of a CORESET is determined by monitoring occasions configured by a search space set. Slots where monitoring occasions occur are determined on the basis of search-space-set s periodicity k, offset o, and duration d. This is illustrated in Figure 6.27, where a common (CSS) set $s = 0$ is configured with periodicity $k = 5$, offset $o = 1$, and duration $d = 2$ corresponding to grey blocks. In addition, a search space set is configured with number of PDCCH candidates for ALs 1, 2, 4, 8, and 16, which are randomized using a hashing function similar to LTE EPDCCH.

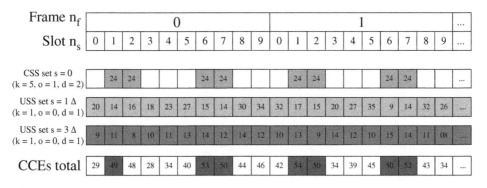

Figure 6.27 Search space sets.

In a monitoring occasion, a UE blindly decodes the configured PDCCH candidates to find a transmitted PDCCH. Monitoring occasions within a slot are determined by a bitmap. NR supports features where multiple monitoring occasions are also allowed within a slot, to reduce latency.

In general, a gNB may configure up to 10 SS sets per BWP. However, UEs capabilities are limited by the number of blind decodings (BDs) UEs are able to perform and the number of CCEs UEs are able to channel-estimate in a slot of a given subcarrier spacing. For example, with 60 kHz subcarrier spacing (SCS), a UE is capable of performing 22 blind decodes and estimating 48 CCEs. If the number of CCEs a UE is configured to monitor in a slot exceeds 48, the UE will not monitor the SS sets with the highest indices; in other words, a UE drops monitoring for PDCCH candidates of SS sets with the highest indices one by one until number of CCEs to be monitored by the UE in the slot is less than the limit. This is illustrated in Figure 6.27, where three SS sets are configured and the total number of CCEs to be monitored in slots is shown. In red slots, the limit of 48 is exceeded, and as a consequence the user-specific (USS) set $s = 3$ is not monitored by the UE. The situation when gNB configures more BDs and CCEs in a slot than the UE is capable of monitoring is called overbooking. Only a special cell (primary cell of a cell group) can be overbooked, and a gNB can never overbook CSS. Overbooking has been introduced to enable maximum usage of the UE's capabilities across the slots.

The actual payload a PDCCH carries is called DCI, and several formats of DCI can be transmitted by a gNB, as listed in Table 6.5. The formats starting with 0 are used for uplink (UL) unicast scheduling, and formats starting with 1 are used for DL unicast scheduling. Formats starting with 2 are group-common formats. A UE is capable of monitoring for at most three unicast formats and at most four DCI formats in a slot. If a gNB configures a UE with more DCI formats to monitor, certain truncation and padding rules apply to align multiple DCI formats in size.

Out of group-common 2_× DCI formats, the 2_0 slot format indicator (SFI) and 2_1 pre-emption indicator are the new features of NR compared to LTE. SFI in DCI format 2_0 indicates a slot combination from a set of preconfigured slot combinations of different slot formats. As shown in Figure 6.25, two slot combinations are indicated, each by one DCI 2_0. The SFI = 0 has been preconfigured using slot formats: Format 0 (DL symbols only) Format 1 (UL symbols only) and Format 8 (13 flexible symbols + 1 UL symbol). The SFI = 1 contains slot Format 2 (flexible symbols only). A flexible symbol, if indicated by a gNB in DCI 2_0, is no-transmit no-receive symbol unless gNB schedules DL or UL

Table 6.5 Downlink Control Information (DCI) formats.

DCI format	Usage
0_0	Scheduling of PUSCH in one cell
0_1	Scheduling of PUSCH in one cell
1_0	Scheduling of PDSCH in one cell
1_1	Scheduling of PDSCH in one cell
2_0	Notifying a group of UEs of the slot format
2_1	Notifying a group of UEs of the PRB(s) and OFDM symbol(s) where UE may assume no transmission is intended for the UE
2_2	Transmission of TPC commands for PUCCH and PUSCH
2_3	Transmission of a group of TPC commands for SRS transmissions by one or more UEs

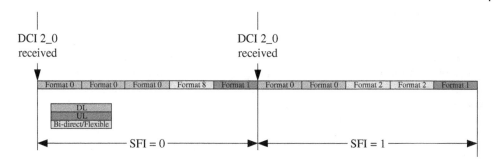

Figure 6.28 Slot Format Indicator (SFI).

transmission overlapping with the symbol. By indicating flexible symbol overlapping with a preconfigured DL/UL transmission or CORESET, a gNB may cancel the preconfigured transmission or prevent a UE or groups of UEs from monitoring the PDCCH in the CORESET. NR currently tabulates 64 slot formats. Format X in [3], however, the table is designed to carry up to 256 entries for future compatibility (Figure 6.28).

DCI format 2_1, a so-called pre-emption indication or interrupted transmission indication DCI, is another broadcast format that is transmitted with specific RNTI (INT-RNTI) that can be common to a group (or all) UEs in the cell. This DCI format is used for notifying the UEs a set of time-frequency resources (RBs and OFDM symbols) that will not carry any information to the UE. The overall idea is that the UE may be first scheduled with a PDSCH, but later the gNB needs to take the already assigned resource to some other, more urgent use, for example, to transmit a URLLC data packet to someone. In this case, part of the already scheduled PDSCH is dropped, and the resources are used to transmit the more urgent PDSCH to someone else. The gNB can use the DCI format 2_1 to broadcast such a pre-emption case to the cell so that the affected UEs can mark the pre-empted resources as erasures in the PDSCH decoding process and avoid trying to use the REs that carried some other UE's more urgent data when decoding the PDSCH that got partially punctured. There is no specified UE behavior for DCI format 2_1 reception; the UE is free to ignore the message, or use it any way it is able to when decoding the affected PDSCH. The usage of the pre-emption indication by the UE is a trade-off between an increased PDCCH decoding load in the UE as it has to check if the indication was sent or not at every possibility configured for it, and the possibility for increased throughput in the event of pre-emption. Similarly, for the network, the trade-off is between sending the pre-emption indication whenever pre-emption actually occurred versus rescheduling the packets that were affected using the regular HARQ procedure.

6.11 Physical Layer Procedures

The physical layer procedures with 5G have many new elements over LTE. One of the new areas is beamforming, which is addressed in the next section in more detail. The other key physical layer procedures are as follows [3, 4]:

- Physical layer retransmission handling (HARQ)
- Power control and power sharing with dual connectivity
- Timing advance

6.11.1 HARQ Procedure

The HARQ procedure is based on the stop-and-wait HARQ operation. Once the packet is transmitted from the gNB, the UE tries to decode it, and if the decoding fails, the UE keeps the soft samples in the buffer and requests a retransmission. Compared to LTE, the UE decoding time is clearly faster, and in an ideal case, a retransmission could occur just three scheduling intervals later, as shown in Figure 6.26. In the case of higher bands and when using a shorter scheduling time interval, the operation could be even faster, but then with the TDD case, the uplink (or downlink) resource is not always available, and thus the selected uplink/downlink allocation will impact the HARQ loop latency experienced. There is support for up to 16 HARQ procedures, which allows operation also with different backhaul latencies in case part of the functionality is located further away from the RF site behind the backhaul connection.

The example timing in Figure 6.29 works with FDD operation when there is always the possibility of uplink transmission. This allows, with the 1 ms allocation period, the retransmission to reach the UE 4 slots later; thus, there is an extra delay of 4 ms from a single retransmission. In the case of TDD operation, the slots are shorter, but uplink is not necessarily available in all cases, and thus there is more variance in the exact delay. With the 2 ms TDD structure, as shown in Figure 6.7, there is anyway an uplink resource available every 2 ms, and thus at most the extra delay experienced would be on the order of 2 ms, but typically less. Thus, the average delay experienced would be comparable to the FDD case due to the shorter slot duration period with 30 kHz SCS.

6.11.2 Uplink Power Control

With 5G, there are initially two types of power control operation, open loop and closed loop. Open loop refers to setting the power based on the received downlink signal, while closed loop refers to the explicit signaling from the gNB to increase or decrease the power. The power control is important to reduce the interference created for the network as well as to minimize the UE power consumption.

The first step with power control occurs on the PDCCH with the scheduling message for the Random Access Response (RAR) in the −6 to 8 dB range. During the operation, the PDCCH can be used to further control the uplink TX power of the UE.

With dual connectivity with LTE and 5G, there are two possible approaches to power control and sharing:

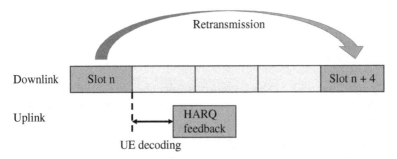

Figure 6.29 Timeline for HARQ operation with FDD case and 15 kHz subcarrier spacing.

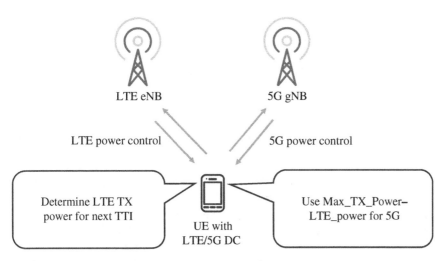

Figure 6.30 Power control with LTE-5G dual connectivity.

- Dynamic power sharing between LTE and 5G. In this, the master of the connection (in the case of option 3 architecture, as covered in Chapter 5, this is LTE) is first allocated the power it needs, and the remaining power is left for 5G operation. This ensures that LTE as the connection anchor can be maintained in all cases, while all the remaining power is available for 5G. This is mandatory in Release 15 (with capability signaling to allow testing the feature first against practical networks). The operation is shown in Figure 6.30.
- Semi-static power sharing between LTE and 5G. In this case, the LTE side reserves some amount of power resources in a semi-static way, and the remaining power is available for the 5G side. This is challenging as LTE side power reservation has to be done with some margin, and on the other hand the unused LTE power is not made available for the 5G side. This is expected to be the case only for some early implementations for the very first UEs and is not foreseen for large commercial use.

6.11.3 Timing Advance

The timing advance procedure is very similar to the LTE approach, with the actual timing advance command being provided in the MAC layer. After the RACH preamble has been received, the gNB will respond with the timing advance in the RAR, and it is then maintained during the connection as frequently as is needed. The step of the timing advance is a function of the SCS. The larger the SCS, the smaller the timing advance adjustment step. Compared to LTE, the timing advance was adjusted to allow very large cells in special deployments beyond 100 km limit. Now the theoretical limit for cell size (radius) is up to 300 km.

6.12   5G MIMO and Beamforming Operation

6.12.1   Downlink MIMO Transmission Schemes

5G NR downlink transmission schemes have been streamlined to a single transmission scheme, compared to LTE where we have CRS- and DM-RS-based transmission schemes. One of the reasons for such a simplification was the fact that the data transmission in NR operates solely on dedicated reference signals. While closed loop transmission schemes are the main design point, these being based on some forms of CSI feedback from the UE, transparent transmit diversity is possible, for example, precoder cycling based on a set of rules, like using the same precoder on the frequency allocation where the same DM-RS is used.

Having set the type of reference signals used for data transmission, the design of 5G targets the number of layers which are transmitted to a single UE and the rules needed for performing multi-user MIMO transmission in an orthogonal way between the users. Indeed, the two main types of transmissions are single- and multi-user MIMO. While single-user MIMO can support a total of 8 transmission layers to a single UE, in the multi-user domain, the network can multiplex up to 12 orthogonal DM-RS ports, meaning that a total of 12 layers can be orthogonally multiplexed between a number of UEs, for example, 12 UEs if each is receiving a single layer or 3 UEs if each is receiving 4 layers. Nevertheless, in the multi-user domain, the network can multiplex even more than 12 layers, if, for example, there is better spatial separation between the users or some forms of orthogonalizations are performed at the transmitter, such as zero forcing.

It is worth noting that codeword-to-layer mapping has the following rules:

- One to four layers are mapped to a single codeword.
- If there are more than five layers, these are mapped on two codewords: the first floor (L/2) layers are mapped on codeword 0, while the remaining layers are mapped on codeword 1.

6.12.2 Beam Management Framework

Beam management is one of the most important components of 5G NR. These mechanisms become necessary when 5G is operated in higher carrier bands, such as above 6 GHz, where radio propagation is more challenging, and beamforming is mandatory for both the gNB and UE in order to compensate for the path loss. The good news is that as the carrier frequency is increasing, the antenna sizes at both the transmitter and receiver become smaller, and hence a high-frequency-antenna solution can accommodate a larger number of antenna elements. The antenna architectures are in any case different: in low carriers, one is using digital transmission where power amplifiers are present for each antenna element, whereas in higher carriers, this is not possible, and hence one power amplifier serves a set of antenna elements, which may be placed on a so-called antenna panel. Both the gNB and UE can have multiple panels: while multiple panels at the gNB enable the multiplexing of multiple UEs, as each panel can transmit toward one UE, multiple panels at the UE would enable the reception of signals in a challenging environment where UE movement, rotation, or beam blockage would benefit from receive signal diversity. Hence, there is a need and motivation for the placing of multiple panels on the device.

One of the fundamental elements of beam management is the concept of a beam pair between the gNB and UE. Both the transmitter and receiver can form beams in different directions; it is said that when the beams "face each other," we can have a beam pair. One should not restrict this possible only to line of sight, as such beam

Figure 6.31 Beam pair between gNB and UE.

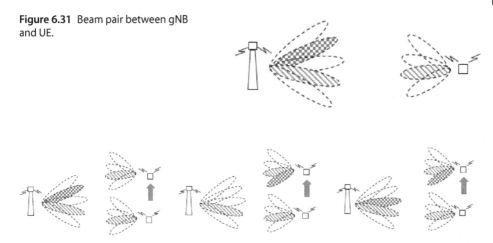

Figure 6.32 Downlink beam training use cases for linear movement of UE (diagonal hash beam = old beam, square dots beam = new beam).

pairs can also occur due to reflections from scatterers. In Figure 6.31,we depict a transmitter having multiple beams and a UE which also operate with multiple beams. Note that these beams can be used for both transmission and reception. If the UE utilizes the same beam for transmitting and receiving, it is said to be capable of beam correspondence.

There are a few challenges in the process of keeping alive beam pairs. Figure 6.32 shows three different intra-cell downlink beam training use cases for UE movement. In the all cases, it is assumed that the UE moves its position linearly with respect to the location of a base station. In the left example, a linear movement introduces the need to change the TX beam at the base station while using the same RX beam at the UE. In the middle example, only the RX beam needs to be changed while the same TX beam is used. In the right, both TX and RX beams need to be changed.

Figure 6.33 presents intra-cell beam training use cases for UE rotation and beam blockage. In the left example, we have UE rotation, which causes the RX beamformer at the UE to change while the TX beam at the gNB is unchanged. In the case of beam blockage, the signal path is blocked by an obstacle, leading to a significant drop of signal quality, for example, tens of decibels in received power, at the receiver. An alternative path, for example, due to reflections (blue beams), may become the best link between the base station (BS) and UE.

Figure 6.33 Necessity for beam change: movement of the UE (left figure), UE rotation (middle figure), and beam blockage (right figure).

Beam management consist of a set of procedures used to assist the UE to set and maintain its receive spatial filter used to transmit and receive. It can be categorized by a few steps:

- Initial beam acquisition
- Beam measurement, refinement, and reporting
- Beam recovery

6.12.2.1 Initial Beam Acquisition

The process of finding the beam pairs is based on beam sweeping and it is done by both the gNB and UE, as shown in Figure 6.34. This means essentially that the transmitter (being the gNB or UE) sweeps a set of beams, while the receiver measures these receive beams. As both the gNB and UE need to perform this procedure, there are ways in which one end sweeps through beam while the other performs measurements, and the other way around. The process ends when a beam pair is established between the transmitter and receiver.

Initial beam acquisition occurs already during the initial access. The gNB transmits the synchronization signal in such a sweeping manner; in fact, all the transmission points from the network perform this operation while the UE performs measurements. As the beams are spatially oriented, just a few of these, if at all any, will be received by the UE. This also depends on the beamwidth with which the initial access beam is transmitted, and in practice this may be wider than the actual data (PDSCH) beam as the signal needs to reach the UE in order for the UE to get connected to the transmission point and hence to the network. For the initial beam acquisition, it is likely that also the UE would utilize a wider beam and for data reception. It is worth noting that the beamwidth is not something which is written in the specification, the gNB and UE being free to utilize whatever beam width they see fit. However, as initial access is a very important procedure, it is likely that the beamwidth used in this situation is wider than the ones used for data transmission of the PDSCH and PUSCH.

6.12.2.2 Beam Measurement and Reporting

Beam measurement and reporting consist of acquiring the beam information for data transmission.

While initial beam acquisition may be done on a wider initial access beam containing the SS Block, this beam can be further refined (narrowed) for data transmission. However, the same beam width may be used for data transmission as it has been used for initial access transmission. In Figure 6.35, we depict an example of such beam refinement. The wide beam of initial access is further narrowed when transmitting CSI-RS. The transmission of CSI-RS is necessary as all the CSI components, such as the PMI,

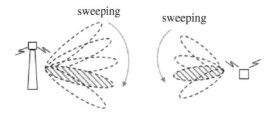

sweeping   sweeping

Figure 6.34 Sweeping beams at gNB and UE.

Figure 6.35 Beam refinement.

SSB wide beam

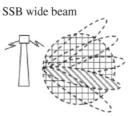

CSI-RS narrow beam

rank, and CQI, are measured on this reference signal. In addition, the so-called TRS (tracking reference signal) is also transmitted via such beams. The TRS is needed for establishing fine time and frequency synchronization at the UE. In some sense, the functionality of LTE CRS has been now split into TRS and CSI-RS, this decoupling bringing more flexibility to the system.

The measurements the UE performs on nonzero power CSI-RS or the SS Block are basic L1-RSRP computations. In this way, the UE finds out the strength of the best beams and reports that to the gNB. The UE is not expected to be configured with more than 64 nonzero power CSI-RS resources for channel measurement for a CSI report. The reported L1-RSRP values are defined by a 7-bit value in the range [−140, −44] dBm with a 1 dB step size.

A UE having one or multiple panels may receive the CSI-RS/SS Blocks with the same spatial filter, and hence with the same beam. It is important for the gNB to know this information, as the transmission on beams received by the same UE Rx beam can occur with more dynamic scheduling, as the RX beam change at the UE is not needed. In Figure 6.36, we have an example where the UE is equipped with two receive panels. Several beams from the same Transmission Reception Point (TRP) may be received by the same beam, and hence the striped Tx beams are received by the striped Rx beam, and the same is true for the square beams. If the UE detects such a configuration, the reporting occurs under the so-called group-based bream reporting.

6.12.2.3 Beam Indication: QCL and Transmission Configuration Indicator (TCI)
The network can indicate to the UE how the signals are transmitted with respect to each other in time. To be more specific, even from the LTE times, the system evolved to a flexible configuration where signals from the UE may arrive from any direction as transmission points can send information to the same UE at the same time (as in

Figure 6.36 Beam grouping.

Beams on panel 2

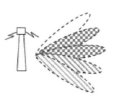
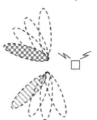

Beams on panel 1

coordinated joint transmission) or they can dynamically alternate in time (as in dynamic point selection). To make such operation possible, the rules of QCL (quasi-colocation) has been introduced. These rules link reference signals to each other with respect to how they have been transmitted. Why is this needed? Because the UE performs channel estimation, and it needs to know how the estimation of various signals needs to be done, as some of the calculated channel statistics may be reused in the baseband processing if the UE is certain that the measured signals originated from the same transmission point. The NR defines four QCL types as follows:

- QCL Type A: Doppler shift, Doppler spread, average delay, delay spread
- QCL Type B: Doppler shift, Doppler spread
- QCL Type C: Average delay, Doppler shift
- QCL Type D: Spatial Rx

The QCL types A, B, and C are applicable for both below- and above-6-GHz transmission, but the spatial QCL Type D is applicable only in higher carriers where the UE is essentially forming beams.

The TCI framework defines pairs of source and target reference signals. The following RS can be used as a source RS to indicate TX beam for downlink:

- SS/PBCH block: RX beam used to receive a certain SS/PBCH block is used as the RX beam for the DL transmission.
- CSI-RS for beam management.
- CSI-RS for CSI computation.

The following RS can be used a source RS to indicate the TX beam for uplink:

- SS/PBCH block: The RX beam used to receive a certain SS/PBCH block is used as the TX beam.
- CSI-RS for beam management: The RX beam used to receive a certain CSI-RS resource is used as the TX beam.
- SRS: The TX beam used to transmit a certain SRS resource is used as the TX beam.

The TCI framework assists the reception of the CSI-RS for CSI acquisition, the CSI-RS for beam management, the DM-RS for PDCCH decoding, and the DM-RS for PDSCH decoding. The logic for using the QCL information is based on the fact that the UE is able to accumulate QCL information from multiple signals. In the tables below, this is represented by the existence of the two reference signals columns (RS1 and RS2). When QCL Type D is configured, as in the above 6 GHz scenarios, the second RS would also indicate QCL Type D. The signaling of TCI states is enabled by using RRC, MAC-CE, and DCI.

For a periodic TRS signal, there are two possible TCI configurations, as described in Table 6.6.

Let us discuss in detail the above two configurations. For configuration 1, in scenarios without QCL Type D such as below 6 GHz, that is, QCL Type C, the average delay and Doppler shift can be inferred from the SS Block for the utilization of the periodic TRS. If QCL Type D is configured and the UE can use the same Spatial Rx filter for the reception of the periodic TRS that was used for the reception of the SS Block, then the DL RS 2 is configured, and hence the UE can infer the average delay, Doppler shift, and Spatial Rx properties from the SS Block. The case in configuration 2 is more interesting:

Table 6.6 TCI states for periodic TRS.

Valid TCI state configuration	DL RS 1	*qcl-Type1*	DL RS 2 (if configured)	*qcl-Type2* (if configured)
1	SS Block	QCL-TypeC	SS Block	QCL-TypeD
2	SS Block	QCL-TypeC	CSI-RS for beam management	QCL-TypeD

below 6 GHz, it is clear that the UE infers the average delay and Doppler shift from the SS Block for utilization of the periodic TRS; above 6 GHz, the UE gets the average delay and Doppler shift from the SS Block while the Spatial Rx properties from a CSI-RS configured to be used for beam management. This means that the beamwidth used for the transmission of the CSI-RS for beam management is used for the transmission of the periodic TRS, but it is different from the beamwidth used for the transmission of the SS Block. This scenario is also the one in which the SS Block can be transmitted with a wider beam, while the CSI-RS for beam management and the periodic TRS are transmitted with a narrower beam.

There is also the case for aperiodic TRS, which simply follows the periodic TRS configuration as the aperiodic TRS cannot exist in the system without a periodic TRS (Table 6.7).

For a CSI-RS used for CSI measurements, the possible TCI configurations described in Table 6.8:

For a CSI-RS used for beam management, the possible TCI configurations described in Table 6.9:

For a DM-RS for PDCCH decoding, the possible TCI configurations described in Table 6.10:

For a DM-RS for PDSCH decoding, the possible TCI configurations described in Table 6.11:

Table 6.7 TCI states for aperiodic TRS.

Valid TCI state configuration	DL RS 1	*qcl-Type1*	DL RS 2 (if configured)	*qcl-Type2* (if configured)
1	Periodic TRS	QCL-TypeA	Periodic TRS	QCL-TypeD

Table 6.8 TCI states for CSI-RS for CSI.

Valid TCI state configuration	DL RS 1	*qcl-Type1*	DL RS 2 (if configured)	*qcl-Type2* (if configured)
1	TRS	QCL-TypeA	SS Block	QCL-TypeD
2	TRS	QCL-TypeA	TRS	QCL-TypeD
3	TRS	QCL-TypeA	CSI-RS for beam management	QCL-TypeD
4	TRS	QCL-TypeB		

Table 6.9 TCI states for CSI-RS for CSI.

Valid TCI state Configuration	DL RS 1	*qcl-Type1*	DL RS 2 (if configured)	*qcl-Type2* (if configured)
1	TRS	QCL-TypeA	TRS	QCL-TypeD
2	TRS	QCL-TypeA	CSI-RS for beam management	QCL-TypeD
3	SS Block	QCL-TypeC	SS Block	QCL-TypeD

Table 6.10 TCI states for CSI-RS for CSI.

Valid TCI state configuration	DL RS 1	*qcl-Type1*	DL RS 2 (if configured)	*qcl-Type2* (if configured)
1	TRS	QCL-TypeA	TRS	QCL-TypeD
2	TRS	QCL-TypeA	CSI-RS for beam management	QCL-TypeD
3	CSI-RS for CSI	QCL-TypeA	CSI-RS for CSI	QCL-TypeD
4[a]	SS Block[a]	QCL-TypeA	SS Block[a]	QCL-TypeD

a) Before TRS configured. Note: This is not a TCI state, rather a valid QCL assumption.

Table 6.11 TCI states for CSI-RS for CSI.

Valid TCI state configuration	DL RS 1	qcl-Type1	DL RS 2 (if configured)	*qcl-Type2* (if configured)
1	TRS	QCL-TypeA	TRS	QCL-TypeD
2	TRS	QCL-TypeA	CSI-RS for beam management	QCL-TypeD
3	CSI-RS for CSI	QCL-TypeA	CSI-RS for CSI[a]	QCL-TypeD[a]
4[b]	SS Block[b]	QCL-TypeA	SS Block[b]	QCL-TypeD

a) Note: QCL-TypeD parameter may not be derived directly from CSI-RS (CSI).
b) Before TRS configured. Note: This is not a TCI state, rather a valid QCL assumption.

6.12.2.4 Beam Recovery

Beam recovery is used for rapid link reconfiguration against sudden blockages. The beam recovery procedure involves several stages. In the first instance, the UE needs to detect that it is in a state in which the beam on which is operating/monitoring is not available anymore; this is called beam failure detection (BFRD). Following this, the UE needs to identify new beams for transmission, and hence the candidate beam identification stage takes place on the configured reference signals (BFD-RS). Once this is done, the UE indicates the beam failure to the gNB; this takes place during the recovery request transmission (BFRQ).

BFRD

A beam fails when in principle the control channel of that particular beam cannot reach the UE. But since that information is not available, the UE needs to monitor some other quantity resembling the transmission of the PDCCH, and while that measurement is below a threshold, the PDCCH and hence that particular beam is considered as not available. Such principles are not new as such; they have been used in LTE for radio link failure (RLF) and also in NR for the same RLF. Indeed, the UE monitors the so-called hypothetical PDCCH block error rate (BLER), which is determined from the quality of the BFD-RS. The BLER threshold value for the hypothetical PDCCH BLER needs to be above 10%.

Beam failure RS (BFRS)

The reference signals on which this monitoring is done are the CSI-RSs, and these need to be spatially QCL-ed with the PDCCH DM-RS. The identification of new beams is done on configured resources representing the CSI-RS and SS Blocks. Each of these can be transmitted with their own beamwidths, in the sense that the CSI-RS can be narrow beams while the SS Block can be wide beams; what matters in this case is that these reference signals would characterize the control channel which is going to be transmitted once the beam has recovered. As an example, if narrow CSI-RS beams have failed, it is very likely that a wider SS Block beam will be the recovery beam; it is not necessary that a failed CSI-RS beam should be replaced by a CSI-RS beam. The identification of the new beams is done by similar measurements as when monitoring the BLER target of the beams, that is, by computing the L1-RSRP of that configured reference signals and applying that against a threshold. NR supports the configuration of the CSI-RS or SS Block only, as BFRS, but also both the CSI-RS and SS Blocks can be configured to be used together.

Beam failure recovery request (BFRQ)

Once the beam(s) have been detected as failed, the gNB needs to know about the failure so that appropriate actions are taken, likely the scheduling of the control and data traffic on new beams. A beam recovery request can be transmitted if the number of consecutive detected beam failure instances exceeds a configured maximum number. The BFRQ is done by transmitting signals on the uplink, and one of the handy signals to use are the PRACH resources, which are available to the UE and which provide contention-free random access (CFRA) toward the gNB, as no collisions will occur for these resources. The UE receives a configuration for PRACH transmission, and these resources are associated with periodic CSI-RS or SS Blocks, which are used as Beam-Failure Reference Signal (BFRS). The association between the CSI-RS/SS Blocks and PRACH resources also implies that the detected DL beam will be used for UL transmission; hence, the concept of beam correspondence is applied. The BFRQ may also be transmitted in a contention-based manner (CBRA), a situation in which the UE is utilizing the contention-based RACH procedure where preamble resources are mapped to downlink signals. In case both CSI-RS and SS Blocks are configured for the UE, the PRACH preambles are associated with the SS Blocks only; in this case, the CSI-RS resources used for new beam identification are found based on the QCL association with the SS Block. On the other hand, there is also the possibility of associating PRACH preambles with either CSI-RS or SS Blocks.

Once the UE transits the BFRQ, it also monitors for an answer from the gNB so that it is acknowledged on the receipt of this information. The monitoring for the recovery request response implies the monitoring of the PDCCH with the assumption that this is spatially QCL-ed with the RS of the identified beam. This monitoring is done during a preconfigured time window, and if the UE does not receive any answer from the gNB, it retransmits the recovery request.

6.12.3 CSI Framework

The CSI framework consists of flexible configurability such that the network can obtain the various forms of CSI information in a straightforward manner. A set of key components are defined for this scope:

o *Reporting settings* refer to what CSI to report and when to report it.
o *Resource settings* configure what signals to use to compute the CSI.

The two types of settings are employed on the basis of *reporting configurations*, which associate what CSI to report and when to report it in conjunction with the signals used to compute the CSI.

In the following, we detail the three CSI framework components.

6.12.3.1 Reporting Settings

The reporting settings consist of an indication of what quantities are to be reported, these being CSI or L1-RSRP related, and configuration information on the quantities to be reported. For example, in terms of codebooks, the parameterization of these codebooks covers both the time and frequency domain behavior. For the time behavior, the UE parameterization consists of information about the periodic, aperiodic, or semi-persistent transmission, while in the frequency domain, it consists of the reporting bandwidth information and frequency granularity for PMI and CQI. In addition, time-domain restrictions for channel and interference measurements, or codebook subset restrictions, may be configured, similar to LTE. In terms of the applicability of such information, each reporting setting is associated with a single downlink BWP.

6.12.3.2 Resource Settings

A resource setting configures a number of CSI resource sets, each CSI resource set consisting of CSI-RS resources, which can be either non-zero power (NZP) CSI-RS resource set(s) or CSI-IM or SS/PBCH block resources which are used for L1-RSRP computation. CSI-RS resource sets also exhibit a time-domain behavior which can be periodic, aperiodic, or semi-persistent. If such time-domain behavior is periodic or semi-persistent, the number of CSI-RS resource sets is limited to one.

There is a set of rules enabling a homogeneous configuration; for example, when a UE is configured with multiple CSI resource configurations consisting of the same NZP CSI-RS resource, the same time-domain behaviors shall be configured for the CSI resource configurations while all the CSI resource settings linked to a CSI report setting should have the same time-domain behavior. Such configurations help in reducing the UE complexity and avoiding possible misconfigurations of the network.

Resources which may be configured for one or more CSI resource settings for channel and interference measurement are the following:

o For interference measurement: CSI-IM or NZP CSI-RS
o For channel measurement: NZP CSI-RS

When, for example, CSI is computed, the UE would assume that the resources used for the computation of the channel and interference can be processed with the same spatial filter; in other words, they can be quasi-co-located with respect to "QCL-TypeD."

6.12.3.3 Reporting Configurations

The reporting configuration links the report settings with the resource settings and contains a list of associated CSI-RS report configurations.

The reporting configurations support a wide range of scenarios and operating modes. In the following, we present an example of such configurations which may be used for non-coherent JT where three transmission points are cooperating.

In Figure 6.37, we have the following configurations:

– *CSI Report Setting 0*: CQI for TRP1
  - Measures channel from TRP1; all other TRPs are quiet
  - Measures interference from other TRPs while TRP1 is quiet
– *CSI Report Setting 1*: CQI for TRP2:
  - Measures channel from TRP2; all other TRPs are quiet
  - Measures interference from other TRPs while TRP2 is quiet
– *CSI Report Setting 3*: CQI for TRP2:
  - Measures channel from TRP3; all other TRPs are quiet
  - Measures interference from other TRPs while TRP3 is quiet

The CSI reporting configurations can be periodic (where the information is transmitted using the PUCCH), aperiodic (where the information is transmitted on the PUSCH), or semi-persistent (where the information is transmitted using the PUCCH and SCI-activated PUSCH). Table 6.12 summarizes the transmission depending on the codebook types, periodicity, and UL channels.

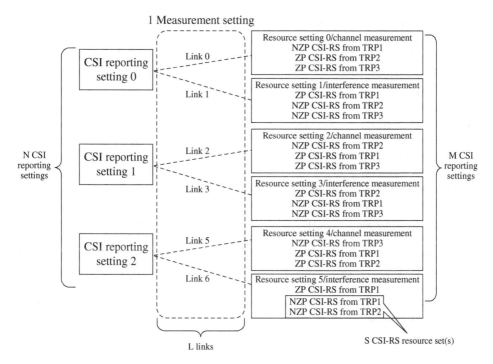

Figure 6.37 CSI reporting configurations.

Table 6.12 Transmission as function of codebook type.

	Periodic	Semi-Persistent	Aperiodic
Type I WB	PUCCH Format 2, PUCCH Format 3,4	PUCCH Format 2 PUSCH	PUSCH
Type I SB		PUCCH Format 3 or 4 PUSCH	PUSCH
Type II WB		PUCCH Format 3 or 4 PUSCH	PUSCH
Type II SB		PUSCH	PUSCH
Type II Part 1 only		PUCCH Format 3 or 4 (UE capability)	

Two important characteristics are defined with respect to the transmission of single or multiple CSI reports:

- CSI reports can be multiplexed on the PUSCH with or without uplink data form the UE. This is a straightforward mechanism, similar to LTE, in terms of efficient UL transmission.
- If the physical channels scheduled to carry the reports overlap in time and frequency, it is understood that such CSI reports collide, and priority rules are defined so that the lower-priority report is not transmitted.

When CSI reporting is performed on the PUSCH, because of the large payload which is envisioned for such a feedback mechanism, a two-part CSI report is used. Two issues need to be mentioned for this design: the information which is mapped on the two parts depends on the CSI feedback type, as we will describe next, and there are also some rules on how the two parts are transmitted. In terms of rules, Part 1 shall be transmitted in its entity before Part 2, one reason being the fact that Pat 1 can be used to identify the number of information bits in Part 2. The following information rules are used for the two types of CSI feedback:

- For Type I CSI feedback, Part 1 contains RI (if reported), CRI (if reported), and CQI for the first codeword. Part 2 contains PMI and the CQI for the second codeword when the rank is larger than four.
- For Type II CSI feedback, Part 1 contains RI, CQI, and an indication of the number of nonzero wideband amplitude coefficients per layer for Type II CSI. Part 2 contains the PMI and the Type II CSI.

For such part-like reporting, the fields of Part 1 and 2 are separately encoded while the fields of Part 1 are also separated encoded. If Report Quantity is configured with CRI/RSRP or SSBRI/RSRP, the CSI feedback consists of a single part. And in case of collisions, omission rules are defined to enable the UE to omit a portion of the Part 2 CSI according to a defined priority order.

When CSI parameters are calculated at the UE, some dependencies between the CSI parameters need to be followed:

o Layer Indicator (LI) shall be calculated based on the reported CQI, PMI, RI, and CRI.
o CQI shall be calculated based on the reported PMI, RI, and CRI.
o PMI shall be calculated based on the reported RI and CRI.
o RI shall be calculated c based on the reported CRI.

Table 6.13 Subband sizes.

Bandwidth used by the UE (PRBs)	Subband size (PRBs)
< 24	N/A
24–72	4, 8
73–144	8, 16
145–275	16, 32

Another important aspect of reporting configurations is the frequency granularity of the CSI reports, which is dependent on the so-called subband sizes. A subband size consists of a number of PRBs which are used in a contiguous manner for processing/transmission. Naturally, the sizes of the subband sizes become larger as the bandwidth on which the UE operates becomes larger, as shown in Table 6.13.

A CSI Reporting Setting configuration defines a CSI reporting band as a subset of subbands of the BWP, where the following reporting configurations are indicated:

o The CSI reporting band as a contiguous or non-contiguous subset of subbands in the BWP for which the CSI shall be reported.
o Wideband or subband CQI, where, if configured, a wideband CQI is reported for each codeword for the entire CSI reporting band, or if subband CQI is configured, one CQI for each codeword is reported for each subband in the CSI reporting band.
o Wideband or subband PMI, where a wideband PMI can be reported for the entire CSI reporting band, or if subband PMI reporting is configured, a single wideband PMI drawn from the W1 codebook is reported for the entire CSI reporting band and one subband PMI drawn from the W2 codebook is reported for each subband in the CSI reporting band, such configuration not being considered when two antenna ports are used.

6.12.3.4 Report Quantity Configurations

CSI components are reported in a variety of combinations, these being suited for various transmission strategies of the gNB. Such CSI components can be grouped into the following sets of components: "none," "cri-RI-PMI-CQI ," "cri-RI-i1," "cri-RI-i1-CQI," "cri-RI-CQI," "cri-RSRP," "ssb-Index-RSRP," or "cri-RI-LI-PMI-CQI." In the following, we describe the meaning of performing such set-based reporting:

o "none" implies that the UE is not required to report any CSI information. This is a mode which obviously turns off any CSI reporting function.
o "cri-RI-PMI-CQI" or "cri-RI-LI-PMI-CQI" implies that the UE shall report a preferred precoder matrix for the entire reporting band, or a preferred precoder matrix per subband.
o "cri-RI-i1" implies that the UE is configured with single-panel Type I codebook reporting while the requested PMI is wideband.
o "cri-RI-i1-CQI" is similar to the previous case, but in addition to the wideband PMI drawn from a single-panel Type I codebook, the UE also is requested to report CQI.
o "cri-RI-CQI" implies a non-PMI port indication report where r ports are indicated in the order of layer ordering for rank r.

o "cri-RSRP" or "ssb-Index-RSRP" implies that the UE needs to measure and report L1-RSRP, with such measurements being performed on either NZP CSI-RS or SS blocks. The performed measurements are limited to no more than 64 CSI-RS and/or SS Block resources. In addition, the measurements may be done independently or may be grouped according to how the CSI-RS or SSB are received simultaneously by the UE, either with a single spatial domain receive filter or with multiple simultaneous spatial domain receive filters.

6.12.4 CSI Components

The CSI components have the function of providing channel information, as seen at the UE, to the transmitter (gNB). For this purpose, various components are available that describe various pieces of information representing the channel, from more detailed to more complex information, as described in the following sections.

6.12.4.1 Channel Quality Indicator (CQI)

The CQI principles of NR are very similar to those of LTE. CQI reporting is based on QPSK, 16-QAM, 64-QAM, and 256-QAM. It is worth noting the definition, which is based on the CQI calculation at the UE, with respect to the utilization of the measurements. The 5G NR specification stipulates that the CQI is calculated *based on an unrestricted observation interval in time.* However, the UE may be subjected to a measurement restriction imposing the performance and reporting of instantaneous measurements. In summary, 5G supports both instantaneous and averaged measurements. The target BLER for eMBB services if 0.1, while for higher-reliability services it is 0.00001. As in LTE, there exists wideband and subband CQI, where for the subband case, a two bit differential CQI is used.

6.12.4.2 Precoding Matrix Indicator (PMI)

Precoded transmission is a well-known technique where the transmitted data is precoded based on the actual channel on which it is going to be transmitted, such that he energy is radiated toward the location of the receiver. While the previous statement is more of an ideal case, in practical systems, the precoder is obtained based on the feedback from the UE, while one of the forms of precoding is based on searching among a set of PMI forming a so-called a codebook, and sending this PMI back to the transmitter (a gNB in a 5G system). Similar to LTE, 5G selected the use of PMI-based operation; however, it diversified the codebook options such that various MIMO use cases are met. Indeed, NR has specified two types of codebooks: the Type I codebook, which better fits single-user transmission, has a standard resolution in terms of CSI feedback, and has a lower payload which needs to be fed back to the transmitter. The Type II codebook is envisioned for multi-user application and has in this respect a higher-resolution CSI feedback, as it is known that in MU MIMO, the CSI resolution has a critical impact, as orthogonality between the users needs to be achieved. The Type II codebook is based on a linear combination of multiple beams and has DFT beams and port selection options. Both Type I and II codebooks are designed for cross-polarized antennas.

The basic principle of precoded transmission is the following: the UE measures the channel based on the CSI-RS (as shown in Figure 6.38), it computes the CSI

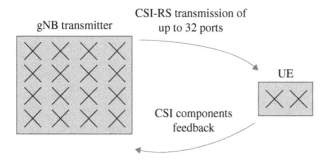

Figure 6.38 Obtaining the CSI feedback: measurements and reporting based on CSI-RS port estimation.

(consisting of the rank, PMI, CQI, etc.), and then sends such CSI information back to the transmitter.

The number of CSI-RS ports the UE can measure is limited. In practice, however, it evolved step by step in LTE, and the first release of NR supports estimation of up to 32 antenna ports. This means that it is possible to transmit a maximum of 32 CSI-RSs in the digital domain. This operation is similar to LTE Class A operation.

Another form of obtaining CSI feedback is based on beam selection, where multiple CSI-RSs are beamed and CRI (CSI-RS Resource Indicator) is fed back. This operation is similar to LTE Class B operation (Figure 6.39).

The gNB arrays may be formed of single or multiple panels. A single-panel array performs the combination of RF beamforming and digital precoding at baseband. Different form factors of such panels may exist, from square, horizontal, or vertical placement of the cross-polarized antennas (Figure 6.40).

Multiple panels are a combination of RF beamforming and digital precoding at baseband (Figures 6.41 and 6.42).

6.12.4.2.1 Type I Codebooks As mentioned in the previous section, the NR Type I codebook consists of single- and multiple-panel designs.

Type I Single-Panel Codebook The design of the Type I codebook supports the configuration of 4, 8, 12, 16, 24, and 32 ports, while for 2 ports it is a simpler structure that we describe next.

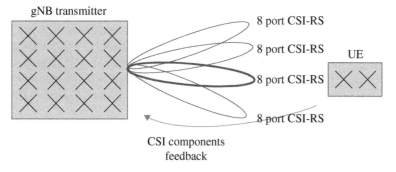

Figure 6.39 Obtaining the CSI feedback: measurements and reporting based on CRI and CSI-RS.

Figure 6.40 Single-panel antenna arrangements: square, horizontal, or vertical placement.

gNB transmitter

Figure 6.41 Multiple-panel SU MIMO transmission.

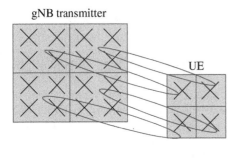

UE

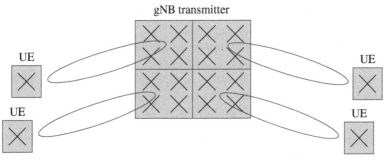

Figure 6.42 Multiple-panel MU MIMO transmission.

For 2 ports, the Type I codebook for rank 1 is given by $\mathbf{W} \in \left\{ \frac{1}{\sqrt{2}} \begin{bmatrix} 1 \\ e^{j\frac{\pi n}{2}} \end{bmatrix}, n = 0, 1, 2, 3 \right\}$, while for rank 2 it is $\left\{ \frac{1}{\sqrt{2}} \begin{bmatrix} 1 & 1 \\ j^n & -j^n \end{bmatrix}, n = 0, 1 \right\}$.

For 4 and more transmission ports, the Type I codebook supports up to rank 8 transmission. The Type I codebook for single panel uses a dual-stage structure similar to LTE codebooks, where $W = W_1 W_2$ and W_1 contains the beam selection and is reported as a wideband component, while W_2 contains the QPSK co-phasing between the two polarizations and possibly also beam selection and can be reported as a subband component.

The design as such follows the LTE block diagonal structure containing the same 2D DFT beam(s) to the two polarizations:

$$W_1 = \begin{bmatrix} B & 0 \\ 0 & B \end{bmatrix}$$

where B is composed of L oversampled 2D DFT beams $B = [b_0 \; b_1 ... b_{L-1}]$.

For ranks 1 and 2, the value of L is configurable as $L = 1$ and $L = 4$. When $L = 1$, beam selection is wideband, whereas further subband beam selection is possible when $L = 4$. The corresponding rank-1 W_2 matrices have the form $W_2 = \begin{bmatrix} 1 \\ e^{j\alpha} \end{bmatrix}$ in the case of $L = 1$ and $W_2 = \begin{bmatrix} e_i \\ e^{j\alpha}e_i \end{bmatrix}$ in the case of $L = 4$, where $e^{j\alpha}$ is a QPSK polarization co-phasing coefficient, and e_i is a size-L selection vector.

For $L = 4$, several beam patterns are supported, similar to LTE Class A (Figure 6.43).

Several 1D/2D antenna ports layouts (N_1, N_2) and oversampling factors (O_1, O_2) are supported as listed in Table 6.14.

Type I Multi-Panel Codebook The multi-panel codebook is supported for ranks 1 to 4. The codebook construction is based on the single-panel design by utilizing a co-phasing between the single-panel precoders:

$$\mathbf{w}_{p,r,l} = \mathbf{b}_{k_1+k'_{1,l},k_2+k'_{2,l}} \cdot c_{p,r,l} \left(\text{normalized by } \frac{1}{\sqrt{N_g R 2 N_1 N_2}} \right)$$

where

- $p = 0, 1, ..., N_g - 1$ are the panels, with N_g being the number of panels
- $r = 0, 1$ are the two polarizations
- $l = 0, 1, ..., R - 1$ are the number of layers, with R being the rank
- $\mathbf{b}_{k_1+k'_{1,l},k_2+k'_{2,l}}$ is taken from the single-panel codebook with $L = 1$
- $c_{p,r,l}$ is the co-phasing for panels and polarizations

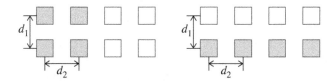

Figure 6.43 Beam patterns for Type I codebook. Left: 2D antenna port layout, right: 1D antenna port layout.

Table 6.14 Antenna power layouts and oversampling factors.

Number of CSI-RS ports	(N_1, N_2)	(O_1, O_2)
4	(2,1)	(4,-)
8	(2,2), (4,1)	(4,4), (4,-)
12	(3,2), (6,1)	(4,4), (4,-)
16	(4,2), (8,1)	(4,4), (4,-)
24	(6,2), (4,3), (12,1)	(4,4), (4,-)
32	(8,2), (4,4), (16,1)	(4,4), (4,-)

Table 6.15 Antenna power layouts and oversampling factors.

Number of CSI-RS ports	(N_g, N_1, N_2)	(O_1, O_2)
8	(2,2,1)	(4,-)
16	(2,2,2), (2,4,1), (4,2,1)	(4,4), (4,-)
32	(2,4,2), (4,4,2), (2,8,1), (4,4,1)	(4,4), (4,-)

The inter-panel co-phasing could be configured in wideband mode for 2 or 4 panels or in subband mode for 2 panels. In subband mode, a maximum of 4 bits PMI payload size is supported per subband.

Several 1D/2D antenna ports layouts (N_g, N_1, N_2) and oversampling factors (O_1, O_2) are supported as listed in Table 6.15.

The inter-panel co-phasing payload is configurable with two modes. Mode 1 consists of a wideband inter-panel co-phasing, has a lower payload, and supports 2- and 4-panel operation. In this mode, the total intra- and inter-panel co-phasing payload is of 2 bits per subband. Mode 2 consists of subband inter-panel co-phasing, has a higher payload, and supports only 2-panel operation. In this mode, the total intra- and inter-panel co-phasing payload is 4 bits per subband.

6.12.4.2.2 Type II codebooks The Type II codebook is a higher-resolution codebook compared to the Type I design. The Type II codebook is a dual-stage codebook structure up to rank 2, $\mathbf{W} = \mathbf{W1} \times \mathbf{W2}$.

For rank 1, the following precoding structure is assumed: $\mathbf{W} = \begin{bmatrix} \widetilde{\mathbf{w}}_{0,0} \\ \widetilde{\mathbf{w}}_{1,0} \end{bmatrix} = \mathbf{W}_1\mathbf{W}_2$, where $\mathbf{W}$ is normalized to 1. For rank 2, the following precoding structure is assumed: $\mathbf{W} = \begin{bmatrix} \widetilde{\mathbf{w}}_{0,0} & \widetilde{\mathbf{w}}_{0,1} \\ \widetilde{\mathbf{w}}_{1,0} & \widetilde{\mathbf{w}}_{1,1} \end{bmatrix} = \mathbf{W}_1\mathbf{W}_2$, where the columns of $\mathbf{W}$ are normalized to $\frac{1}{\sqrt{2}}$. The first stage $\mathbf{W1}$ consists of a set of L orthogonal beams ($L \in \{2, 3, 4\}$) selected from the predefined oversampled 2D DFT beams for a single polarization, and beam selection is realized in wideband (WB). The second stage, $\mathbf{W2}$, consists of $2L-1$ beam combining coefficients for L selected beams and 2 polarizations in each layer. Generally, beam combining coefficients can be divided into phase combining and amplitude scaling quantization separately. Phase combining is configured for subband (SB) reporting, and amplitude scaling is configured to report the WB amplitude with or without the SB differential amplitude. The weighted combination of L beams is given by the following coefficients:

$$\widetilde{\mathbf{w}}_{r,l} = \sum_{l=0}^{L-1} \mathbf{b}_{k_1^{(i)} k_2^{(i)}} \cdot p_{r,l,i}^{(WB)} \cdot p_{r,l,i}^{(SB)} c_{r,l,i}$$

where

- $\mathbf{b}_{k_1,k_2}$ is an oversampled 2D DFT beam
- r are the two polarizations, and l can go up to two layers
- $p_{r,l,i}^{(WB)}$ represents the wideband beam amplitude scaling factor for beam i and on polarization r and layer l

- $p_{r,l,i}^{(SB)}$ represents the subband beam amplitude scaling factor for beam i on polarization r and layer l
- $c_{r,l,i}$ represents the phase combiner for beam i on polarization r and layer l. This is configurable between 2 and 3 bits using QPSK or 8PSK.

The Type II codebook benefits from a configurable amplitude scaling mode between wideband and subband (with unequal bit allocation) and wideband only. The beam selection is wideband only and consists of an unconstrained beam selection from an orthogonal basis. As was the case for Type I codebooks, the Type II cookbook has several 1D/2D antenna ports layouts (N_1, N_2), and oversampling factors (O_1, O_2) are supported as listed in Table 6.16.

The Type II CSI feedback requires a large reporting overhead to acquire enhanced spatial channel information.

6.12.4.2.3 Port selection codebooks

NR also supports a codebook for beamformed CSI-RS. This comes as an extension of Type II CSI for rank 1 and 2, where $\mathbf{W}_1$ is given as follows:

$$\mathbf{W}_1 = \begin{bmatrix} \mathbf{E}_{\frac{X}{2} \times L} & 0 \\ 0 & \mathbf{E}_{\frac{X}{2} \times L} \end{bmatrix},$$

where X is the number of CSI-RS ports and $L \in \{2, 3, 4\}$.

The possible values of X follow the Type II single-panel codebook:

$$\mathbf{E}_{\frac{X}{2} \times L} = \left[e^{\left(\frac{X}{2}\right)} \mathrm{mod}\left(md, \frac{X}{2}\right) \quad e^{\left(\frac{X}{2}\right)} \mathrm{mod}\left(md+1, \frac{X}{2}\right) \cdots \right.$$
$$\left. \times e^{\left(\frac{X}{2}\right)} \mathrm{mod}\left(md+L-1, \frac{X}{2}\right) \right]$$

where

- $e^{\left(\frac{X}{2}\right)}_i$ is a length $\frac{X}{2}$ vector with the i-th element equal to 1 and 0 elsewhere
- The port selection is given by $m \in \left\{0, 1, \ldots, \left\lceil \frac{X}{2d} \right\rceil - 1\right\}$ where m is wideband
- The values of d is configurable $d \in \{1, 2, 3, 4\}$ under the condition that $d \leq \frac{X}{2}$ and $d \leq L$

Table 6.16 Type II codebook antenna power layouts and oversampling factors.

Number of CSI-RS ports	(N_1, N_2)	(O_1, O_2)
4[a]	(2,1)	(4,-)
8[a]	(2,2), (4,1)	(4,4), (4,-)
12	(3,2), (6,1)	(4,4), (4,-)
16	(4,2), (8,1)	(4,4), (4,-)
24	(6,2), (4,3), (12,1)	(4,4), (4,-)
32	(8,2), (4,4), (16,1)	(4,4), (4,-)

a) Beam selection is not used for 4-port L = 2 and for 8-port L = 4.

The amplitude scaling and phase combining coefficients, along with their configurations, follow the Type II single-panel codebook.

6.12.4.3 Resource Indicators: CRI, SSBRI, RI, LI

The UE can be configured with a set of nonzero-power CSI-RS resources out of which it may be asked to report a subset. The identification of such a nonzero-power CSI-RS is done by a CSI-RS resource indicator. When a UE is configured with more than one nonzero-power CSI-RSs, it can report a set of N UE-selected CSI-RS resource-related indices. This is a very useful indicator as it can quickly point to the N best CSI-RS resources the network can use further.

As we have seen for the indication of the CSI-RS, a similar indication is possible for SS Blocks; the indicator is called SSBRI.

The LI indicates which column of the precoder matrix of the reported PMI corresponds to the strongest layer of the codeword corresponding to the largest reported wideband CQI. If two wideband CQIs are reported and have equal value, the LI corresponds to the strongest layer of the first codeword.

The RI is a well-known indicator in MIMO, and it is a measure of the total number of layers that can be transmitted.

6.12.5 Uplink MIMO Transmission Schemes

NR supports two transmission schemes for the PUSCH: codebook-based transmission and non-codebook-based transmission. These are described in the next subsections.

6.12.5.1 Codebook-Based Uplink Transmission

When codebook-based transmission is used, the UE precodes the PUSCH based on a precoder which is indicated by the gNB. The following steps are performed:

- The UE transmits SRS resources to the gNB. The gNB estimates the UL channel based on these resources and hence determines the UL transmission rank and also identifies the best SRS, each of the SRS resources being tagged with an SRI if multiple SRS resources are configured. In addition to the SRI and rank, the gNB estimates the precoder to be applied over the layers that corresponds to that SRS resource. This precoder is selected from a predefined UL codebook over the SRS ports in the selected SRS resource by the SRI. At the end of all these processing stages, the gNB has the SRI, Transmit Precoder Matrix Indicator (TPMI) (in the uplink direction, the PMI is denoted by Transmit PMI to differentiate it from the downlink PMI), and rank, which can be used by the UE for the UL transmission. If a single SRS resource is configured, no SRI needs to be indicated to the UE; the TPMI is used to indicate the preferred precoder over the SRS ports in the configured single SRS resource. The TPMIs are supposed to be wideband, and while frequency-selective TPMIs might provide further performance improvement, the cost of signaling such information would be tremendously high as the baseline wideband information would be multiplied by the number of subbands.
- The gNB indicates the SRI and TPMI to the UE, and hence the UE has now the necessary information for performing codebook-based uplink transmission.

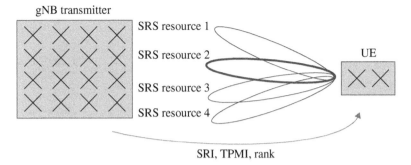

gNB transmitter

SRS resource 1
SRS resource 2
SRS resource 3
SRS resource 4

UE

SRI, TPMI, rank

Figure 6.44 Codebook-based UL transmission procedure.

What we have been describing above is depicted in Figure 6.44. In this example, the UE transmits four SRS resources, and the gNB identifies the second resource as the best. Further, the rank and precoder are computed for this SRS resource. The gNB indicates to the UE the SRI (which means that the second SRS resource is to be used), the computed TPMI, and rank (Figure 6.44).

Under the codebook-based operation, three forms of transmission are possible in the uplink:

○ Full coherence: All the ports can be transmitted coherently.
○ Partial coherence: Port pairs can be transmitted coherently. This can happen when the UE has calibrated antenna ports within a panel but also has multiple panels not calibrated to one another. Coherent transmission occurs inside a panel where an SRS resource is configured as such.
○ Non-coherent: No port pairs can be transmitted coherently. In this case, the transmit chains of the UE operate in a more independent fashion, and there is no phase adjustment between them.

6.12.5.2 Non-Codebook-Based Uplink Transmission

When non-codebook-based transmission is used, the UE would precode the PUSCH transmission, but only SRI information is used, and hence no Transmit PMI (TPMI) is indicated by the gNB, as in codebook-based transmission. Again, as in codebook-based transmission, only wideband operation is possible. The logic of non-codebook-based UL transmission is to exploit channel reciprocity more. The following steps are performed:

• The UE estimates the channel based on CSI-RS, and determines a precoder.
• The UE transmits one or multiple SRSs, which are precoded with the precoder obtained during the previous step.
• The gNB selects the best-received SRS and indicates this to the UE, based on the SRI. Further PUSCH transmission occurs based on the indicated SRI (Figure 6.45).

6.13 Channel Coding with 5G

Channel coding is one of the main hardware critical components in 5G, and codes used in 5G are capable of providing good performance gains with a lower implementation

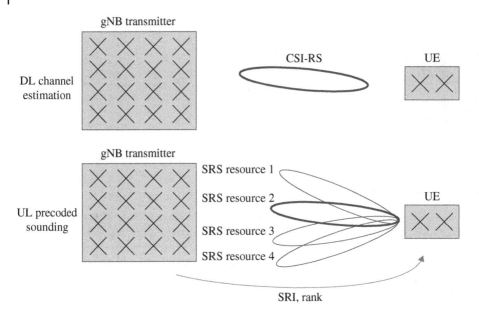

Figure 6.45 Non-codebook-based UL transmission procedure.

complexity and processing delays compared to its predecessors. In 3GPP Rel-15, different service scenarios were considered in the initial phase of the channel coding discussion; however, the final selection of the channel coding scheme was mainly decided based on the eMBB requirements, which were identified as performance of the coding scheme, implementation complexity, the latency of encoding and decoding, and the flexibility (e.g. variable code length, code rate, HARQ) of the coding scheme. The coding schemes were separately discussed for the data channel and control channel considering differences in the supported block sizes and code rates.

As a result, for the first time in 3GPP, data channels use LDPC codes, and for the first time in any standard, the control channel design is based on polar coding.

6.13.1 Channel Coding for Data Channel

LDPC is a very mature coding scheme which has been considered by many other standards. The flexibility that the coding scheme provided in implementations together with competitive performance can be identified as one of the key advantages that LDPC possesses when compared to all other coding schemes. If the implementation requirement is to support very low energy consumption, LDPC has the capability of realizing that with physically optimized circuits. Likewise, the same LDPC code can be used with full parallelized implementations if the objective is to achieve high throughputs. LDPC implementations can also support finer granularity of block sizes and code rates.

LDPC has different design flavors, and each offers different benefits. It is well understood that quasi-cyclic (QC) LDPC codes are the most realistic type and provides lower encoding/decoding complexities compared to other variants. This variant is

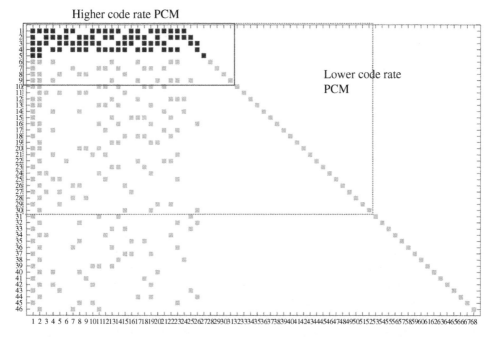

Figure 6.46 Support for different rates in 5G LDPC base graph #1

already used in standards like IEEE 802.11ac, IEEE 802.11ad, IEEE 802.11a/b/g/n, IEEE 802.15.3c, IEEE 802.1e, DVB C2/S2/T2, etc. 5G channel coding is also based on these QC-LDPC codes. However, compared to many other variants of LDPC codes standardized in the past, 5G supports finer granularity of block sizes, code rates, and incremental redundancy (IR) HARQ. For example, IEEE 802.11n only defines fixed code rates and block sizes in the LDPC design and supports only 12 combinations, but 5G LDPC design construction supports 1-bit granularity of block sizes, IR-HARQ, and all practical code rates by utilizing two base graphs. LDPC design in 5G is a rate-compatible design, where higher to lower code rates are supported by extending the same base graph. Figure 6.46 shows the structure of the first base graph used in 5G, where the part shown in a smaller box is represents a higher code rate transmission, and the part in the larger box represents a lower code rate transmission. In this type of design, incremental parity bits can be generated based on the information bits with the encoding with a lower code rate parity check matrix (PCM) and using different parts of the encoded bits in each transmission; those bits can be decoded with the matching PCM at the decoding side. In the following section, we provide a quick overview of basic code construction details and the 5G coding chain.

6.13.1.1 5G LDPC Code Design

The 5G LDPC design [5] is mainly based on two base graphs, where each base graph has eight different shift coefficient designs. Each of these shift coefficient designs is further extended to support multiple lifting sizes. To explain this in more detail, we have to consider the fundamentals of QC-LDPC codes.

The representation of an QC-LDPC PCM, **H**, can be as follows:

$$
H = \begin{bmatrix}
P_{1,1} & P_{1,2} & P_{1,1} & & P_{1,N} \\
P_{2,1} & P_{2,2} & P_{2,1} & & P_{2,N} \\
\cdot & \cdot & \cdot & & \cdot \\
\cdot & \cdot & \cdot & \cdots\cdots & \cdot \\
\cdot & \cdot & \cdot & & \cdot \\
\cdot & \cdot & \cdot & & \cdot \\
P_{(N-K_b),1} & P_{(N-K_b),2} & P_{(N-K_b),3} & & P_{(N-K_b),N}
\end{bmatrix},
$$

where $P_{i,j}$ is a cyclic-permutation matrix obtained from the zero matrix and z by the z cyclically shifted identity matrix to the right (z is the lifting size). Also, $P_{i,j}$ often represented as a numerical entry which is the value of the shift. All the nonzero entries of **H** define the connections between check and variable nodes, and this is generally known as the base graph. For base graph #1 in NR, $N = 68$ and $K_b = 22$, which in turn provides the dimensions mentioned in Figure 6.46.

In general, the base graph defines the structure of the code and defines the basic framework for the hardware dimensions of the encoding/decoding. 5G LDPC base graphs contain specific design components that can be illustrated as in Figure 6.47 The 1's in the matrix represent the nonzero elements of the shift coefficient designs, where each shift coefficient design has a numeric value that defines the cyclic shift used in the parity PCM. The 0 represents a zero matrix with the dimension used in the lifting

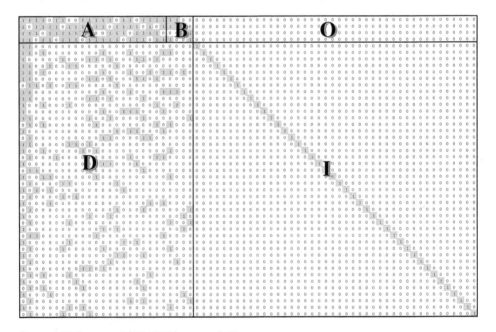

Figure 6.47 Structure of 5G LDPC base graph #1

size. The parts of the base graph #1 (base graph #2 also has similar parts with different dimensions) are as follows:

o Matrix **A** and **B** correspond to the kernel part, where the highest code rate is defined.
o Matrix **A** further corresponds to the systematic bits.
o Matrix **B** is a square matrix and corresponds to the parity bits; it has a dual diagonal structure (i.e. main diagonal and off-diagonal).
o Matrix **O** is a zero matrix.
o Matrix **D** corresponds to the extended part of the base graph.
o The first rows of matrix **D** have a quasi-row orthogonality.
o The last rows of matrix **D** have a row orthogonality.
o Matrix **I** corresponds to the extended part of the base graph and is an identity matrix.

As mentioned earlier, in 5G LDPC design, two base graphs are introduced such that the code provides good performance at a broader range of block sizes and code rates, and also improves the latency and performance for lower block sizes and code rates. The base graph #1 can support a maximum code block size of 8448 bits and code rates from 8/9 to 1/3 without any puncturing and repetition, while base graph #2 can support a maximum code block size of 3840 bits and code rates 2/3 to 1/5 without any puncturing and repetition.

Eight shift coefficient designs per base graph are defined in 5G LDPC codes, and each shift coefficient design includes a set of lifting values as in Table 6.17. The lifting values of each shift coefficient design can be changed such that different code block sizes are supported with 5G LDPC codes.

In LTE turbo, the same code is applied across all the block sizes and code rates. However, that principle is changed in 5G due to the use of two base graphs. The use of the base graphs is defined on the basis of the transport block size and target code rate indicated in the modulation and coding scheme (MCS). The base graph #2 is used when the transport block size (TBS) $\leq$292 for all rates, or TBS $\leq$3824 and Rate $\leq$2/3, or Rate $\leq$1/3. LDPC BG #1 is used otherwise (illustrated in Figure 6.48).

6.13.1.2 5G LDPC Coding Chain

The basic channel coding chain in 5G is illustrated in an earlier section with Figure 6.18. Here, we provide a summary of the design details used in each block of the coding chain.

Table 6.17 Sets of LDPC lifting size.

Set number	Set of lifting sizes (Z)
1	{2, 4, 8, 16, 32, 64, 128, 256}
2	{3, 6, 12, 24, 48, 96, 192, 384}
3	{5, 10, 20, 40, 80, 160, 320}
4	{7, 14, 28, 56, 112, 224}
5	{9, 18, 36, 72, 144, 288}
6	{11, 22, 44, 88, 176, 352}
7	{13, 26, 52, 104, 208}
8	{15, 30, 60, 120, 240}

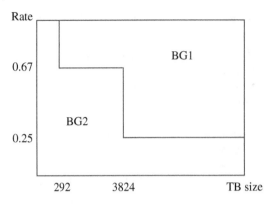

Figure 6.48 Use of the base graphs.

First, CRC bits are attached to the transport block. This is done separately for smaller TBS and larger TBS. The main motivation for the use of different CRCs is the overhead that can be saved by assuming a shorter CRC length for the smaller TBS. In general, the false alarm rate (FAR) at the receiver side is defined by the length of the CRC attached to the information block. The understanding is that an FAR of 2^{-L} can be achieved by using an L number of CRC bits. However, unlike the LTE turbo codes, LDPC codes are associated with a parity check mechanism which helps to detect if the code blocks are received without any error or not. In other words, parity check also provides FAR improvements in addition to the CRC bits. This provides a reduction in CRC bits compared to LTE (24 CRC bits). However, the reduction in CRC length is useful only in smaller TBSs where the overhead of CRC is significant. Therefore, 16-bit CRC is used when the TBS is lower than 3824, and 24-bit CRC is used for all other TBSs.

After CRC attachment, the input bit sequence can be segmented into more than one code block when certain conditions are satisfied. If the base graph 2 is used, especially when the rate is below ¼ and the input bit sequence exceeds 3840, the code blocks are segmented assuming 3816 as the maximum code block size (without code block CRC). Otherwise, 8424 is used as the maximum code block size without the code-block-level CRC. In summary, 5G allows two different maximum dimensions for the code block sizes depending on the selection of the base graph. Each code block after the segmentation has a code-block-level CRC of 24 bits (Figure 6.49).

After segmentation, the selection of the lifting size to be used in the LDPC encoding is decided based on predefined criteria where one value from the set of lifting sizes in Table 6.17 is selected. This lifting value also provides the shift coefficient design to be used in the encoding. Since the number of bits in a code block may not perfectly match the dimension of the encoding matrix, some amount of padding is done before

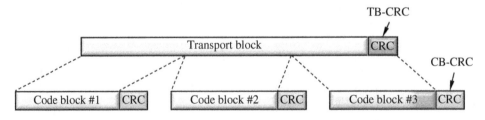

Figure 6.49 CRC attachment (TB-level and CB-Level) and code block segmentation.

the encoding procedure. However, these padding bits are removed after the encoding. In channel encoding, the LDPC design explained in the earlier section is used for each code block.

After the encoding, the coded, bits of each code block are copied to a circular buffer and rate matching is performed. The rate matching in 5G has two different components: the first is bit selection from the rate matcher, and the second is the bit interleaver of the selected bits. The length of the circular buffer is defined on the basis of the encoded block size or the limited buffer rate matching (LBRM). The former is applied for large TBSs, where a limit in the TBS is used as the LBRM TBS, which is determined from the code rate (2/3), resource allocation (depending on the active BWP), and other assumptions which are required in the transport block determination procedure in 5G. As LBRM is not the main topic of discussion in this section, we do not go into the details here. The selection of the bits is made sequentially from predefined positions in the circular buffer, and these predefined starting positions usually provide different redundancy versions. Redundancy versions are defined as RV0, RV1, RV2, and RV3. In 5G, RVs are not uniformly separated as in LTE. For base graph #1, the starting positions of RVs are 0, 17/66, 33/66, and 56/66 in fractions of the circular buffer. For base graph #2, they are 0, 13/50, 25/50, and 43/50 in portions. An illustration of different RVs and retransmission options is given in Figure 6.50.

After the bit selection, each coded block goes through a bit interleaver to increase the robustness of the performance in wireless channels. The bit interleaver in 5G channel coding is quite similar to the LTE bit interleaver. In principle, both have block interleavers, and the writing is row-wise left to right, whereas reading is from column-wise top to bottom. One difference is that in 5G the number of rows in the block interleaver

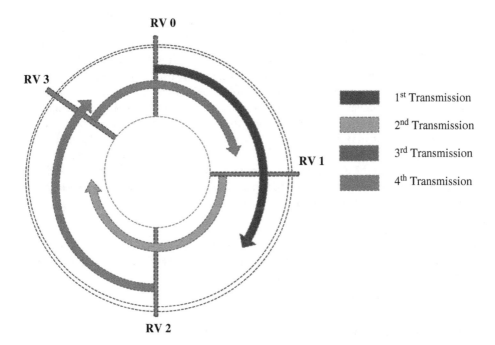

Figure 6.50 Circular buffer rate matching in 5G.

is defined by the modulation order used for the transmission. The 5G block interleaver with dimensions defined by the modulation order is called the systematic bit priority ordering interleaver for RV0, where systematic bits of the coded block get somewhat higher reliability compared to parity bits as they are mapped to the most significant bits in the modulation mapping.

After the rate matcher, the bits of each coded block are concatenated together, and output is sent to the next level of the physical layer processing.

6.13.2 Channel Coding for Control Channels

In 5G, polar coding is adopted as the main coding scheme for both DL and UL control channels except for the very small block lengths. For the very short block sizes, the same codes used by LTE are adopted, which are repetition for 1 bit, simplex codes for 2 bits, and Reed–Muller codes for 3–11 payload bits.

The polar code is a relatively new channel coding scheme compared to other codes like LDPC and Turbo, and shows good performance especially when the code block sizes are smaller. The better performance is obtained with code concatenation with CRC and also with the help of successful cancelation list decoders. At least from the algorithm complexity point of view, polar codes also showed a lower number such as $O(L * N * log_2(N))$, where N is the encoded block length, and L is the list size. These features made polar codes more suitable for control channels and led them to be adopted as the main coding scheme in both downlink and uplink control.

The concept of channel polarization is used in polar coding construction and is well discussed in the literature. When the information block length increases, the codeword tends to have two different regions, one error free and the other with zero capacity. In such cases, the error-free part can be used for data transmission, and the other region can be predefined to a known value, which is also called frozen bits. In polar encoding, the bit positions are carefully treated as they may belong to one of these two categories depending on the reliability levels of each bit position. In 5G polar codes, each bit position has a reliability ranking where higher-reliability bits are used to map the information bits, and the rest are set to zero. Also, the construction of polar coding makes the code always operate with powers of two as the output bits. Due to this, polar encoding and decoding always assumes a mother polar codeword size, which is always a power of two. For example, 512 is the maximum mother codeword size in 5G DL control channels, and 1024 is the maximum mother codeword size in 5G UL control channels. Even though these sizes are used in the encoding or decoding, the actual transmissions often require rate matching, and polar coding has good rate matching schemes that can be still applied without loss of significant performance.

6.13.2.1 5G Polar Coding Design
There are three different types of polar coding designs in 5G due to the different requirements identified in 3GPP for DL and UL control channels and for performance in regions with different block sizes. These designs are as follows:

- DL control: Distributed CRC polar code.
- UL Control:
  - Parity check (PC) and CRC concatenated polar code: 12–19 bits of UCI
  - CRC concatenated polar code: Above 19 bits of UCI

In summary, all these designs are not state-of-art polar coding schemes defined in academia. 5G control channel coding schemes are concatenated codes which use polar coding with CRC and/or PC bits, and these provide much better performance compared to traditional polar coding designs.

In DL control channels, the CRC bits are distributed among information bits considering the benefits of using the CRC bits for decoding and also to support early termination, which is considered essential due to the number of blind decodes associated with the DL control. To distribute the CRC bits inside the information bits, 5G uses a simple interleaving mechanism. The interleaver itself is defined by using the CRC generator matrix, where the CRC generator matrix has the bottom-to-top unidirectional growth property of CRC (with the increase in information block size). Using this property of the CRC generator matrix, a single interleaving pattern for different block sizes is defined in 5G. Figure 6.51 shows a toy example of the principle used here, where P1-P4 are CRC bits and 1–12 are information bits, the distribution of CRC bits is accomplished by using a single interleaver for all information block sizes, and a simple nulling operation is used to derive the interleaved block sizes for smaller block size (10 bits) with the corresponding CRC bits. In 5G, a 164-bit long interleaver is used with this principle to support CRC interleaving.

In the distributed CRC polar code used in DL control channels, 8 CRC bits are distributed, and 16 bits are appended at the end. The benefits of the distributed CRC polar code are as follows:

- Early termination without impacting FAR performance. This saves UE energy consumption and reduces decoding latency.
- Flexibility of using the same code for different purposes. CRC distribution provides the capability of different decoding choices such as using distributed bits when pruning the lists, selecting fewer or more bits for error correction versus detection depending on the service requirement, and using the traditional CRC-aided (CA)-polar coding scheme.
- Reduction in the FAR. The rate at which incorrectly received messages are decoded as correct, that is, the FAR, is reduced to obtain targets low as 2−21 by careful selection of the distribution pattern.

For the UL control channel, CRC bits are appended at the end of the information payload. To circumvent higher overheads at smaller payloads, different regions of the payload use different CRC lengths. When the payload is between 12 to 19 bits, 6 CRC

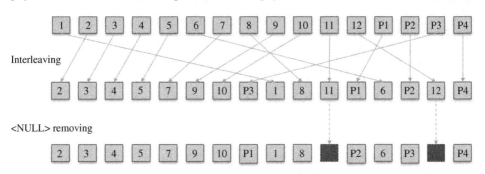

Figure 6.51 Distributed CRC polar coding design with nested property.

bits are appended with three PC bits. Above 19 bits of information payload, 11 CRC bits are appended.

For distributed CRC (DL control) and CRC concatenation (UL control) designs, the same procedure is applied at the polar encoder with the selected mother codeword size. At the polar encoder, the concatenated bits are mapped to the most reliable position (from the ranked positions), and the remaining positions are set to zero. This ranking order is known as the polar sequence, which is a nested pattern supporting up to a 1024-bits-long polar codeword. Even though the length can be up to 1024 in UL, the maximum supported polar codeword length for DL is 512. The PC and CRC concatenated polar codes have minor variations in the parity check bit mapping, which uses a specific mapping mechanism considering the structure of the Kronecker matrix and polarization weights.

The polar codeword length used for encoding may not always be the maximum mother codeword size that is supported by 5G. The selection of the polar codeword size depends on the information payload size, rate matching output (i.e. available resources for control channel), and minimum threshold code rate (1/8). A predefined method is used when deriving the exact polar codeword length based on these parameters.

Unlike many other coding schemes, rate matching in the 5G control channel has several sub-steps. First, a sub-block interleaving is performed on the output bit stream to arrange the coded bit stream; next, the coded bits are copied to a circular buffer, and finally, the bit selection is performed depending on the code rate supported in the control channel. If the rate matching output length (based on the code rate) is greater than the polar codeword size, simple repetition is applied. Otherwise, puncturing and shortening are used depending on the code rate. When the code rate is lower than or equal to 7/16, puncturing is used, whereas shortening is used for other rates.

A bit interleaver is only applied to the UL control channel. The interleaver is known as the triangular interleaver, where the write-read operation is similar to traditional block interleaving.

6.14 Dual Connectivity

The use of dual connectivity with either LTE or another 5G carrier is an essential part of 5G operation. The very first wave of 5G devices need to support the dual connectivity operation between LTE and 5G. This requires some extra considerations, especially for the resource sharing between the two uplink transmissions. The dual connectivity between 5G and LTE does not assume tight coordination between the eNB and gNB. Also, co-siting is not required for dual connectivity operation, unlike with traditional carrier aggregation.

Dual connectivity was also introduced in LTE Release 12, but dual connectivity was not deployed in real networks and is not supported on the UE side. The challenge was the requirement of two independent transmitters and receivers, as only a limited set of UEs supported uplink carrier aggregation even if the uplink throughput had clearly improved.

The key element is the uplink power sharing between LTE and 5G. If the power sharing is semi-static, this would mean that even if, for example, LTE would not be transmitting anything, the UE would reserve part of the transmission power for the LTE side, thus reducing coverage for the 5G side. With dynamic power sharing, the 5G uplink transmission power is reduced only momentarily if transmission power is really needed on the LTE side. The uplink power control was addressed in further detail in Section 6.11.

Release 15 (part of the late drop) also includes 5G–5G dual connectivity, which assumed synchronization between different gNBs. This was necessary to limit the impact to the uplink power control, as the uplink frame timing has some alignment between the two transmitters. In Release 16, this limitation is removed, and the Release 16 UE does not require time synchronization between different 5G gNBs in case of dual connectivity, as covered in Chapter 17. The example use case for 5G–5G dual connectivity, which has been discussed in 3GPP as well, has been the dual connectivity operation with the 3.5 GHz and 28 GHz bands, as shown in Figure 6.52. This would allow the full benefit of the 5G core features to be obtained as well as low latency and high reliability with 5G while at the same time benefiting from the very high data rates with mmW spectrum. With this example, one could get on the order of 2 Gbps from the 3.5 GHz band (with 100 MHz carrier available), and a single 400 MHz channel available in the 28 GHz band could provide a user data rate on the order of 3 Gbps. Thus, combined, this would enable a downlink peak data rate of 5 Gbps. The detailed peak rate depends further, for example, on the uplink/downlink allocation ratio, as addressed further in Section 6.15.

With standalone 5G operation, naturally all the uplink transmission power can be dedicated for 5G operation without the need to coordinate between the LTE and 5G transmitters.

A special case can arise for difficult band combinations, when the UE would be operating in the time division multiplexing (TDM) mode between LTE and 5G to avoid interference. This could be needed, for example, if the actual carriers chosen on the LTE side would cause interference with the actual carriers used on the 5G band. In Europe, the problem could arise with some of the allocations on the 1800 MHz band if the allocation on the 3.5 GHz spectrum coincides exactly with the second-order harmonics. The TDM operation would also impact the network side and reduce LTE and 5G performance due to the longer delay, as the uplink is not always available due to the TDM operation. Also, the performance on the LTE side is reduced, and one needs to delay uplink HARQ feedback compared to the normal operation. An example of

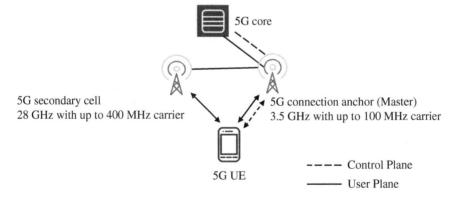

Figure 6.52 5G–5G dual connectivity operation.

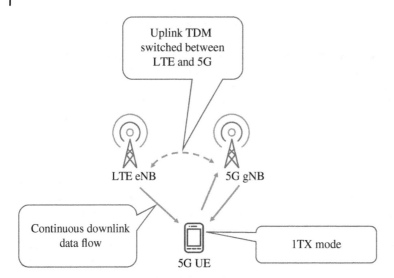

Figure 6.53 LTE and 5G dual connectivity in difficult band combination configuration (with exact second-order harmonics coinciding with the part of the 5G band in use).

TDM uplink is shown in Figure 6.53. The different dual connectivity options from the architectural perspective were covered in Chapter 5.

In the downlink direction, the support for dual connectivity requires more from the receiver when compared to the LTE downlink intra-site carrier aggregation. One needs to be able to keep synchronization to two independent base stations which have independent timing, and a key challenge is also how to share decoding resources between LTE and 5G as discussed in more detail in Chapter 13.

There is also an additional option that could be considered with dual connectivity, called Supplemental Uplink (SUL). The SUL principle employs an additional uplink carrier which would be transmitted typically on a lower frequency band than 5G, and this could be transmitted only when the 5G uplink is not used on the regular band. There are also different alternatives for SUL on the lower band, either with TDM or FDM with the LTE uplink operation. As SUL is an optional UE feature and SUL elements in the 3GPP specification are still experiencing corrections in 2019, it is not foreseen to be deployed in the near term.

6.15 5G Data Rates

With 5G, the supported data rate by the UE is not dependent on a particular UE category, but rather it is dependent on the supported frequency band, at least in Release 15. For example, with the 3.5 GHz band, the UE supports the following:

- 100 MHz bandwidth
- Four receiver antennas
- 256-QAM modulation (downlink)

These basically define the UE-supported data rate to be on the order of 2 Gbps. It is likely that in a later phase new UE types will be defined (for example, for IoT use cases)

Table 6.18 Sub-1 GHz, 3.5 GHz, and 28 GHz downlink peak data rate calculation example.

	Sub-1 GHz	3.5 GHz	28 GHz
Bandwidth	10 MHz	100 MHz	800 MHz (2 × 400 MHz)
Subcarrier spacing	15 kHz	30 kHz	120 kHz
Number of PRBs	52	273	528
Modulation	Downlink 256-QAM Uplink 64-QAM	Downlink 256-QAM Uplink 64-QAM	Downlink 64-QAM Uplink 64-QAM
MIMO	Downlink 2 × 2 Uplink no MIMO	Downlink 4×4 Uplink no MIMO	Downlink 2×2 Uplink no MIMO
PDCCH symbols	1 symbol for full bandwidth	1 symbol using 50% bandwidth	1 symbol using 50% bandwidth
DM-RS symbols	1	1	1
SSB	Not considered	Not considered	Not considered
TDD radio	Downlink 1.0 (FDD) Uplink 1.0 (FDD)	Downlink 0.74 Uplink 0.23	Downlink 0.74 Uplink 0.23
Coding rate	0.93	0.93	0.93
Peak rate downlink	**111 Mbps**	**1.80 Gbps**	**5.2 Gbps**
Peak rate uplink	**42 Mbps**	**104 Mbps**	**0.80 Gbps**

which will support much lower data rates. These are also the minimum values; there may be UEs that could support even more antennas and thus be capable of an even higher data rate.

With low frequency bands, such as 600 or 700 MHz, the UE is required to support only two receiver antennas, and thus the resulting data rates are naturally lower.

With the bands in the range 24–29 GHz, the minimum UE-supported bandwidth is 200 MHz; thus, the resulting data rate with 4-stream MIMO would be on the order of 4 Gbps. However, the expected number of diversity streams is less, typically 2, so the higher data rates are created by using a larger bandwidth instead of using 2-stream MIMO. With an expected allocation on the order of up to 800 or even 1000 MHz per operator, per-UE data rates on the order of 4–5 Gbps could be achieved (with multiple 200 or 400 MHz carriers).

When calculating the achievable data rate, the uplink/downlink configuration will also have an impact in the case of TDD deployment. For a configuration with a large downlink allocation, a higher peak downlink data rate is naturally achieved. Table 6.18 presents an example data rate calculation that is valid for a Release 15 UE supporting the 3.5 GHz frequency band (Band n78) as well as an example with the 28 GHz band.

6.16 Physical Layer Measurements

The physical layer measurements reported by the UE toward the network are typically based on either the Secondary Synchronization (SS) signal or on the CSI-RS. Naturally, for the use with mobility toward LTE, additional LTE measurements based on LTE RSRP are supported.

For measurements within the 5G network, the following measurement are defined in [6]:

- Synchronization Signal reference signal received power (SS-RSRP). This is based on the REs carrying the SSS.
- Synchronization Signal reference signal received quality (SS-RSRQ), which can be defined as the SS-RSRP/NR carrier Received Signal Strength Indicator (RSSI).
- Synchronization Signal signal-to-noise and interference ratio (SS-SINR).

The linear average over the power contribution (in [W]) of the REs carrying the SSSs divided by the linear average of the noise and interference power contribution (in [W]) over the REs carrying the SSSs within the same frequency bandwidth.

- SS reference signal received power per branch (SS-RSRPB), which is the linear average over the power contributions of the REs that carry the SSSs.
- CSI reference signal received quality (CSI-RSRQ).
- CSI signal-to-noise and interference ratio (CSI-SINR):

BTS side: SSS transmit power.

During network planning, one should avoid not only using the same physical cell ID in close proximity, but also avoid using the same SSS sequence in the near-by cell as the UE measurements may get corrupted due to the same sequence coming from two synchronized base stations.

6.17   UE Capability

Unlike in LTE, 5G does not define specific UE categories to determine the maximum data rate capabilities. The approach chosen in Release 15 was to let the supported frequency band determine what kind of data rates to be supported. This is derived basically by taking into account the following:

- Minimum bandwidth supported for a given band
- Number of receiver/transmitter antennas supported
- Coding rate (minimum)
- Modulation supported

For example, the widely supported 3.5 GHz band gives us the following (Release 15):

- Minimum bandwidth necessary for a UE is 100 MHz.
- Minimum number of receiver antenna and MIMO streams is 4.

With a coding rate of 0.95 and using 256-QAM modulation (not band specific) as well as assuming the downlink-heavy configuration, we obtain a data rate of approximately 2 Gbps.

In addition, there are many other physical-layer-related UE capabilities, as described in [7], grouped into the following categories:

- Waveform-, modulation-, subcarrier-spacing-, and cyclic-prefix-related capabilities. Most of them are mandatory, except the 60 kHz SCS and support for the extended cyclic prefix. Naturally, if the UE supported, for example, only FR1, then it would be mandated to support only 15 and 30 kHz SCSs.

- Initial-access- and mobility-related capabilities address, for example, whether a UE is using the SS Block as a basis radio link monitoring (mandatory) or whether one also considers CSI-RS measurements for radio resource management (optional). MIMO-related capabilities deal with the MIMO and beamforming operation, and there are a large number of different capabilities (more than 60 different capability entries in this area).
- Downlink-control-channel- and procedure-related capabilities address whether the UE supports, for example, more than one TCI state.
- Uplink-control-channel- and procedure-related capabilities define, for example, what kind of PUCCH formats the UE supports.
- Scheduling and HARQ operation capabilities address, for example, whether the UE supports more than one UL/DL switching point in a slot (optional) or whether inter-slot frequency hopping is supported for the PUSCH (also optional) or whether the UE supports rate matching around LTE CRS (relevant when sharing the same band), which is defined as mandatory with capability signaling. In general, the capabilities defined as "mandatory with capability signaling" are features which 3GPP expects the UEs to support once they become available on the network side.
- Carrier-aggregation- and dual-connectivity-related capabilities, as well as BWP and SUL. Dual connectivity and carrier aggregation are naturally linked to the supported band combinations. The basic BWP operation is mandatory (without capability signaling), while all SUL aspects are optional for a UE.
- Channel coding does not have any specific capabilities. The data channel LDPC is naturally mandatory as well as the different alternatives for control channel encoding/decoding.
- Uplink power control, with basic power control operation, is mandatory. Interestingly, dynamic power sharing between LTE and 5G is also mandated for UEs. This allows (in contrast to semi-static UL power sharing) the coverage to be maximized on the 5G side also when compared to semi-static power sharing between LTE and 5G uplink transmissions.

The large number of parameters may at the start of the network cause some confusion as to what is actually available on the UE side. The 3GPP specification finalization for the UE capability was only concluded in December 2018, and thus UEs need to follow that version of the specifications to be able to tell the network correctly which features they support. It cannot be guaranteed that UEs based on an earlier specification version will always be supported properly in the network.

6.18 Summary

The 5G L1 design provides a capable and versatile radio design for different use cases and a large range of frequency bands. The flexibility with a large number of different parameters and options likely will cause some challenges until the implementations in the marketplace gain more clarity on the features actually supported in the field.

The new L1 design with support for wideband operation and low complexity, yet high performance LDPC channel coding, enables higher bit rates than the earlier-generation designs. The new capabilities with beamforming allow the performance to be boosted and get most of the higher frequency bands as well. The flexibility of the L1 structure, including how frequently the gNB needs to transmit synchronization/reference signals, allows the network power consumption to scale much better as a function of the traffic compared to LTE. The use of pipeline processing together with a shorter slot duration or mini-slot allows low-latency communications to be supported.

Some of the Release 15 defined features are tailored for the use of non-traditional use cases, for example, URLLC, and thus are not foreseen in the first-phase implementations but likely later when 5G has established itself outside the classical mobile broadband domain.

References

1 Holma, H. and Toskala, A. *LTE for UMTS*, 2e. Wiley.
2 3GPP Technical Specification TS 38.211. NR; Physical hannels and Modulation.
3 3GPP Technical Specification TS 38.213. NR Physical Layer Procedures for Control.
4 3GPP Technical Specification TS 38.214. NR Physical Layer Procedures for Data.
5 3GPP Technical Specification TS 38.212. NR Multiplexing and Channel Coding.
6 3GPP Technical Specification TS 38.215. NR Physical Layer Measurements.
7 3GPP Tdoc RP-182866 (2018). RAN1 UE Feature List. December 2018.

7

5G Radio Protocols

Tero Henttonen[1], Jarkko Koskela[1], Benoist Sébire[2], and Antti Toskala[1]

[1] *Nokia, Finland*
[2] *Nokia, Japan*

7.1 Introduction

The Long-Term Evolution (LTE) radio protocols were primarily designed for the provision of PS services through a flat architecture. They were a major improvement over the previous generations, eliminating the complexity inherent in the support of circuit-switched (CS) services and in a convoluted architecture. Many of the original principles of LTE remained untouched since Release 8 and have proved to be a solid baseline for over a decade. Thus, in the early days of 5G standardization, it was agreed to use the LTE radio protocols as a baseline for 5G and to enhance them for the support of very high data rates with low latency, dynamic spectrum usage, and flexible Quality of Service (QoS).

This chapter describes the 5G radio protocols, with a focus on the modifications and enhancements brought to the LTE baseline. Unless otherwise mentioned, it can be assumed that the descriptions apply to both 5G with dual connectivity with LTE toward Evolved Packet Core (EPC) (non-standalone), and 5G with connection to 5GC (standalone).

5G Technology: 3GPP Evolution to 5G-Advanced, Second Edition.
Edited by Harri Holma, Antti Toskala, and Takehiro Nakamura.

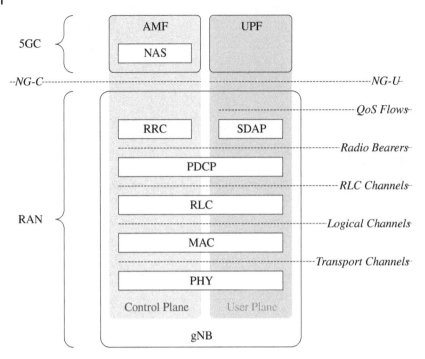

Figure 7.1 3GPP radio protocols.

7.2 5G Radio Protocol Layers

The 5G radio protocols consist of a user plane (UP) and a control plane (CP). The UP is located between the IP layer and the physical layer: in the OSI model of computer networking, the UP corresponds to the Data Link Layer and is therefore commonly referred to as Layer 2. Layer 2 of New Radio (NR) consists of four different sublayers, as depicted in Figure 7.1:

- Service Data Adaptation Protocol (SDAP, specified in 3GPP TS 37.324), which offers QoS flows to the UPF in 5GC (see Chapter 5).
- Packet Data Convergence Protocol (PDCP, specified in 3GPP TS 38.323), which offers radio bearers to SDAP.
- Radio Link Control (RLC, specified in 3GPP TS 38.322), which offers RLC channels to PDCP.
- Medium Access Control (MAC, specified in 3GPP TS 38.321), which offers logical channels to RLC.

Note that the SDAP layer is only needed when interfacing with 5GC. In other words, SDAP is not needed for non-standalone 5G with dual connectivity with LTE toward EPC but is required for standalone 5G with connection to 5GC.

The CP consists of the Radio Resource Control (RRC) protocol, as shown in Figure 7.1. On top of RRC is the Non-Access Stratum (NAS) protocol of the AMF in 5GC (see Chapter 5).

In addition to the mechanisms required to handle the new aspects of the 5G physical layer, the key drivers for making changes over the LTE baseline for the radio protocols were (1) the need to deal with a combination of even larger data rates and lower latency than in earlier generations; (2) the provision of dynamic spectrum usage; and (3) the support for a new QoS framework from 5GC. In the following, the main changes introduced for those key drivers are summarized.

First, in order to support a combination of large data rate and low latency, the 5G radio protocols were optimized for UE processing. The processing requirements are constrained by real-time operations. In the uplink, this means the operations that need to be performed by the UE to build an uplink transport block once it receives a grant from the network. In order to relax the processing requirements, real-time operations were pushed down the radio protocols and located as closely as possible to the physical layer to allow as much processing as possible without knowing the grant size, that is, to allow as many non-real-time operations and as much pre-processing as possible. Maximizing the number of non-real-time operations also makes parallel processing easier, which is essential to reduce battery consumption. The main changes introduced in 5G radio protocols compared to LTE to ease UE processing are as follows:

- RLC concatenation moved to MAC (the building of RLC PDUs becomes a non-real-time operation – see Section 7.5).
- Asymmetric placement of MAC CEs in Uplink and Downlink (the processing of MAC CEs becomes easier for the UE – see Section 7.6.13).
- Interlacing of subheaders in MAC (pipeline processing of MAC PDUs is made possible – see Section 7.6.13).

Furthermore, reordering operations in the radio protocols were streamlined with only PDCP reordering packets (reordering windows were previously located in both RLC and PDCP in LTE). Without having to reorder packets in RLC, PDCP can now decipher packets as they reach the UE (see Section 7.4). A gap in the sequence of acknowledged packets in RLC no longer stalls deciphering operations in PDCP.

Second, the granularity of spectrum usage in NR goes below that of the carrier frequency of legacy systems: with the addition of Bandwidth Parts (BWPs), it becomes possible to constrain the UE to operate within a limited portion (i.e. a part) of the frequency bandwidth of a cell (hence, the name "bandwidth part" – as covered in Section 7.6.10). Furthermore, the BWP used by the UE can be ordered to change within the same cell, and all BWPs need not have the same numerology.

Third, when connecting to 5GC, the QoS is not tied to the radio access bearer model of legacy systems. With 5GC, each packet is marked by the Core Network with a Quality of Service Flow ID (QFI), and the gNB can use that information to select an appropriate radio bearer. The one-to-one mapping between radio access bearer from the core network and the radio bearer from the radio access network (RAN) disappears, and the gNB is entrusted with configuring and selecting appropriate radio bearers to satisfy the QoS requirements (see Section 7.3).

7.3 SDAP

7.3.1 Overview

The SDAP layer handles the mapping of QoS flows from a PDU session onto Data Radio Bearers (DRBs). Unlike in previous generations, there is no one-to-one mapping between Radio Access Bearers (RABs) from the Core Network and Radio Bearers in the RAN. In 5G, the QoS model is not based on RABs but on QoS flows; it supports QoS flows that require a guaranteed flow bit rate (GBR QoS Flows) and QoS flows that do not require a guaranteed flow bit rate (non-GBR QoS Flows). The model gives a greater freedom to the RAN: 5GC only marks packets with a QFI, and the RAN is entrusted to use that information for configuring and selecting appropriate DRBs for the corresponding QoS flows. Not all QoS flows are expected to require a separate DRB, and several QoS flows can be multiplexed by the gNB onto the same DRB by the RAN. The QFI is carried in an encapsulation header over NG-U and serves as a pointer toward a QoS profile describing a number of QoS parameters to characterize the flow, for instance, a Guaranteed Flow Bit Rate (GFBR), Maximum Flow Bit Rate (MFBR), and Maximum Packet Loss Rate (see 3GPP TS 23.501). The QoS profile also contains a 5G QoS Identifier (5QI), which points toward QoS characteristics, for instance, the priority level, Packet Delay Budget (PDB), Packet Error Rate (PER), and Maximum Data Burst Volume (MDBV; see Section 7.3.3). This is summarized in Figure 7.2 below. The correspondence between 5QI and QoS characteristics is either fixed in the specifications or signaled to the gNB by 5GC (when the PDU session is established or later with a PDU session modification procedure).

The mapping of QoS flows by the gNB onto DRBs is based on the QFI and the associated QoS profiles (i.e. QoS parameters and QoS characteristics signaled from the Core Network) as depicted in Figure 7.3. The mapping of Service Data Flows (SDFs) onto QoS flows is governed by the CN.

In the uplink, the mapping of QoS flows by the UE onto RBs is governed by a set of mapping rules signaled to the UE by the gNB as depicted in Figure 7.4. The mapping of SDF onto QoS flows is governed by the CN and signaled to the UE at the NAS level.

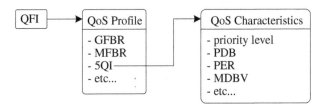

Figure 7.2 5G QoS parameters.

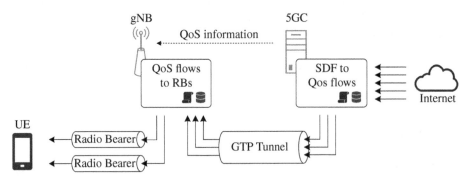

Figure 7.3 Mapping of QoS flows – downlink.

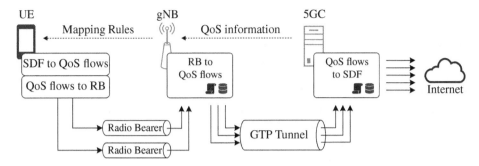

Figure 7.4 Mapping of QoS flows – uplink.

The mapping rules are configured either explicitly via RRC signaling or implicitly via Reflective Quality of Service (RQoS). With RQoS, the UE derives the uplink mapping rules from the information available in the downlink. For instance, a QoS flow x received on DRB y in the downlink implicitly tells the UE to use that same DRB for sending packets from that same QoS flow in the uplink; that is, a mapping rule for QoS flow x to DRB y is implicitly configured. Furthermore, all QoS flows for which no mapping rules are configured are mapped by default on a default DRB. As a result, the 5G QoS framework allows a mode of operation where (1) a default DRB with radio protocols configured to provide a default QoS is used to carry the bulk of traffic; and (2) only QoS flows with specific requirements are dynamically (re)mapped onto dedicated DRBs as they appear, without involving 5GC. Together with RQoS, this greatly minimizes the amount of signaling required and reduces CP latency: not all DRBs need to be configured when transiting to CONNECTED from IDLE, and the relocation of QoS flows from one RB to another does not always require CP signaling. As a matter of fact, in a typical usage scenario, the number of flows to deal with is very high, and they come and go frequently. For example, when browsing the Internet, one click can initiate a dozen TCP/IP flows. With RQoS, the allocation of radio bearers is still controlled by the network, but CP signaling is limited to the setup of the bearers: explicit signaling of the mapping rules for the plethora of flows is not required.

The configuration of the SDAP sublayer to handle the QoS flows of one PDU session is called an SDAP entity. In single connectivity (when only one gNB is used), there are as many SDAP entities as PDU sessions established by 5GC.

7.3.2 QoS Flow Remapping

QoS flow remapping from one bearer to another is needed whenever the mapping rule of a QoS flow changes, for instance, at handover when the target gNB has a different mapping policy than the source gNB, or when the gNB moves a new QoS flow away from the default bearer (as explained in Section 7.3.1. above).

When remapping a QoS flow from an old bearer to a new one, some packets from that QoS flow are likely still awaiting transmission on the old bearer – with the large amount of pre-processing allowed in 5G (see Section 7.2), such a scenario is indeed expected to be common. After updating the mapping rule, packets from the QoS flow will therefore reach the receiver from both the old bearer and the new bearer in parallel for as long as the old bearer contains packets from that QoS flow (please note that a solution requiring the UE to pull all pre-processed data on the old bearer from the QoS flow to be remapped was not considered feasible). In-order delivery then necessitates buffering of fresh data on the new DRB for as long as data remains on the old DRB. Such buffering can either take place at the receiver or at the transmitter.

Buffering in the transmitter relies on the following principle: transmission of fresh data on the new bearer starts only after having transmitted all packets from the relocated QoS flow on the old bearer. It is transparent to the receiver but requires the transmitter to buffer fresh data from the relocated QoS flow.

Buffering in the receiver relies on the following principle: fresh data from the new bearer is only delivered to the upper layers once all the packets from the relocated QoS flow on the old bearer have been received and delivered in sequence to the upper layers. It is transparent to the transmitter but requires the receiver to buffer fresh data from the QoS flow.

In order to minimize the buffering requirements on the UE, buffering in the transmitter (i.e. gNB) is used in the downlink while buffering in the receiver (i.e. also gNB) is used in the uplink. However, in order to help the gNB confirm that all data from the relocated QoS flow has been sent on the old bearer, an end marker is introduced. The end marker is always transmitted by the UE on the old bearer after updating a mapping rule.

An example of QoS flow relocation in the downlink is depicted in Figure 7.5, where a QoS flow A initially mapped onto RB1 together with a QoS flow B is remapped to RB2 (Step 1 in the figure). After updating the mapping rules, fresh data from QoS flow A remains buffered in the transmission queue of RB2 as long as RB1 contains data from QoS flow A (Step 2 in the figure). Once RB1 has no remaining data from QoS flow A, delivery of data from QoS flow A on RB2 can start (Step 3 in the figure).

An example of QoS flow relocation in the uplink is depicted in Figure 7.6, where a QoS flow A initially mapped onto RB1 together with a QoS flow B is remapped to RB2 (Step 1 in the figure). After updating the mapping rules, fresh data from QoS flow A on DRB2 is buffered in the receiver as long as the end marker (depicted with M) is not received (Step 2 in the figure). Once the end marker is received, the buffered data from QoS flow A on DRB2 is delivered to the upper layers and buffering ends (Step 3 in the figure).

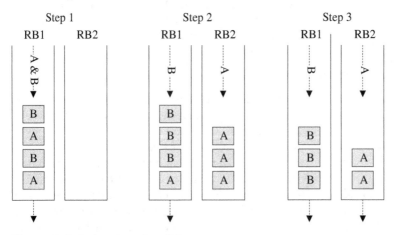

Figure 7.5 QoS relocation – downlink.

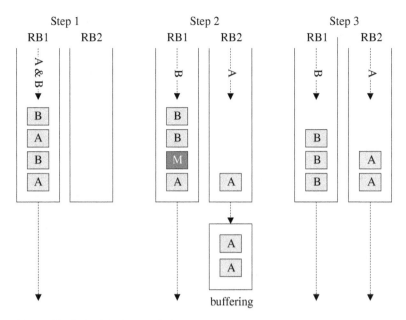

Figure 7.6 QoS relocation – uplink.

7.3.3 MDBV

One of the new QoS characteristics introduced in 5G is the MDBV. It denotes the largest amount of data that is required to be served within a given delay and is one of the characteristics of the QoS profile in 5G (see Section 7.3.1 above).

MDBV is used in radio admission control to assess how many delay-critical GBR bearers can be supported in parallel in a cell. If delay-critical GBR bearers started sending more data than initially indicated by the MDBV, the gNB would have no choice but to take the statistical variations into account when performing admission control, thereby reducing not only the number of delay-critical GBR bearers it allows, but any type of bearers for which resources need to be guaranteed. For instance, considering a delay-critical GBR service requiring 160 octets to be sent within 5 ms every second, a PDB of 5 ms, an MDBV of 160 octets, and a GBR of 1280 bit/s (160 octets/s) would be configured. Without the MDBV, the gNB would have to cope with a bit rate up to 256 000 bit/s (160 octets/5 ms), i.e. 200 times larger. This simple example underscores the importance of the MDBV for maximizing cell capacity for GBR bearers in 5G.

7.3.4 Header

In the downlink, an SDAP header is configured for a DRB when RQoS is required. In the uplink, an SDAP header is configured for a DRB when multiple QoS flows are mapped onto that DRB:

- The QFIs are needed in the uplink for the gNB to identify the QoS flows and mark the UP PDUs on NG-U.
- The end markers are needed to handle QoS flow remapping as described in Section 7.3.2 above.

Going back to the example given in Section 7.3.1 where a default bearer is used to handle the bulk of the traffic, that default bearer would then be configured with SDAP headers in both the uplink and the downlink, and the separate bearers additionally configured to individually handle specific QoS flows would not have any SDAP header configured.

7.4 PDCP

7.4.1 Overview

The PDCP layer handles DRBs and guarantees in-order delivery without duplicates to the upper layers, performs header compression when required, and is the layer responsible for enforcing security in the RAN through ciphering and integrity protection. In 5G, the PDCP layer also takes care of duplication for increased reliability and reduced latency.

The configuration of the PDCP sublayer for handling one radio bearer is called a PDCP entity. There are as many PDCP entities as there are radio bearers to handle.

In the case of LTE and 5G dual connectivity, the 5G-based PDCP layer is always used, as shown in Chapter 5. This allows the actual data routing (if via eNB or gNB) to be transparent to the UE implementation, since the user plane operation remains the same with both routing options.

7.4.2 Reordering

The design process of the PDCP reordering window in LTE is interesting. First, a reordering window at handover was agreed upon in order to ensure in-sequence delivery during handovers. Then, to cope with possible forwarding losses during handover, the reordering window was changed to a *duplicate discard window*. Finally, to simplify the behavior, it was agreed to apply the window always (i.e. not only at handover) and remove the flush timer. As a result, when the UE receives a PDCP SDU in LTE, it delivers it to a higher layer together with all PDCP SDUs with lower sequence numbers (SNs), regardless of possible gaps and in ascending order of the SNs. Further simplifications were considered, but with Release 8 coming to an end, no further modifications were agreed upon, and the principles of the LTE PDCP reordering window were left untouched since.

5G opened the door for further simplifications: instead of employing reordering in both RLC and PDCP, it was decided to perform reordering only in PDCP. Without having to reorder packets in RLC, PDCP can now decipher packets on the fly: a gap in the SN sequence of acknowledged packets in RLC no longer stalls delivery toward PDCP, and deciphering operations no longer have to deal with large bursts of data. Furthermore, unlike in LTE, it is also possible to configure PDCP not to reorder packets. This is aimed at applications that tolerate out-of-sequence delivery.

Another simplification agreed upon in 5G was to rely on the COUNT variable to specify UE behavior (instead of relying on the SN, as in LTE) and to assume that it never wraps around. COUNT is still determined based on the received SN and the state variables as in LTE, but the simplification makes the descriptions lighter and much easier to understand (one can compare Section 5.1.2.1.2 from 3GPP TS 36.323 for LTE with

Section 5.2.2.1 from 3GPP TS 38.323 for 5G). With 32 bits for COUNT, approximately 4.3 billion packets can be exchanged before a wraparound occurs. This corresponds to 6442 Gb of data, or 994 days of continuous phone calls (respectively assuming 1500 bytes payload and one packet every 20 ms). There are therefore no practical limitations associated with the assumption that COUNT does not wrap around. Note that if it did wrap around, the gNB could then simply move the QoS flow to a new radio bearer.

7.4.3 Security

Security is guaranteed in the RAN via ciphering and integrity protection. As in LTE, the configuration of integrity protection is mandatory for signaling radio bearers (SRBs) and for emergency calls of UEs in a limited service state, the NULL algorithm is configured. Unlike in LTE, though, integrity protection is not limited to SRBs (see Section 7.7); it can also be configured for DRBs, albeit up to a bit rate given by the UE capabilities.

The integrity protection function of PDCP includes integrity protection in the transmitter and integrity verification in the receiver. Integrity protection results in a 32-bit MAC-I, which is appended at the end of the PDCP PDU. The integrity protection algorithms and their inputs are defined in 3GPP TS 33.501.

The ciphering function of PDCP includes ciphering in the transmitter and deciphering in the receiver. Only the data portion of the SDAP PDU is ciphered, together with the MAC-I when present. Leaving the SDAP header un-ciphered allows data routing to take place in the receiver before de-ciphering. If the SDAP header were also ciphered, this would have imposed certain hardware architecture restrictions on both the UE and the network by precluding such routing. The ciphering algorithms and their inputs are defined in 3GPP TS 33.501.

Figure 7.7 below summarizes security in PDCP and to which portion of the PDCP PDU integrity protection and ciphering apply.

When after deciphering, integrity verification fails in the receiver for a radio bearer, only the corresponding PDU is discarded and RRC is notified. If this happens on an SRB, RRC then triggers a reestablishment. Not triggering a reestablishment when integrity verification fails on a DRB limits the impacts of packet injection and reduces the chances of Denial of Service (DoS).

7.4.4 Header Compression

Header compression in PDCP is based on the Robust Header Compression (RoHC) framework defined in RFC 5795, which defines multiple header compression algorithms

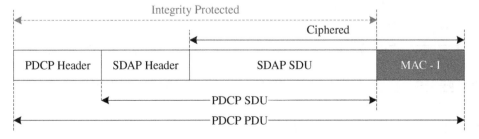

Figure 7.7 Security in PDCP.

(also known as profiles), specific to the layers located above the Layer 2 of the RAN, for example, TCP/IP or RTP/UDP/IP. Header compression was motivated by, and is commonly used for, voice services to reach comparable coverage as legacy CS services for which no RTP/UDP/IP header was present.

Once configured for a DRB, RoHC produces two kinds of packets: PDCP PDU with a compressed header and RoHC feedback packets. The former kind is always associated with a PDCP SN and can be ciphered, whereas the latter is categorized as a PDCP control PDU, does not have an associated SN, and is neither ciphered nor integrity protected.

7.4.5 Duplicates and Status Reports

Duplicates can be generated by the protocol layers below PDCP when the acknowledgments governing retransmissions fail to reach the transmitter. This typically happens at handover, when there is no time to synchronize the transmission and receive windows of RLC in acknowledged mode (AM) before resetting the RLC entities: the latest PDUs correctly received by the receiver are not known to the transmitter. To guarantee lossless mobility, those PDUs are always retransmitted after handover, thereby generating duplicates.

In order to avoid delivering such duplicates to the upper layers, PDCP discards them based on the received SN: if a PDU with the same SN has already been received, it is discarded. Then, to reduce the number of duplicates occurring after handover, PDCP status reports can be exchanged in the target cell after handover. It is important to understand that PDCP status reports are not needed to guarantee lossless mobility; they only help reduce the overhead generated by the duplicates over the air interface. Also, because retransmissions in the target cell can already start in one direction (e.g. downlink) before receiving and processing the PDCP status report from the reverse direction (e.g. uplink), not all duplicates might be eliminated. The overhead of a couple of duplicates over the air interface is, however, acceptable to guarantee low latency for lossless mobility (waiting for status reports to be received and processed might unnecessarily delay the transmission). Figure 7.8 illustrates an example of how status reports work.

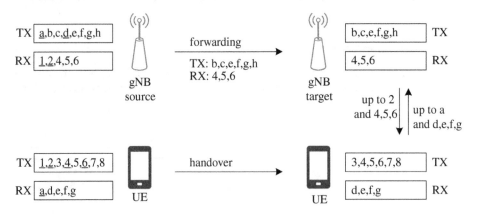

RX: *x* delivered to upper layers/TX: *x* acknowledged by lower layers

Figure 7.8 PDCP status reports.

As depicted, in the downlink, before handover, the source gNB transmits PDUs {a, b, c, d, e, f, g, h}, but only {a, d} are acknowledged by the UE. In the receive window of the UE, PDUs {b, c} are known to be missing, blocking {d, e, f, g} from being delivered to upper layers while PDU {a} was already delivered to upper layers. In the uplink, before handover, the UE transmits PDUs {1, 2, 3, 4, 5, 6, 7, 8}, but only {1, 2, 4, 6} are acknowledged by the gNB. The first missing PDU is therefore {3}. In the receive window of the gNB, PDU {3} is known to be missing, blocking {4, 5, 6} from being delivered to upper layers while PDUs {1, 2} were already delivered to upper layers. At handover, the source gNB forwards the contents of its receive and transmit windows to the target gNB, that is, PDUs {b, c, e, f, g, h} for the downlink and {4, 5, 6} for the uplink, while the UE prepares the retransmission of all PDUs starting from the first known missing one, that is, PDUs {3,4,5,6,7,8}. After handover, the uplink status report generated by the gNB signals to the UE that PDUs up to {2} were correctly received, as well as PDUs {4, 5, 6}. The UE can then clear those PDUs from its transmission buffer and only retransmit PDUs {3, 7, 8} instead of {3, 4, 5, 6, 7, 8}. The downlink status report generated by the UE signals to the gNB that PDUs up to {a} were correctly received, as well as PDUs {d, e, f, g}. The gNB can then clear those PDUs from its transmission buffer and retransmit only PDUs {b, c, h} instead of {b, c, e, f, g, h}.

7.4.6 Duplication

For the support of Ultra-Reliable and Low Latency Communications (URLLC) services, duplication is introduced in PDCP in 5G. With duplication, in addition to sending the original PDUs on a "primary leg," another leg, the "secondary leg," is used to transmit duplicates of the original PDUs as depicted in Figure 7.9.

As long as the two legs are independent of one another, duplication increases the chances of getting the PDU through in the shortest time possible: the fastest of the two legs at any point of time always pulls the transmission forward. To guarantee that the two legs are independent of one another and are not transmitted on the same radio path, different cells or different cell groups are used for diversity. The former is called carrier aggregation (CA) duplication, and the latter is called DC duplication. In both cases, it is guaranteed that the original PDUs and their duplicates will not end up on the same transport block. Note that CA duplication relies on LCP restrictions to ensure that different transport blocks are used (see Section 7.6.6).

Once a PDU is known to have reached the receiver, if the corresponding PDU still awaits transmission, it is discarded (the original PDU on the primary leg if the duplicate was received via the secondary leg, or the duplicate on the secondary leg if the original

Figure 7.9 Duplication.

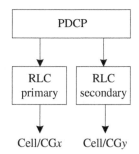

PDU was received on the primary leg). At the receiver, the second time the same PDU is received, it is discarded using the same mechanisms that were described in Section 7.4.5.

7.5 RLC

7.5.1 Overview

The main tasks of the RLC sublayer are segmentation and error correction. In the transmitter, segmentation consists in cutting RLC SDUs (i.e. PDCP PDUs) into segments that fit into the payload allocated for transmission by the lower layers. This naturally occurs only if the SDU is too large to fit. In the receiver, the segments are then put back together through a process called SDU reassembly. Error correction is achieved by ARQ and relies on status reports to trigger retransmissions. RLC offers three modes of operations: AM, unacknowledged mode (UM), and transparent mode (TM).

The RLC sublayer of 5G differs from the RLC sublayer of 4G on a number of points: no concatenation, no reordering, no sequence numbering, and offset-based segmentation always. Concatenation is a real-time operation which can only be performed once the grant size is known. Removing concatenation from RLC and relying on the lower layers to concatenate PDUs allows the processing of RLC PDUs to become a non-real-time operation. Without a reordering function for RLC PDUs (see Section 7.4.2), the need for SNs in UM mode becomes limited to the reordering of segments of a single PDU, thus reducing overhead. This is especially useful for services with a stable packet size for which the grant can be tailored to accommodate at least one packet always, for instance, a voice service. Finally, using an offset for segmentation always simplifies RLC operations (see Section 7.5.2).

The configuration of the RLC sublayer to handle one radio bearer from PDCP is called an RLC entity. When duplication is not configured, there is one RLC entity per radio bearer. When it is configured, there are two.

7.5.2 Segmentation

Segmentation occurs when an RLC SDU is too large to fit into the payload allocated for transmission by the lower layers and consists in cutting it into segments in the transmitter and putting these segments back together in the receiver through a process called SDU reassembly.

In LTE, an offset was only used for re-segmentation, when an already-transmitted PDU would need to be segmented to fit into the payload available for a retransmission. Although it decreases the overhead, it complicates RLC operations by mixing different kinds of PDUs. Thus, for 5G, it was decided to rely on an offset always, but only transmit it from the second segment to help UE processing. Let us explain why.

First, the RLC PDU header is dynamic, and the offset is only transmitted when required. In other words, when there is no segmentation, the RLC header is minimized. Second, because RLC SDUs are transmitted in sequence and segmentation only occurs when needed, what fits into the payload indicated from lower layers can be described as follows:

- $\{0;1\}$ last segment of SDU_i + $[0;n]$ complete SDUs + $\{0,1\}$ first segment of SDU_{i+n+1}; *or*
- 1 segment of SDU.

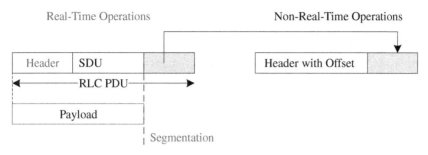

Figure 7.10 Segmentation.

That is, the first segment of an SDU always comes last and is always followed by the next segment. Third, segmentation is a real-time operation, which can only be performed once the size of the grant is known. If the offset were added to the first segment, the RLC PDU header size would be fixed only after knowing whether segmentation has occurred. Not having an offset in the first segment makes the RLC PDU header size stable from the viewpoint of real-time operations (see Note 1). When segmentation occurs, only the one-bit flag in the header of the RLC PDU needs to be changed and the PDU cut to produce a first segment. The second segment can then be built with the offset, but in non-real time. This is depicted in Figure 7.10.

7.5.3 Error Correction

ARQ in 5G is based on the same principles as in LTE: the retransmissions of RLC PDUs or RLC PDU segments in the transmitter are based on status reports from the receiver; status reports can be polled by the transmitter or triggered by the receiver; reaching the maximum number of retransmissions triggers radio link failure (RLF) (see Section 7.5.5 for one exception).

The main difference with LTE lies in the details of the RLC status report. Without concatenation in RLC, there can be a large number of RLC PDUs sent in a single transport block, and then lost. The status report in 5G was therefore enhanced to handle multiple SNs: a range of lost PDUs can be signaled to keep overhead under control.

7.5.4 Transmissions Modes

In TM, an RLC entity does not do anything and remains transparent: the transmitter side does not segment, nor add any headers, and the receiver side simply delivers the PDU as is to the upper layers. TM mode is meant to be used for SRB0 (see Section 7.7.1). In UM, an RLC entity provides segmentation service to the upper layers. In AM, an RLC entity provides segmentation and error correction services to the upper layers.

7.5.5 Duplication

Duplication for URLLC services (see Section 7.4.6) only impacts the handling of RLF in RLC: because the addition of a secondary leg for duplication should not increase the chances of declaring RLF, reaching the maximum number of retransmissions on

---

Note 1: One exception is when a 12-bit SN is used in RLC UM, in which case the header grows by one byte (from one to two) for the first segment.

the secondary RLC entity does not trigger RLF. If the secondary leg also triggered RLF, the network would have to be very conservative when configuring additional cells/CGs to minimize the risk failure of the secondary path. Having such a requirement on the configuration of additional resources would limit the usefulness of duplication.

7.6 MAC Layer

7.6.1 Overview

The MAC layer is the lowest sublayer of Layer 2 and is responsible for interfacing with the physical layer. As such, it is the UP layer which had to evolve the most to cope with the new properties of the 5G radio (the sublayers of the UP above MAC being fairly isolated).

The main functions of the MAC layer include random access, mapping between logical channels and transport channels, multiplexing and demultiplexing of MAC SDUs (i.e. RLC PDUs), error correction with HARQ, scheduling information reporting, discontinuous reception (DRX), BWP operation, and beam failure detection (BFD) and recovery. The MAC layer also conveys the signaling required for functions like timing advance and recommended bit rate. Lastly, the MAC layer is used to control the selection, activation, and deactivation of features such as PDCP duplication (see Section 7.4.6), and SCells in CA.

In the following, after describing the logical channels, the MAC functions and services are explained, with emphasis laid on the main differences compared to LTE.

7.6.2 Logical Channels

The MAC sublayer provides services to the upper layers through logical channels. Logical channels can be categorized into two different groups: Control Channels (CCHs), used exclusively for the transport of CP signaling, and Traffic Channels (TCHs), used for carrying UP data. Different logical channels are defined for different data transfer services:

- Broadcast Control Channel (BCCH), used for broadcasting system information in the downlink (see Section 7.7.2).
- Paging Control Channel (PCCH), used in the downlink to transfer paging information and notify UEs of system information changes and public warning system (PWS) transmissions (see Section 7.7.3).
- Common Control Channel (CCCH), used by SRB0 to transfer control information between the UE and the network (see Section 7.7.1), in both the downlink and uplink.
- Dedicated Control Channel (DCCH) used by SRB1, SRB2, and SRB3 as a point-to-point channel for dedicated control information between the UE and the network (see Section 7.7.1), in both the downlink and uplink.
- Dedicated Traffic Channel (DTCH,) used as a point-to-point channel for transferring UP data, in both the downlink and uplink.

Mapping of the logical channels to the transport channels and further to the physical channels is shown in Figure 7.11, where:

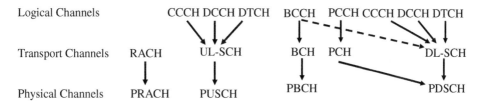

Figure 7.11 Mapping of the uplink and downlink logical and transport channels.

- In the uplink, all the logical channels are mapped to the Uplink Shared Channel (UL-SCH), and there is no logical channel mapped on the Random Access Channel (RACH), as it does not carry any information above the MAC layer.
- In the downlink, BCCH uses the Broadcast Channel (BCH) as the transport channel for the Master Information Block (MIB), while the actual System Information Blocks (SIBs) are mapped on the DL-SCH (see Section 7.7.2); the PCCH is mapped onto the PCH; and the remaining logical channels (CCH, DCCH, and DTCH) are mapped onto the DL-SCH.

There are no multicast-related channel defined in 5G specifications in Release 15, nor is there work in progress to define those in Release 16.

7.6.3 Random Access Procedure

The random access procedure is used by UEs in the IDLE and INACTIVE states to gain access to the network (see Section 7.7.1) and is used by UEs in the CONNECTED state when no valid PUCCH resources are available, for instance, at handover, after SR failure (see Section 7.6.5) or when beam failure occurs (see Section 7.6.11). A random access procedure can be contention based (CBRA) or contention free (CFRA), the latter being used only when the UE is already known to the network and has preamble(s) allocated. The physical layer details of the random access procedure are covered in Chapter 6. From a MAC sublayer perspective, the main differences compared to LTE are related to beam-based operation and uplink selection:

- The resources for the random access procedure are always linked to particular beams, and the MAC sublayer in the UE first needs to select beams before transmitting a preamble in the uplink. If beam failure occurs during a random access procedure, beam selection is initiated once more. This implies that a UE may switch from CFRA to CBRA if the contention-free resources of the CFRA procedure are associated with failed beams.
- When Supplemental Uplink (SUL) is configured, the MAC sublayer in the UE selects the uplink carrier to use for the random access procedure (either NUL or SUL). The selection is done based on the RSRP threshold, and once a carrier is selected, it is used throughout the random access procedure until it either succeeds or fails, in which case another carrier can be selected.

7.6.4 HARQ and Transmissions

As in LTE, multiple parallel HARQ processes are maintained in 5G [1], and the MAC sublayer instructs the physical layer whether to flush the soft buffer of an HARQ process or to fill it with new symbols. But unlike in LTE, asynchronous HARQ is used in both the uplink and downlink for increased flexibility. Synchronous HARQ and non-adaptive retransmissions are therefore not supported, and retransmissions always rely on explicit scheduling on the PDCCH (see Chapter 6) for simplified operations in the MAC sublayer.

Initial transmissions are either explicitly scheduled via the PDCCH or pre-configured via RRC. The pre-configuration of grants was already introduced in LTE as *semi-persistent scheduling* (SPS) to reduce the scheduling overhead for services with fixed transmission properties such as voice services (one voice frame every 20 ms). Once an SPS grant is activated by the PDCCH, it follows a fixed pattern configured by RRC. In 5G, the same mechanism is used, but has been given a different name in the uplink: *configured uplink grant type 2*. Why type 2? Because 5G also introduces a *configured uplink grant type 1*.

Configured uplink grant type 1 allows quick transmissions of small data on contention-based resources signaled by RRC. Unlike type 2, no PDCCH is required to activate the grant. Type 1 is especially suited for URLLC services.

Finally, the downlink was also enhanced in 5G to make it possible for the gNB to pre-empt ongoing downlink transmissions on the PDSCH with latency-critical transmissions, not necessarily for the same UEs (as discussed in Chapter 6). This is achieved by using a different RNTI on the PDCCH.

7.6.5 Scheduling Request

The scheduling request procedure is triggered by the UE whenever resources are needed for a new transmission on the UL-SCH. When PUCCH resources are available, an SR is sent on the PUCCH; otherwise, a random access procedure is used (see Section 7.6.3). An SR is considered as pending until resources are granted by the gNB and is repeated on the PUCCH as long as it is pending. When the number of SRs sent on the PUCCH reaches a maximum configured by the higher layer, SR failure is declared, and a random access procedure is triggered.

The scheduling request procedure of 5G differs from that of LTE by allowing multiple SR configurations. Each SR configuration consists of a set of PUCCH resources tied to a numerology, BWP, and cell. Whether a particular logical channel can use a particular SR configuration is also explicitly configured by RRC. With multiple SR configurations, it then becomes possible to reserve more SR opportunities for URLLC services and identify the logical channel(s) requesting resources directly from the SR used to request them.

7.6.6 Logical Channel Prioritization and Multiplexing

The mapping of RLC PDUs from the logical channels onto transport blocks to be transmitted by the physical layer is governed in the MAC sublayer by a process called logical channel prioritization, or LCP for short. LCP in 5G builds upon the basic principles introduced in LTE. Two loops are used to determine how many bits can be transmitted for each logical channel: the first loop relies on a token bucket algorithm to serve the logical channels in order of priority up to their configured prioritized bit rate (PBR), and the second loop serves logical channels in strict order of priority only. The first loop is used to guarantee that logical channels of low priority will not starve as long as they have a PBR configured.

After LCP, the MAC sublayer indicates to each RLC entity how many bits are available for transmission for the corresponding logical channel. But unlike in LTE, the MAC sublayer is responsible for the multiplexing of RLC PDUs from each logical channel (see Section 7.5.1), and an RLC entity produces as many PDUs and PDU segments as allowed by the number of bits indicated by MAC (see Section 7.5.2). When a logical channel is empty, others will be transmitted, and if none have data available for transmission, padding will be used to fill the transport block.

Another main difference in 5G lies in the addition of LCP restrictions. With LCP restrictions, it becomes possible to limit the usage of a particular grant, cell, or numerology to a subset of logical channels. The three main benefits are as follows: (1) with LCP restrictions, a logical channel can be linked to the physical properties of the resources used for uplink transmission; for instance, the numerology with the largest subcarrier spacing and/or shortest PUSCH transmission duration can be reserved for URLLC services; (2) with LCP restrictions, possible collisions on contention-based resources (configured grant type 1; see Section 7.6.4) can be minimized by restricting their usage to services which really need them, for instance, URLLC services; and (3) with LCP restrictions, duplicates are guaranteed to be sent in different transport blocks (see Section 7.4.6).

7.6.7 BSR

A BSR signals the amount of data awaiting transmission in the UE buffers. The reporting is done per Logical Channel Group (LCG), and RRC configures the mapping between the logical channels and an LCG. An LCG can contain one or several logical channels, and not all LCGs need to be configured. With BSR, the gNB can assess how many bits the UE has to transmit.

The same types of BSR as in LTE are used in 5G: *regular BSR*, which triggers a scheduling request procedure if there are no uplink resources to transmit them; *periodic BSR*, which is only sent if there is an available grant in the uplink (and does not trigger a scheduling request procedure if there is none), and *padding BSR*, which replaces padding. Typically, when new data becomes available for transmission, a regular BSR is triggered, which in turns triggers a scheduling request to signal the gNB that an uplink grant is needed for the UE to send something (CP signaling or UP data). Once the data exchange between the UE and the gNB has started, periodic BSRs can be used to update the gNB on a regular basis, while padding BSRs can be used to determine when the transmission buffers are empty or identify which LCG cannot be transmitted due to LCP restrictions (see Section 7.6.6) without increasing the overhead.

As in LTE, a Short and a Long format are defined for the BSR. The Short format is used to report only one LCG, while the Long format allows multiple LCGs to be reported. In 5G, the number of LCGs is increased to 8 (instead of 4 as in LTE). The motivation for this increase lies in the QoS framework (see Section 7.3.1). Indeed, by freeing the RAN from the CN constraints to support multiple RABs, 5G allows flexible QoS in the RAN with a large number of radio bearers. With eight LCGs, it became necessary to minimize the overhead for the Long BSR, and only the LCGs which have data available for transmission are now reported. In other words, the Long BSR is dynamic in size. The Short BSR remains 1 byte long, though, to minimize the overhead for low-bit-rate services and avoid fragmenting UP PDUs when a grant is tailored for a given service (e.g., a configured grant type 2 – see Section 7.6.4).

Finally, as in LTE, the truncated BSR is used when the number of padding bits is not large enough to signal all the LCGs having data available for transmission. The truncated BSR is essential for the gNB to know that there is more data in the UE buffers than what is being reported in lieu of padding. Without the truncated BSR, the gNB could conclude that the LCGs that are not being reported have no more data to transmit and stop granting additional resources to the UE.

7.6.8 PHR

A PHR signals the difference between the nominal maximum transmission power and the estimated UL-SCH transmission power. With PHR, the gNB can take the transmission power into account in packet scheduling.

As in LTE, the first type of PHR is used for PUSCH transmission, and the third one for SRS transmission on SCells configured with SRS only. Also, a virtual PHR is used in CA when no real transmission takes place on an SCell and a reference format is used (e.g. when the SCell is de-activated). The second PHR type, originally introduced in LTE for simultaneous PUSCH and PUCCH transmission, is not applicable to the 5G radio link.

A PHR is triggered when the path loss seen by the UE changes abruptly (up to a threshold configured by the gNB), when a configured timer expires (to allow for periodic reporting), or when an SCell is activated (in the case of CA). A PHR is only useful when data exchange between the UE and the gNB is taking place, and thus – unlike a BSR – a PHR never triggers a scheduling request procedure. Consequently, a triggered PHR remains pending until an uplink grant allows it to be sent.

7.6.9 DRX

Discontinuous reception (DRX) provides sleeping opportunities to the UE by mandating the time periods during which it must be awake to receive (and decode) the PDCCH for possible downlink transmissions and uplink grants. Even without transmissions, keeping the RF ON for PDCCH reception and performing blind decoding attempts to figure out if any of the PDCCH transmissions within the PDCCH search space are intended for the UE consumes power. By allowing the UE to switch its RF OFF and not perform any PDCCH decoding, DRX greatly enhances battery life.

The DRX framework of 5G is built upon the same principles as in LTE: the DRX cycle defines how often the UE needs to wake up, and once awake, for how long the PDCCH monitoring activity needs to last. The main difference compared to LTE is related to the HARQ operation in the uplink (see Section 7.6.4). With asynchronous HARQ, the retransmission opportunities do not occur at a fixed point in time, and the monitoring activity is more dynamic.

7.6.10 Bandwidth Parts

While DRX reduces UE activity in the time domain, BWP aims at reducing it in the frequency domain. With BWP, it becomes possible to constrain the UE to operate within a limited portion (i.e. a part) of the frequency bandwidth of the cell. Consider a voice service in the 3.5 GHz spectrum and a cell bandwidth of 100 MHz; obviously the data flow

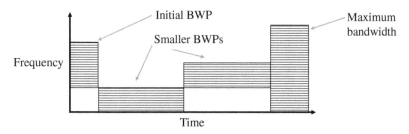

Figure 7.12 BWP example.

only requires a small fraction of the allocated spectrum. With BWP, the UE does not have to monitor the full channel bandwidth and can therefore reduce its power consumption.

Several BWPs can be configured by RRC, but only one is active at a time (the *active* BWP). The configured BWPs can differ in width (to save power in periods of low activity), in location within the frequency domain (to increase scheduling flexibility), and in the subcarrier spacing offered (to allow different services). Switching between configured BWPs for a UE can be triggered by RRC signaling, physical layer signaling (DCI), or by the MAC sublayer itself when an inactivity timer expires, or upon initiating a random access procedure.

We distinguish between two kinds of BWPs: the *initial* one and the *default* one. The initial BWP is the BWP selected by the MAC sublayer when performing a random access, or configured by the network for an SCell in CA. The default BWP is the BWP associated with the inactivity timer by the network: when an inactivity timer is configured, the expiry of the timer switches the active BWP to the default BWP. An example of BWP operation is depicted in Figure 7.12. Note that BWP also impacts system information reception (see Section 7.7.2) as well as the UE measurement operations.

7.6.11 BFD and Recovery

BFD relies on the same principles as RLF: the physical layer notifies the upper layers of failure events and once too many such events are detected (as measured on the Beam Failure Detection Reference Signal for the beams in questions), the UE initiates beam failure recovery with a random access procedure: it starts with beam selection, and upon completion of the random access procedure, beam failure recovery is considered successful. Note that the random access procedure can be contention free when random access resources were provided by the gNB for the selected beam (see Section 7.6.3).

The process is controlled by many parameters, for instance, how many beam failures are detected with a given time window and what is the threshold (in the RSRP level) for considering SSB reception good enough to start the beam recovery process. If the SSB level is too weak, then beam recovery procedure is not started, and RLF follows. The process is illustrated in Figure 7.13.

7.6.12 Other Functions

The MAC sublayer supports a large number of additional functions by means of MAC Control Elements (MAC CE) signaling, including the following :

- Timing Advance commands to maintain uplink timing in large cells (this was a basic LTE functionality already).

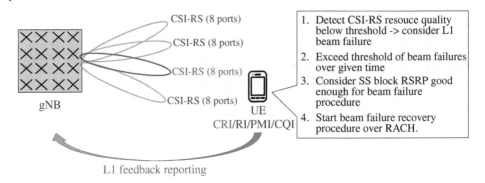

Figure 7.13 Beam failure detection and recovery.

- Activation and deactivation of SCells in CA to reduce power consumption in periods of low activity, for instance, when the PCell is enough to fulfill the QoS requirements or when the UE has no data to transmit (the functionality was introduced in Release 10 of LTE).
- Activation and deactivation of PDCP duplication (new 5G functionality described in Section 7.4.6).
- Bit rate recommendations (in DL) and recommendation queries for RAN-assisted codec adaptation (as introduced in the Release 14 of LTE).
- Further physical layer parameter or parameter set configurations and selections.

7.6.13   MAC PDU Structure

As in LTE, a MAC PDU consists of MAC subPDUs, with the content of each MAC subPDU being indicated with a MAC subheader, pointing toward either control data (a MAC CE), user data (a logical channel), or padding.

Unlike in LTE though, the MAC subheaders are not grouped together at the beginning of the MAC PDU. In 5G, MAC subPDUs are concatenated one after another by LCP (see Section 7.6.6) in order to ease pipeline processing in both the receiver and the transmitter.

Another difference compared to LTE is the asymmetric structure: in the downlink, the MAC CEs are multiplexed at the front of the PDU, while in the uplink they are appended at the end. This gives more time for the UE to process the MAC CEs it receives (in the downlink), and compute the MAC CEs it needs to send (in the uplink).

Figure 7.14 summarizes the PDU structure in the downlink, while Figure 7.15 depicts the uplink.

7.7   The RRC Protocol

7.7.1   Overview

The main task of the RRC protocol is to configure the UE with the parameters needed by other protocol layers, and to maintain connectivity between the UE and the network. The RRC protocol configures UEs in all RRC states and is responsible for the connection

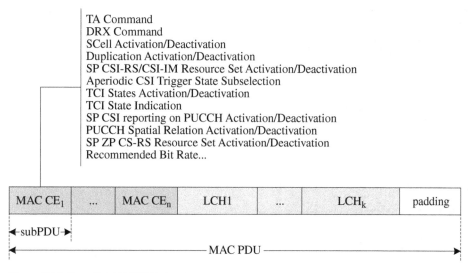

Figure 7.14 Downlink MAC PDU structure.

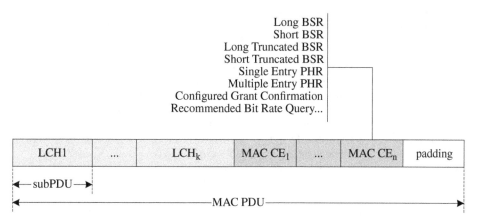

Figure 7.15 Uplink MAC PDU structure.

control, broadcast of system information (SI), measurement configuration and reporting, and other related functions (e.g. inter-RAT mobility, security activation, transfer of NAS protocol information).

There are three RRC states in NR: IDLE, INACTIVE, and CONNECTED. All other state transitions except IDLE to INACTIVE are allowed states based on various triggers, as illustrated in Figure 7.16.

The UE operations in each RRC state are different: in CONNECTED, the UE is fully controlled by the network, and in IDLE, the UE is mostly functioning autonomously without (explicit) network control. Unsurprisingly, the INACTIVE state falls somewhat in between these, although it is closer to the IDLE than to the CONNECTED state.

The various differences and similarities for each RRC state summarized in Table 7.1

Inter-RAT mobility is also supported between various RRC states, with some exceptions: Mobility between NR and UTRAN/GERAN is not supported in Release 15, nor

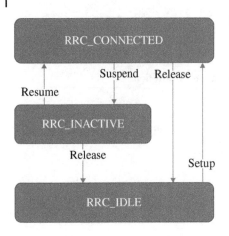

Figure 7.16 RRC states and allowed state transitions.

Table 7.1 UE operation in different RRC states.

	RRC state		
UE operation	**RRC_IDLE**	**RRC_INACTIVE**	**RRC_CONNECTED**
Connection control	**UE-based** (based only on received SI)	**UE-based** (based only on stored RAN area and received SI configuration)	**Network-based** (via RRC messages between UE and network)
Mobility	**UE-based:** Reselection based on SI and stored RRC release configuration	**UE-based:** Reselection based on SI and stored RRC release configuration	**Network-based:** Network decides and triggers any cell changes
System information maintenance and paging	UE monitors **SI and CN-based paging** from network	UE monitors **SI, CN-based paging, RAN area – based notifications** from network	UE monitors **SI** (with paging used for SI changes)
Measurement reporting	**N/A** (used for cell reselection but not reported)	**N/A** (used for cell reselection but not reported)	**As configured by RRC** (Periodic or event-triggered reporting)
RRC context stored at UE	SI from serving cell	RRC context configured by network, SI from serving cell and stored RAN area configuration	RRC context configured by network

is a direct transition from NR RRC_INACTIVE to LTE RRC_INACTIVE state. This is partly because the initial NR deployments were envisioned to be done where LTE coverage existed already, and support for older systems was to be done later if seen necessary. Having fewer inter-RAT transitions also simplifies the inter-RAT handling and allows for more focused inter-operability testing. The allowed inter-RAT state transitions are depicted in Figure 7.17 below.

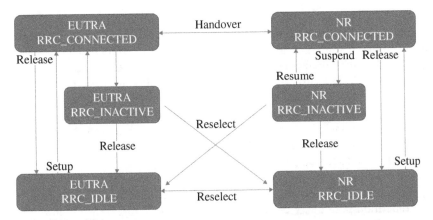

Figure 7.17 Mobility to and from NR.

Table 7.2 SRB types in 5G.

SRB usage	SRB type			
	SRB0	SRB1	SRB2	SRB3
Main purpose	Connection establishment, resumption, and reestablishment (i.e. messages requiring UE identification)	Most RRC messages (both before and after AS security establishment); may include NAS information before security activation	NAS and other lower-priority messages; only configured after security activation	Only for specific EN-DC messages: Measurement reporting and SCG reconfigurations not involving MN
Channel	CCCH	DCCH	DCCH	DCCH
Ciphering	No	Before security activation: No After security activation: Yes	Yes	Yes
Integrity protection	Yes	Yes	Yes	Yes
Used only for EN-DC	No	No	No	Yes

The RRC messages are sent using the SRBs, which always have higher priority than DRBs to ensure they are sent as quickly as possible. The SRBs 0-3 can be configured in NR for the following purposes, as shown in Table 7.2.

7.7.2 Broadcast of System Information

System information contains both NAS and access stratum (AS)-related information. Based on the characteristics and usages of the information, the system information elements are grouped together into the *MasterInformationBlock* (MIB) and different SIBs. Generally, system information provides the UE the parameters to operate in RRC_IDLE and RRC_INACTIVE states and also provides sufficient information to move to RRC_CONNECTED. For a UE in RRC_CONNECTED, the network can provide system information through dedicated signaling using the *RRCReconfiguration* message, for example, if the UE has an active BWP with no common search space configured. Nevertheless, the UE shall acquire the MIB of the PSCell to get the SFN timing of the Secondary Cell Group (SCG) (which may be different from the Master Cell Group (MCG)).

The MIB is always transmitted on the BCH with a periodicity of 80 ms and repetitions within 80 ms, and it includes parameters that are needed to acquire SIB1 from the cell. The first transmission of the MIB is scheduled in subframes as defined in [2], and repetitions are scheduled according to the period of the SSB.

SIB1 is transmitted on the DL-SCH with a periodicity of 160 ms and variable transmission repetition periodicity as specified in TS 38.213 [13, Section 13]. The default transmission repetition periodicity of SIB1 is 20 ms, but the actual transmission repetition periodicity is up to the network implementation. For the SSB and CORESET multiplexing pattern 1, the SIB1 repetition transmission period is 20 ms. For the SSB and CORESET multiplexing pattern 2/3, the SIB1 transmission repetition period is the same as the SSB period.

SIBs other than SIB1 are carried in *SystemInformation* (SI) messages, which are transmitted on the DL-SCH. Only SIBs having the same periodicity can be mapped to the same SI message. Each SI message is transmitted within periodically occurring time domain windows (referred to as SI-windows having the same length for all SI messages). Each SI message is associated with an SI-window, and the SI-windows of different SI messages do not overlap. That is, within one SI-window, only the corresponding SI message is transmitted.

Figure 7.18 shows how the UE can find each SI message to read the actual SIBs in it.

SIB1 includes information regarding the availability and scheduling (e.g. mapping of SIBs to the SI message, periodicity, and SI-window size) of other SIBs with an indication

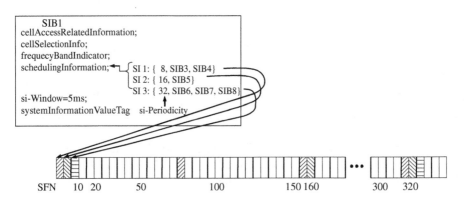

determine integer value $x = (n - 1)*w$, where n is the order of entry in the list of SI message and w is the *si-WindowLength*

si-Window starts at subframe #a, where a = x mod 10, in the next radio frame for which SFN mod T = FLOOR $(x/10)$ where T is the si-Periodicity of the concerned SI message

h

Figure 7.18 Acquisition of SI message.

whether one or more SIBs are only provided on demand and, in that case, the configuration needed by the UE to perform the SI request. SIB1 is a cell-specific SIB.

SIB2 includes cell reselection information common for intra-frequency, inter-frequency, and/ or inter-RAT cell reselection.

SIB3 includes neighboring-cell-related information relevant only for intra-frequency cell reselection. The IE includes cells with specific reselection parameters as well as blacklisted cells.

SIB4 includes information relevant only for inter-frequency cell reselection, that is, information about other NR frequencies and inter-frequency neighboring cells relevant for cell reselection.

SIB5 includes information relevant only for inter-RAT cell reselection, that is, information about E-UTRA frequencies and E-UTRAs neighboring cells relevant for cell reselection.

SIB6, SIB7, and *SIB8* contain an Earthquake and Tsunami Warning System (ETWS) primary notification, ETWS secondary notification, and Commercial Mobile Alert System (CMAS) notification, respectively. Commonly these are referred as the PWS and can be used to warn users about disaster situations, for example, earthquakes and tsunamis.

SIB9 contains information related to GPS time and Coordinated Universal Time (UTC). The UE may use the parameters provided in this SIB to obtain the UTC, GPS, and local time.

7.7.2.1 Validity and Change of System Information

The UE may use a stored version of the system information for PCell except the *MIB* and *SIB1*; for example, after cell reselection, upon return from out of coverage or after the reception of SI change indication. Any SIB except SIB1 can be configured to be cell specific or area specific, using an indication in the SIB1 scheduling information. The cell-specific SIB is applicable only within a cell that provides the SIB, while the area-specific SIB is applicable within an area referred to as the SI area, which consists of one or several cells and is identified by the *systemInformationAreaID* broadcasted in SIB1.

Whenever the network needs to update the SI, one needs to ensure that both the network and the UE apply the updated information roughly at a same time. This is ensured so that the UE is informed about the SI change prior to the actual change of information and the updated SI is broadcasted in the modification period following the one where the SI change indication is transmitted. Nevertheless, due to nature of the information of the PWS SIBs, the update mechanism has to be faster, and thus the UE needs to immediately acquire any PWS information whenever it is indicated as changed. The UE receives indications about SI modifications and PWS modifications using Short Message transmitted with P-RNTI over DCI.

A UE in RRC_IDLE or in RRC_INACTIVE shall monitor for SI change indication in its own paging occasion every DRX cycle. A UE in RRC_CONNECTED shall monitor for SI change indication in any paging occasion at least once per modification period if the UE is provided with a common search space to monitor paging. Otherwise, the network needs to update the SI by dedicated signaling for the UE in RRC_CONNECTED.

An ETWS- or CMAS-capable UE in RRC_IDLE or in RRC_INACTIVE shall monitor for indications about PWS notification in its own paging occasion every DRX cycle.

An ETWS- or CMAS-capable UE in RRC_CONNECTED shall monitor for indication about PWS notification in any paging occasion at least once per modification period if the UE is provided with common search space to monitor paging.

When the relevant SI for the SCell is changed, the RAN releases and adds the concerned SCell. For the PSCell, SI can only be changed with Reconfiguration with Sync.

7.7.3 Paging

The main purpose of the paging message is to page a UE in the RRC_IDLE and RRC_INACTIVE mode for a mobile-terminated call. In addition, the UE may be informed about SI changes by the paging short message, that is, via the PDCCH using P-RNTI with or without an associated paging message. The same short message may be used to also indicate any updates to emergency SI, that is, ETWS or CMAS. The UE identity in the paging message is either S-TMSI (for NAS paging) or I-RNTI (for RAN paging). NAS paging is in normal scenarios used in RRC_IDLE only, but in some error scenarios where the network loses track of the UE, it may need to send S-TMSI paging to the UE in RRC_INACTIVE; if this occurs, a state transition to RRC_IDLE would be triggered.

A UE may use DRX for paging reception in order to reduce the power consumption. So that the network can reach a UE utilizing DRX, there needs to be a common understanding regarding when a UE has to listen to a paging message. This is achieved by a standardized calculation giving Paging Frame (PF) and Paging Occasion (PO) as described in [3].

In multi-beam operations, the length of one PO is one period of beam sweeping, and so the UE can assume that the same paging message is repeated in all beams of the sweeping pattern; thus, the selection of the beam(s) for the reception of the paging message is up to the UE implementation.

It is not practical for all UEs listen to paging at the same time points, as then paging message sizes grow impractically large because many UE identities will need to be included to indicate which UE is actually paged. Thus, the paging calculation uses UE identity to distribute UEs to different PFs and POs. Even with this it is likely that in the same PF/PO there will be multiple UEs listening, so the network has to convey UE identity in the paging message for the UE to know that the paging message is for the UE. Also, another benefit of distributing UEs to multiple POs is to increase the paging capacity of the system.

In NR, the UE identity used for the paging calculation was changed from IMSI to S-TMSI (temporary UE ID allocated by AMF) in order to enhance security, as with IMSI it is possible to partially decode UE IMSI if a malicious gNB sends fake paging messages with false IMSIs.

The paging cycle, which is the periodicity with which the UE needs to monitor a PO, can be configured to be common for all UEs in a cell, or the network can configure an NAS-level UE-specific paging cycle used in RRC_IDLE as well as for RAN paging; that is, paging in an RRC_INACTIVE network can use a UE-specific paging cycle configuration. The UE will follow always the shortest among the paging cycles configured for it, regardless of whether it arrives via dedicated or broadcast signaling. This ensures that it is possible to reach the UE regardless of the initiating entity for paging.

The PF is calculated quite similarly as in LTE, but in addition one can configure the PF to align to RMSI transmissions. In this way, it is possible to minimize UE and network power consumption, or alternatively, in a capacity-limited system, the network may want to un-align POs so as not to collide with RMSI occasions to increase capacity for a paging message.

In NR, UE_ID is used to distribute UEs between different POs of the PF – somewhat similarly as in LTE. Though a PO associated with a PF does not need to occur in the same radio frame the PF started in, it should occur before the next possible PF of any UE. The actual slot determination of the PF is a little different compared to LTE – first, in LTE we had fixed subframes within radio frames that could be used for paging. But in NR, the TDD UL/DL slot configuration is very flexible, and we cannot used fixed subframes as in LTE.

Possible PDCCH monitoring occasions for paging are determined according to paging search space [4]. If the paging search space equals search space zero, then the configuration is called default association, otherwise it is called non-default association. A PO lasts for the length of the beam sweep (i.e. for as many PDCCH monitoring occasions as there are transmitted SSBs in the cell).

In the case of default association, the PDCCH monitoring occasions for paging are the same as for RMSI, as defined in [4].

For non-default association, the network may configure each PO to start on any allowed slot (e.g. not UL slot, should be part of the paging search space). Alternatively, POs are situated on consecutive PDCCH monitoring occasions.

7.7.4 Overview of Idle and Inactive Mode Mobility

Public Land Mobile Network (PLMN) selection is a part of the NAS procedures, and AS helps NAS to perform the procedure by searching available PLMNs if requested to do so by NAS – the PLMN is available if the measured RSRP value of the cell is greater than −110 dBm.

A priority value can be assigned to PLMNs by the home PLMN operator. The UE shall search for higher-priority PLMNs at regular time intervals and search for a suitable cell if another PLMN has been selected. For example, the operator may have configured preferred roaming operators for the USIM (Universal Subscriber Identity Module) card. When the UE is roaming and not camped to the preferred operators, it tries to periodically find the preferred operator.

If the UE is unable to find a suitable cell to camp on or if the location registration failed, it attempts to camp on a cell irrespective of the PLMN identity and enters a "limited service" state in which only emergency calls can be made.

A USIM must be inserted in the UE in order to perform the registration. As in LTE, NR uses only USIM as it can provide stronger protection against identity theft compared to SIM (Figure 7.19).

Once NAS has selected a PLMN, AS tries to select a cell of the selected PLMN based on the radio measurements and starts receiving the BCHs of that cell and find out if the cell is suitable for camping, which requires that the cell not be barred and adequate radio quality . After this procedure, called cell selection, the UE must register itself to the network, thus promoting the selected PLMN to the registered PLMN with corresponding NAS procedures. If the UE is able to find a cell that is deemed as a better

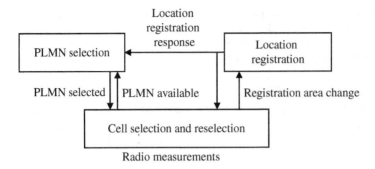

Figure 7.19 Idle mode overview.

candidate for reselection according to the reselection criteria (Section 7.7.4.1), it reselects onto that cell and camps on it and again checks if the cell is suitable for camping. If the cell where the UE camps in RRC_IDLE does not belong to at least one tracking area to which UE is registered, location registration needs to be performed. The overview is shown in Figure 7.19. Similarly, if the cell where the UE camps in RRC_INACTIVE does not belong to the RAN-based notification area to which the UE is registered, a RAN area update needs to be performed. By performing location registration procedures, the network can limit paging to a smaller area and avoiding paging overload.

7.7.4.1 Cell Selection and Reselection Process

When the UE is first powered on, it will initiate the Initial Cell Selection procedure. The UE will scan all RF channels according to its capabilities to find a suitable cell. On each carrier frequency, the UE needs only search for the strongest cell. Once a suitable cell is found, this cell is selected. Initial cell selection is used to ensure that the UE gets into service (or back to the service area) as fast as possible.

The UE may also have stored information about the available carrier frequencies and cells in the neighborhood. This information may be based on the SI or any other information the UE has acquired in the past – 3GPP specifications do not exactly define what kind of information the UE is required or allowed to use for SI cell selection. If the UE does not find a suitable cell using the stored information, the Initial Cell Selection procedure shall be started in order to ensure that a suitable cell is found.

For the cell to be suitable, it has to fulfill the S-criterion:

$$S_{rxlev} > 0 \text{ AND } S_{qual} > 0$$

where:

$$S_{rxlev} = Q_{rxlevmeas} - (Q_{rxlevmin} + Q_{rxlevminoffset}) - P_{compensation} - Qoffset_{temp}$$

$$S_{qual} = Q_{qualmeas} - (Q_{qualmin} + Q_{qualminoffset}) - Qoffset_{temp}$$

$Q_{rxlevelmeas}$ is the measured cell received level (RSRP), $Q_{rxlevelmin}$ is the minimum required received level [dBm], and $Q_{rxlevelminoffset}$ is used when searching for a higher-priority PLMN. $Qoffset_{temp}$ is only applied if the connection setup fails consecutively multiple times. A quality-based, that is, RSRQ-based S-criterion ($Squal$), may also be configured by the network if deemed necessary.

Whenever the UE has camped to a cell, it will continue to find a better cell as a candidate for reselection according to the reselection evaluation process (Section 7.7.4.2). In order to do this, the UE needs to measure neighbor cells, which are indicated in the neighbor cell list in the serving cell. The network may also ban the UE from considering some cells for reselection, which is also known as blacklisting of cells. Performing measurements consume UE battery, and to limit the power consumption of the UE, various rules to limit the impact have been defined (Section 7.7.4.4).

7.7.4.2 Intra-frequency and Equal-Priority Reselections

The cell ranking is used to find the best cell for the UE camping in the case of intra-frequency reselection or on reselection to equal-priority frequency. The ranking is based on the criterion R_s for the serving cell and R_n for the neighboring cells.

$$Rs = Qmeas, s + Qhyst - Qoffsettemp$$

$$Rn = Qmeas, n - Qoffset - Qoffsettemp$$

Where Q_{meas} is the RSRP measurement quantity, Q_{hyst} is the power domain hysteresis to avoid ping-pong, and Qoffset is an offset value to control different frequency specific characteristics (e.g. propagation properties of different carrier frequencies) or cell specific characteristics. $Qoffset_{temp}$ is only applied if the connection setup fails consecutively multiple times. In the time domain, $T_{reselection}$ is used to limit too frequent reselections. Reselection occurs for the best-ranked neighbor cell if it is better ranked than the serving cell for a longer time than $T_{reselection}$. Q_{hyst} provides hysteresis by requiring any neighbor cell to be better than the serving cell by an RRC configurable amount before reselection can occur, and $Qoffset_{s,n}$ and $Qoffset_{frequency}$ provide the possibility to bias the reselection toward particular cells and/or frequencies. The cell reselection parameters are illustrated in Figure 7.20.

Normally, ranking is based on a comparison of the measurement quantities derived as described in Section 7.7.4.4, but in a multi-beam environment, it is possible to enable performance of cell reselection for the cell with the highest number of beams above

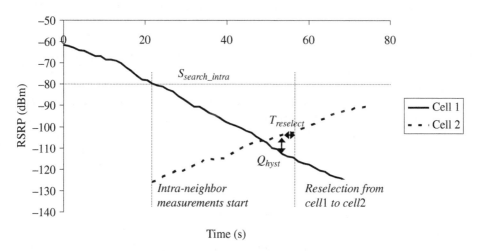

Figure 7.20 Idle mode intra-frequency cell reselection algorithm.

the threshold (i.e. *absThreshSS-BlocksConsolidation*) among the cells whose R value is within *rangeToBestCell* of the R value of the highest-ranked cell. If there are multiple such cells, the UE will perform cell reselection for the highest-ranked cell among them. In this way, the UE prefers reselection for a cell with more high-quality beams, and degradation of a beam does not trigger immediate reselection.

7.7.4.3 Inter-Frequency/RAT Reselections

Similarly to E-UTRAN in NR, priority-based reselection evaluation was chosen to be the main reselection method between different frequencies and RATs (called a layer from now on), as it was seen that operators would like to control how the UE prioritizes camping on different RATs or frequencies. In priority-based reselection, each layer is given a priority each layer is given a priority, and based on this information, the UE tries to camp on the highest-priority frequency/RAT if it can provide acceptable service. So that the UE can decide if acceptable service can be provided, the network allocates to each frequency/RAT a threshold ($Thresh_{x, high}$) which has to be fulfilled before reselection to such a layer is performed. A similar $T_{reselection}$ as in intra-frequency reselections is utilized; that is, the new layer needs to fulfill the threshold for consecutive time intervals of $T_{reselection}$ before reselection is performed. This is used to eliminate reselections if just temporary fading for the evaluated frequency occurs. In order to make reselection to a lower-priority layer, the UE will not reselect to that if the higher-priority layer is still above the threshold or if the lower-priority frequency is not above another threshold ($Thresh_{x, low}$) in order to ensure that the lower-priority layer provides sufficient quality.

The main parameters for controlling the idle mode mobility are shown in Table 7.3.

7.7.4.4 Cell Selection and Reselection Measurements

In order to limit the need to perform reselection measurements, if the serving cell reception level or quality is high enough, i.e. above the configured threshold, the UE does not need to make any intra-frequency measurements. In this way, if the serving cell is good and as the ranking is used to trigger reselection, the UE can omit neighbor cell measurements. Similarly, for equal-priority and lower-priority inter-frequency measurements, another threshold can be configured to omit these measurements if the serving cell is good enough.

But in order to ensure that the UE camps on the highest-priority layer, the UE cannot omit measurements of higher-priority layers regardless of the serving cell quality.

For cell selection in multi-beam operations, the measurement quantity of a cell is left to the UE implementation.

For cell reselection in multi-beam operations, the UE derives a cell measurement quantity as the linear average of the power values of up to the maximum number of the highest beam *(nrofSS-BlocksToAverage)* measurement quantity values above the threshold (*absThreshSS-BlocksConsolidation*). If not a single beam is above the threshold, the UE considers the cell measurement quantity to be equal to the highest beam measurement quantity.

7.7.4.5 Reselection Evaluation Altered by UE Mobility

If the UE moves fast, the network can adjust cell reselection parameters. The high (or medium) mobility state is entered based on the number of cell reselections N_{CR} within a predefined time T_{CRmax}. In order to avoid adjusting reselection parameters when the UE

Table 7.3 Main parameters for idle mode mobility.

Parameters	Description
Q_{hyst}	This parameter specifies the hysteresis value for ranking criteria. The hysteresis is required to avoid ping-ponging between two cells.
$T_{reselection}$	This parameter gives the cell reselection timer value. The cell reselection timer together with the hysteresis is applied to control the unnecessary cell reselections.
$Q_{rxlevmin}$, $Q_{qualmin}$	This parameter specifies the minimum required received level in the cell in dBm. Used in the S-criterion, i.e. in the cell selection process.
$S_{intrasearchP}$, $S_{nonintrasearchP}$, $S_{intrasearchQ}$, $S_{nonintrasearchQ}$	These parameters specify the threshold (in dB) for intra-frequency, inter-frequency, and inter-RAN measurements. The UE is not required to make measurements if $S_{servingcell}$ is above the threshold.
N_{CR}, T_{CRmax}	These parameters specify when to enter the medium- or high-mobility state: if the number of cell reselections within time T_{RCmax} is higher than N_{CR}, the UE enters the medium- or higher-mobility state and modifies Q_{hyst} and $T_{reselection}$ based on configured scaling factors.
$Thresh_{x, high}$	This specifies the threshold used by the UE in reselection toward the higher-priority frequency X than the currently serving frequency.
$Thresh_{x, low}$	This specifies the threshold used in reselection toward frequency X from a higher-priority frequency.
nrofSS-BlocksToAverage	This specifies number of SS blocks to average for cell measurement derivation.
absThreshSS-BlocksConsolidation	This specifies the minimum threshold for the beam which can be used for cell measurement derivation.
rangeToBestCell	This specifies the R value range which the cells whose R value is within the range can be a candidate for the highest-ranked cell.

is ping-ponging between two cells, those cell reselections are not counted in the mobility state calculations. As the "speed" estimation is based on the count of reselections, it does not give the exact speed, but just a rough estimate of UE movement; nevertheless, this provides a means for the network to control UE reselection behavior in accordance with the UE movement by making the UE modify *Qhyst* and/or *Treselection* based on the UE mobility state. Different from E-UTRA, the speed-dependent scaling method is not applied in the RRC_CONNECTED state for connected mode mobility parameters.

7.7.5 RRC Connection Control and Mobility

7.7.5.1 RRC Connection Control

Connection control can be considered as the primary task of RRC: Ensuring that the UE is connected to the network and is able to transmit and receive data. Thus, RRC

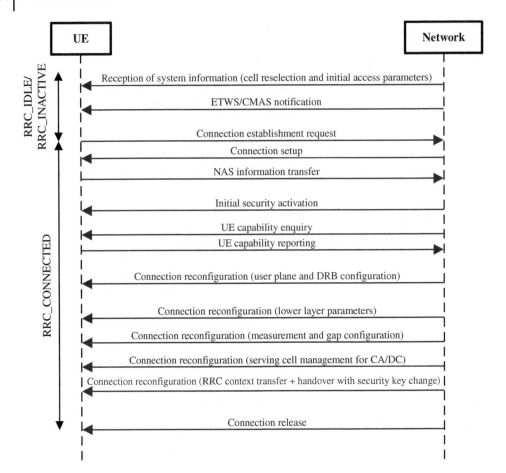

Figure 7.21 Example of RRC connection control tasks during lifetime of an RRC connection.

connection control is used for setting up, modifying, and releasing the configured UE RRC state and the corresponding parameters in each RRC state via broadcast or dedicated signaling. The mobility procedures (which are further detailed in Section 7.7.5.3) are also accomplished using the umbrella of RRC connection control procedures, by reusing the RRC connection modification procedure. These basic tasks of connection control are illustrated in a sequence diagram in Figure 7.21 (showing most connection control actions, depicted in an exemplary manner to show how they could occur during the lifetime of a single connection).

Most RRC connection control actions are atomic: Only one connection control task can operate at the same time. There are some exceptions (e.g. reporting certain failures and providing UE assistance information to the network), but the main principle is that the UE executes one RRC command at a time, and each RRC command must be completed before executing the next one. This obviously also means that if the network wishes to perform two successive commands that modify the UE RRC context, the second one has to take the results of the first one into account. It is still possible to send multiple RRC messages to be "buffered" for the UE, but since they are processed in the order they are received, the network has to ensure that nothing would happen to create inconsistency in the referred RRC context.

Certain RRC connection control messages carry RRC transaction identifiers (RTIs), which are used by the UE to ensure that no duplicate RRC messages are acted upon (if the same RRC message is repeated due to errors), as well as by the network to identify when the UE has finished applying a particular RRC message. Thus, the RTI is always first inserted by the network in a downlink RRC message, and then echoed back to the network in the corresponding uplink RRC message acknowledgment (if that is sent – there are some cases when there is no UL response). The rules determining when the RTI is included are quite straightforward: for the downlink, RRC messages typically include the RTI, and the UE always uses the same RTI when responding to the message. The exceptions to this are messages that push UE to IDLE/INACTIVE, or broadcast messages. For the uplink, an RRC message that would require a network response uses the RTI as does the UE response to a downlink RRC message. Otherwise, uplink messages do not include the RTI.

Tables 7.4 and 7.5 illustrate the RRC messages that do and do not include the RTI, with explanations of why the RTI is or is not included.

7.7.5.2 RRC Connection Setup from IDLE and INACTIVE

The most basic task of the RRC protocol is establishment of the RRC connection for a UE that needs to transfer data. While the network may initiate the UE connection attempt

Table 7.4 RRC messages that include RRC transaction identifier.

RRC message	Comments
RRCReestablishment, RRCResume, RRCSetup	Downlink message that has a response (setting up UE connectivity)
RRCReconfiguration	Downlink message that has a response (modifying RRC configuration)
CounterCheck, SecurityModeCommand, UECapabilityEnquiry	Downlink message that has a response (requesting the UE to provide information or acknowledge the network procedure)
DLInformationTransfer	Downlink message (the default rule)
MobilityFromNRCommand, RRCRelease	Downlink message (the default rule)

Table 7.5 RRC messages that do not include RRC transaction identifier.

RRC message	Comments
RRCReject	Moving UE to IDLE via RRC connection rejection
MIB, SIB1, Paging	Broadcast messages
MeasurementReport, FailureInformation, LocationMeasurementIndication, UE assistanceInformation, ULInformationTransfer	UE-originating messages with no response in DL
RRCSetupRequest, RRCResumeRequest, RRCResumeRequest1, RRCSystemInfoRequest, RRCReestablishmentRequest	UE-originating messages that may not have a direct downlink response

by paging the UE, the procedure always starts with the UE accessing the network via a random access procedure, which establishes the necessary uplink and downlink timing and identification protocols.

RRC involvement in the connection setup procedure is rather simple: Once the handshake via random access is completed, the RRC layer identifies the UE and configures it with the basic (initial) RRC configuration that allows starting basic information exchange between the UE and the network. Once the RRC connection is set up, the UE RRC acknowledgment message also delivers the NAS information to the network to enable setting up of the remaining UP details (e.g. services and their configuration details).

For this purpose, the basic RRC connection setup includes a configuration that ensures that (1) UE capabilities can be queried if necessary, (2) further reconfigurations can be done, and (3) information usable for scheduling (e.g. CQI) can be provided. This enables the network to perform the so-called "first RRC reconfiguration" that finally initiates the UP communication using the desired bandwidth based on the UE capabilities.

Once the UP configuration is ready and the UE is configured according to its capabilities, the connection establishment is fully completed, and data transmissions in the uplink and downlink can proceed.

The connection setup conceals one hidden aspect that is not readily apparent: the BWP usage. Since the MIB is sent using semi-fixed resources, and reception of the MIB then tells the UE how to receive SIB1, which then allows the UE to perform initial access and receive the rest of the SI. However, to simplify the initial access, the UE only utilizes a downlink bandwidth equal to the MIB bandwidth in the initial access, which means it does *not* utilize the so-called initial DL BWP during the initial access. Only once the RRC connection setup is received by the UE will it start using the initial DL BWP (or another BWP configured by the network in the initial RRC configuration sent in the connection setup message). However, since the network does not know the exact UE capabilities at this stage yet (see Section 7.7.6.3), the initial configuration has to be such that any UE can utilize it, and since the UE bases its cell suitability criterion only on the initial (DL) BWP bandwidth, the network may not have much choice but to use the initial BWP in the initial RRC connection establishment.

7.7.5.3 Mobility and Measurements in Connected Mode

Mobility procedures in the connected mode can be divided to two basic tasks: the cell change procedures (which are typically called "handovers") and measurements (which are mostly performed by UEs and reported to the network). Mobility is fully controlled by the network in the connected mode: the UE only acts according to the network configuration and makes an autonomous decision only in case the radio link starts to fail. This means that whatever the UE does is configured by the network: which frequencies to measure, the reporting triggers, thresholds and hysteresis, and the type of measurement reporting (i.e. one-shot, event-triggered, and event-triggered periodic or periodic). The network can utilize the UE measurement result in various ways (e.g. for generic connection control purposes), but most typically the measurement information is used for triggering handovers or triggering other measurements that can lead to handovers.

The UE measurement model in NR is more complicated than in LTE due to the beam measurements: though the beam measurements are first processed and filtered in L1

Table 7.6 Measurement events in NR RRC.

Measurement event	Example use case
A1: Serving cell > threshold	Deactivating (e.g. neighbor cell or inter-RAT) measurements since current serving cell is sufficient
A2: Serving cell < threshold	Activating (e.g. neighbor cell or inter-RAT) measurements since current serving cell may become insufficient
A3: Neighbor cell + offset > SpCell	Triggering (intra-frequency) handover to neighbor cell
A4: Neighbor cell > threshold	Considering neighbor cell as HO target or triggering further measurements
A5: SpCell < threshold1 AND Neighbor cell > threshold2	Triggering (inter-frequency) handover to neighbor cell
A6: Neighbor cell + offset > SCell	Triggering SCell change or release
B1: Inter-RAT neighbor cell > threshold	Considering inter-RAT neighbor as inter-RAT HO candidate
B2: SpCell < threshold1 AND Inter-RAT neighbor cell > threshold2	Triggering (inter-RAT) handover to neighbor cell

(the details of which are left to the UE implementation), there is additional L2 processing to consolidate the input from (potentially) multiple beams and select the best beam, after which there is still L3 filtering (to smoothen the instantaneous variations in the measurements) to process the final measured value that can be reported to the network. If beams are not used, then process simplifies to the same model as in LTE (i.e. L1 filtering followed by L3 filtering) (Table 7.6).

7.7.6 RRC Support of Upper Layers

7.7.6.1 NAS Message Transfer

RRC supports NAS message transfer, which is required for, for example, connection setup to exchange the required DRB information between the UE and the network. The NAS messages may be either transferred with dedicated RRC messages designed for NAS message transfer in the uplink/downlink or piggybacked within other RRC messages. The piggybacking can be performed upon RRC reconfiguration (in the downlink direction) or upon the RRC setup/resume completion (in the uplink direction).

The information within the NAS messages is always fully transparent to the RRC, which merely sends or receives the messages and passes them directly to the NAS layer without attempting to decode or comprehend them.

7.7.6.2 Network Slicing

Network slicing (as described in Chapter 5) is an important concept in NR but is almost invisible to the RRC layer. It is only visible in the completion of RRC setup when the UE passes its current S-NSSAI list to the network. This information is the same information that is included in the NAS protocol also passed in the same message, but since NAS is terminated in the AMF, the gNB is not required to comprehend the NAS protocol.

Signaling the UE S-NSSAI list over RRC allows the gNB to perform faster AMF selection toward the correct network slice without requiring it to comprehend the NAS protocol.

7.7.6.3 UE Capability Transfer

With the proliferation of different UE vendors and new models being released continuously, the network needs to understand the UE capabilities to configure the UE correctly. The RRC defines a procedure that allows the UE to indicate this information to network, along with the ASN.1 encoding format that allows the capabilities to be comprehended by the gNB regardless of the version of the UE and network (i.e. the capabilities of a UE supporting Release 17 RRC can still be understood by network supporting only Release 15 RRC).

The UE capabilities are typically only requested when it first registers in a network, for example, at power-up or upon return from flight mode to active usage. After the capabilities are known, they are stored in the AMF using the identity assigned to the UE at the registration. This allows the next gNB to fetch them based on the UE identity given at the RRC connection request when the NAS connection is established between the UE and the AMF. Once the capabilities are known, the network can configure the UE with the supported features.

The UE capabilities involve rather simple signaling, mostly consisting of rather compressed bitmap-like strings whose positions indicate what the UE supports. They are encoded via ASN.1 to preserve backward and forward compatibility, and the basic principles are much the same as with LTE networks. However, because most UEs support CA with multiple bands, the amount of information can grow to be very large as many capabilities are dependent on the exact band combination. Several optimizations were done already in LTE to combat this problem, and to keep the UE capability size manageable, the NR capabilities are requested and processed slightly differently than the LTE UE capabilities.

First, the network always indicates for which frequency bands it wishes to obtain the UE capabilities: for example, assume that the UE supports CA between any two carriers from bands n77, n78, n80, and n81. This means that the UE supports a total of six CA band combinations. When requesting the UE capabilities, the network could indicate that it is only interested in the UE capabilities of bands n78 and n79, which would mean that the UE only indicates a single CA band combination between n78 and n79, thus reducing the size to 1/6th of the full capabilities.

Second, heavy use of index-based referencing is used to avoid duplication of information. The so-called Feature Sets have been defined to store the combination of capabilities supported within one band in a band combination: each feature set is reported only once but can be associated with any band. By combining multiple such feature sets, it is then possible to indicate support for multiple alternative feature sets, as shown in Figure 7.22. This allows that a single Feature Set can be reused in multiple band combinations, thus avoiding duplication.

7.7.7 Different Versions of Release 15 RRC Specifications

Release 15 NR was released in three different waves, as was described in Chapter 2. The "early drop" with non-standalone NR (only supporting E-UTRA – NR Dual Connectivity), the NR standalone drop (supporting NR but not NR DC), and the "late drop" NR standalone (supporting NR DC and additional architecture options).

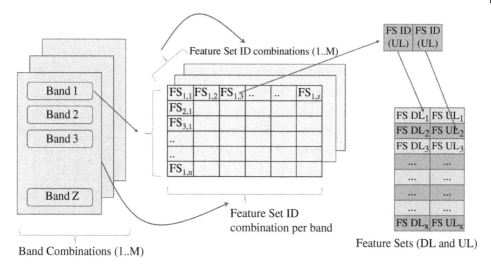

Figure 7.22 UE capability feature set and feature set combinations.

While the traces of the "RRC versioning" are visible in the RRC specifications, they are not obvious without understanding how ASN.1 works. There are several different ways to extend signaling in ASN.1: using the so-called extension markers (denoted with ellipsis, i.e. three consecutive dots), using encapsulated extensions (via OCTET STRINGs that hide the content from the decoders through which parts of the message might pass), and the so-called "empty SEQUENCEs" (which can be added only at the end of a message and are signified by leaving an undefined container at the end of the message so that the container contents are defined only in later versions). All of these mechanisms are used in RRC (for various reasons), but each one signifies that a specification has been extended at some point of time.

7.8 Radio Protocols in RAN Architecture

The division of the radio protocols in 3GPP 5G RAN architecture is shown in Figure 7.23. The DU, as introduced in Chapter 5, covers time critical functionality which includes the retransmission handling (this including besides L1 also the MAC and RLC layers) while the PDCP, SDAP, and RRC are in the Central Unit (CU). As shown in Figure 7.23, the CU can be further divided into the control and user plane parts (CU-CP and CU-UP, respectively). The protocol stacks on different interfaces inside the RAN and between the RAN and the core are covered in Chapter 5.

7.9 Summary

The objectives of the 5G radio protocols are to enable on the one hand easy implementation of high rate connectivity and on the other hand to support both power-efficient as well as high-capacity network operation with solutions like the BWP. The introduction of new elements like the flow-based QoS service, RRC_inactive state, and network slicing ensures that 5G can address different needs in line with the service evolution on top 5G networks.

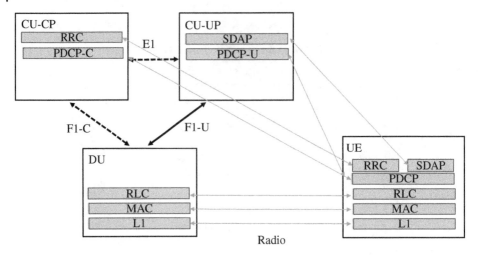

Figure 7.23 5G control and UP protocols in 5G architecture.

5G provides inter-working with LTE in Release 15, and will provide eventually in Release 16 inter-working with 3G as well to ensure voice call continuity, as discussed in Chapter 15. The dual connectivity operation is built around using 5G-based PDCP, which helps the UE implementation as then data routing is not visible to the UE.

References

1 3GPP Technical Specification TS 38.214 (2019). NR Physical Layer Procedures for Data, Version 15.6.0. June 2019.

2 3GPP Technical Specification TS 38.211 (2019). NR Physical Channel, Version 15.6.0. June 2019.

3 3GPP Technical Specification TS 38.304 (2019). NR User Equipment (UE) Procedures in Idle Mode and in RRC Inactive State, Version 15.4.0. June 2019.

4 3GPP Technical Specification TS 38.213 (2019). NR Physical Layer Procedures for Control, Version 15.6.0. June 2019.

8

Deployment Aspects

Harri Holma[1], Riku Luostari[2], Jussi Reunanen[3], and Puripong Thepchatri[3]

[1] *Nokia, Finland*
[2] *Nokia, New Zealand*
[3] *Nokia, Thailand*

CHAPTER MENU

8.1 Introduction

The capabilities of 5G networks do not depend only on 3GPP specifications and on network and user equipment but also on the practical network deployment constraints. 5G technology alone cannot provide great performance if there is not enough spectrum or enough base stations or if interference is not minimized by coordination. Some of those deployment aspects are discussed in this chapter. Section 8.2 considers spectrum resources, Section 8.3 discusses the density of the existing base station network, Section 8.4 illustrates the usage of mobile data and the asymmetry between the downlink and uplink, Section 8.5 discusses base station site solutions, Section 8.6 describes Electromagnetic Field (EMF) considerations, Section 8.7 presents network synchronization and coordination requirements, Section 8.8 discusses the 5G overlay solution, and Section 8.9 summarizes the chapter.

5G Technology: 3GPP Evolution to 5G-Advanced, Second Edition.
Edited by Harri Holma, Antti Toskala, and Takehiro Nakamura.
© 2024 John Wiley & Sons Ltd. Published 2024 by John Wiley & Sons Ltd.

8.2   Spectrum Resources

Spectrum is the key resource for any wireless communication. The amount of spectrum defines to a large extent the amount of capacity and the data rates that can be achieved by the mobile operator. If the amount of spectrum is low, it is difficult to provide high capacity and high data rates. The amount of spectrum per operator varies considerably between countries and even within countries. Figure 8.1 illustrates the amount of spectrum per operator in a few example countries. This figure shows the total amount of spectrum downlink and uplink combined between 700 and 2600 MHz in 2019. The total spectrum resources are more than 700 MHz between those frequencies if the entire spectrum is available for the mobile operators. That allows in the best case more than 200 MHz per operator with three operators. We can make a few observations from this figure:

- There are a few operators having more than 200 MHz spectrum at 700–2600 MHz.
- The typical amount of spectrum per operator is approximately 150 MHz.
- The highest amount of available spectrum is typically in Europe, in advanced Asian markets like Japan, Korea, and Australia, and in China.
- The lowest amount of available spectrum is typically in India, in developing Asian markets, like Indonesia, and in Africa and Latin America. The reason is that part of the spectrum is not available for mobile operators. Another reason is that the number of operators may be more than three.

This entire spectrum is not available for Long-Term Evolution (LTE) yet because legacy 2G and 3G networks need to be maintained in most markets, and they eat part of the spectrum. This is especially true in developing markets, where the share of legacy devices is still substantial.

The available spectrum will increase considerably with the allocations of 3.3–5.0 GHz bands. More spectrum at 700–2600 MHz will also be made available gradually in those markets where part of the spectrum is unallocated. Operator consolidation will also increase the amount of spectrum per operator. When all these factors are considered, the typical available spectrum will increase to 300 MHz per operator in the near future.

8.2.1   Spectrum Refarming and Dynamic Spectrum Sharing

Spectrum refarming refers to the process of migrating blocks of spectrum from one technology to another. Spectrum refarming has been done from 2G to 3G and to LTE, and refarming will continue to 5G. Typical spectrum strategy evolution is illustrated in Figure 8.2. The target has been to minimize the spectrum allocation for legacy 2G and 3G technologies. Some operators have been able to close down legacy networks, while many operators still keep 2G and 3G running to maintain legacy devices, old Internet of Things (IoT) devices, and to provide circuit-switched (CS) voice support. 2G and 3G systems are mostly maintained at the 900 MHz band. The other bands between 700 and 2600 MHz have been refarmed to LTE to provide more capacity and higher data rates. 5G introduction in many markets happens with the 3.5 GHz band and with some low bands like 700 MHz. The next step will be gradual refarming from LTE to 5G. The refarming from LTE to 5G can utilize Dynamic Spectrum Sharing (DSS), where both

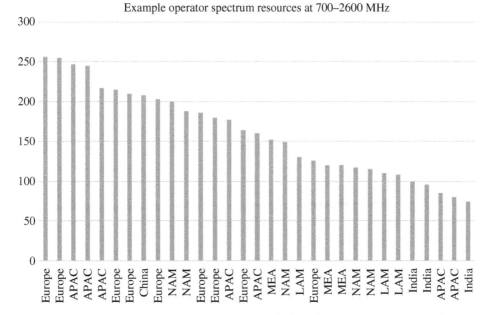

Figure 8.1 Typical operator spectrum resources at 700–2600 MHz downlink + uplink combined (APAC = Asia Pacific, NAM = North America, MEA = Middle East Africa, LAM = Latin America).

Figure 8.2 Typical spectrum strategy for sub-6-GHz bands.

MHz	Before 2019	2019/20	2020 and beyond
3500		5G	5G
2600	LTE	LTE	
2300	LTE	LTE	
2100	3G+LTE	LTE	
1800	2G+LTE	LTE	LTE/5G
1500		LTE	
900	2G+3G	2G+3G	
800	LTE	LTE	
700		LTE/5G	

LTE and 5G control signals occupy the full channel, and the resource allocation can be done in a dynamic way in the time domain depending on the instantaneous device distribution and service requirements. DSS uses close coordination between LTE eNodeB and 5G gNodeB packet schedulers for jointly optimizing the resource allocations. DSS can make the refarming to 5G a smoother process compared to earlier technologies. DSS is discussed in Chapter 10.

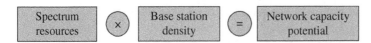

Figure 8.3 Network capacity potential defined by amount of spectrum and base station density.

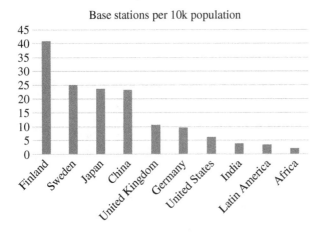

Figure 8.4 Base station density in different countries.

8.3 Network Density

The maximum achievable network capacity is defined by the spectrum resources and by the network density. Even if spectrum resources are great, the network capacity remains low if the network density is low. This basic principle is shown in Figure 8.3. Network density refers here to the number of base stations compared to the population. There are major differences between countries. Figure 8.4 illustrates the approximate macro network densities in terms of base stations per 10 000 persons. The calculation includes the combined number of base stations of all operators divided by the country's population. The highest network densities are found in the Nordic countries, and in Japan, Korea, and China. The medium network densities can be found in the European and North American networks. The lowest densities are typical in Latin America, Africa, and India. When we combine the spectrum resources and site densities together, we can conclude that there are likely challenges with network capacity in the United States, India, and in many developing countries in Asia and in Africa.

8.4 Mobile Data Traffic Growth

8.4.1 Mobile Data Volume

Early technology visions for the year 2020 indicated that the expected mobile traffic may be very high – even 1 GB/person/day. Such high mobile data traffic was considered a very bold target 10 years ago, but the growth was even higher in many networks. Mobile data traffic exceeded 1 GB/person/day in some countries already in 2018 and will exceed it in many countries by 2020. Figure 8.5 shows the mobile data usage per person per day

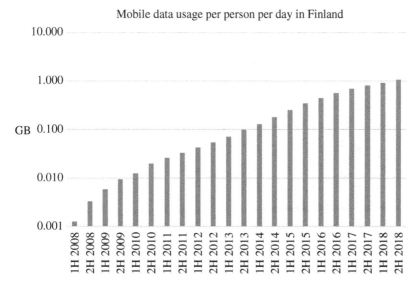

Figure 8.5 Mobile data usage per person per day in Finland.

in Finland from 2008 until 2018. The usage exceeded 1 GB per day during 2018. The total growth in 10 years has been 300×, corresponding to a nearly 80% annual growth rate on average. Mobile data usage is nearly as high in Taiwan and Saudi Arabia as well. The very high data usage requires two conditions: first, the networks need to have good quality in terms of coverage and capacity, and second, the data pricing needs to be flat rate. When these conditions are fulfilled, the customers tend to use lot of LTE mobile data because it is so convenient.

The total global mobile data during 2018 was approximately 600 petabytes (PB) per day, that is 600 000 terabytes (TB) or 600 000 000 gigabytes (GB). Most of that data consumption is created in China, India, the United States, and Europe. Figure 8.6 illustrates the estimated total mobile data usage per day in those four leading regions. The United States was the leading market in terms of mobile data until 2016, while all four areas had roughly similar data volumes during 2017. The fast growth in China and India made those countries larger in terms of data usage during 2018 simply because of the very large population in both countries. The fast data growth in India was enabled by the launch of a new LTE network in 2016 combined with attractive data pricing.

8.4.2 Traffic Asymmetry

The traffic asymmetry between the downlink and uplink should be considered for network optimization. The spectrum allocation in Frequency Division Duplex (FDD) systems is symmetric between the downlink and uplink, while Time Division Duplex (TDD) systems allow one to configure more capacity in the downlink direction. The average asymmetry collected from a large number of LTE networks globally is illustrated in Figure 8.7: the average asymmetry is 11:1, and it has been slowly increasing during the last few years. Therefore, TDD systems are inherently more spectrally efficient than FDD systems. The high asymmetry is mainly driven by streaming video, which is the most

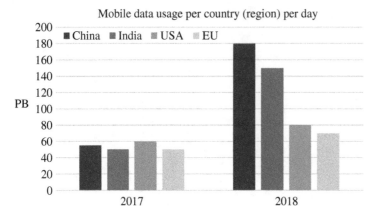

Figure 8.6 Mobile data usage per country or region per day.

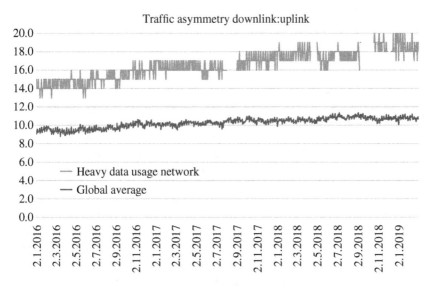

Figure 8.7 Downlink–uplink asymmetry in mobile data in LTE networks.

typical service in LTE networks. The asymmetry can be even more extreme, as shown in Figure 8.7, in a network with heavy data usage. The asymmetry in that network is 19:1. The reason is that when customers use a lot of data, the traffic is even more dominated by video streaming, which further increases the asymmetry. If we assume that TDD uses an 80:20 split between the downlink and uplink, and if the downlink spectral efficiency is two times higher than that for the uplink, then the ratio between the downlink and uplink capacity is 10:1, matching the typical traffic asymmetry.

8.5 Base Station Site Solutions

Base station sites have experienced major changes during the last few years and will experience further changes during the next few years. It will have an impact on the way

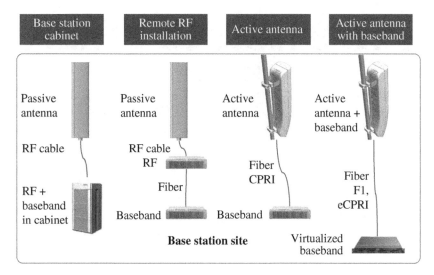

Figure 8.8 Evolution of base station site.

operators are building the sites. Figure 8.8 illustrates evolutionary steps in the base station site. The traditional solution on the left has a large base station cabinet housing RF and baseband units. RF cable connects RF units to passive antenna. The drawbacks of this solution are long RF cables with high losses and bulky base station cabinets with complex installation. The next phase brought RF units close to the antenna, which minimizes RF cable loss. This solution is also known as remote RF installation. Also, the baseband unit can be located within a small casing, which makes the installation work more flexible. The antenna has been a passive component in the first two evolutionary phases. The third phase made the antenna an active component by integrating RF components inside the antenna. An active antenna and a baseband unit are connected with optical fiber, where the interface is typically Common Public Radio Interface (CPRI). An active antenna brings a number of benefits: RF losses can be minimized, beamforming can be enabled, and installation can be simplified as RF cables or connectors are not required. The fourth step in the site solution evolution is the integration of part of the baseband processing with the active antenna and the rest in the virtualized local cloud. Low-latency baseband is integrated with the antenna, and more latency-tolerant baseband can be centralized. The detailed functionality split depends on the fiber capability. The interface options between baseband sections are described in Chapter 5, and the main options are called F1 and eCPRI. The fourth step makes the base station site extremely lean: it is just an antenna with a fiber connection to the virtualized baseband.

The evolutionary steps in the base station site show that the antenna is a key element in the future base station site because most functionalities will be integrated with the antenna. Figure 8.9 describes the typical steps in antenna evolution. The traditional passive antenna approach has two separate cross-polarized (x-polarized) antennas: one antenna covering low bands below 1 GHz and another antenna covering the high band around 2 GHz. The second phase shows dual x-polarized antennas, which has become a common solution lately, especially at the 1800–2600 MHz bands. It allows four transmitters and four receivers (4T4Rs), which improves coverage and data rates. The peak data

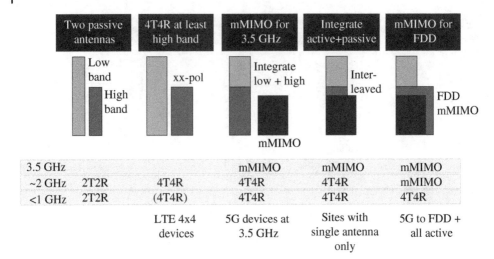

Figure 8.9 Base station antenna evolution.

rate can be doubled by using 4×4 MIMO, which is supported by high-end LTE devices at the 1800–2600 MHz bands. 4T4R is sometimes also used at low bands even if devices do not support 4×4, because 4R uplink can improve coverage. The coverage improvement is beneficial especially for Voice over LTE (VoLTE) in order to match the coverage of CS voice service. The next step in the evolution is the addition of massive Multiple Input Multiple Output (MIMO) antennas for 5G at 3.5 GHz. The first massive MIMO antenna is typically a separate antenna. Many base station sites have limitations in terms of the number of antennas. It may be beneficial to integrate all the passive antennas into a single antenna to make room for the new massive MIMO antenna. A further evolutionary step integrates all the passive antennas with 3.5 GHz massive MIMO antenna, which enables supporting all bands with just a single antenna: 700–2600 MHz with a passive antenna and 3.5 GHz with a massive MIMO antenna. The last step brings the massive MIMO antenna also to FDD bands at 1800–2600 MHz, which makes sense, especially when 5G device penetration on those bands increases.

8.6 Electromagnetic Field (EMF) Considerations

Base station installations must comply with the local EMF exposure limitations. Most countries have adopted the RF exposure limits developed by International Commission on Non-Ionizing Radiation Protection (ICNIRP). This limit corresponds to 10 W/m^2 measured over 6 min moving average. Some countries enforce more stringent local rules. EMF limits must be considered as part of 5G network deployment planning. A new challenge in 5G is the usage of beamforming from the EMF point of view. Massive MIMO provides high beamforming gain, which is excellent for coverage and capacity but can pose new issues regarding RF exposure and EMF limitations. Beamforming increases the Equivalent Isotropically Radiated Power (EIRP) and creates larger EMF exclusion zones than conventional non-beamforming antennas if we assume the theoretical peak value for the beamforming gain. The peak EIRP value with a 3.5 GHz

Table 8.1 Compliance distances with traditional and with statistical approaches Adapted from [1].

	Probability and averaging time	Compliance distance
Traditional		16.1 m
Statistical	95%/10 s	6.4 m
	99%/10 s	6.6 m
	95%/60 s	6.9 m
	99%/60 s	7.3 m

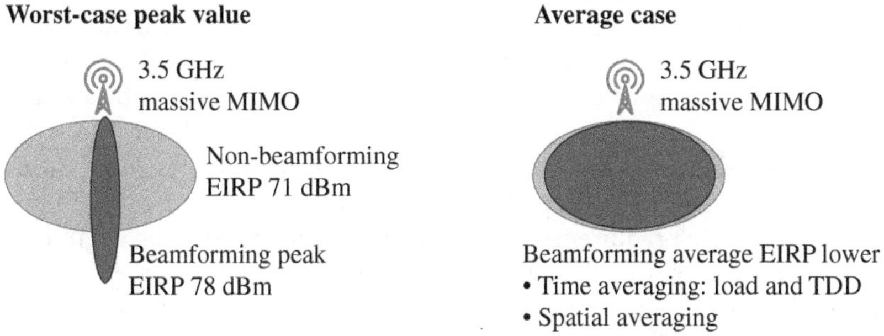

Figure 8.10 Peak and average EIRP values with beamforming antenna.

beamforming antenna can be 78 dBm with 200 W and a 25 dBi beamforming gain, which is substantially more than that obtained with the traditional solution with an 18 dBi sector antenna. This worst case, however, does not represent the typical case with beamforming, where the average value is lower due to time averaging and spatial averaging. The cell is not fully loaded all the time, and the downlink transmission does not happen all the time in TDD. Also, the spatial beam direction changes constantly. Therefore, a statistical model would provide a more practical measurement of the real EMF values. We performed 3GPP-compliant system simulations considering that the EMF is averaged in the time and spatial domains assuming a full buffer traffic model. The results show that the averaged beamforming gain is much less when compared to the maximum achievable gain. The averaged transmitted power is 20–30% of the peak value. The required compliance distances are shown in Table 8.1 and are nearly half the distances computed with the traditional method. More details can be found from [1–3] (Figure 8.10).

8.7 Network Synchronization and Coordination Requirements

Most LTE networks globally use FDD technology. The importance of TDD is increasing, with a growing number of TD-LTE networks and with 5G utilizing TDD technology at 2.6 and 3.5 GHz. TDD technology brings new requirements for the deployments in terms of network synchronization and inter-operator coordination. This section addresses the requirements and the solutions needed in TDD networks. While FDD networks have downlink and uplink channels on separate frequencies relatively far apart and well outside each other's band pass filters, in TDD networks they are located on the same frequency. Also, in the case of adjacent frequencies, uplink, and downlink frequencies are typically within the bandpass of each other's filters. Downlink and uplink resources are multiplexed in the time domain rather than in the frequency domain; band pass filters cannot provide the protection from downlink-to-uplink interference as they do in FDD networks.

As base transceiver station (BTS) antennas are typically located on rooftops, there is a good chance for Line of Sight (LoS) between various other BTS antennas exist, and the path loss between antennas is expected to be very small. As the transmission and receiving takes place within the band pass of each other's filters, downlink transmission occurring at the same time that the neighboring BTS is receiving would cause very severe BTS-to-BTS interference. Even though the user equipment (UE)-to-UE interference scenario is also possible, they are considered much less damaging with a lower likelihood.

For the above reasons, downlink and uplink need to take place simultaneously in all base stations, and switching should take place very accurately at the same time. Therefore, TDD networks have strict requirements for synchronization and timing alignment of downlink and uplink symbols for all networks operating on the same band. All operators should preferably synchronize their networks and use the same TDD frame configuration between the uplink and downlink. The TDD base station synchronization can be obtained from Global Positioning System (GPS) or via the transport network with the Timing over Packet (ToP) solution.

8.7.1 Main Interference Scenarios in TDD System

The main interference scenarios in TDD networks are BTS-to-BTS interference and UE-to-UE interference. Figure 8.11 illustrates both interference cases when Cell 1 and Cell 2 are not synchronized. In interference case A, Cell 2 transmission to UE2 overlaps with the reception of Cell 1 from UE1. If the two BTSs are close enough, then the interference level can be so high that the uplink of Cell 1 is completely blocked by the simultaneous Cell 2 transmission.

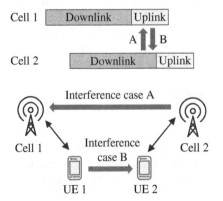

Figure 8.11 BTS-to-BTS and UE-to-UE interference due to timing misalignment within operator's own band.

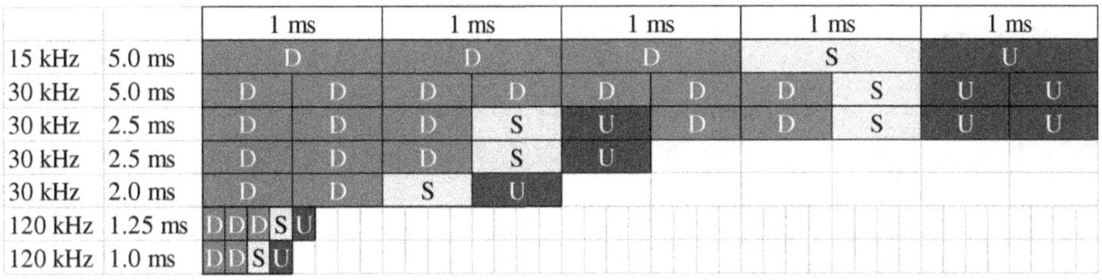

		1 ms		1 ms		1 ms		1 ms		1 ms	
15 kHz	5.0 ms	D		D		D		S		U	
30 kHz	5.0 ms	D	D	D	D	D	D	D	S	U	U
30 kHz	2.5 ms	D	D	D	S	U	D	D	S	U	U
30 kHz	2.5 ms	D	D	D	S	U					
30 kHz	2.0 ms	D	D	S	U						
120 kHz	1.25 ms	D D D S U									
120 kHz	1.0 ms	D D S U									

Figure 8.12 Typical TDD frame configuration options (D = Downlink slot, S = Special slot, U = Uplink slot).

In interference case B, the UE1 transmission interferes with the UE2 reception, and if UE1 and UE2 are close enough, the reception of UE2 is blocked by the UE1 transmission. UE-to-UE interference depends heavily on the UE locations and on the UE transmission power levels. The worst-case interference can be challenging, but the probability of it occurring is very low.

These interference scenarios can also occur if the cells are synchronized, but different asymmetries between downlink and uplink are used in adjacent cells. Therefore, the base stations need to be synchronized in TDD networks, and the same asymmetry must be used if we want to completely avoid these interference cases.

8.7.2 TDD Frame Configuration Options

While LTE specifications limit the uplink and downlink frame configurations to seven different options, 5G specifications allow a large number of options for splitting the frame between the uplink and downlink: the TDD frame size can be selected, and the split between the downlink and uplink can be selected. Typical 5G frame configurations are shown in Figure 8.12 for the downlink, uplink, and special slots. The special slot includes a guard period (GP) for switching between the downlink and uplink. The special slot is typically configured mostly for the downlink with a GP of a few symbols. The 15-kHz option with the 5 ms frame is designed for coexistence with TD-LTE. There are a few different options for the 30 kHz case with 5.0, 2.5, and 2.0 ms frame sizes that are typically used for mid-band cases without TD-LTE coexistence. All cases have asymmetric allocation where the downlink has larger share of the time than the uplink. mmWave cases use a 120 kHz subcarrier spacing, and the TDD frame is normally 0.625 or 0.5 ms.

8.7.3 Cell Size and Random Access Channel

TDD frame configurations have an impact on the relative capacity split between the downlink and uplink, on the latency, and also on the maximum cell radius. For example, if we have 30 kHz subcarrier spacing and just one uplink slot in the frame, only the short Physical Random Access Channel (PRACH) format can be utilized but not the long PRACH format. Short PRACH limits the cell radius. The maximum cell radius with short PRACH with a 30 kHz subcarrier spacing is less than 5 km and with a 15 kHz subcarrier spacing less than 10 km. A larger cell radius may be needed, especially in sub-urban and rural cases where high base station antennas and open propagation can enable larger cells from the propagation point of view. Long PRACH allows very large cell sizes. Long PRACH is possible with two consecutive uplink slots, for example, with the DDDSUDDSUU frame shown in Figure 8.13.

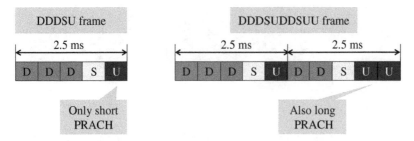

Figure 8.13 Frame configurations with 30 kHz subcarrier spacings and PRACH options.

8.7.4 Guard Period and Safety Zone

Radio waves travel at the speed of light, and hence from the BTS-to-BTS interference point of view, it is also worth noting that the duration of the GP defines the safety zone distance beyond which the downlink and uplink will start overlapping even if the network is perfectly synchronized. The duration of the GP, measured in microseconds, defines the interference safety zone distance. At the speed of light, in 1 μs, the radio waves travel 300 m. To achieve, for example, a 15 km safety range, about a 50 μs GP would be required from when transmitter switches off to when receiver is switched on. If a 40 km safety zone is required, a 134 μs GP would be necessary – but this also emphasizes the requirement for good-quality physical layer design and the importance of reducing overshooting. In LTE, the GP is defined by the selection of the Special Sub-Frame (SSF), of which nine different options are specified in [4]. In 5G, there are no SSFs as in TD-LTE; however, the slot formats defined in [5] (Tables 4.3.2–3) can contain a certain number of guard symbols defined by the networks. Typical settings for guard symbols are two or four in the case of a 30 kHz subcarrier spacing. Calculation of the safety distance should also consider factors like the transit period for switching off the transmitter as well as the Timing Advance Offset (TAO), which reduces the practical safety range. These are included in the following picture showing the safety zone ranges for all LTE and 5G GP configurations (Figures 8.14–8.16).

Having a greater GP allows for greater safety zones and helps to better cope with small timing alignment errors, but comes at the cost of capacity and peak throughput.

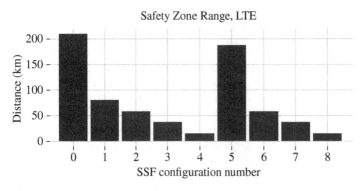

Figure 8.14 Safety zone ranges with different special subframe configurations for LTE.

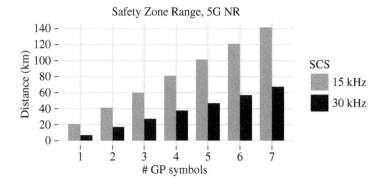

Figure 8.15 Safety zone ranges with different GP configurations for 5G with 15 and 30 kHz subcarrier spacings.

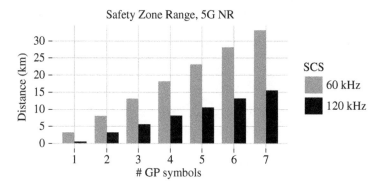

Figure 8.16 Safety zone ranges with different GP configurations for 5G with 60 and 120 kHz subcarrier spacings.

For example, a GP of five symbols with 30 kHz allows a 50 km safety zone, but it takes 7% of the system capacity since there are a total of 70 symbols in the 2.5 ms frame.

8.7.5 Intra-Frequency Operation

As there is often LoS between BTS antennas located on rooftops, the RF path between BTS sites provides very little isolation. Hence, BTS-to-BTS interference can be considered the worst kind of interference that can occur in a TDD network. Figure 8.17 show the increase in dB to the noise floor assuming the worst-case scenario where transmitting and receiving antenna are pointing toward each other and there is no obstruction in between. The distance calculation assumes the worst-case scenario where both antennas are on rooftops and visible to each other, and hence free space path loss calculation is used. With these modest power and antenna gain assumptions, the potential interference over the thermal noise in the victim cell receiver would be a massive 40 dB as far as 50 km away. This is clearly an unwanted scenario, and it is obvious that BTSs on the same frequency are required to switch between transmit and receive modes simultaneously.

The interfering BTS antenna gain here and in the following cases is assumed to be 27 dBi and victim antenna 9 dBi per connector. The receiving end antenna gain is the gain

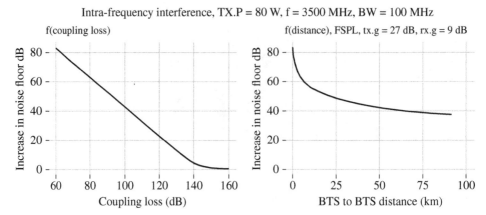

Figure 8.17 Intra-frequency BTS-to-BTS interference increase as a function of coupling loss between BTSs and as a function of distance assuming free space loss between BTSs.

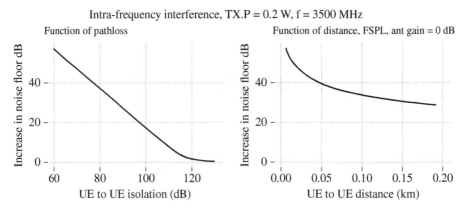

Figure 8.18 Intra-frequency UE-to-UE interference.

of a single transceiver array boundary connector. The coupling loss is the loss between the transmitter and receiver connectors.

Similarly, when two UEs are near each other, for example, in the same room and served by two different BTSs with timing differences, they can interfere with each other. Figure 8.18 show the noise floor increase assuming there is LoS between the two UEs.

The victim UE could potentially suffer from interference of tens of dBs above the thermal noise even if the separation is tens to hundreds of meters. In special event conditions where the UE density is high, this would quite likely result in severely degraded service.

A relatively rare mode of radio propagation which is often called *atmospheric ducting* is also worth mentioning. During certain weather conditions such as inversion layers, radio waves are bent between different horizontal atmospheric layers of different densities, forming a waveguide. This can result in a radio channel in which the radio waves can follow the curvature of Earth and have much reduced attenuation. Due to the reduced attenuation, interference can occasionally be experienced even hundreds of kilometers away. Some geographical areas are more likely to experience such conditions than others.

Figure 8.19 BTS-to-BTS interference due to timing misalignment between adjacent operator bands.

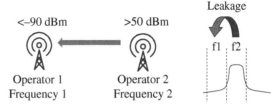

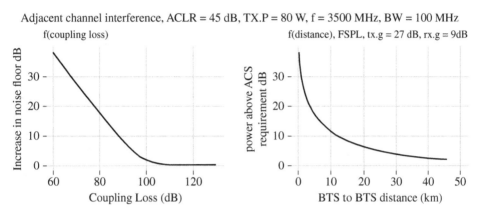

Figure 8.20 Adjacent channel BTS-to-BTS interference.

8.7.6 Inter-Operator Synchronization

The synchronization and coordination requirements also need to be considered between different operators in the same area. The inter-operator interference case is illustrated in Figure 8.19. Operators 1 and 2 have dedicated frequencies that are adjacent. The frequency domain filtering is not ideal, and there is some leakage between the frequencies. Figure 8.19 illustrates a case where two operators have their base stations co-sited and use adjacent frequencies. The base station transmission power can be more than 50 dBm, while the thermal noise level with 100 MHz bandwidth is less than −90 dBm. If we wish to have uncoordinated timing between the operators, the required isolation should be more than 140 dB. The adjacent channel leakage and selectivity can provide approximately 40 dB attenuation between frequencies, and we would need about 100 dB additional coupling loss between the two operator antennas. If the antennas are co-sited or located close to each other, it is not possible to guarantee such a high coupling loss. Therefore, the assumption is that adjacent operators need to synchronize their networks and apply the same asymmetry in macrocell deployments below 6 GHz bands.

We will next consider more detailed calculations about inter-frequency interference cases. The adjacent channel leakage ratio (ACLR) and adjacent channel selectivity (ACS) performance are specified separately both for BTS and UE transmitters and receivers by 3GPP. For LTE, it is defined in [6] and [7] and for 5G in [8] and [9]. Figure 8.20 shows the BTS-to-BTS increase in the noise floor assuming the ACLR is 45 dB, and the UE-to-UE increase in the noise floor assuming the ACLR is 30 dB.

The potential interference even with these rather modest assumptions in the victim BTS can be tens of dB above the thermal noise floor even several kilometers away. Similarly, below, the UE-to-UE interference can be severe even if the UEs are separated by

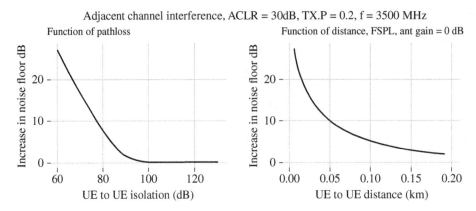

Figure 8.21 Adjacent channel UE-to-UE interference.

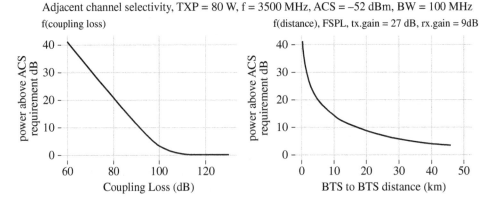

Figure 8.22 Adjacent channel receiver selectivity in BTS.

hundreds of meters if they are located in open space. At first sight, this may seem not too serious, but if, for example, a special event is considered where there can be thousands of UEs concentrated in small area, the service would be severely impacted. As the ACS performance requirement in 3GPP is very similar to the ACLR requirement, only ACLR pictures are shown here (Figure 8.21).

The ACS is a measure of the receiver's ability to receive a wanted signal at its channel frequency at the antenna connector in the presence of an adjacent channel signal, and these are also specified in [8] and [9]. Like the ACLR, the ACS would be a potential cause of degraded performance in BTSs up to several kilometers away in case there is no synchronization and coordination between one's own and adjacent channel operation. Interference considering ACS in BTS reception is shown in Figure 8.22 and in UE reception is shown in Figure 8.23.

UE-to-UE ACS poses a smaller problem but would cause issues, for example, in stadiums where a high concentration of devices would have LoS between each other.

8.7.7 Synchronization Requirements in 3GPP

There are a number of requirements defined in 3GPP that lead to the necessity for phase synchronization in 5G TDD networks.

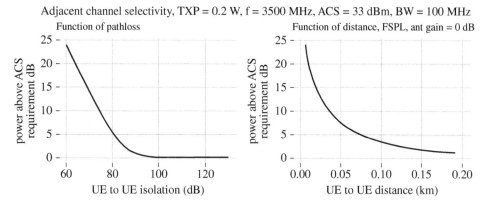

Adjacent channel selectivity, TXP = 0.2 W, f = 3500 MHz, ACS = 33 dBm, BW = 100 MHz

Figure 8.23 Adjacent channel receiver selectivity in UE.

8.7.7.1 Cell Phase Synchronization Accuracy

This is probably the most important requirement because most of the early 5G systems will be based on TDD frequency bands where the downlink and uplink share the same frequency. The 3GPP specification [10] defines the cell phase synchronization accuracy for TDD as the maximum absolute deviation in the frame start timing between any pair of cells on the same frequency that have overlapping coverage areas. The minimum requirements of the cell phase synchronization accuracy measured at base station antenna connectors is defined as better than 3 μs. The cell phase synchronization accuracy was defined based on the different downlink/uplink interference scenarios mentioned in the aforementioned chapters, where several important factors such as the cell size, timing advance, BTS switching time, UE switching time, and the GP duration were taken into consideration. The value of 3 μs is applicable for all subcarrier spacings defined in 3GPP Release 15, that is, 15, 30, 60, and 120 kHz. It is important to note that due to the short symbol duration of 60 and the 120 kHz subcarrier spacing, a minimum of two GP symbols is required to avoid downlink–uplink interference.

8.7.7.2 Maximum Receive Timing Difference (MRTD) for LTE–5G Dual Connectivity

The 3GPP specification [10] defines the MRTD as the relative timing difference between the subframe timing boundary of an E-UTRA Pcell and the slot timing boundaries of a PScell to be aggregated for EN-DC (E-UTRA-NR Dual Connectivity) that the UE is capable of handling. Though the MRTD is defined as a UE requirement, it indirectly defines the phase synchronization requirement for base stations as well because any phase synchronization error between LTE and 5G base stations will directly impact the MRTD measured at the UE. The 3GPP specification [10] defines the MRTD for inter-band asynchronous EN-DC to be 500, 250, 125, and 62.5 μs for 15, 30, 60, and 120 kHz subcarrier spacings, respectively. These MRTD values are half of the slot duration of their corresponding subcarrier spacings hence they will always be met for any kind of deployment. Therefore, phase synchronization between LTE and 5G base stations are not required for inter-band EN-DC. For intra-band EN-DC, [10] defines support of synchronous EN-DC in the UE as mandatory with an MRTD requirement of 3 μs, assuming that the LTE and 5G base stations are collocated, whereas asynchronous EN-DC needs to be explicitly signaled by the UE to the network via Radio Resource Control (RRC) signaling . It is important to note that LTE may still require certain timing and phase accuracy in performing an aligned measurement gap to ensure good performance for the EN-DC setup.

8.7.8 Synchronization from Global Navigation Satellite System (GNSS)

The GNSS is a general term describing any satellite constellation that provides positioning, navigation, and timing services on a global basis. It is probably the most accurate timing system that is available globally and is economical for use by telecom operators. Different countries may want to have their own GNSS which can be operated independently of other countries. The most popular system today is GPS, but there are a few other global options as well:

- GPS is operated by the US military and is the most popular GNSS system worldwide.
- GLONASS is a global GNSS owned and operated by the Russian Federation.
- Beidou is a regional GNSS owned and operated by the People's Republic of China. China is currently expanding the system to provide global coverage by 2020.
- Galileo is a global GNSS owned and operated by the European Union. The system is planned to be completed by 2020.
- Quasi-Zenith Satellite system (QZSS) is a regional GNSS owned by the Government of Japan and operated by Quasi-Zenith Satellite (QZS) System Service (QSS) Inc. Japan plans to have an operational constellation of four satellites by 2018, expanded to seven satellites by 2023.

The GNSS solution involves connecting the GNSS antenna/receiver to the base station. The GNSS receiver is a device that receives and digitally processes the signals from a GNSS satellite constellation in order to provide position and accurate time information to the base station. The GNSS receiver can be a single device that integrates both the antenna and receiver in itself or it can be separated from the antenna. In the latter case, the GNSS receiver is connected to the antenna by a coaxial cable and provides power directly to antenna. The GNSS receiver needs to be installed with clear satellite visibility and can be directly connected to the Synchronization Input interface of the base station. Power for the GNSS receiver is supported through the combined power and control cable connected to the Sync Input interface of the base station. The base station provides an integrated power feeding to the GNSS receiver enabling cable lengths of up to 300 m (Figure 8.24).

The GNSS synchronization solution has many benefits. It can provide the highest level of synchronization accuracy required by the most demanding applications in mobile networks such as Observed Time Difference of Arrival (OTDOA), which is required by the regulator in the United States to meet E911 requirements. OTDOA is used to determine the position of the UE with an accuracy in the range of 50 m. To achieve this high level of precision, it is required that the base station must achieve a minimum phase accuracy of 100 ns. This can be effectively provided by the GNSS solution. The GNSS solution therefore can easily provide ±1.5 μs phase accuracy to each base station.

The accuracy and stability of the GNSS solution are also independent of other operational aspects in mobile networks. They are not impacted by the downtime of other network elements, traffic congestion in the radio interface or transmission network,

Figure 8.24 GNSS synchronization solution.

GNSS receiver

Power and ↑   *Synchronization*
control to   *signal to base*
GNSS   ↓ *station*

Base station
system module

and the size of the mobile network. The GNSS synchronization solution is therefore a very attractive solution for providing a highly accurate synchronization solution with performance that is independent of the operation of mobile network.

Nevertheless, there are also a number of considerations that can be considered as disadvantages of the GNSS solution. They can, however, be viewed as subjective because they may depend on the operator's environment, which differs from one operator to another.

- The GNSS solution is viewed as an expensive solution because of the additional cost of the GNSS receiver and the installation costs that the operator need to incur for every base station site.
- The deployment of the GNSS receiver can sometimes be difficult because it often requires clear satellite visibility. This means that it may not be possible to install the GNSS receiver in locations such as an indoor building, urban canyon, or tunnel, where clear satellite visibility will be an issue.
- The accuracy and stability of the GNSS solution may be impacted by factors that are outside the operators' control such as bad weather condition, GPS jamming, or GPS spoofing.
- Problems such as GPS jamming can be very common in some countries, for example, via criminal activities or via military exercises. In the worst case, it can bring the whole network down, as the base station may have to shut itself down to avoid further interference to the system.

8.7.9 Synchronization with ToP

The disadvantages of the GNSS solution mentioned above create a demand for an alternative synchronization solution that should be cheaper, have greater site deployment flexibility, and can be controlled fully by the operator. ToP is a solution where phase synchronization can be obtained from the transmission network or backhaul network and can meet the aforementioned requirements. The ToP with phase synchronization solution is based on the IEEE1588-2008 standard and follows the ITU-T profile. The relationship between IEEE1588-2008 and ITU-T requires some explanation. IEEE1588-2008 is a standard which defines a set of methods and attributes to be used to achieve timing accuracy in a packet network. It is not designed specifically for any application. It can be used as well in industry for automation purposes. For IEEE1588-2008 to be usable in real-life networks, the standard requires that a "profile" be created to meet the requirement of each application. ITU-T, which is a telecommunications standardization body, then takes the IEEE1588-2008 standard and defines a set of profiles to be used for telecommunications purposes. There are some telecom vendors that do not support ITU-T profiles and instead use their proprietary profile and claim IEEE1588-2008 compliance. Such equipment may have interoperability issues with equipment that support ITU-T telecom profiles. Operators should therefore strictly follow ITU-T profiles for the deployment of the ToP with phase synchronization solution. Apart from the set of functionalities defined in ITU-T that shall be implemented in network elements such as the base station, transmission equipment, and Grandmaster, ITU-T also defines network recommendations for phase synchronization deployment that operators can follow to ensure successful deployment of ToP with phase synchronization.

An overview of the ToP solution is shown in Figure 8.25. The ToP with phase synchronization solution is an end-to-end solution which basically consists of following functionalities/network elements:

- The grandmaster clock is a network element that is connected to the GNSS receiver and provides precision timing protocol (PTP) packets over the IP/Ethernet interface toward the transmission network. The grandmaster clock is able to maintain timing even during quite long GPS jamming.
- The boundary clock is a network element which consists of both master and slave functionalities. At the slave side, it receives a PTP packet from the grandmaster or another boundary clock and terminate the PTP protocol and tries to estimate the correct time from the master. At the master side, a new PTP packet is created based on the timing information of the boundary clock and sent out to the next boundary clock in the chain or slave. In short, each boundary clock tries to obtain the accurate time information from the master and pass it to the next boundary clock/slave in the chain. Boundary clock functionality can be integrated into the switch or router in the transmission network.
- The transparent clock is a network element that transparently forwards the PTP packet it receives from one port to another without terminating the PTP session. The transparent clock, however, adds residence delay (i.e. the duration of the PTP packets residence in the transparent clock) to the PTP packets so that the next slave in the chain can use this information to calculate the correct time. Transparent clock functionality can be integrated into the switch or router in the transmission network.
- The slave is a network element that terminates the PTP protocol and tries to estimate the correct time from the master. A BTS is a network element that consists of both slave functionality and the end application. An end application takes the final time information from the slave for its usage.
- Full on-path support – every transmission hop between the grandmaster and slave (BTS) must be equipped with either boundary clock functionality or transparent clock functionality.
- Partial on-path support – Some of the transmission hops between the grandmaster and slave (BTS) can be equipped with boundary clock functionality.

Figure 8.25 ToP with phase synchronization solution.

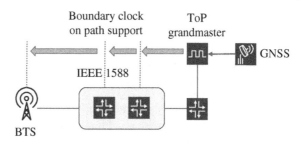

ITU-T defines the following two sets of telecom standards to be used for full on-path support and partial on-path support, respectively:

- ITU-T G.8271.1: Network limits for time synchronization in a packet network with full timing support from the network [11]
- ITU-T G.8275.1: Telecom profile for phase synchronization with full on-path support [12]
- ITU-T G.8271.2: Network limits for time synchronization in a packet network with partial timing support from the network [13]
- ITU-T G.8275.2: Telecom profile for phase synchronization with partial on-path support [14]

The first thing that the operator must know is the phase synchronization accuracy requirement of the end application. Only then can the operator start detailed planning of the solution and decide which solution can be used. In TD-LTE, the phase sync requirement is ±1.5 μs. The ITU-T has correspondingly designed ITU-T G.8271.1 and ITU-T G.8271.2 to meet ±1.5 μs requirement at the end application. ITU-T G.8271.1 designs, for example, the maximum number of boundary clocks that can be supported between the grandmaster and slave to meet the ±1.5 μs requirement. The operator can therefore easily follow this network limit definition to roll out the TD-LTE network. Fortunately, the 5G phase synchronization requirement is the same as for TD-LTE, that is, ±1.5 μs, and hence both ITU-T G.8271.1 and ITU-T G.8271.2 can be fully reused for 5G.

Second, the operator needs to decide if it will provide full on-path support or partial on-path support. Full on-path support provides a more reliable, more accurate phase synchronization solution but comes at probably higher cost of deployment, as every transmission hops must support the boundary clock/transparent clock. It may also be easier to deploy this solution as the accuracy achievable at the slave (BTS) can be concretely estimated from the number of hops. The details of how each hop contributes to the total phase error are clearly described in ITU-T G.8271.1.

A partial timing support is a cheaper solution because of not all network elements need to support the boundary clock. However, it is indeed not straightforward to design a network with partial timing support: for example, what is the maximum number of hops we can have in the chain without the boundary clock? ITU-T G.8271.2 defines a metric called "pktSelected2wayTE" to measure the phase error of the partial timing support network and help the operator design their network. However, to measure this metric, the operator will require quite detailed analysis of the end-to-end delay in both the uplink and downlink directions of the transmission network and hence a good level of competency is required. This task can be accomplished with dedicated equipment which provides the pktSelected2wayTE measurement; however, the equipment is considered quite expensive, and its correct handling will require some technical knowledge. It is also important to note that the design may need to be done more than once, for example, due to transmission network changes or traffic profile changes.

For these reasons, it is recommended to deploy full on-path support if possible. It will greatly help the operator to ease the deployment and operational effort. It is important to note that there would, of course, be the case where partial timing support may be unavoidable, for example, a leased line transmission network.

Third, to ensure full interoperability of all network elements and to meet the designed limit provided by ITU-T G.8271.1/ITU-T G.8271.2, the operator should respectively use equipment and configurations that are compliant with ITU-T G.8275.1/ITU-T G.8275.2.

Finally, the operator needs to ensure that the required accuracy can be achieved and maintained. There could still be a number of problems that could happen in the ToP with phase synchronization solution that operators need to be aware of:

- IEEE1588-2008 works based on the assumption that the network is symmetric; that is, the uplink constant delay and downlink constant delay must be the same. The operator needs to ensure that if asymmetric transport equipment like a microwave radio or Gigabit Passive Optical Network (GPON) is used, they must be equipped with the boundary clock solution to solve the asymmetry issue.
- Even if we are using symmetric transmission technology like fiber in IP routing, asymmetry could be caused by different fiber lengths in the uplink and downlink directions. This could happen because normally two fibers are required for the uplink and downlink directions. The operator needs to ensure that such "constant" asymmetry will be properly measured and compensated in the boundary clock or slave.
- The operator should avoid fully utilizing the transmission interface as a congested interface will cause additional queueing time in the transmission equipment, and that could impact the synchronization accuracy. It is recommended to dimension the interface such that less than 80% of the interface capacity will be used for all traffic.
- The operator should properly plan the backup transmission route to be ready to support ToP with phase synchronization as well so that the synchronization will not be impacted when transmission re-routing takes place.
- It is highly recommended to check the synchronization accuracy at the slave. This can be easily done by connecting the BTS with both the GNSS receiver and ToP with phase synchronization.
- It is important to note that 5G is still in the early phase, and the phase sync requirements may change, for example, when more advanced features are introduced. In that case, the existing ToP with phase synchronization transmission network planned for the ±1.5 μs phase error budget may need to be re-planned.

8.7.10 Timing Alignment Between Vendors

Even though each vendor can use the same source of timing, the processing delays between equipment may differ, and it is necessary to align the timing for each carrier on the same band. This can be done in the laboratory, where the BTSs provided by all vendors are set up. A power probe is installed in each BTS transmitting antenna port and

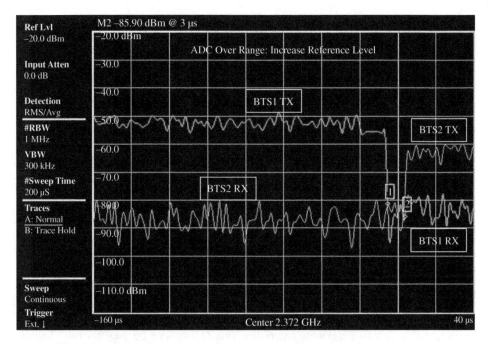

Figure 8.26 Two nearly "perfectly" misaligned TDD BTSs on oscilloscope screen.

connected to an oscilloscope. The timing is then adjusted by tuning a BTS parameter that offsets the timing.

The following figure shows an example of poor alignment where BTS2 starts transmission when BTS1 has transferred to the receiving mode. This would result in massive interference in both directions. Transmissions and switching to the receiving mode should start at the same time (Figure 8.26).

The synchronization process between eNodeB's, for both the intra- and inter-frequency cases, could also be automated in the future by utilizing the receiver in the BTS. The BTS could be switched to UE-receive-only mode for a period during which it could measure the timing differences on the same frequency and on other frequencies on the same band and potentially automate the timing alignment in TDD networks.

8.8 5G Overlay with Another Vendor LTE

This section discusses options how to implement the 5G network using radio vendors other than the existing LTE network. The target is to introduce a new radio network vendor with 5G without significant support from an incumbent LTE vendor, and without significantly impacting the resultant network performance and user experience.

The first phase of 5G is deployed with a non-standalone architecture (NSA), which couples the LTE and 5G radio layers together. The two layers are logically joined via an X2 connection, which allows for dual connectivity between the LTE and 5G radios. While this X2 connection is specified by 3GPP, LTE and 5G interworking between two

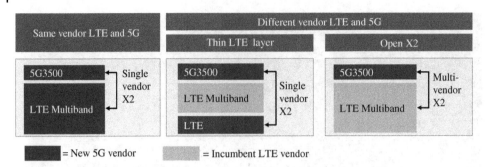

Figure 8.27 Two options for 5G overlay from a new vendor on top of LTE network from an existing incumbent vendor.

different vendors may be challenging due to the prevalence of vendor-specific implementations, resulting in the need to exchange vendor-specific algorithms. If the operator would like to avoid inter-vendor X2 testing, then another option is to bring a thin layer of LTE (in the current 700, 1800, or 2100 MHz bands, for example) that can be deployed together with a new mid-band 5G carrier. This thin layer provides the anchor point for the LTE coverage network and ensures that many of the benefits of NSA architectures can be realized by new 5G-enabled mobile devices, without compromising the user experience of legacy LTE devices. The redirection feature can be used to move legacy LTE mobiles to the LTE multiband network, while 5G-capable mobiles are moved to the thin layer of LTE. These two multivendor options are illustrated in Figure 8.27.

A traditional NSA-based solution can combine LTE and 5G data rates together in dual connectivity mode, allowing the user to leverage the full LTE+5G network capacity. This benefit can be obtained with a single vendor solution and with a multivendor solution based on an open X2 interface. If a new vendor with a thin LTE layer is introduced, LTE resources cannot be aggregated between the two vendors, and the aggregated data rate will be lower than with a single vendor. The thin layer solution may still be attractive since a large 5G carrier alone can bring very high data rates without LTE aggregation.

New 5G services require a new 5G core network to fully realize the flexibility and capability of 5G including network slicing. The 5G core is designed with standalone (SA) radio architectures as the primary focus. In order to provide complete coverage for standalone 5G, low frequency bands must be utilized for 5G. The thin layer of LTE can be converted into a 5G layer using DSS. This need to upgrade radio units to support 5G and DSS also affords the opportunity to completely replace the incumbent vendor with another vendor's solution that will allow legacy LTE users and 5G users and both NSA and SA modes to be fully supported with a simplified cell site configuration, as shown in Figure 8.28.

8.9 Summary

This chapter presented a few important factors that have a major impact on 5G network performance and deployment complexity. Spectrum resources and base station density are relevant factors in defining how much capacity the radio network is able to provide. More spectrum and more base stations lead to higher capacity. That is also visible in the

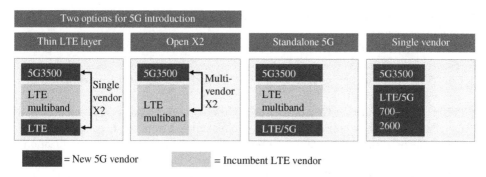

Figure 8.28 Long-term migration plans for a new vendor 5G overlay.

current LTE networks where the highest data volumes tend to occur in those countries where the network density is relatively high and spectrum resources are relative large. The chapter discussed practical base station and site solutions. The evolution leads to close integration of RF and part of baseband functionalities to the antenna, which can improve the radio performance and minimize the site complexity. The chapter presented network synchronization requirements and solutions. All TDD networks, including 5G TDD, require accurate phase synchronization between base stations and also between operators running their networks in the same frequency band. In addition, the downlink and uplink must have the same asymmetry. Network synchronization can be obtained by GPS and by ToP via a transport network.

References

1 Baracca, P., Weber, A. Wild, T., Grangeat, C. 2018. A Statistical Approach for RF Exposure Compliance Boundary Design in Massive MIMO Systems, in 22nd International ITG Workshop on Smart Antennas (WSA 2018), Bochum, Germany, March 2018.

2 IEC62232:2017 (2017). *Determination of RF Field Strength, Power Density and SAR in the Vicinity of Radiocommunication Base Stations for the Purpose of Evaluating Human Exposure.* International Electrotechnical Commission.

3 IEC TR62669:2019 (2019). *Case Studies Supporting IEC 62232 - Determination of RF Field Strength, Power Density and SAR in the Vicinity of Radiocommunication Base Stations for the Purpose of Evaluating Human Exposure.* International Electrotechnical Commission.

4 3GPP 36.213 Technical Specifications (2019). Evolved Universal Terrestrial Radio Access (E-UTRA), Physical layer procedures.

5 3GPP 38.211 Technical Specifications (2019). NR; Physical channels and modulation.

6 3GPP 36.101 Technical Specifications (2019). Evolved Universal Terrestrial Radio Access (E-UTRA); User Equipment (UE) radio transmission and reception.

7 3GPP 36.104 Technical Specifications (2019). Evolved Universal Terrestrial Radio Access (E-UTRA); Base Station (BS) radio transmission and reception.

8 3GPP 38.101-1 Technical Specifications (2019).NR; User Equipment (UE) radio transmission and reception; Part 1: Range 1 Standalone.

9 3GPP 38.104 Technical Specifications (2019). NR; Base Station (BS) radio transmission and reception.

10 3GPP 38.133 Technical Specifications (2019).NR; Requirements for support of radio resource management.

11 ITU-T Recommendation G.8271.1. (2017). "Network limits for time synchronization in packet networks".

12 ITU-T Recommendation G.8275.1. (2016). "Precision time protocol telecom profile for phase/time synchronization with full timing support from the network".

13 ITU-T Recommendation G.8271.2. (2017). "Network limits for time synchronization in packet networks with partial timing support from the network".

14 ITU-T Recommendation G.8275.2. (2016). "Precision time protocol telecom profile for time/phase synchronization with partial timing support from the network".

9

Transport

Esa Markus Metsälä and Juha Salmelin

Nokia, Finland

CHAPTER MENU

9.1 5G Transport Network

9.1.1 5G Transport

Transport networks for 5G will be more versatile than today and characterized by the different 5G use cases and different network architectures (as described in Chapters 1 and 5) and also by different kinds of operator networks and different operating models.

In addition to reducing latency and increasing capacity and user peak rates of the mobile network to a new level, there is a strong tendency to simplify the mobile backhaul and make it robust against failures and abnormal situations, so that mobile services run uninterrupted over the 5G transport, while minimizing transport operational needs. Security is vital as well as resiliency.

From a networking technology standpoint, there at first sight is no dramatic change from LTE where already all the 3GPP logical interfaces were IP (Internet Protocol) based. A special case is the low layer split point, which is now with 5G evolving from the legacy Common Public Radio Interface (CPRI)/Open Base Station Architecture Initiative (OBSAI) to standard networking technologies.

However, a major evolution in technology is still underway as connectivity relies more and more on IP, due to the availability of IP services for resiliency, security, flexibility, and scalability, with operational and troubleshooting tools. IP (natively or with Multi-protocol Label Switching MPLS) is a common building block for transport services for

5G Technology: 3GPP Evolution to 5G-Advanced, Second Edition.
Edited by Harri Holma, Antti Toskala, and Takehiro Nakamura.
© 2024 John Wiley & Sons Ltd. Published 2024 by John Wiley & Sons Ltd.

mobile backhaul and is needed even more for 5G. At the same time, low speed links, old transport nodes, and technologies may gradually be retired from the network during the upgrade phase of 5G and replaced with modern solutions.

The optical layer, besides supplying large bandwidth for communications, is becoming to a greater extent capable of solving connectivity needs directly in the photonic domain, with wavelength routing and wavelength-selective switching technologies, protection capabilities, and OAM (Operation, Administration, and Maintenance).

Solutions for high capacity links are also offered by wireless media. Traditional microwave frequencies have limited bandwidth, but higher bands (e.g. E-band, W-band, D-band) have enough spectrum so that wireless links with 10G and even higher capacities are possible.

Combining optical and wireless domains with IP services works very well as a networking solution.

In IP networking, a recent development has extended IP MPLS VPN (virtual private network) principles to Ethernet services with E-VPN (Ethernet VPN), bringing to users of Ethernet service capabilities from the IP domain like multi-homing and load balancing. In another networking area, deterministic network behavior is progressing in IEEE (Institute of Electrical and Electronics Engineers) and IETF (Internet Engineering Task Force), with tools for covering new use cases, for example, in the industrial area and also in 5G low layer fronthaul.

With cloud deployments, 5G Core and radio network functions are virtualized, and these functions may be located more freely than previously. Connectivity in the transport network must be flexible so that these functions may be relocated dynamically, with the help of SDN (Software Defined Networking).

This all makes the 5G transport network more heterogeneous and flexible, as the transport network should support not only the first phases of 5G, but also include capabilities to adapt to future needs and 5G use cases, as transport network redesigns are costly and time-consuming.

Some building blocks here relate to having the following:

1. Physical media with adequate bandwidth for the links, in order to be able to increase capacity without excessive lead times
2. A networking connectivity model which is flexible and resilient in order to arrange connectivity between any locations as wished; and in order to support service availability in midst of a failure of any critical link or node
3. Protected network resources and network domain communications, in order to block any unintended access and cryptographically protect communications on links
4. A culture of consistently documenting, following, and enforcing sound design principles

Against all the diverse needs and different cases, a range of physical media and networking technologies and solutions exist, and those need to be assessed against the intended target state, with a time scope of at least some years ahead.

9.1.2 Types of 5G Transport

The majority of the sub-6-GHz macro-gNBs will be deployed to the same site as LTE. It is possible to use the existing LTE transport if the capacity is sufficient, but quite often the transport capacity need to be upgraded, for example, from 1 Gbps to 10 Gbps. Copper or microwave radios (MWRs) need to be changed to fiber or millimeter wave radios for more bandwidth.

Small cells are needed to add a capacity layer under the macro base stations. A new transport link is needed for the new cells. If the target is not to increase capacity but to fill coverage holes, Integrated Access and Backhaul (IAB) may be used.

Small cells using mmW frequency have a very small cell size. A completely new and dense transport network is needed. Those gNBs that are not in the hottest spots may use IAB.

IAB is a connection without any cost, but it consumes part of the access capacity. Copper is the second-lowest cost solution, but capacity needs are limiting the distance to meters or tens of meters. mmW E-band radio needs line of sight (LoS) and a quite stable installation, but capacity reaches 10 Gbps. In the future, D-band (140 GHz) radios will be beam-steerable, enabling easy alignment, tolerating installations on swaying light poles and easily supporting 25 Gbps and more, but only with limited distances like below a couple of hundred meters. The highest capacities are reached with optical cabling, but often the digging or leasing costs are limiting usage.

MEC (Multiaccess Edge Computing) can be used if low latency is needed and/or the transport capacity is limited. MEC can be divided into core, edge, and even to far edge like uMEC in the 5G light poles.

For private networks of factories or harbors, completely new transport networks are needed, but the investment is not huge as only a couple of gNBs are needed in the limited factory area in a typical case. For ultra-reliability, protected transport links are used.

For city networks with mmW gNBs, a complete new transport network is needed, and the cost may be huge. One solution is to use light poles for sites and create a shared network together with mobile operators and a city. In this case, only one shared transport network is needed. The needed transport capacity is bigger than just the traffic from gNBs since typically IoT (Internet of Things) traffic from the smart light poles are using the same transport. The city with a need to boost the digital ecosystem in the city area may share the costs of sites and fiber transport. The city anyway needs a network for the sensors and cameras to meet the UN (United Nations) sustainable development goals (SDGs) in the future (see [1]). Combining multiple networks into a single one benefits mobile operators as well as the city.

Indoor networks have only a limited possibility of using existing LAN (local area network) networks. Quite often, there is not enough capacity technically available, but the new multi-Gbps LAN technology enables at least the low end of 5G capacities. Otherwise a new transport network is needed. IAB can be used to connect those gNBs that have LoS to each other. Window CPE (Customer Premises Equipment) can be used to get the signal from an outdoor gNB.

9.1.3 Own Versus Leased Transport

Transport network strategies are often broadly categorized into two, according to whether the mobile operator owns and operates its own transport network, or whether the transport links are leased from a third party connectivity service provider. Traditionally, incumbent operators have more of fixed infrastructure like optical or copper cabling and related sites, while mobile-only operators have far less or none of such infrastructure and have relied on leased lines and possibly on self-built and self-operated microwave radio links.

In practice, mobile transport networks often partly use their own infrastructure, and partly leased services.

One technical impact is that with leased services monitoring of the SLA (service level agreement), any violation of it is of special importance, as is fault localization: it is required to be able to detect which side of the UNI (user to network interface) the fault is on.

In mobile networks, the transport network may be shared along with the shared radio network infrastructure with two or more mobile operators. Sharing the infrastructure in some way is expected to grow. Furthermore, many networks are operated as a service. All this blurs the otherwise clear distinction own and leased transport.

9.1.4 Common Transport

Common transport here refers to the fact that often multiple radio technologies like 2G, 3G, 4G, and 5G are supported partly or (almost) completely by the same transport network. On the base station site, several base stations may be deployed and combined by a transport device to a common uplink, or alternatively, a single multi-radio capable base station supports many radio standards with a common transport link.

In these cases, transport links and nodes are shared for all technologies. The common links will fulfill the needs for all of the radio technologies, meeting the specific demands for capacity, quality of service (QoS), availability, and security. With 5G, typically the existing network links need to be upgraded at least for capacity, and also assessed for other 5G demands like latency, connectivity, security, and availability.

When the access link or node is shared, failure on this common facility impacts all radio technologies. This needs to be considered in the availability analysis and design and addressed, especially if it is essential that at least one (or some) air interface service from the site continues its operation specifically if another technology experiences an outage.

9.1.5 Mobile Backhaul Tiers

A transport network may be categorized into separate network tiers: access, distribution, and core, which all have distinctive characteristics.

Figure 9.1 shows three tiers of mobile backhaul, partitioning the e2e network into tiers with distinct characteristics, based on tiered network architecture, see e.g. [2].

In the access tier, individual base stations (macro- or small cell gNBs), distributed (gNB-DUs) units, as well as remote radio units (RUs) are connected to the network via first mile[1] access links. The access tier may consist of just a single link, a few links, and also possibly of some site switches/routers functioning as small access hubs. Access links may be – and often are – supplied by wireless transmission, otherwise fiber is the first option, wherever available.

The access tier is very important, as a major part of the transport network cost originates from here. Access links are needed in volumes, and even if the access link is shared

---

1 The term "last mile" is used as well. The terms "first mile" and "last mile" both refer to the link connecting the access device (base station in the 5G case) to the network.

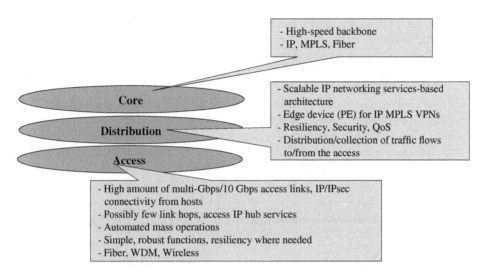

Figure 9.1 Mobile backhaul network tiers.

for multiple base stations on the same site, there are still a large number of sites in any practical network that need to have connectivity to the network. The high capacity demanded by 5G translates into high bandwidth need on the access links, which in turn drives up the cost of those links. A single access link needs to supply at least the single user peak rate, which can be high with 5G. Wavelengths and WDM (wavelength division multiplexing) will be used more and more in the access to reduce the amount of fiber needed. With small cell deployments (mmWave 5G bands), many of the requirements are even more demanding. Access tier infrastructure will also be shared more and more for multiple usages, due to the cost and difficulty of supplying the access links.

In the access tier, operability is of specific importance. Configuration and operation of the access tier has to be automated and as simple to operate as possible with troubleshooting capabilities.

In the distribution tier, traffic flows from the access hub sites, and access links are collected. The architecture is scalable and resilient, and basically an arbitrary number of new access regions and sites can be added with connectivity services based on IP defined as needed between sites, for example, with IP MPLS VPN service (PE, provider edge device). In many cases, IP services will be supported already in the access tier for horizontal X2/Xn traffic flows, resiliency, and security.

Central IPsec GW (SEG) and gnB-CU (Centralized Unit) from the mobile network elements would also often be located in the distribution tier (with central SEG alternatively on the core tier). QoS-, security-, and resiliency-related configurations need to be implemented on the edge devices as well as on any access tier hubs. WDM access links may also terminate here and interface the IP service router for horizontal connectivity and locally processed traffic flow (gNB-CU). Statistical multiplexing is possible, with the condition that not all access links need full bandwidth at the same instant.

The backbone/core tier is a high capacity, resilient, and secure fast backbone optimized for wide area/countrywide connectivity to the core network elements such as the UPF (User Plane Function) and AMF (Access Management Function). Non-mobile network traffic is often carried by the same backbone: enterprise traffic, fixed residential Internet, and so on.

For 5G there is more variability in the use cases and in deployment options than ever before, and the location of the 5G network elements in the mobile backhaul tiers needs to be adapted so that it fits the targeted use case, and vice versa, some functions on the mobile backhaul tiers may need to be placed deeper toward, or into the access tier.

For example, low-latency applications require the core network function (UPF) to be located close to the UE, which can be supported by having the Mobile Edge Computing (MEC) function in the distribution or even in the access tier. For ultra-reliable communications, it may be necessary to have full resiliency already in the access tier and in the first mile access links, which in turn requires that the connectivity solution in the access be capable of resilient service. A basic prerequisite for Ultra-Reliable Low-Latency Communication (URLLC), but essential for the non-URLLC case as well, is that failure domains be limited, failures contained, and the network kept in control under all circumstances.

Every mobile transport network is a specific case, as the infrastructure on the current network often introduces some constraints or at a minimum provides a starting point for the 5G evolution. The intended target state may also differ. Many candidate technologies exist that may all be technically viable. Which alternative provides the best fit to the needs is very much case specific.

9.1.6 Logical and Physical Transport Topology

Logical connectivity, which refers to which 5G mobile elements need connectivity with each other, is different from the physical transport topology. Logical connectivity needs to follow 3GPP definitions for the interfaces between the mobile network elements.

The physical transmission topology is mostly irrelevant as long as logical (IP) connectivity exists between elements, for example, between the gNB and the core. The physical topology becomes interesting in certain cases, like optimizing propagation delay (since the actual cable distance on the signal path matters) and resiliency (since different paths should avoid fate-sharing with each other).

9.1.7 Standards Viewpoint

In regard to transport, standardization in 3GPP is concentrated on the definition of the protocol stacks for the logical interfaces of the mobile system, with references to the respective standards of that protocol, with those protocols typically standardized by IETF and IEEE.

How connectivity should be arranged between the mobile nodes, with what physical media or with what technology, is not defined by 3GPP, as long as the logical interface specification (protocol stack) is met. Since the logical interfaces rely on well-known networking protocols, IPv6, IPv4, and IPsec (IP security suite of protocols), networks, links, and equipment do not need to be mobile network specific. However additional functional requirements may be imposed from synchronization.

Figure 9.2 shows the scope of 3GPP work in the transport and networking areas. The division into RNL (Radio Network Layer) and TNL (Transport Network Layer) is shown. Not all relevant standards or forums are included.

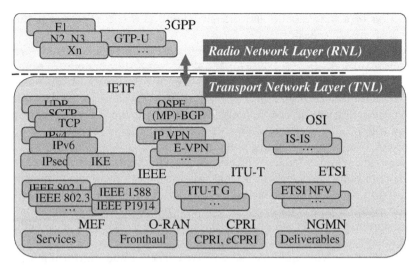

Figure 9.2 5G radio and transport standardization.

The clear division of standards into RNL and TNL allows RNL to be mostly agnostic to the TNL layer. This makes it simpler to adopt new TNL technologies or otherwise modify TNL behavior.

Especially in the area of transport network implementation and solutions for physical media, multiple standardization activities exist, for example, in ITU-T and IEEE, that should be referred to. On the networking layer, most of the standards are developed in IETF as RFCs (Request for Comments). Since the TNL layer apart from the logical interface definition is to a great extent an implementation topic, candidate solutions can be – and are – reused from IP enterprise networks, fixed residential Internet networks, optical transmission systems, and other transport and networking domains.

The low layer split point is a special case also in standardization: it is so far not defined at the protocol level by 3GPP, but instead there are industry forum standards and the IEEE task force. It also has special demands for capacity and latency.

9.2 Capacity and Latency

9.2.1 Transport Capacity Upgrades

The very first transport question for the 5G system is, how much capacity is needed for the transport network, for the first mile link, and from any hub point upward?

Different strategies for dimensioning are used, depending on the amount of information available, business strategies and targets, with the final result of the method used also in practice double-checked and possibly impacted by the resulting cost of the transport service. Typically, a major upgrade or strategic redesign cycle of the transport network occurs every once in a while, complemented with smaller incremental designs taking place more often.

During a major redesign, technology platforms possibly change, while the smaller cycles rather enhance the existing production network with changes of limited scope. Introduction of a new mobile network generation is a trigger to assess the future-proofness of current transport.

Overdimensioning is costly and should be avoided; on the other hand, any future capacity upgrade requiring site visits (e.g. for installing higher-bit-rate optical pluggables) becomes costly, too, especially if the upgrade concerns access links which are numerous. With the introduction of a new system, the traffic volume in the network grows with the penetration rate of the terminals. The take-up can, however, be rapid, as was experienced with LTE. The transport network has to be prepared in advance.

In the absence of a traffic profile (traffic volume and mix of services), the transport capacity need for a single (first mile) access link is estimated from the air interface capacity and from the requirement to support the single user peak rate. The single user peak rate is important, as it is easily observable for the user. With 5G, that rate is high, and some access links easily become bottlenecks for that service even for a single user, meaning that the maximum bit rate is not available through these sites unless access link capacity is first expanded.

At hub sites, traffic flows from multiple cell sites are collected, and statistical multiplexing is exploited to reduce capacity on the uplinks.

9.2.2 Access Link

For the single access link, a common way to establish the transport capacity based on air interface capacity is shown in Figure 9.3. The benefit of this approach is that the estimate can be done without any traffic mix assumption and that the access link then reflects the air interface capacity. The difficulty is that it is not in any direct way tied to real traffic demand and also that the air interface in practical networks is seldom in ideal conditions.

In Figure 9.3, for each gNB sector, a peak bit rate and an average bit rate is given. In addition, the single user peak rate is shown. Often the transport network is at minimum

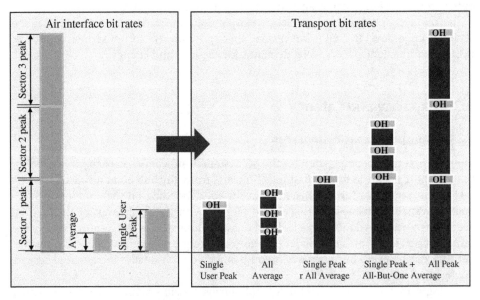

Figure 9.3 Dimensioning: example gNB configuration with three sectors.

required to support the maximum single user peak rate that the system and terminals offer. The single user peak rate then depends on terminal capabilities, and those may increase significantly over time.

To arrive at a transport capacity, protocol overhead (OH) is added, which is logical interface and traffic mix dependent.

All-average refers to a case where all sectors of the gNB transmit at an average rate (say, 0.2 of the peak rate) simultaneously. Single-peak, all-average takes the higher requirement from the two: either one sector peak or the sum of all sector average rates. Single-peak, all-but-one average refers to a single sector transmitting at the peak rate and all the others at an average rate. All sectors transmitting simultaneously at the maximum rate results in the all-peak scenario, which is the rather theoretical maximum rate.

9.2.3 Distribution Tier

The previous method gives an estimate for the capacity needed for the first mile link (single site). When traffic is combined from multiple access links in the distribution tier (or on any hub site), the statistical multiplexing gain reduces the capacity needed on the common uplink. The amount of gain depends on the traffic patterns, number of sites, and whether those sites experience peak traffic condition at the same time. Typically, sites in business areas are busy during working hours, while residential areas produce traffic peaks in the evenings.

When multiple sources queue for service such as egress port transmission, the delay increases rapidly when the utilization on the link grows above 70–80%. Thus, network links cannot be loaded up to full nominal capacity without causing delay, which is experienced especially by flows in the lowest priority queues.

9.2.4 Backhaul and High Layer Fronthaul Capacity

For 5G TDD duplexing, the capacity in the downlink (DL) or uplink (UL) direction also depends on the DL/UL ratio; here, 80% DL and 20% UL is assumed, so any symmetrical transport link will be downlink limited. The transport protocol overhead in the examples include IPv6, IPsec, and Ethernet overhead with VLANs (virtual local area networks).

The overhead in the downlink direction is 15–20% when the packet mix is dominated by large user packets, which is typically the case. In the uplink direction, the overhead is larger, assuming smaller user packets (such as TCP acknowledgements). Nevertheless, the downlink imposes higher requirements for transmission link capacity.

Figure 9.4 shows the transmission bit rates for the single-peak/all- average (Alternative A) and the single user peak rate requirement (Alternative B). Each has its merits; in practice, one consideration is how costly it is to carry out possible capacity upgrades in the future, if dimensioning follows the lower end.

With the assumption of a sub-6-GHz band, a macro-cell site (three-sector configuration, 100 MHz bandwidth) needs around 2.0–4.3 Gbps capacity. The single user peak rate (2 Gbps) requirement is lower than the single sector peak rate in this case. 1 Gbps Ethernet port speed is inadequate, and this makes the 10 Gbps port the preferred choice for F1 and NG interface links.

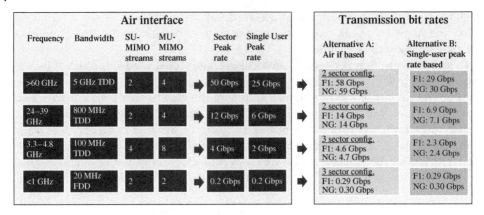

Air interface								Transmission bit rates	
Frequency	Bandwidth	SU-MIMO streams	MU-MIMO streams		Sector Peak rate	Single User Peak rate		Alternative A: Air if based	Alternative B: Single-user peak rate based
>60 GHz	5 GHz TDD	2	4	➡	50 Gbps	25 Gbps	➡	2 sector config. F1: 58 Gbps NG: 59 Gbps	F1: 29 Gbps NG: 30 Gbps
24–39 GHz	800 MHz TDD	2	4	➡	12 Gbps	6 Gbps	➡	2 sector config. F1: 14 Gbps NG: 14 Gbps	F1: 6.9 Gbps NG: 7.1 Gbps
3.3–4.8 GHz	100 MHz TDD	4	8	➡	4 Gbps	2 Gbps	➡	3 sector config. F1: 4.6 Gbps NG: 4.7 Gbps	F1: 2.3 Gbps NG: 2.4 Gbps
<1 GHz	20 MHz FDD	2	2	➡	0.2 Gbps	0.2 Gbps	➡	3 sector config. F1: 0.29 Gbps NG: 0.30 Gbps	F1: 0.29 Gbps NG: 0.30 Gbps

Figure 9.4 Backhaul capacity (NG, F1) requirement.

On higher frequency bands, capacities of the air interface grow rapidly, and tens of Gbps rates are needed to match the air interface, going to more than 50 Gbps if 5 GHz air bandwidth is available in the future with >60 GHz frequencies. The single user peak rate requirement on the 24–39 GHz band is 6 Gbps, which can be fulfilled by around 7 Gbps transport capacity and again with a 10 Gbps port.

In addition to user plane flows, transport links carry the mobile network control plane, management plane, some transport control flows, and possibly synchronization plane flows. While important in other respects, these traffic types typically require orders of magnitude less than what is required for the user plane. See an analysis of the bit rate of these flows in LTE in [3].

With dual connectivity (5G-5G or 5G-4G), the gNB may divert part of the traffic to the other base station, via the Xn/X2 interface. With X2, the 4G air interface may be used for additional throughput, in case 5G air interface conditions are poor. For this traffic, the X2 transport link needs to have the amount of bandwidth that 4G is estimated to be able to support. If the base stations are co-sited, that transport link is site internal. With 5G-5G, dual connectivity is supported by the Xn interface. The behavior of this interface is similar.

9.2.5 Low Layer Fronthaul Capacity

At the low layer split interface, the transport capacity depends on the exact split, how many bits are utilized for the sample data, whether compression is used, and on which protocol stack is deployed on that interface. This causes a range of results when calculating the required bit rates. Figure 9.5 shows a comparison between legacy CPRI and the O-RAN standard [4] with split option 7-2 (see Chapter 5, Figure 5.8 and the related discussion in Chapter 5 for split options). The transport overhead is included. In the given value range, the low value approximates the floor that can be reached, the value being very close to the backhaul requirement. The bit rates become even higher than the high value in the range if a large sample width (e.g. 16 bits) is deployed and compression is omitted.

The bold text in Figure 9.5 represents the user plane bit rate with compression of sample data. The CPRI rate is shown for comparison, and a significant reduction in the bit rate is evident.

	2 streams	4 streams	8 streams	16 streams
20 MHz	0,2–0,5 Gbit/s 16 Gbit/s (CPRI)	0,4–1 Gbit/s 32 Gbit/s (CPRI)	0.8–2 Gbit/s 64 Gbit/s (CPRI)	1.6–4 Gbit/s 128 Gbit/s (CPRI)
100 MHz	1–2,5 Gbit/s 80 Gbit/s (CPRI)	2–5 Gbit/s 160 Gbit/s (CPRI)	4–10 Gbit/s 320 Gbit/s (CPRI)	8–20 Gbit/s 640 Gbit/s (CPRI)
200 MHz	2–5 Gbit/s 320 Gbit/s (CPRI)	4–10 Gbit/s 320 Gbit/s (CPRI)	8–20 Gbit/s 640 Gbit/s (CPRI)	16–40 Gbit/s 1280 Gbit/s (CPRI)
800 MHz	8–20 Gbit/s 640 Gbit/s (CPRI)	16–40 Gbit/s 1280 Gbit/s (CPRI)	32–80 Gbit/s 2560 Gbit/s (CPRI)	64–160 Gbit/s 5120 Gbit/s (CPRI)

Figure 9.5 Low layer split point capacities.

Even with this, the RU capacity can be very high if the air interface bandwidth is large and there are many streams. The remote antenna (RU) needs n × 10G/25G Ethernet ports to connect to the network. For a site with multiple RUs, the capacities need to be supplied for each RU.

While CPRI and OBSAI use a constant bit stream continuously between the antenna and the baseband, the O-RAN specification is frame/packet based. What is common is that radio L2/L1 scheduling takes place in the baseband over the low layer split point interface, and all scheduled transmissions should be received in time by the RU, calling for quite deterministic behavior on the transport link.

9.2.6 Latency

One key 5G objective is reducing the e2e latency of user communication. Transport contributes significantly to this, since for long distances, propagation delay alone becomes a major factor. Wireless transmission as a media has an inherent advantage here over fiber/copper, since the signal propagation delay is smaller in the air (3.3 vs. 5 µs/km)

In addition to the e2e latency, any transport link on a 3GPP logical interface has to also meet the system requirement: how long a delay can be tolerated with the 5G system interface still remaining operational. Both the e2e latency and the system operation aspect are relevant, and which one becomes more stringent is interface dependent.

The e2e latency targets for 5G are documented, for example, in 3GPP TR38.913, "Study on Scenarios and Requirements for Next Generation Access Technologies" [5], and in "NGMN 5G Extreme Requirements: End-to-End Considerations" [6]). The performance chapter (Chapter 10) includes further analysis. 3GPP documents 8 ms (with only the radio stack considered), while 10 ms is assumed for the radio access network as an example in NGMN. For the URLLC service, the target is 1–2 ms.

The E2E latency is important as a reduction in the latency improves the performance of many applications, and some of them are not feasible unless the latency is low enough. Propagation delay is unavoidable in transport links (although paths can be optimized), while low speed or congested links, excessive buffers, and complex operations in the data plane of transport and network devices can – and should – be completely avoided. Even a single low-performance node or link in the transport network may in the worst case introduce significant unnecessary latency for a large number of users, as every transport link serves an aggregate of end users.

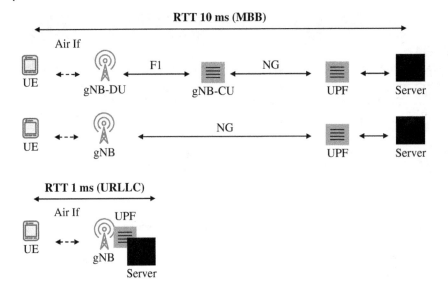

Figure 9.6 RTT components.

The round trip time (RTT) budget is divided into the air interface, UE processing, baseband processing, transport legs, core function, and connectivity to the server, as shown in Figure 9.6.

For the 10 ms e2e target of mobile broadband (MBB), the transport legs on F1 and NG may not contribute more than a few milliseconds (e.g. 1–4 ms RTT) to the latency, depending on the latency consumed by the other components. Transport needs to be assessed as part of the whole system.

As in Figure 9.6, for the e2e latency, the low-latency service necessitates an architecture where the UPF, gNB, and server are collapsed and present in the same site or anyway close to the radio, and also the transport architecture is different from that of the MBB.

From the system operation viewpoint, the F1 and NG interfaces tolerate longer than 1–4 ms RTT. All the delay-critical radio functions (fast retransmissions, radio scheduling, Radio Link Control [RLC] protocol) are located in the gNB-DU. That makes the F1 and NG interfaces non-delay-critical. An increase in the delay on these interfaces mainly impacts the e2e delay and performance.

Low-latency fronthaul instead is delay critical from the system operation aspect. Since now the delay-critical radio layers and functions (scheduling, L1 retransmissions, etc.) operate over the low layer fronthaul, the delay tolerance is limited to around 50–100 μs or up to a few hundreds of microseconds RTT, depending on the frequency band (subcarrier spacing), TTI (transmission time interval), and the number of Hybrid-ARQ processes assumed at the terminal.

9.2.7 QoS Marking

In the 5G Core, QoS is flow based, based on packet filters and individual flows within a PDU session. The related identifiers are the QoS flow identifier QFI, for the QoS profile, and the 5G QoS identifier 5QI, for the QoS characteristics of a flow. The N2 (NG-C) interface signals these identifiers to the 5G RAN.

In transport, QoS is mapped to the traffic class field of the IPv6 packet, or the type of service field of the IPv4 packet, and further when needed into the MPLS traffic class and into Ethernet 802.1Q priority bits, to ensure that the flow is correctly handled in the transport network. For stronger guarantees, traffic engineering techniques are required, for example, with Multiprotocol Label Switching Traffic Engineering (MPLS TE).

Since the traffic volume is typically very downlink dominated, the downlink direction is especially important for addressing capacity bottlenecks, primarily with a combination of link bandwidth upgrades while also ensuring proper QoS configuration.

9.3 Technologies

9.3.1 Client Ports

At the 5G nodes (gNB, gNB-DU, gNB-CU, etc.), physical interfaces that connect the mobile node to the network are in practice Ethernet ports. Figure 9.7 illustrates the Ethernet port speeds against the capacities calculated in the previous section.

Requirements are collected from Figure 9.4 of the previous section: Cases A1–A3 dimension the F1 and NG interfaces based on one sector peak, which was assumed to be higher than the sum of the average rates of all sectors. Cases B1-B3 use the single user peak rate as the basis, leading to a lower capacity requirement. Cases A1 and B1 are for sub-6-GHz frequencies, A2 and B2 for mmWave, and A3 and B3 for future high bands over 60 GHz. Sub-6G-Hz assumed a three-sector configuration, while mmWave and higher bands assumed two sectors.

Standard Ethernet interfaces of 10G, 25G, 40G, and 100G are utilized. On the client side, 25G is currently the highest single-lane rate supported. 40G and 100G are based on multiples of 10G and 25G, for example, $4 \times 10G$ and $4 \times 25G$.

Copper interfaces are limited to site internal cabling. Multimedia fiber can reach 100–300 m and has been commonly used in data centers. Single-mode fiber extends to 10 or 40 km and can be used to directly connect to the peer element. Otherwise the client ports interface a transmission device (e.g. WDM) or some other networking device (a switch, router) for connectivity to remote sites.

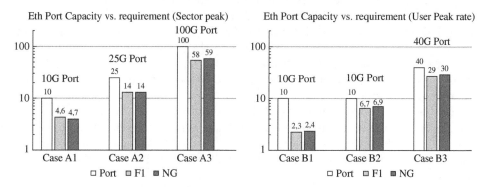

Figure 9.7 Ethernet port speeds and 5G requirements.

9.3.2 Networking Technologies Overview

An overview of some key networking technologies and protocols is included in Table 9.1, as candidate building blocks for the 5G networking service.

Optical technology has introduced a new layer, Layer-0 (Photonic Layer), that has the highest capacities available as physical media, is agnostic to higher layer protocols, supports photonic layer multiplexing and switching with high amount of channels, and optionally supports the OTN (Optical Transport Network) hierarchy. Wavelength multiplexing increases fiber utilization and can provide an optical path through a long distance, with OAM and protection capabilities where needed. CWDM (Course Wavelength Division Multiplexing) and DWDM (Dense Wavelength Division Multiplexing) grids are standardized by ITU-T.

In Passive Optical Networks (PONs), currently 10G is the highest bit rate supported for symmetrical PON (like Next Generation NG-PON2, XGS-PON), with NG-PON2 supporting an aggregate of 40 G downstream (with four wavelengths). With evolution, 25G PONs can even better match 5G speeds, and targets also low layer fronthaul application with the development of cooperative DBA (Dynamic Bandwidth Assignment). See, for example, [7].

Wireless is the primary alternative to fiber, where fiber is not available. Traditional microwave bands provide limited bandwidth and can support capacities up to around 1–2 Gbps. For 10 Gbps links and more, higher bands (E-band and D-band) are used. Wireless remains a very viable option, with low latency and a path for even higher throughputs such as 25 and 40 Gbps. Protection methods are commonly supported.

Copper technologies appear limited in both bandwidth and reach (despite huge developments in the technology), except for site internal interconnects and indoor Ethernet cabling.

IP connectivity service is the de facto solution in 5G for connectivity, either as a native IP service or, for example, as an IP MPLS VPN service. The IP forwarding plane is usually complemented with the IP control plane (routing protocols) offering flexible connectivity service in any scale.

The IPv6 protocol (RFC 8200) [8] is attractive for new deployments like 5G as it improves the network by having a large address space with end-to-end addressing, avoidance of broadcasts, avoidance of fragmentation in intermediate nodes, address autoconfiguration, concept of link local addressing, and further services.

The MPLS data plane with the IP control plane is an essential networking technology that supports services like IP MPLS VPNs [9], E-VPNs [10], and MPLS traffic engineering including fast reroute. For IP MPLS VPN and E-VPN, the control plane relies on MP–BGP (Multiprotocol Border Gateway Protocol) [11] for carrying customer network prefixes/addresses. The IP MPLS VPN service relies on a common MPLS/IP architecture where a single network can be shared for multiple uses and applications very flexibly and in almost any scale.

MEF (Metro Ethernet Forum) [12] defines Ethernet services, such as E-Line and E-LAN, that are widely used for mobile backhaul. With the service definitions, the service is abstracted from the implementation, allowing different underlying technologies that are not visible to the user. Recently, MEF has extended the work to cover IP services.

Table 9.1 Networking technologies overview.

Technology	Description, Application, Comment
CWDM	Point-to-Point optical link, ITU-T CWDM grid for 18 channels
DWDM	ITU-T DWDM grid 40, 80, 160 channels, possibilities for topologies, OAM, long range
Wavelength routing	Routing a wavelength based on G-MPLS or SDN control
OTN	Sonet/SDH type of hierarchy extended to 100G and beyond
PON	Passive Optical Network using shared media, TDM uplink with DBA (Dynamic Bandwidth Assignment) algorithm
Copper	G. Fast supports 1 Gbit/s (DL+UL combined) for 50 m distance
Wireless	Link lengths of a few hundred meters, 10 Gbps (E-band) and more (D-band)
IPv6	Native IPv6 connectivity service, several improvements, coexistence with IPv4 by dual stack
IPv4	Native IPv4 connectivity service
IP MPLS VPN	IP VPN service specific to a customer, on a common MPLS/IP infrastructure
E-VPN	Ethernet VPN, Leverages IP MPLS VPN principles to Ethernet services
E-Line	MEF definition for point-to-point Ethernet service
E-Tree	MEF definition for point-to-multipoint Ethernet service
E-LAN	MEF definition for multipoint service
MPLS/IP	Key technology building block for multiple usages, MPLS data plane, IP control plane
MPLS TE	MPLS traffic engineering for path and capacity reservation, fast reroute, etc.
MPLS FRR	MPLS Fast reroute for 50 ms recovery on a back-up LSP Label Switched Path
MPLS-TP	MPLS Transport Profile for transport network behavior
G-MPLS	Generalized MPLS for routing of wavepath and other LSPs
Native	Native Ethernet bridging (IEEE802.1Q), Local area technology
Ethernet TSN	Time-Sensitive Networks, IEEE Task Group, support for deterministic behavior
DetNet	Deterministic Networks, IETF Working Group, support for deterministic behavior
Segment routing	Source routing paradigm, MPLS or IPv6 based
SDN	Software defined networking, central control of network flows
IPsec	3GPP selected technology for cryptographic protection on network domain
MACSEC	Link layer (MAC) security, complementing IPsec

Segment routing is source routing, where the list of intermediate nodes to be visited is given as a segment list [13]. The segment list can be populated by, for example, by SDN control/network management, or by a path computation element. Segments are identified by MPLS labels or by IPv6 addresses. Segment routing can be used, for example, as a form of traffic engineering or to realize service function chaining.

Native Ethernet bridging best suits limited LANs. If no IP service is possible, an alternative to native bridging is to have the flows programmed by an SDN controller. The Time-Sensitive Networking (TSN) task group is discussed in Section 9.5.

The IP security suite of protocols [14–16] is for cryptographically securing the network domain interfaces of 5G, consisting of services for data origin authentication, integrity protection, encryption, and anti-replay protection. For IPsec, both tunnel mode and transport mode are referred to by 3GPP, as well as addition of DTLS (Datagram Transport Layer Security) for the Stream Control Transmission Protocol (SCTP)-based control plane [17]. MACSEC is a link layer security service which can complement IPsec with solutions that scale with the port count.

With time, cryptographic algorithms that were previously considered safe may become breakable, which is why it is essential to keep the algorithms up to date with the latest standard.

SDN and the concept of centralized control applies not only to switches but also to network nodes in general: optical elements and devices, switches, wireless transmission products, MPLS and IP routers, cloud infrastructure/data centers, and site solutions. This allows the possibility of controlling heterogeneous network layers and nodes unifying the operation, with SDN controllers and a Network Orchestrator.

9.4 Fronthaul and Backhaul Interfaces

9.4.1 Low Layer Fronthaul

Low layer fronthaul refers to the low layer split point, which is not defined by 3GPP in detail.

Figure 9.8 illustrates the O-RAN Alliance low layer fronthaul protocol stack and a low layer split, where part of the 5G L1 is in the O-RU (O-RAN-Radio Unit), while the rest of the radio protocols are in the 5G baseband. The baseband can be further split into O-CU (O-RAN-Control Unit) and O-DU (O-RAN-Distributed Unit) with a high layer split point.

O-RAN specifies the transport protocols to be used in carrying the low layer split interface data, with two types of transport headers, either enhanced CPRI (eCPRI) header or IEEE-1914.3 header.

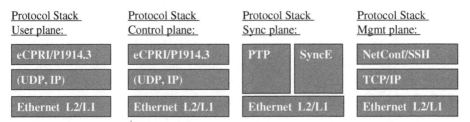

Figure 9.8 Low layer split point interface.

For user plane transport, sample data is mapped either into UDP/IP or directly into Ethernet frames (without the IP layer) [4]. IPv6 and IPv4 are both included as alternatives. The control plane flow consists of real-time control data. The control plane stack is identical to the user plane stack. Due to the real-time control information, retransmissions of control messages are not feasible, and a reliable transport protocol is not defined.

Synchronization flow carries timing information to the O-RU, and the standard includes the Precision Timing Protocol (PTP) along with Synchronous Ethernet. The management plane is defined with the NetConf/SSH/TCP/IP protocol [18, 19].

Instead of transmitting time-domain samples as with CPRI, O-RAN with 5G transmits samples in the frequency domain. This reduces the bit rate required on the transport interface, as well as the possibility of compressing sample data.

Compared to other 5G network transport segments, the low layer fronthaul is very different from the backhaul or from the high layer fronthaul, due to requirements for low latency and high capacity.

IEEE 802.1CM [20] has defined TSN for fronthaul applications. Non-TSN flows and TSN flows can be supported by the same network. Profile A mostly relies on IEEE802.1Q, while an optional Profile B introduces additional capabilities, frame preemption, and interspersing express traffic [21]. The standard also specifies requirements for synchronization.

9.4.1.1 Network Solutions

The physical distance from the RUs to the baseband is capped to some tens of kilometers due to the propagation delay: 25 km on fiber alone means 250 µs RTT, which may be, depending on a number of factors, already excessive. On top of this, some allowance needs to be given for the network nodes for processing.

One of the targets for remote RU deployment is simplification of the network by locating baseband computing at a central site. The RUs should be simple appliances with minimal manual operation and no – or rare – site visits. This target drives also the split point to include more of the baseband computing in the baseband unit, provided that the resulting transport rates remain reasonable, so it is a compromise between these two goals.

The simplest transmission configuration is with point-to-point fiber: each RU is directly attached to the respective baseband unit using dedicated fiber. The only facility in between is the fiber cabling.

For fiber count savings, traffic flows from the RU sites may be optically (passively or actively) multiplexed into a single fiber. The additional element involved now is the optical device. In the digital domain, an alternative includes introducing a device to first perform the multiplexing at the digital level (Ethernet frames or IP packets) before converting the signal to an optical signal for transmission. This adds the switch element and the digital layer for the operations.

While deployments are expected to be very fiber dominated, wireless media need not be ruled out, provided that enough capacity exists and that the radio product does not introduce too much processing delay as the latency budget is tight. Both of these prerequisites are met with wireless media on the higher bands such as the E-band, and wireless may complement fiber at least in some of the cases, for example, for <6 GHz deployment, where the capacity requirement is not extremely high.

9.4.1.2 Security

User signals on the low layer split point are cryptographically protected by the PDCP (Packet Data Convergence Protocol) layer between the UE and the gNB over the low layer fronthaul transport. Due to the PDCP protection, sensitive information is not easily available, even if the transport layer has no cryptographic protection. Legacy CPRI/OBSAI did not define further security services, and neither does the O-RAN specification demand this. High bit rates would also mandate high-performance cryptographic engines.

The management plane to the RU needs cryptographic protection as it is otherwise lacking all protection. In the O-RAN standard, NetConf over SSH (Secure Shell) is included.

9.4.2 NG Interface

The NG interface is defined for connectivity between the 5G RAN and the 5G Core network. NG2 is the control plane interface (NG-C, carrying NG-AP), and NG3 is the user plane interface (NG-U). In 5G RAN, the element is either the gNB-CU (cloud deployment) or the classical gNB.

The protocol stack on the user plane consists of GTP-U/UPD/IP (IPv6 or IPv4) and IPsec. The control plane stack relies on (DTLS)/SCTP/IP (IPv6 or IPv4) and IPsec. The protocol stack is included in Figure 5.3 of Chapter 5.

9.4.2.1 Connectivity

Multiple NG2 and NG3 interfaces may exist in the same 5G RAN node. This allows resiliency to a failure in a single core network node (or site) and also load balancing of traffic in the core network nodes.

Basically, any connectivity service that supports IP flows between the peer nodes is feasible.

In Figure 9.9, classical gNBs use the first mile access service of a wireless link or fiber. An IP router is located at the access hub site acting also as a CE (customer edge) router, providing an attachment circuit to the PE device and to the IP MPLS VPN service.

With IP services, dual-homing/multi-homing is supported, and a resilient service can be provided. For the customer (gNB), the IP MPLS VPN appears as a router providing

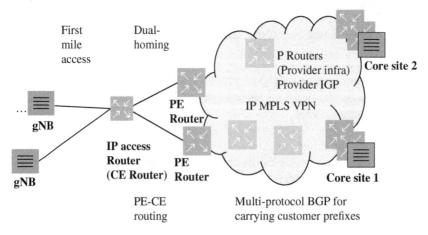

Figure 9.9 NG interface example (classical gNB).

IP connectivity service to customer networks (UPFs and other gNBs) on different sites in that specific VPN. The customer's routes are not shared with other customers, and service provider's routes are as well separate.

9.4.2.2 Security

The NG interface lacks cryptographic protection both on the user plane and on the control plane, except for NAS (Non-Access Stratum) signaling from the UE to the core network. Cryptographic protection with IPsec is required.

9.4.3 Xn/X2 Interfaces

The Xn/X2 interface is defined for connectivity between the 5G RAN and 5G RAN or LTE nodes, for mobility between 5G RAN nodes (like X2 in LTE), and for dual connectivity, either for the 5G-5G case or for the 5G-LTE case. The interface includes both the user plane and the control plane.

The user plane protocol stack consists of GTP-U/UDP/IP and IPsec. The control plane protocol stack consists of (DTLS)/SCTP/IP and IPsec. The IP protocol may be IPv6 or IPv4. The Xn interface and protocol stack are shown in Figure 5.10 of Chapter 5.

9.4.3.1 Connectivity

Xn introduces parallel connectivity between 5G nodes, basically a full mesh between neighbor nodes. The link is seldom physically a direct link, but instead the logical connectivity is arranged with links connecting to a central hub point, which provides the IP connectivity service.

In the case of a high layer split architecture, it is the gNB-CU that interfaces other RAN nodes, 5G or 4G.

9.4.3.2 Security

The user plane is PDCP protected cryptographically (PDCP is carried over the GTP-U), while the control plane is unprotected (except for the RRC Radio Resource Control). IPsec (and DTLS) is needed to protect the control plane, and IPsec may provide additional protection for the user plane.

With IPsec protection using a central security gateway (SEG), the signal must visit the SEG, so the SEG becomes the hub point. Alternatively, IPsec tunnels may be created directly between the 5G nodes (gNBs), and the tunnel then routed via the first common IP hub point.

9.4.3.3 Dual Connectivity

The above applies as well for the dual connectivity case (LTE-5G, 5G-5G), with the difference that there may be higher capacities involved, depending on the scheduler decisions. When LTE and 5G nodes share a site, the link is site internal. For dual connectivity, a flow control procedure is defined over the Xn/X2 interface, which limits the traffic flow if congestion occurs.

9.4.4 F1 Interface

In the CU-DU split architecture, the F1 interface connects the gNB-DU to the gNB-CU. Since the latency-critical protocols terminate at the gNB-DU, the F1 interface is not as demanding as the low layer fronthaul interface in this respect.

The protocol stack at the F1 interface is shown in Figure 5.12 of Chapter 5. The transport layers in the control plane are (DTLS)/SCTP/IP and in the user plane are GTP-U/UDP/IP, IP being either IPv6 or IPv4. IPsec is the protocol defined for cryptographic protection.

With the gNB-CU, part of the computing functions of 5G radio are centralized, and as an example, implemented as VNFs (virtual network functions) in the cloud infrastructure. Since the real-time critical functions are implemented in the gNB-DU, computing requirements on the cloud infrastructure are not as demanding as would be the case if the lower layers of the 5G radio protocols were also deployed in the cloud. As a possible further evolution, these real-time functions can also be moved to the cloud, which then changes the F1 interface to the low layer fronthaul interface.

The gNB-DUs only interface the gNB-CU, with one gNB-CU serving, for example, hundred gNB-DUs, depending on the implementation. The data center interfaces the transport network via a data center edge switch/router.

With the CU-DU higher layer split architecture, it is often assumed that the centralized unit relies on cloud computing on a data center; however, this is an implementation topic from the 3GPP standards viewpoint.

9.4.4.1 Security on F1

The F1 interface consists of the user plane and control plane (F1 AP, Application Protocol). On the user plane, since PDCP terminates at the gNB-CU, end user IP flow is cryptographically protected by PDCP over the F1, between the UE and the gNB-CU. On the control plane, NAS signaling as well as RRC signaling are as cryptographically well protected over the F1. Otherwise, the F1 control plane messaging is clear text, which is why cryptographic protection with IPsec in the network domain is required.

On the user plane, the end user IP flow is cryptographically protected even if the transport IP layer is not. In addition, IPsec may be used to protect the F1 user plane.

9.5 Specific Topics

9.5.1 Network Slicing in Transport

Following the principle that different slices (and the customers there) are tenants on the shared 5G network infrastructure, logical separation of the tenants calls for a network slice to be attached to a specific transport service (a transport slice) in the otherwise common transport network. The approach depends on the use case and needs, as in some cases it is sufficient to support the network slice on the core network and the air interface only and share the same transport service for all of the slices, meaning that the transport layer is agnostic to slices. Other cases require a dedicated transport service for the slice, with logical separation and possibly in some way slice specific transport service.

The example network diagram in Figure 9.9 for the NG interface with IP MPLS VPNs fits transport network slicing nicely. A network slice may on the transport network side now be attached to its own VPN service. Each VPN is as the name implies its own virtual network with no connectivity to other VPNs (other slices), and yet all slices rely on a common, shared MPLS/IP infrastructure, which is beneficial from the network implementation, operation, and cost viewpoints, and also aligns with the target of improving network efficiency. Figure 9.10 illustrates the operation.

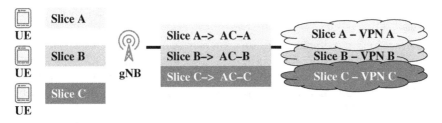

Figure 9.10 Network slicing in transport example.

AC stands for attachment circuit; each AC is associated with a VRF (virtual private network routing and forwarding) instance and customer A's (Slice A's) routes are distributed only for sites of VPN-A.

On top of the logical separation (different VPNs) for slices, further services may be configured when needed, like guaranteed capacity (e.g. with MPLS Traffic Engineering), specific connectivity service to a local UPF or Mobile Edge Computing (MEC) function, higher availability, and so forth.

SDN can help greatly in provisioning the transport services needed for the slices, especially if there are many slices and slice-specific configurations are needed over a heterogeneous transport network.

9.5.2 URLLC Transport

Ultra-reliability and low latency introduce a new level of performance requirements for transport, but only for a specific, limited area. Since the environment and the use cases for the URLLC services vary, transport design alternatives are also very much case specific.

9.5.2.1 Latency

For the low-latency service, server and content are implemented physically close to the cell site, implying a local UPF function and possibly collapsing radio and core elements to the same site. The resulting transport network architecture is very different from the traditional macro 5G network.

In order to guarantee 1–2 ms latency, deterministic behavior is needed from the transport network. The related IEEE task groups are the TSN and Deterministic Networking (DetNet) in IETF [22, 23].

9.5.2.2 Reliability

For ultra-reliability (5 nines to even 8 nines), any transport leg from the terminal via cell site up to the core network and to the server or other peer needs to be taken into account. In 3GPP 22.261, reliability is defined as the "percentage value of the amount of sent network layer packets successfully delivered to a given system entity within the time constraint required by the targeted service, divided by the total number of sent network layer packets" [24].

Reliability is thus not equivalent to availability. The first mile transport link availability in a traditional macro transport network is planned, for example, for four nines (99.99%), still allowing for 52 min of outage in a year. What helps is that radio network cells often partly overlap and service is not lost even when an individual site or the access link fails. Access links are seldom duplicated.

For recovery, protection switching or restoration may be utilized, or a combination of the two, for example, with IP and optical services. Replicated signals or packet streams (1 + 1 type of protection) enable the receiving end to select the signal from the working path at any time. A number of solutions exist (mobile backhaul related in [25]), which can be considered for URLLC transport.

Specifically with respect to URLLC, in a 5G radio network, dual (or multi) connectivity to the terminal is studied in 3GPP, with packet stream duplication at the PDCP layer [26]. See Section 7.4.6 in Chapter 7 for PDCP layer duplication. Since the PDCP layer terminates at the gNB (or gNB-CU), this packet duplication also covers the F1 and Xn transport links. Underlying transport services are then designed for sufficient transport layer availability taking into account the possible higher layer replication, so that the system target is met.

IEEE TSN proposes multiple enhancements to the local area bridging standards, for example, replication of streams over multiple disjoint paths [27]. The deterministic networks working group of IETF addresses the same field but more from the IP networking standpoint, in collaboration with IEEE.

9.5.3 IAB (Integrated Access and Backhaul)

Provisioning the first mile access link to the base station is in many cases difficult. Fiber may not necessarily be available right at the desired location, even when it is available in a cable conduit underground.

Using a separate wireless radio equipment is also difficult if the site space is limited. IAB solves these issues without needing either extra site space or equipment, provided that part of the UE access bandwidth can be used for backhaul. This is true especially in the higher frequency bands where bandwidth is not as scarce resource.

IAB relies on the CU-DU split architecture. Each DU is logically connected to the CU, but the F1 interface may be relayed through intermediate IAB nodes as self-backhaul instead of traditional fiber or other media as shown in Figure 9.11 [28]. The IAB node shares the air interface resources and services both the access UEs attached to it and the backhaul link for the tail-end DU(s) for their connectivity to the CU.

The IAB connectivity consists of multiple hops (chaining) and possibly a tree topology. Intermediate IAB nodes provide a relay function with an adaptation layer.

The standard work is in progress in 3GPP. In further 3GPP releases, IAB should also support moving IAB nodes, for example, as the backhaul for moving vehicles.

9.5.4 NTNs (Non-Terrestrial Networks)

3GPP has a study item on NTNs [29, 30], with the target of integrating satellite-based communication links and services to the users into the 5G system. Architectural aspects are studied in TR23.737, "Study on Architecture Aspects for using Satellite Access in 5G" [31]. The benefit is the extension of 5G communication service to remote areas, or ships in the middle of the ocean, airplanes, and further cases.

With LEO (Low Earth Orbit) platforms, communication latency is drastically reduced (e.g. from 500–800 ms RTT to 10–40 ms) compared to GEO (Geostationary Earth Orbit)

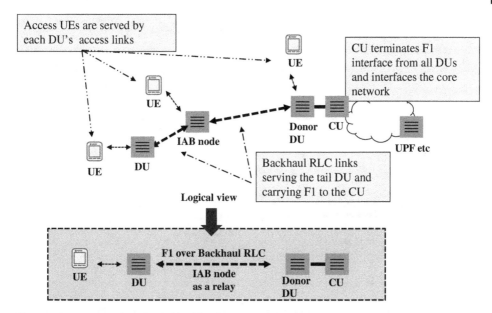

Figure 9.11 Integrated access and backhaul.

services, and at the same time the possibility of using standard 5G technology (with some adaptations) promises lower cost of communications.

Key components of the NTN architecture candidates are illustrated in Figure 9.12. Satellite or other high-altitude platforms may consist of only a relay function (radio frequency conversion) from the access side to the feeder link to the Ground Station GS, or it may include part or all of 5G radio baseband processing (gNB-DU or full gNB). The remaining part of the 5G radio and core network is then located on the ground with the GS, or anyway reachable via the GS. Performance depends on the selected functional split and whether retransmissions to the terminal are supported by the satellite or via the GS.

Terminals are specific satellite terminals requiring larger antenna, or alternatively, regular 5G UEs connect to the satellite service via a satellite gateway, which then communicates with the satellite. With high-altitude platforms, regular UEs may be feasible.

Satellites may also have a link to another satellite (Inter-Satellite Link, ISL) which may be needed if the satellite cannot reach any GS directly.

The architecture reuses existing 5G logical interfaces, protocols, and concepts as much as possible, with adaptations to cope with longer latencies and other NTN specifics.

9.5.5 Time-Sensitive Networks

Originally called the IEEE Audio and Video Bridging (AVB) task group, the group was renamed to TSN TG (Time-Sensitive Network Task Group) [22], as other application areas emerged.

Many of the targeted use cases are today served by vendor- or application-specific protocols and solutions, in factory automation, automotive, and other areas. With the enhancements under definition, these use cases are targeted to be served by Ethernet

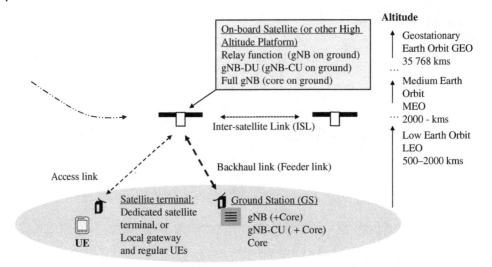

Figure 9.12 5G non-terrestrial networks.

and IP technology, and further, such that the new use cases and these flows can coexist with the non-TSN flows.

With TSN, applications rely on a common accurate time base, obtained, for example, by PTP with a profile defined in IEEE802.1AS [32]. Low latency, ultra-reliability, and resource management with dedicated resources are addressed with specific enhancements.

An interesting addition is 802.1Qbv (Enhancements for Scheduled Traffic), which essentially introduces TDM (time division multiplexing)-type of slots into Ethernet while allowing other traffic types in the same segment [33]. This capability is difficult to implement with standard networking without dedicating a wire for the purpose.

Possible application areas of TSN in 5G are in the low layer fronthaul and with the URLLC and IOT use cases.

References

1 Sustainable Development Goals. https://www.un.org/sustainabledevelopment
2 Froom, R., Sivasubramanian, B. and Frahim, E. (2007). *Building Cisco Multilayer Switched Networks*, 4th Edition. Cisco Press. ISBN 978-1587052736.
3 Metsälä, E.M. and Salmelin, J. (eds.) (2015). *LTE Backhaul: Planning and Optimization*. Wiley. ISBN: 1118924649.
4 O-RAN (2019). Fronthaul Control, User and Synchronization Plane Specification Version 1.0. March 2019.
5 3GPP TR38.913 V1.0.0 (2016). Study on scenarios and requirements for next generation access technologies.
6 NGMN (2018). 5G Extreme Requirements: End-to-End Considerations, 2.0. 16-Aug-2018.
7 ITU-T Supplement 66 (2018). 5G wireless fronthaul requirements in a passive optical network context. 10/2018.

8 IETF RFC 8200 (2017). Internet Protocol, Version 6 (IPv6) Specification.

9 IETF RFC 4364 (2006). BGP/MPLS IP Virtual Private Networks (VPNs).

10 IETF RFC 7432 (2015). BGP MPLS-Based Ethernet VPN.

11 IETF RFC 4760 (2007). Multiprotocol Extensions for BGP-4.

12 MEF, https://www.mef.net

13 IETF RFC 8402 (2018). Segment Routing Architecture.

14 IETF RFC 4301 (2005). Security Architecture for the Internet Protocol.

15 IETF RFC 4303 (2005). IP Encapsulating Security Payload (ESP).

16 IETF RFC 7296 (2014). Internet Key Exchange Protocol Version 2 (IKEv2).

17 3GPP TS 33.210 V16.1.0 (2019). Network Domain Security (NDS); IP network layer security.

18 O-RAN-WG4.MP. 0-v01.00 (2019). Management Plane Specification.

19 IETF RFC 6242 (2011). Using the NETCONF Protocol over Secure Shell (SSH).

20 IEEE Std 802.1CM™ (2018). Time-Sensitive Networking for Fronthaul.

21 IEEE Std 802.3br™ (2016). Interspersing express traffic specification.

22 Time-Sensitive Networking (TSN) Task Group. https://1.ieee802.org/tsn

23 Deterministic Networking (detnet). https://datatracker.ietf.org/wg/detnet/about

24 3GPP TS 22.261 V16.0.0 (2017). Service requirements for the 5G system.

25 Metsälä, E. and Salmelin, J. (eds.) (2012). *Mobile Backhaul*, 204–248. Wiley. ISBN: 978-1-119-97420-8.

26 3GPP TR 23.725 V16.0.0 (2018). Study on enhancement of Ultra-Reliable Low-Latency Communication (URLLC) support in the 5G Core network (5GC).

27 IEEE 802.1CB-2017 (2017). Frame Replication and Elimination for Reliability.

28 3GPP TR 38.874 V16.0.0 (2018). Study on Integrated Access and Backhaul.

29 3GPP TR 38.811 V1.0.0 (2018). Study on New Radio (NR) to support non-terrestrial networks.

30 3GPP TR 38.821 V0.8.0 (2019). Solutions for NR to support non-terrestrial networks.

31 3GPP TR23.737 V1.0.0 (2019). Study on architecture aspects for using satellite access in 5G.

32 IEEE Std 802.1AS-2011 (2011). Timing and Synchronization for Time-Sensitive Applications in Bridged Local Area Networks.

33 IEEE P802.1Qbv IEEE Std 802.1Qbv™ (2015). Enhancements for Scheduled Traffic.

10

5G Performance

Harri Holma[1], Suresh Kalyanasundaram[2], and Venkat Venkatesan[3]

[1] *Nokia, Finland*
[2] *Nokia, India*
[3] *Nokia Bell Labs, United States of America*

10.1 Introduction

This chapter considers 5G technology from the performance point of view by illustrating 5G network capabilities to the operators and to the end users in terms of data rates, capacity, coverage, energy efficiency, connectivity, and latency. 5G networks need to fulfill a number of new performance targets since they target diverse use cases that can be categorized into three different areas:

- Extreme mobile broadband for high data rates and capacity for smartphones, tablets, and laptops. That is the traditional use case of mobile broadband networks.
- Massive Internet of Things (IoT) communication for tens of billions of connected devices like sensors and control units. This use case requires low-cost devices with low power consumption and high connectivity capacity.

5G Technology: 3GPP Evolution to 5G-Advanced, Second Edition.
Edited by Harri Holma, Antti Toskala, and Takehiro Nakamura.
© 2024 John Wiley & Sons Ltd. Published 2024 by John Wiley & Sons Ltd.

Table 10.1 End user performance targets.

User data rate	• Indoor: 1 Gbps downlink and 0.5 Gbps uplink
	• Dense area: 300 Mbps downlink and 150 Mbps uplink
	• Crowd: 25 Mbps downlink and 50 Mbps uplink
	• Everywhere: 50 Mbps downlink and 25 Mbps uplink
	• Ultra-low-cost broadband: 10 Mbps downlink and 10 Mbps uplink
	• Airplanes: 15 Mbps downlink and 7.5 Mbps uplink
End-to-end latency	• 1 ms in low latency case
	• 10 ms in most cases
	• 50 ms in ultra-low-cost
Mobile speed	• Airplanes 1000 km/h
	• Mobile and ultra-reliable cases 500 km/h
	• Typical cases up to 120 km/h
	• Indoor and crowd 3 km/h

- Critical communication for remote controlling of machines or even cars. This use case requires extreme reliability and low latency.

More detailed 5G performance requirements were defined by Next Generation Mobile Networks (NGMN) and by International Telecommunication Union (ITU). NGMN requirements are shown in [1] and ITU requirements in [2]. ITU refers to 5G technology as IMT-2020. The reference case for IMT-2020 is IMT-Advanced capabilities. IMT-Advanced requirements were met by LTE-Advanced in 2009. IMT-2020 capabilities must be substantially higher than IMT-Advanced. 5G systems start in 2019–2020, and 5G must carry the wireless traffic far beyond 2030 for a number of new use cases. 5G targets are typically up to 10 times higher than LTE-Advanced capability and up to 100 times higher in traffic density. We present a short summary of the targets here: end user performance targets are shown in Table 10.1, system design targets in Table 10.2, and other targets in Table 10.3. More detailed 5G targets are described in Chapter 2. The target is to provide a minimum of 1 Gbps practical data rates in indoor cases, 300 Mbps in dense areas, and 50 Mbps everywhere. The target is also to provide ultra-low-cost broadband with a data rate of 10 Mbps. The latency target is set to 1 ms for the most demanding cases and 10 ms for other cases. The required traffic density is

Table 10.2 System performance targets.

Connection density	• Indoor: 75 per 1000 m^2 office
	• Dense area: 2500/km^2
	• Crowd: 30 000 per stadium
	• Everywhere: 400/km^2 suburban and 100/km^2 rural
	• Ultra-low-cost broadband: 16/km^2
	• Airplanes: 80/plane
Traffic density	• Indoor: 15 Gbps downlink and 2 Gbps uplink in 1000 m^2 office
	• Dense area: 750 Gbps downlink and 125 Gbps uplink in km^2
	• Crowd: 0.75 Gbps downlink and 1.5 Gbps uplink in stadium
	• Everywhere: 20 Gbps in downlink and 10 Gbps uplink in km^2
	• Ultra-low-cost broadband: 16 Mbps/km^2
	• Airplanes: 1.2 Gbps downlink and 0.6 Gbps uplink per plane

Table 10.3 Other targets.

Device power efficiency	• Smartphone battery life three days • Low-cost IoT device battery life 15 years
Network energy efficiency	• 1000 times more traffic than today with 50% lower energy usage
Resilience and reliability	• Network availability 99.999%
Ultra low cost networks	• Low-cost mobile broadband for low Average Revenue Per User (ARPU) markets • Low-cost solution for low ARPU IoT services

15 Gbps for a $1000\,\text{m}^2$ office area and 750 Gbps per km^2 in dense areas. Further targets are also defined in terms of energy efficiency both for devices and for networks. IoT device battery lifetime should be up to 15 years, and network energy consumption should be lowered by 50% while carrying 1000 more traffic. That means 2000 times lower energy consumption for each transferred data bit.

10.2 Peak Data Rates

The data rate can be increased with larger bandwidth, with more antennas using multi-stream transmission and with higher-order modulation. All these solutions are part of the 5G specifications. The peak data rate can be calculated with the transmission bandwidth, with the maximum number of multiantenna transmissions, and with number of bits per symbol with the maximum modulation level.

$$Datarate = Bandwidth\ Antenna\ streams\ \frac{bits}{symbol}$$

Assuming 100 MHz bandwidth, four antennas with four parallel streams and 256-QAM modulation with 8 bits per symbol gives a theoretical peak rate of 3.2 Gbps. The practical peak rate is lower for a number of reasons: guard bands, control channel overhead, channel coding, and downlink–uplink split in the Time Division Duplex (TDD) operation. 5G with 100 MHz, four antennas, and TDD give practically up to 2 Gbps in the downlink. Low bands typically have 10–20 MHz spectrum and two to four antennas with Frequency Division Duplex (FDD) giving a peak rate of 100–400 Mbps. Millimeter waves with up to 800 MHz bandwidth and two antennas enable up to 8 Gbps. The long-term target could be even 5 GHz of spectrum with above-50-GHz frequencies and four antennas giving a peak rate of 100 Gbps. The approximate peak rates are shown in Figure 10.1.

More detailed peak data rate calculations with typical 5G parameters are illustrated in Table 10.4. Three options are presented: sub-1-GHz with 10 MHz FDD, 3.5 GHz with 100 MHz TDD, and 28 GHz with 800 MHz TDD. The corresponding subcarrier spacings are 15, 30, and 120 kHz. The downlink modulation is assumed to be 256-QAM for sub-1-GHz and 3.5 GHz cases and 64-QAM for the mmWave case. The uplink modulation is assumed to be 64-QAM. Multiple Input Multiple Output (MIMO) capability is assumed to be four streams in the downlink for mid-bands, dual stream in the downlink at low FDD bands and at mmWave, and single stream in the uplink. Each slot is

$$Data\ rate = Bandwidth \cdot Antenna\ streams \cdot \frac{bits}{symbol}$$

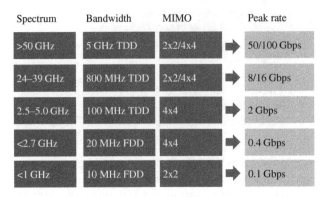

Figure 10.1 Approximate peak data rates as a function of bandwidth and antennas.

Table 10.4 Peak data rate calculations.

	Sub-1-GHz FDD	2.5–5.0 GHz TDD	24–39 GHz TDD
Bandwidth	10 MHz	100 MHz	800 MHz (2 × 400 MHz)
Subcarrier spacing	15 kHz	30 kHz	120 kHz
Number of PRBs	52	273	528
Modulation	Downlink 256-QAM Uplink 64-QAM	Downlink 256-QAM Uplink 64-QAM	Downlink 64-QAM Uplink 64-QAM
MIMO	Downlink 2 × 2 Uplink no MIMO	Downlink 4 × 4 Uplink no MIMO	Downlink 2 × 2 Uplink no MIMO
PDCCH symbols	1 symbol for full bandwidth	1 symbol using 50% bandwidth	1 symbol using 50% bandwidth
DMRS symbols	1	1	1
SSB	Not considered	Not considered	Not considered
TDD radio	Downlink 1.0 (FDD) Uplink 1.0 (FDD)	Downlink 0.74 Uplink 0.23	Downlink 0.74 Uplink 0.23
Coding rate	0.93	0.93	0.93
Peak rate downlink	**111 Mbps**	**1.80 Gbps**	**5.2 Gbps**
Peak rate uplink	**42 Mbps**	**0.104 Gbps**	**0.80 Gbps**

PRB, Physical Resource Block; MIMO, Multiple Input Multiple Output; PDCCH, Physical Downlink Control Channel; SSB, Synchronization Signal Block.

assumed to carry one Physical Downlink Control Channel (PDCCH) symbol and one Demodulation Reference Signal (DMRS) symbol. PDCCH is assumed to use the whole bandwidth for the sub-1-GHz case and 50% of the bandwidth for higher frequencies. Broadcast and synchronization channels are not considered in this calculation. TDD is assumed to have three downlink slots followed by one slot with a 10-symbol uplink,

2-symbol guard and 2-symbol uplink part, and one full uplink slot. The downlink peak rates with these assumptions are 111 Mbps for sub-1-GHz, 1.80 Gbps for 3.5 GHz, and 5.2 Gbps for 28 GHz with a coding rate of 0.93. The corresponding uplink rates are 42, 104 and 803 Mbps. A coding rate of 1.0 means no coding, and the downlink peak rates would be 120 Mbps, 1.95 Gbps, and 5.6 Gbps. The relative data rates in the downlink and in uplink can be adjusted according to the downlink-uplink ratio in TDD.

10.3   Practical Data Rates

10.3.1   User Data Rates at 2.5–5.0 GHz

Practical user data rates will be lower than the theoretical peak rates. The data rates depend on a number factors including bandwidth, signal levels, inter-cell interference levels, and on the number of simultaneous users. We illustrate example data rates with a 3.5 GHz urban deployment using 100 MHz bandwidth and massive MIMO three-sector antennas with 100 W maximum output power. The inter-site distance is 500 m, and the base station (BTS) antennas are located on rooftops 25 m above ground level. The time allocation between the downlink and uplink in TDD is 4:1. UEs are assumed to have four receive antennas. 80% of UEs are located indoors and 20% outdoors. The traffic model is 0.5 MB file transfer. The results are illustrated in Figure 10.2 for the downlink and in Figure 10.3 for the uplink both for average data rates and for cell edge data rates. Three different loading levels are shown: 16, 32, and 48 UE arrivals per second. The rate of 16 UEs each with 0.5 MB download corresponds to a cell throughput of 64 Mbps, which is relatively low loading. The average data rate in downlink is 400–500 Mbps with low loading with 16TRX and 64TRX. The corresponding cell edge data rate is 130–140 Mbps. When the loading increases, the data rates decrease. With 48 UE arrivals per second,

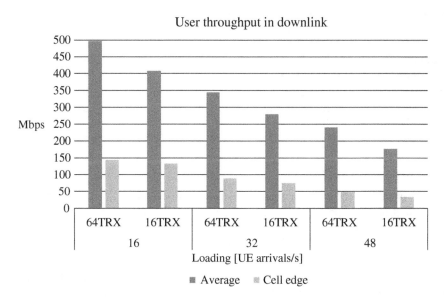

Figure 10.2 User data rates in the downlink with 3.5 GHz spectrum and 100 MHz bandwidth.

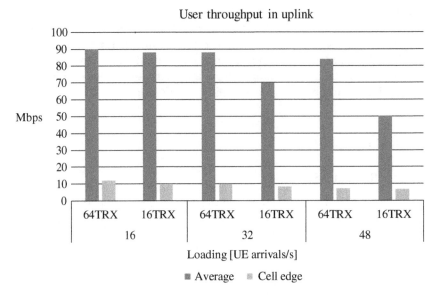

Figure 10.3 User data rates in the uplink with 3.5 GHz spectrum and 100 MHz bandwidth.

the average user throughput is 180–240 Mbps, which is an approximately 50% lower data rate than with 16 UE arrivals per second. The cell edge throughput drops relatively more with a higher loading than does the average throughput. With 48 UE arrivals per second, the cell edge throughput is 35–50 Mbps, which is a 70% lower data rate than with 16 UE arrivals per second. The reason for this is that a higher inter-cell interference has a relatively larger impact on the cell edge data rates than on the average data rates.

Uplink data rates are substantially lower than downlink data rates mainly because the downlink is allocated four times more time than the uplink in this simulation. The average uplink data rate is 90 Mbps with 64TRX and drops to 50 Mbps with 16TRX and higher loading. We note that the difference between the average and cell edge throughput in the uplink is 8–10 times while in the downlink it is 3–5 times. The reason for this is that the uplink data rates are mainly noise limited in macro cells, and the UE transmission power is the limiting factor in the uplink cell edge data rates. The downlink data rates are mainly interference limited.

10.3.2 User Data Rates at 28 GHz

The data rate discussion above focused on frequencies below 6 GHz. The following simulations in urban micro cells consider data rates with 28 GHz spectrum. The assumption is that all UEs are outdoors, the inter-site distance is 50–100 m, the base station antenna height is 10 m, and the base station power is 6 W with an 8 × 8 antenna configuration. The results are shown in Figure 10.4. The simulations show a median data rate of 1.6 Gbps, a peak rate of 6 Gbps, and at the cell edge 0.75 Gbps with 800 MHz bandwidth. The results illustrate that the 28 GHz band can be used for providing very high user data rates in dense networks. When the inter-site distance was increased to

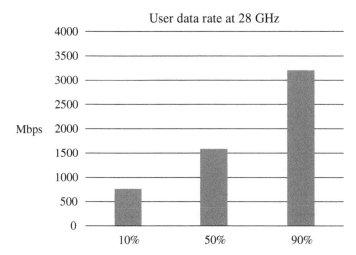

Figure 10.4 User data rate distribution in 28 GHz simulations with 800 MHz bandwidth and two users per cell with an inter-site distance of 50–100 m.

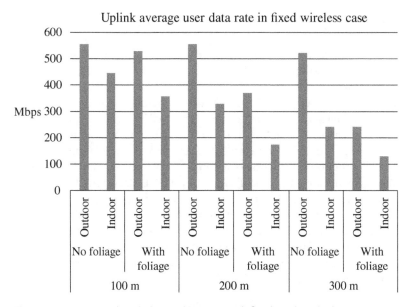

Figure 10.5 Average downlink user data rates with fixed wireless deployment at 28 GHz.

300 m, the throughput was lower due to coverage limitations. Only those UEs with a line-of-sight connection were able to access a high-data-rate connection.

10.3.3 User Data Rates with Fixed Wireless Access at 28 GHz

5G radio can also be utilized for providing high speed fixed connection to homes. 5G may be able to provide a fast enough connection with a potentially lower cost compared to fiber installation. This section illustrates radio performance simulations for the fixed

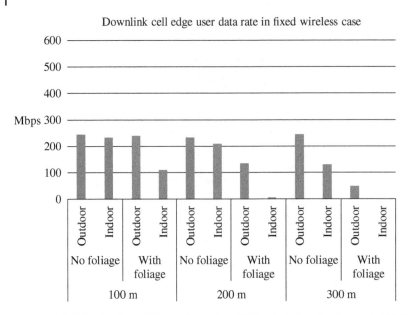

Figure 10.6 Cell edge downlink user data rates with fixed wireless deployment at 28 GHz.

wireless use case using an 800 MHz system bandwidth at 28 GHz. The radio frame is split between the downlink and uplink in the ratio 50:50. The simulations consider a residential environment with three-sector 5G base stations at 6 m height using a beamforming antenna and a total power of 20 W. The base station antenna includes 16×16 elements each with cross-polarization. A maximum of two MIMO streams per UE and four multiuser MIMO streams per cell is assumed. The UE antenna is assumed to have beamforming as well with a total of 4×2 elements each with cross-polarization. UEs at homes are mounted either on the outside of a house or are located inside the house. The simulations illustrate the impact of the following factors: UE placement, foliage, and cell size. Two different foliage assumptions are considered: no foliage and heavy foliage. Foliage refers to the leaves of trees and plants which cause additional attenuation at high frequencies. Three different inter-site distances are considered: 100, 200, and 300 m. Simulation results are illustrated in three separate plots: average downlink data rates in Figure 10.5, cell edge downlink in Figure 10.6, and cell edge uplink in Figure 10.7. We can make the following notes from the simulations:

- An average downlink data rate of 500 Mbps and cell edge data rate of 100 Mbps can be achieved with an inter-site distance of 300 m without foliage with outdoor installations, 200 m with heavy foliage with outdoor installations, and 100 m with indoor installations.
- Uplink data rates are lower than downlink rates due to the lower UE transmission power. An average rate of 300 Mbps and cell edge rate of 100 Mbps can be achieved only with 100 m inter-site distance and outdoor installations if heavy foliage is considered. An inter-site distance of 200 or 300 m can be achieved for outdoor installations without foliage.

More details can be found in [3].

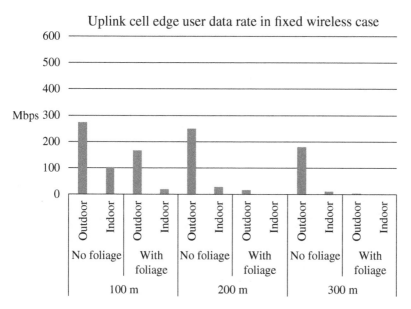

Figure 10.7 Cell edge uplink user data rates with fixed wireless deployment at 28 GHz.

10.4 Latency

10.4.1 User Plane Latency

Low latency communication (LLC) is one of the targets in 5G networks. Latency minimization is considered important for mission-critical communication. Low latency is also beneficial for interactive applications and for protocol performance. Data rate and spectral efficiency have theoretical limits defined by the bandwidth, signal-to-noise ratio (SNR), and the Shannon limit [4]. There are no such theoretical limits for latency. The limit is just a practical protocol design which is a trade-off between performance, efficiency, and complexity. The latency components in HSPA, LTE, and 5G networks are shown in Table 10.5. The main solution for minimizing latency is a shorter transmission time interval (TTI). A shorter TTI makes the transmission time shorter, and it also makes the buffering and processing times shorter. 5G supports the shortest TTI of 0.125 ms, while LTE Release 8 has 1 ms TTI and HSPA has 2 ms TTI. 5G latency can be decreased further by using mini-slot transmission. It is considered later in this section. Uplink and downlink transmission takes one TTI. There is also a frame alignment time needed in the transmission buffer until the next TTI starts. The average buffering time is half TTI both in the uplink and downlink. We assume that at a minimum, PDCCH transmission and waiting time until Physical Uplink Shared Channel (PUSCH) is needed in LTE with a 3 ms delay. We assume that 5G can use grant-free transmission without any scheduling delay. Scheduling delay would be greater if the UE must send a capacity request and the base station must run a packet scheduling decision. Data encoding and decoding take some time in the transmitter and receiver. Processing times will decrease in 5G simply because of the shorter TTI. The UE processing time requirement in LTE is more than 2 ms, and in 5G it is as low as below 0.2 ms. We also assume that the base station processing time correspondingly decreases. In practice, some latency is also caused

Table 10.5 HSPA, LTE, and 5G two-way latency (=round trip time) components.

Delay component	HSPA	LTE Release 8	5G with short TTI transmission
Downlink transmission (ms)	2	1	0.125
Uplink transmission (ms)	2	1	0.125
Downlink frame alignment (ms)	1	0.5	0.07
Uplink frame alignment (ms)	1	0.5	0.0625
Scheduling (ms)	1.3[a]	3–18[b]	—
UE processing (ms)	8	2.3	0.2
BTS processing (ms)	3	2	0.2
Transport and core (ms)	2 (including RNC)	1	0.1 (for local content)
Total (ms)	**20**	**11–28**	**0.9**

RNC, Radio Network Controller.
a)   Shared Control Channel (SCCH) transmissions.
b)   Scheduling period + capacity request + scheduling decision + PDCCH signaling.

Table 10.6 UE processing time requirements in downlink reception data from [5].

UE capability	15 kHz	30 kHz	60 kHz	120 kHz
Capability 1 for mobile broadband UE	8 symbols = 0.57 ms	10 symbols = 0.36 ms	17 symbols = 0.30 ms	20 symbols = 0.17 ms
Capability 2 for low latency UE	3 symbols = 0.21 ms	4.5 symbols = 0.186 ms	9 symbols = 0.186 ms	—

by the transmission network and by the core network. We assume 1 ms transport latency in LTE. The assumption for 5G is local content close to the base station, which makes the transport latency close to zero.

These estimates show that HSPA can achieve a round trip time (RTT) of approximately 20 ms, LTE with pre-allocated scheduling can achieve 11 ms RTT, and 5G even below 1 ms RTT with these processing time assumptions. These latencies are optimistic values without retransmission and with minimum transport delays. The RTT evolution is illustrated in Figure 10.8.

Example live network measurements with smartphones in Helsinki are shown in Figure 10.9, with HSPA showing 19 ms average latency and LTE showing 11 ms average latency. These measurements illustrate that it is possible to get close to the estimated RTTs also in live networks when the core network gateway and server are close to the base station (Tables 10.6 and 10.7).

UE processing time is one of the contributing factors in the latency. 5G UE requirements for PDSCH decoding are defined in 3GPP for mobile broadband UE called Processing Capability 1 and for low latency services called Processing Capability 2. The processing time is dependent on the subcarrier spacing. For low latency UE,

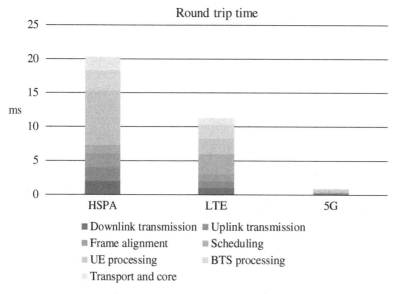

Figure 10.8 Round trip time evolution.

Figure 10.9 Round trip time measurements in live network.

Table 10.7 UE processing time requirements in uplink transmission data from [5].

UE capability	15 kHz	30 kHz	60 kHz	120 kHz
Capability 1 for mobile broadband UE	10 symbols = 0.71 ms	12 symbols = 0.43 ms	23 symbols = 0.41 ms	36 symbols = 0.32 ms
Capability 2 for low latency UE	5 symbols = 0.36 ms	5.5 symbols = 0.19 ms	11 symbols = 0.19 ms	—

the requirement for receiver processing is approximately 0.2 ms and for transmitter processing 0.19–0.36 ms. For mobile broadband UE, the requirement is 0.17–0.57 ms depending on the subcarrier spacing. These processing time requirements are substantially tighter than in LTE Release 8, where more than 2 ms was allowed for UE processing. The practical processing times can be lower than the minimum 3GPP requirements. We assume for simplicity in the following text that 5G UEs are just fulfilling 3GPP processing time requirements. We also assume that the processing time in the base station is half of the processing time in the UE. There are no 3GPP requirements defined for the network element processing times.

5G latency is impacted by the TTI size. Table 10.8 and Figure 10.10 show the latency values with full TTI transmission. TTI size is a function of the subcarrier spacing: a larger subcarrier spacing has a shorter TTI, leading to shorter latency. We assume

Table 10.8 Two-way latency with different subcarrier spacing and full TTI transmission with UE capability 1.

Delay component	15 kHz	30 kHz	60 kHz	120 kHz
Downlink transmission (ms)	1.0	0.5	0.25	0.125
Uplink transmission (ms)	1.0	0.5	0.25	0.125
Downlink frame alignment (ms)	0.5	0.25	0.125	0.0625
Uplink frame alignment (ms)	0.5	0.25	0.125	0.0625
UE processing – reception (ms)	0.57	0.36	0.30	0.17
UE processing –transmission (ms)	0.71	0.43	0.41	0.32
BTS processing – reception (ms)	0.29	0.18	0.15	0.09
BTS processing – transmission (ms)	0.36	0.22	0.22	0.16
Transport and core (ms)	0.10	0.10	0.10	0.10
Total (ms)	**5.0**	**2.8**	**1.9**	**1.2**

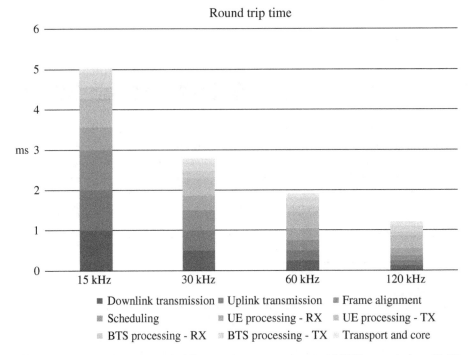

Figure 10.10 Two-way latency with different subcarrier spacings and full TTI transmission with UE capability 1.

here continuous transmission and reception, which means FDD, and UE processing capability 1. These assumptions lead to latency of 5.0, 2.8, 1.9, and 1.2 ms for subcarrier spacings of 15–120 kHz.

Latency can be improved with mini-slot transmission by using only part of the TTI. Mini-slot improves latency especially for low band FDD deployments using 15 kHz subcarrier spacing. Table 10.9 and Figure 10.11 show the latency with 15 kHz mini-slot

Table 10.9 Two-way latency with mini-slot transmission with 15 kHz with UE capability 2.

Delay component	14-symbol	7-symbol	4-symbol	2-symbol
Downlink transmission (ms)	1.0	0.5	0.29	0.14
Uplink transmission (ms)	1.0	0.5	0.29	0.14
Downlink frame alignment (ms)	0.5	0.25	0.14	0.07
Uplink frame alignment (ms)	0.5	0.25	0.14	0.07
UE processing – reception (ms)	0.21	0.21	0.21	0.21
UE processing – transmission (ms)	0.36	0.36	0.36	0.36
BTS processing – reception (ms)	0.11	0.11	0.11	0.11
BTS processing – transmission (ms)	0.18	0.18	0.18	0.18
Transport and core (ms)	0.10	0.10	0.10	0.10
Total (ms)	**4.0**	**2.5**	**1.8**	**1.4**

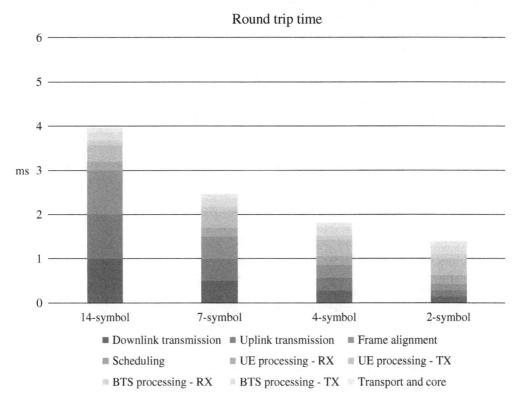

Figure 10.11 Two-way latency with mini-slot transmission with 15 kHz with UE capability 2.

assuming UE processing capability 2. The latency estimate shows that it is possible to achieve below 2 ms two-way latency with 15 kHz FDD and with mini-slot.

FDD has an inherent benefit in terms of latency because uplink and downlink transmissions can run simultaneously. The TDD frame will add some latency because uplink transmission is not available during the downlink, and vice versa. When the packet

Sub-6 GHz with 30 kHz subcarrier spacing

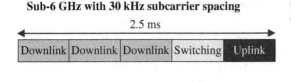

Figure 10.12 TDD frame adds latency compared to FDD.

mmW with 120 kHz subcarrier spacing

D = downlink
S = switching
U = uplink

arrives in the uplink transmission buffer, it has to wait until the uplink transmission part of the TDD frame. The packet in the downlink may also need to wait until the downlink transmission part starts, but the downlink frame alignment time is typically short because the TDD frame is configured mostly for the downlink. Also, if the server response comes quickly, the packet does not need to wait at all for the downlink transmission. If the packet arrives in the uplink buffer in the worst case just when the uplink transmission has started, the packet has to wait for the whole TDD frame before the transmission can start. The average waiting time is half of the TDD frame. Typical TDD frame configurations are shown in Figure 10.12. We assume that the switching subframe is configured mostly for the downlink. If we have a 2.5 ms TDD frame at sub-6-GHz bands, the average waiting time in the uplink is 1.25 ms. The corresponding value in FDD would be half of the 0.5 ms frame, which is 0.25 ms. Therefore, the TDD frame adds on average 1.0 ms latency in the uplink. This value comes on top of the RTT shown in Table 10.8. Therefore, we can expect that the best-case RTT measurements with sub-6-GHz TDD with 30 kHz is 2.8 ms + 1 ms = 3.8 ms when using full TTI transmission. The average waiting time in the uplink with mmW and 120 kHz is approx. 0.3 ms, which makes the best-case RTT approx. 1.5 ms. We can conclude that the sub-6-GHz TDD system will not be able to deliver 1–2 ms latency, but the best-case latency will be more than 3 ms. If we need 1–2 ms latency, then we have two options: a low band FDD solution with mini-slot for wide area low latency services or a mmW TDD solution for local area low latency services. Best-case latency values for example deployment options are shown in Table 10.10 and in Figure 10.13.

An example RTT measurement with a pre-commercial 5G system with 3.5 GHz TDD is illustrated in Figure 10.14. The latency varies between 3.9 and 6.4 ms, which is close to the theoretical values.

Table 10.10 Best case two-way latency summary for example deployment cases.

System assumption	Best-case latency (ms)
Sub-1-GHz, FDD 15 kHz, full TTI, UE capability #1	5.0
Sub-6-GHz, TDD, 2.5 ms frame, 30 kHz, full TTI, UE capability #1	3.8
mmW, TDD, 0.625 ms frame, 120 kHz, UE capability #1	1.5
Sub-1-GHz, FDD 15 kHz, 2-symbol mini-slot, UE capability #2	1.4

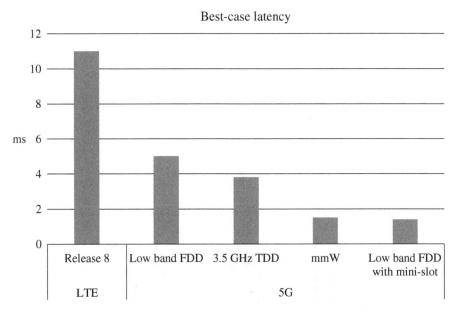

Figure 10.13 Best-case two-way latency with different 5G deployment options.

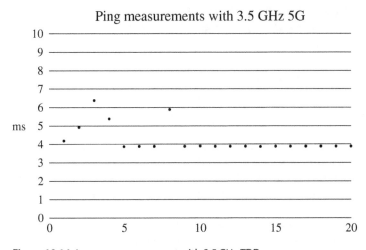

Figure 10.14 Latency measurement with 3.5 GHz TDD.

10.4.2 Low Latency Architecture

When the radio latency becomes lower in 5G, it becomes more important to minimize the total end-to-end latency beyond the radio network. The signal propagation speed in optical fiber is 2×10^8 m per second. The two-way propagation time in 1000 km direct fiber is 10 ms, in 100 km 1 ms, and in 10 km 0.1 ms; see Figure 10.15. If we want to achieve 1 ms end-to-end RTT including radio and transport, the content must be close to the radio, which implies a maximum of a few tens of kilometers or preferably even a shorter

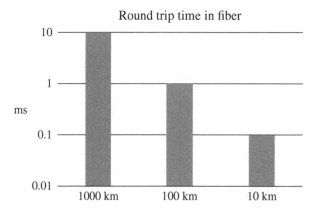

Figure 10.15 Theoretical two-way propagation delay in direct optical fiber.

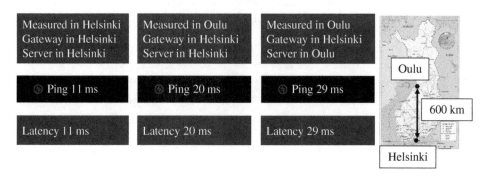

Figure 10.16 Impact of distance on practical latency.

distance. Therefore, local content and a local gateway will be relevant for low latency use cases.

The impact of distance on the latency can be demonstrated with RTT measurements in an LTE network. The measurement case is illustrated in Figure 10.16. The local operator has a core network gateway located in Helsinki. Ping measurements over the LTE network show a latency of 11 ms in Helsinki. The same measurement in Oulu shows a 20 ms latency, that is, an additional latency of 9 ms. The distance from Helsinki to Oulu is 600 km, which in theory corresponds to an additional latency of 6 ms, but the practical latency is always more than the theoretical minimum. The third measurement was done in Oulu while selecting a local server in Oulu. The measured latency was 29 ms because the signal travels to the gateway in Helsinki, back to the server in Oulu, again to Helsinki, and back to the UE in Oulu. There is no gateway in Oulu. It is typically the case in mobile networks today that the core network architecture is centralized. These simple measurements illustrate that millisecond-level latency requires changes in the network architecture, and more gateways are required.

Two example solutions for low latency are Multi-Access Edge Computing (MEC) and the Distributed Core network. MEC brings the content close to the subscribers. Distributed Core allows access to the local Internet and intranet. Figure 10.17 illustrates the architecture evolution from a centralized core network to the distributed approach.

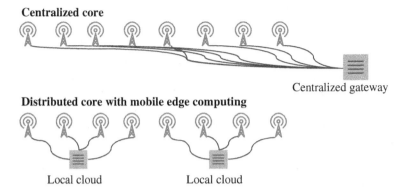

Centralized core

Centralized gateway

Distributed core with mobile edge computing

Local cloud   Local cloud

Figure 10.17 Distributed architecture with mobile edge computing.

The traditional network architecture includes only a small number of centralized core network gateways, which can cause long propagation delays.

10.4.3 Control Plane Latency

Another part of latency optimization is the control plane setup time. A low latency control plane enables low latency also for the transmission of the first packets. If the control plane latency is high, then the first packets have high latency due to the setup times. The following control plane protocols increase the latency in LTE networks:

- Transition from idle to Radio Resource Control (RRC) connected takes 50–100 ms. The delay is because approximately 10 signaling messages are required between the UE and the network for the connection setup. The typical inactivity timer in LTE networks is 5–10 s, after which the UE is moved to idle. Therefore, RRC connection setup occurs quite frequently for the first packet in LTE networks.
- Resource allocation takes 10–20 ms if there are no PUSCH resources available. The UE needs to send capacity requests and get the capacity allocation on the PDCCH. The inactivity timer value is typically just 0–50 ms before the allocation is released. Therefore, resource allocation delay occurs frequently in LTE networks.

There is clearly room for improving the control plane latency by RRC and protocol design in 5G networks. The main solutions are the RRC connected inactive state and grant-free access; see Figure 10.18. The RRC connected inactivate state allows the RRC connection to be kept alive while minimizing the UE power consumption. Grant-free

LTE design	LTE latency	5G solution	5G latency
Transition from RRC idle to RRC connected	Latency 50–100 ms	RRC connected inactive	<10 ms
Resource allocation and packet scheduling	Latency 10–20 ms	Grant-free access	0 ms

Figure 10.18 Control plane latency solutions in 5G.

access allows sending data with predefined resources without any network resource allocation. These 5G features are described in more detail in Chapters 3 and 7.

The estimated setup delays are shown in Table 10.11. The first case is LTE RRC connection setup from idle, which is estimated to take 76 ms. The next three cases are 5G setup from RRC inactive with different TTI sizes and different UE processing delay assumptions. We can note that the processing delays in UE and BTS take more than 50% of the latency. 5G setup time with 1 ms TTI is 31.5 ms, which is more than 50% faster compared to LTE. The setup time is below 10 ms when using mini-slot and assuming short UE processing delays.

5G non-standalone architecture (NSA) adds an extra latency because of the 5G setup time. The UE is first connected to the LTE network, and adding 5G leg takes some time.

Table 10.11 Estimated control plane latency with 5G connected inactive state data from [6].

Delay component	LTE Release 14 (ms)	5G with 1 ms TTI (ms)	5G with 1 ms TTI and short processing delay (ms)	5G with 1/7 ms TTI and short processing delay (ms)
RACH scheduling period	0.5	0.5	0.5	0.07
MSG1: RACH preamble	1.0	1.0	1.0	0.14
MSG2: Random access response	3.0	3.0	1.5	1.5
UE processing delay	5.0	5.0	1.0	1.0
MSG3: RRC request	1.0	1.0	1.0	0.14
BTS processing delay	4.0	4.0	4.0	3.0
MSG4: RRC setup	1.0	1.0	1.0	0.14
UE processing delay	15.0	15.0	3.0	3.0
MSG5: RRC setup complete	1.0	1.0	1.0	0.14
BTS processing delay	4.0			
MME processing delay	15.0			
BTS processing delay	4.0			
Security mode command	1.5			
UE processing delay	20.0			
Total	76.0	31.5	13.0	9.1

MSG, message; MME, Mobility Management Entity.

Figure 10.19 5G leg setup time measurement in non-standalone architecture.

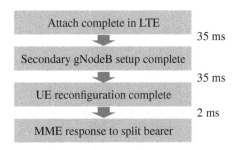

An example measurement is shown in Figure 10.19. The setup time in this case is below 100 ms. If the UE must first make measurements to find out the 5G signal levels, the setup time will be higher.

10.5 Link Budgets

10.5.1 Link Budget for Sub-6-GHz TDD

The 5G link budget is an important tool for estimating the coverage areas of 5G cells. We illustrate here the main factors impacting the maximum path loss values for different physical channels. The path loss can be converted to the cell range with propagation models like Okumura–Hata. We use the following assumptions:

- The base station antenna structure has a major impact on the link budget. Massive MIMO with beamforming gain can improve the link budget considerably. A 3.5 GHz three-sector beamforming antenna can provide a gain of 25 dBi. The full beamforming gain assumes eight beams per sector for both polarizations, which means, 16-64TRX massive MIMO. That gain is available for the user-specific channel PDSCH/PUSCH, PDCCH/PUCCH, and the Physical Random Access Channel (PRACH). We assume a gain of 22 dBi for the broadcast channel PBCH since eight beams can be used for the common channels. For reference, a passive three-sector antenna at the 2 GHz band typically has a gain of 17–18 dBi and a sub-1-GHz antenna 15 dBi. The antenna gain values naturally depend on the size of the antenna and other antenna parameters.
- Device antennas have an impact on the link budget. The link budget assumes four-antenna reception in the UE at 3.5 GHz and two-antenna reception at low FDD bands.
- The base station RF noise figure is assumed to be 2 dB and the UE RF noise figure 7 dB.
- The base station output power is 200 W (53 dBm) and the device power 0.4 W (26 dBm) for TDD and 0.2 W (23 dBm) for FDD.
- The downlink share in TDD is assumed to be 80% and for the uplink 20%. If the 1 Mbps PUSCH data rate is required, the instantaneous data rate must be 5 Mbps during the uplink part. That explains why the PUSCH is the limiting factor for the coverage.
- The PDSCH and PUSCH SNR requirements are taken from Shannon theory with a 4 dB margin. It is further assumed that uplink 2RX gives a 3 dB gain and downlink 4RX gives a 4 dB gain. In theory, 4RX could provide more gain, but designing four antennas in the small form factor in the device is challenging. We assume

that the PUSCH bandwidth (number of physical resource blocks [PRBs]) is large enough so that Quadrature Phase Shift Keying (QPSK) modulation can be used. The PDSCH transmission is assumed to use a maximum of 50% of the downlink PRBs, which corresponds to two simultaneous UEs in the downlink sharing power equally.

- A long PUCCH format with 30 kHz is assumed with an SNR requirement of −6 dB.
- A long PRACH sequence with 30 kHz and three PRBs is assumed with an SNR requirement of −10 dB.
- The Sounding Reference Signal (SRS) is assumed to have a low bandwidth of four PRBs. SRS transmission has flexibility between uplink signaling and downlink beamforming quality. Wideband SRS provides better quality for the downlink beamforming. Narrowband SRS has greater coverage while providing lower quality estimates for the beamforming. We assume that the SNR requirement for SRS is −10 dB. The SRS must be received with each receiver branch separately, which does not give beamforming gain. Therefore, the antenna gain for the SRS is lower than for other channels. The calculation assumes a 16 dBi gain for SRS reception.
- The PDCCH SNR requirement is assumed as −6 dB, and the transmission uses 24 PRBs with an aggregation level of four.
- The PBCH SNR requirement is assumed as −10 dB, and the transmission uses 20 PRBs.

The link budget calculation shows LTE1800 for reference with a maximum path loss of 155 dB including an 18 dBi antenna gain. That path loss typically gives a 500 kbps uplink data rate. The path loss value corresponds to a reference signal received power (RSRP) of −122 dBm with 2×40 W downlink power. The path loss difference between 3.5 and 1.8 GHz is assumed to be 10 dB. Therefore, the LTE1800 path loss is shown as 165 dB in Figure 10.20 when converted to 3.5 GHz.

The results for the uplink link budget are shown in Table 10.12 and for the downlink in Table 10.13. The results are summarized in Figure 10.20. We can make a few observations:

- PUSCH is the limiting factor for coverage for data rates above 100 kbps. 1 Mbps uplink coverage is approximately 15 dB shorter than the downlink 10 Mbps coverage or uplink control channel coverage. The PUSCH coverage issue can be solved by using a lower band uplink transmission.
- The SRS is the next limiting factor for coverage. Even with 4 PRB transmission, the SRS link budget is short because the full beamforming gain cannot be used for the SRS. One option is to use UE feedback for beamforming to get around the SRS coverage limitation or to use pre-filtering for SRS reception to get the beamforming gain.
- PRACH and PUCCH have good enough coverage to take benefit of low band LTE for uplink with the dual connectivity solution.
- The 5G downlink can still provide more than 10 Mbps data rate where the control channel coverage ends. The 5G downlink link budget and data rate are heavily dependent how many users share the downlink power. This calculation assumes that a single user is allocated 50% of 200 W power.

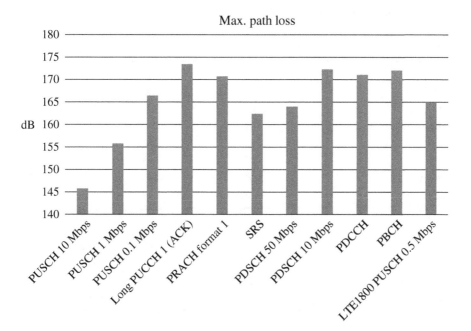

Figure 10.20 Link budget benchmarking for different physical channels.

Table 10.12 Uplink link budget for sub-6-GHz TDD.

System parameter	PUSCH 10 (Mbps)	PUSCH 1 (Mbps)	PUSCH 0.1 (Mbps)	Long PUCCH 1 (ACK)	PRACH format 1	SRS
Data rate (Mbps)	10.0	1.0	0.1			
Tx power (W)						
Tx power (dBm)	26	26	26	26	26	26
Subcarrier spacing (kHz)	30	30	30	30	30	30
PRBs	50	5	2	1	3	4
Uplink share	20 %	20 %	20 %			
Antenna gain (dBi)	25	25	25	25	25	16
Interference margin (dB)	0	0	0	0	0	0
Noise figure (dB)	2	2	2	2	2	2
SNR (dB)	4.7	4.7	−2.0	−6.0	−8.0	−10.0
Sensitivity (dBm)	−94.7	−104.7	−115.4	−122.4	−119.7	−120.4
Max path loss (dB)	120.7	130.7	141.4	148.4	145.7	146.4
Max path loss with beams (dB)	145.7	155.7	166.4	173.4	170.7	162.4

Table 10.13 Downlink link budget for sub-6-GHz TDD.

System parameter	PDSCH 50 (Mbps)	PDSCH 10 (Mbps)	PDCCH	PBCH
Data rate (Mbps)	50.0	10.0		
Tx power (W)	200	200	200	200
Tx power (dBm)	53.0	53.0	53.0	53.0
Subcarrier spacing (kHz)	30	30	30	30
Total PRBs	275	275	275	275
Used PRBs	138	138	24	20
Downlink share	80 %	80 %		
Antenna gain (dBi)	25	25	25	22
Interference margin (dB)	0	0	0	0
Noise figure (dB)	7	7	7	7
SNR (dB)	1.0	−7.3	−6.0	−10.0
Sensitivity (dBm)	−89.0	−97.3	−103.6	−108.4
Max path loss (dB)	139.0	147.3	146.1	150.1
Max path loss with beams (dB)	164.0	172.3	171.1	172.1

10.5.2 Link Budget for Low Band FDD

Low band link budgets have some differences compared to the previous link budget: FDD technology is used instead of TDD, massive MIMO beamforming is not utilized, and antenna gains are lower than at 3.5 GHz TDD bands. No SRS is needed in FDD because uplink sounding cannot be used for downlink beamforming due to different uplink and downlink frequencies. Table 10.13 illustrates an example data channel link budget for sub-1-GHz FDD for 0.5 Mbps uplink and for 10 Mbps downlink with 10 MHz bandwidth. The antenna gain with a three-sector base station is assumed to be 15 dBi. The antenna gain is typically higher, up to 18 dBi at 2 GHz bands, and even higher in the case of narrow-beam six-sector antennas. The link budget assumes a base station power of 80 W and two simultaneous users. The results show that the maximum path loss is approximately 152–154 dB. A base station antenna gain of 18 dBi would mean a path loss of 155–157 dB (Table 10.14).

10.5.3 Link Budget for Millimeter Waves

We consider a relatively high data rate of 50 Mbps uplink and 500 Mbps downlink for the millimeter wave link budget. The limiting factor for coverage for such high data rates is the data channel. We assume a base station transmission power of 4 W and a noise figure of 5 dB. The beamforming antenna gain is assumed to be 26 dBi. The downlink calculation assumes single active user. The link budget for the millimeter wave system is shown in Table 10.15. The results illustrate that the uplink and downlink maximum path loss values are similar at approximately 136 dB. The values are substantially shorter than in sub-6-GHz link budgets because we assume higher user data rates and because of the lower base station transmission power. The resulting cell range at millimeter waves

Table 10.14 Data channel link budget for sub-1-GHz FDD.

5G uplink		5G downlink	
PUSCH 0.5 Mbps		**PDSCH 10 Mbps**	
Data rate (Mbps)	0.5	Data rate (Mbps)	10.0
Tx power (W)		Tx power (W)	80
Tx power (dBm)	23	Tx power (dBm)	49
Subcarrier spacing (kHz)	15	Subcarrier spacing (kHz)	15
		Total PRBs	50
PRBs	4	Used PRBs	25
Uplink share	100 %	Downlink share	100 %
Antenna gain (dBi)	15	Antenna gain (dBi)	15
Interference margin (dB)	0	Interference margin (dB)	0
Noise figure (dB)	2	Noise figure (dB)	7
SNR (dB)	−1.1	SNR (dB)	7.6
Sensitivity (dBm)	−114.5	Sensitivity (dBm)	−92.8
Max path loss (dB)	137.5	Max path loss (dB)	138.8
Max path loss + antenna gain (dB)	152.5	Max path loss + antenna gain (dB)	153.8

Table 10.15 Data channel link budget for mmWave case at 24–39 GHz.

5G uplink		5G downlink	
PUSCH 50 Mbps		**PDSCH 500 Mbps**	
Data rate (Mbps)	50.0	Data rate (Mbps)	500.0
Tx power (W)		Tx power (W)	4
Tx power (dBm)	26	Tx power (dBm)	36.0
Subcarrier spacing (kHz)	120	Subcarrier spacing (kHz)	120
		Total PRBs	100
PRBs	50	Used PRBs	100
Uplink share	20 %	Downlink share	80 %
Antenna gain (dBi)	26	Antenna gain (dBi)	26
Interference margin (dB)	0	Interference margin (dB)	0
Noise figure (dB)	5	Noise figure (dB)	7
SNR (dB)	5.9	SNR (dB)	11.0
Sensitivity (dBm)	−84.5	Sensitivity (dBm)	−74.4
Max path loss (dB)	110.5	Max path loss (dB)	110.4
Max path loss with beams (dB)	136.5	Max path loss with beams (dB)	136.4

will be very short compared to low bands because of a lower maximum path loss and because of a higher propagation loss. The propagation models are presented in Chapter 4 and propagation measurements in Chapter 11.

10.6 Coverage for Sub-6-GHz Band

The first part of the coverage discussion focuses on the 3.5 GHz band: signal propagation, beamforming antenna gain, and uplink solutions are discussed.

10.6.1 Signal Propagation at 3.5 GHz Band

Higher frequency leads to higher signal attenuation and shorter propagation. The Okumura–Hata propagation models show a 10 dB difference in path loss between 3.5 and 1.8 GHz. Nokia has conducted propagation measurements at 3.5 GHz, and the results are summarized in Figure 10.21: the outdoor propagation difference is 5.8 dB in non-line-of-sight (NLOS) conditions and less in the line-of-sight (LOS) case. There is an additional 3–5 dB indoor penetration loss at 3.5 GHz, which makes the total path loss difference 9–11 dB in indoor conditions. For more details, see [7]. We can conclude that there is a propagation difference between 3.5 and 1.8 GHz of approximately 10 dB. The difference is less outdoors and can be more in deep indoors.

10.6.2 Beamforming Antenna Gain

The physical size of the antenna becomes smaller when the frequency increases. Alternatively, the antenna gain can be increased while maintaining the antenna size at higher bands. A 3.5 GHz antenna can therefore provide higher gain than a 1.8 GHz antenna in the same form factor. Figure 10.22 shows typical three-sector four-port passive antennas at 700–900 MHz bands and at 2 GHz bands. The antenna at 2 GHz gives an 18 dBi gain

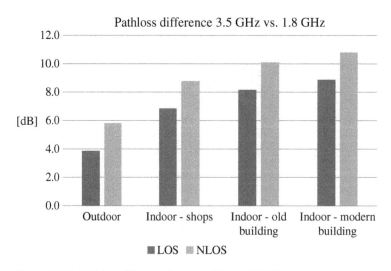

Figure 10.21 Path loss difference between 3.5 and 1.8 GHz.

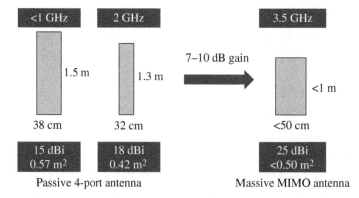

Figure 10.22 Antenna gain and size benchmarking of low band passive antenna with mid-band massive MIMO antenna.

with a frontal area of approximately $0.4\,\text{m}^2$. A massive MIMO antenna at $3.5\,\text{GHz}$ has a similar area while providing 25 dBi antenna gain, that is, 7 dB more than at 1.8 GHz. A higher antenna gain means that the beams become narrower. Therefore, traditional sectorization alone does not work, but beamforming solutions are required in 5G in order to benefit from the higher antenna gain.

The relative link budgets on different frequencies are shown in Figure 10.23. The Okumura–Hata propagation model is applied to obtain the relative path loss values on different frequencies. The same maximum path loss assumed for the different bands. The results show that the coverage becomes shorter on higher frequencies. The gain from massive MIMO at $3.5\,\text{GHz}$ is assumed to be 6 dB. The difference between the downlink and uplink is assumed to be 8 dB. The results show that $3.5\,\text{GHz}$ with massive MIMO has a similar or better downlink coverage compared to the 1.8 GHz uplink. But the 5G uplink coverage even with massive MIMO is 4 dB shorter than with LTE1800. The next section takes a look at the uplink coverage improvements.

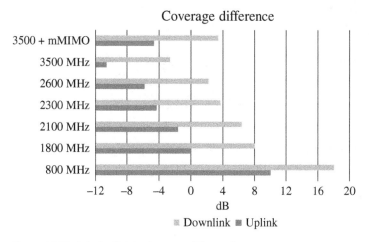

Figure 10.23 Relative link budgets on different frequencies.

10.6.3 Uplink Coverage Solutions

The uplink coverage can be improved by using a lower band. There are a few different options for taking advantage of low band. The solutions are discussed in more detailed in the following subsections.

1. Low band LTE with dual connectivity
2. Low band 5G with carrier aggregation
3. Low band 5G with supplemental uplink

10.6.3.1 Low Band LTE with Dual Connectivity

Dual connectivity is the first solution for improving the uplink coverage. First 5G devices are based on NSA and support dual connectivity between LTE and 5G radios. With split bearer in dual connectivity, the uplink can be mapped to either of the two systems. 5G can be configured as the primary entity in the split bearer case, and 5G radio is used for data transmission as much as possible. If the 5G uplink connection becomes weak and the throughput low, then the data volume in the uplink transmission buffer increases, and the UE will use LTE for the uplink data transmission. The high-level concept is shown in Figure 10.24.

The principles of dual connectivity are as follows:

- The UE decides the uplink data split between LTE and 5G. The threshold for 5G usage is defined by the network parameters. The 5G base station decides the downlink split between LTE and 5G radios.
- The Layer 1 and 2 control plane is needed in both radios.

While LTE is used for the uplink user plane, the 5G uplink must still deliver Radio Link Control (RLC) Acknowledgments (ACK/NACK) on the PUSCH as well as RLC status

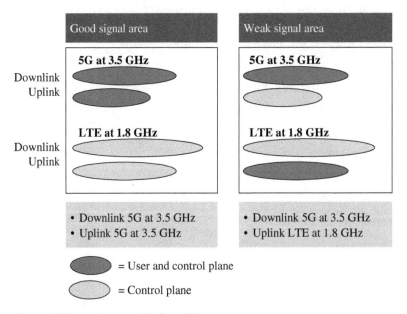

Figure 10.24 Dual connectivity for uplink coverage.

5G at 3.5 GHz

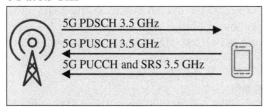

Dual connectivity 5G + LTE

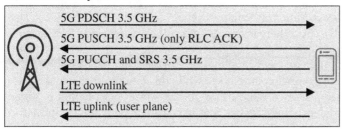

Figure 10.25 Physical channel locations and coverage.

reports, the SRS for 5G downlink beamforming, and the PUCCH for layer 1 hybrid automatic repeat request (HARQ) ACK/NACK. The physical channel locations are shown in Figure 10.25. Fortunately, the 5G uplink control channels have better coverage than data channels. Therefore, it is feasible to use LTE for the user plane while still running the control plane in 5G.

Dual connectivity allows inter-site operation where 5G and LTE connections use different sites. That capability can be useful in the early phase of 5G deployment where single 5G site can cover multiple LTE sites. The concept is shown in Figure 10.26. Note that in theory, coverage is shorter at higher bands. But 5G coverage at 3.5 GHz can be larger than LTE at 1800 MHz because there is no/low inter-cell interference at 3.5 GHz. LTE networks are mostly interference limited because they are designed to provide great indoor coverage. Therefore, one 5G cell at 3.5 GHz could potentially collect traffic from many LTE cells from those users that are in favorable conditions. If we simply assume that the LTE network design includes 20 dB indoor loss, the outdoor coverage area can be over 10× larger. That means one 5G cell can collect all outdoor users, and some indoor users, from the coverage area of 10 LTE cells.

Figure 10.26 Inter-site dual connectivity.

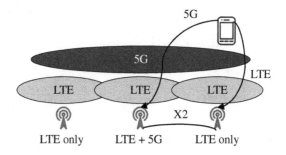

Some synchronization is required between LTE and 5G base stations for asynchronous dual connectivity, but the requirements are relatively relaxed. The maximum timing difference between LTE and 5G is allowed to be 250 μs for the 30 kHz 5G case; see [8]. This requirement is relaxed compared to the TDD network synchronization requirement of 3 μs.

10.6.3.2 Low Band 5G with Carrier Aggregation

Low band usage for 5G is one alternative to make 5G coverage better both in the uplink and in downlink. If there are no empty spectrum blocks available, spectrum sharing between 5G and LTE needs to be considered. The traditional refarming from 2G to 3G and to LTE was a slow process. Similar frequency domain refarming can be utilized also in the transition from LTE to 5G. There is another alternative available as well because LTE and 5G frame structures are so similar. That is dynamic spectrum sharing (DSS) in the time and frequency domains, which allows one to have full bandwidth for LTE and for 5G from the control channel point of view while the data channel resources are shared dynamically between the two radios. DSS requires fast coordination between LTE and 5G packet schedulers. The different low band 5G options are illustrated in Figure 10.27.

DSS brings benefits because it allows the allocation to be changed instantaneously between LTE and 5G. On the other hand, DSS has a lower spectral efficiency due to overhead. Figure 10.28 shows the channel structure in the time and frequency domains for the shared LTE–5G carrier. 5G capacity is impacted by LTE Cell Reference Signals (CRSs) and PDCCH overheads. There is less impact from the 5G control channels on LTE since 5G does not use any CRS. If all LTE UEs support Transmission Mode 9 (TM9), then the shared carrier has a lower overhead because less CRS transmission is required.

Low band 5G can be combined with higher band 5G with carrier aggregation in the same way as carrier aggregation in LTE. Low band 5G can be used for the uplink in the weak signal, while high and low band 5G together can be used for the downlink. The physical channel locations are shown in Figure 10.29 for the case where the downlink uses both low band and high band and the uplink uses just low band.

10.6.3.3 Supplemental Uplink

Supplemental uplink solution uses low band 5G but only for the uplink direction. Part of the LTE uplink spectrum is allocated for 5G, while the LTE downlink spectrum is kept

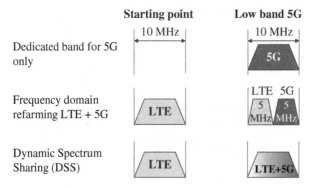

Figure 10.27 Low band 5G deployment options.

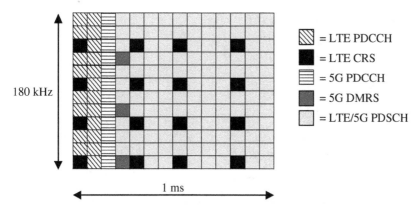

Figure 10.28 LTE and 5G sharing in time domain.

Low band (700 MHz) + high band (3.5 GHz) 5G

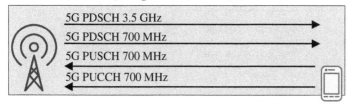

Figure 10.29 Physical channel locations with 5G carrier aggregation.

Figure 10.30 Supplemental uplink concept.

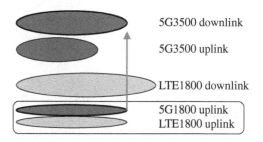

fully allocated for LTE. This approach is quite handy because traffic is highly asymmetric today: downlink traffic is typically 10 times more than uplink traffic. Therefore, the FDD uplink spectrum utilization is lower than the downlink utilization, and it is feasible to allocate part of the uplink spectrum for 5G. The supplemental uplink concept is shown in Figure 10.30.

Uplink sharing between LTE and 5G can be done partly in the frequency domain and partly in the time domain. The center frequency of the LTE carrier cannot be changed because the downlink LTE transmission uses full bandwidth. Therefore, the LTE PDCCH is located at the edge of the carrier. The data channels for LTE and 5G are located in the center of the carrier and switching between LTE and 5G is possible in the time domain, as shown in Figure 10.31.

Operating band combinations for the supplemental uplink are shown in Table 10.16. Many of the existing low band LTE options are included, especially for Europe and Asia.

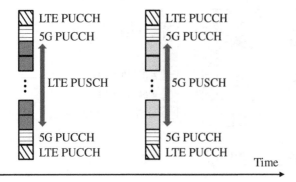

Figure 10.31 Options for frequency sharing between LTE and 5G.

Table 10.16 Supplemental uplink band combinations data from [9].

5G band	Supplemental uplink
5G 3500 (n78)	LTE 1800 uplink (n80)
5G 3500 (n78)	LTE 900 uplink (n81)
5G 3500 (n78)	LTE 800 uplink (n82)
5G 3500 (n78)	LTE 700 uplink (n83)
5G 3500 (n78)	LTE 2100 uplink (n84)
5G 3500 (n78)	LTE AWS uplink (n86)
5G 4500 (n79)	LTE 1800 uplink (n80)
5G 4500 (n79)	LTE 900 uplink (n81)

We should note, however, that Uplink Sharing from UE Perspective (ULSUP) is not mandatory for Release 15 devices. ULSUP refers to the case where 5G uplink and LTE use the same band in the case of dual connectivity. For example, if LTE uses 1800 MHz in dual connectivity, then 5G cannot use 1800 MHz for the supplemental uplink without network explicit resource sharing coordination. The ULSUP discussion is part of 3GPP Release 16. In Release 15, timeframe uplink sharing from the network point of view is the only supported mechanism for uplink sharing. This means that in the case of dual connectivity and supplemental uplink on the same band, it is the anchor LTE eNodeB that is required to control the sharing between LTE and the 5G uplink.

10.6.3.4 Benchmarking of Uplink Solutions

Dual connectivity is the first solution to improve the uplink coverage without any extra functionality, because the first 5G devices will anyway support dual connectivity. Dual connectivity has the additional benefit that it is feasible also for inter-site cases. Dual connectivity will increase latency since the uplink connection uses LTE instead of 5G, but it should not be a problem for mobile broadband services. Dual connectivity requires that 5G uplink control channels be optimized for maximum coverage.

Table 10.17 Benchmarking of uplink coverage solutions.

Solution	Description	Minimum latency (ms)	Architecture option
Dual connectivity	LTE for uplink user plane. 5G control plane at 3.5 GHz.	5–10	Option 3x
Low band 5G	5G uplink and downlink at low band.	1–2	Option 2 and Option 3x
Supplemental uplink	5G uplink at low band.	2–3	Option 2 and Option 3x

When new 5G services are required with the 5G core network, then the Option 2 Standalone architecture will be utilized. Dual connectivity is not available with Option 2, and low band 5G will be needed. Low band 5G can be aggregated with 3.5 GHz for high capacity and data rates.

The supplemental uplink is one option for uplink coverage, but the practical gain over dual connectivity is not obvious, and there are some challenges for the supplemental uplink transmission when LTE and 5G share the same frequency. The benchmarking of uplink coverage solutions are shown in Table 10.17.

10.7 Massive MIMO and Beamforming Algorithms

Massive MIMO is a major booster in radio performance. This section discusses some of the design factors that have an impact on massive MIMO benefits: antenna configurations, beamforming algorithms, and network functionality split.

10.7.1 Antenna Configuration

The main parameters defining the massive MIMO antenna configuration are the number of antenna elements, also known as physical antenna elements, and the number of transceivers (TRX), also known as logical antenna elements. The number of antenna elements defines the physical size of the antenna and the antenna gain, which is relevant for coverage. The typical antenna configuration at the 3.5 GHz band has 192 antenna elements with 12 rows in the vertical direction, 8 columns in the horizontal direction, and cross-polarization. The antenna element spacing is typically half wavelength. The wavelength depends on the frequency and is 8.6 cm at 3.5 GHz, leading to an antenna element spacing of 4.3 cm. The antenna height is then in practice 60 cm and the antenna width 40 cm by assuming a half wavelength (λ) distance between antenna elements and an additional half wavelength at the edges of the antenna. Such an antenna can be deployed at many macro sites from the size point of view. The theoretical antenna width as a function of the antenna columns is shown in Figure 10.32, illustrating that the eight-column solution is feasible from the antenna-width point of

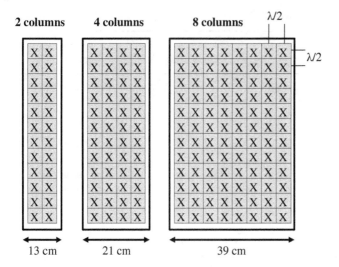

Figure 10.32 Antenna size at 3.5 GHz band with half wavelength spacing between antenna elements.

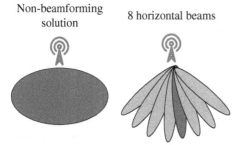

Figure 10.33 Up to eight horizontal beams.

view. The eight-column antenna allows the creation of up to eight independent beams in the horizontal domain. The case is shown in Figure 10.33.

The relationship between physical and logical antenna configurations is shown in Figure 10.34. The leftmost case shows the physical antenna with 12 rows, 8 columns, and cross-polarization. The number of TRXs is typically smaller than the number of antenna elements because some antenna elements can be combined. If there are 16 TRXs, then the logical antenna configuration is shown in the rightmost case with one row, eight columns, and cross-polarization. That case allows the creation of eight horizontal beams and a single vertical beam. The 16TRX antenna allows full exploitation of horizontal beamforming, but does not give any vertical beamforming. If there are 64 TRXs, then the logical antenna configuration is shown in the center case. 64 TRXs allows the creation of eight horizontal beams and additionally four vertical beams.

The gain of 64TRX over 16TRX depends heavily on the environment. If the area is flat and there are no high-rise buildings, the additional gain of vertical beamforming is limited. If there are high-rise buildings or high mountains, the gain from 64TRX can be higher. Figure 10.35 illustrates the case with vertical beamforming for high-rise buildings.

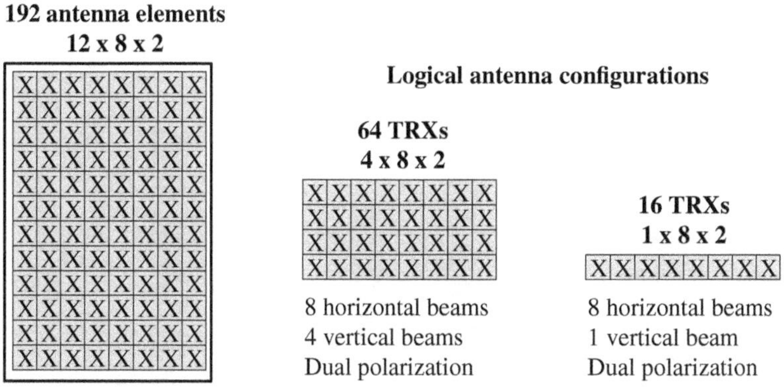

192 antenna elements
12 x 8 x 2

Logical antenna configurations

64 TRXs
4 x 8 x 2

8 horizontal beams
4 vertical beams
Dual polarization

16 TRXs
1 x 8 x 2

8 horizontal beams
1 vertical beam
Dual polarization

Figure 10.34 Physical and logical antenna configurations.

Figure 10.35 Vertical beamforming.

Figure 10.36 summarizes spectral efficiency simulations in the urban macrocell environment using the 3.5 GHz spectrum. The number of studied TRXs is 8, 16, 32, and 64. The number of antenna elements is 64, 128, and 192. We can note the following:

- For both 128 and 192 antenna elements, the spectral efficiency improves with increasing TRXs.
- For 128 antenna elements, 32 and 64 TRXs give around 20% and 25% spectral efficiency gains, respectively, over 16TRX operation.
- For 192 antenna elements, 32 and 64 TRXs give around 26% and 33% spectral efficiency gains, respectively, over 16TRX operation.
- 192 antenna elements with 16 TRXs give similar spectral efficiency when compared with 128 antenna elements. However, 64 TRXs give around 10% better performance with 192 antenna elements compared to 128.

10.7.2 Beamforming Algorithms

10.7.2.1 Grid of Beams and User-Specific Beams

The 3GPP specifications do not define algorithms for beamforming in the base station. Some simulations assume ideal beamforming algorithms in the downlink and in the uplink. The computational complexity of such algorithms may be too heavy for practical implementation. Therefore, we need to study simplified algorithms that can

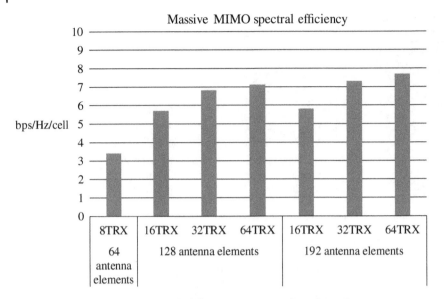

Figure 10.36 Spectral efficiency with different massive MIMO configurations.

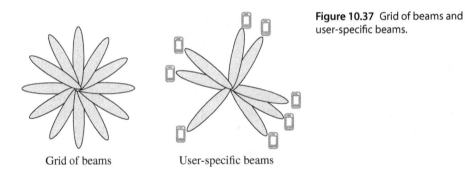

Figure 10.37 Grid of beams and user-specific beams.

deliver nearly the same performance. We take two example solutions in Figure 10.37: a grid of fixed beams and user-specific beams. Figure 10.38 shows the downlink benchmarking of user-specific beamforming with eigen beamforming (EBF) and the simpler grid of beams (GoBs). EBF gives +20% higher average throughput and +10% higher cell edge throughput than GoBs with multiuser Massive Multiple Input Multiple Output (MU-MIMO).

Figure 10.39 shows the uplink data rates with a full-blown Interference Rejection Combining (IRC) receiver and the combination of IRC and beam selection. The IRC receiver combines the signals to maximize the signal to interference and noise ratio (SINR). The alternative solution brings N best beam signals that are used for IRC combining. The simulations assume 64 receiver branches. The performance is very similar when using 8-16IRC with best beam selection: 8IRC receiver with beam selection gives 93% of the 64IRC performance, and 16IRC receiver with beam selection gives 97% of the 64IRC performance.

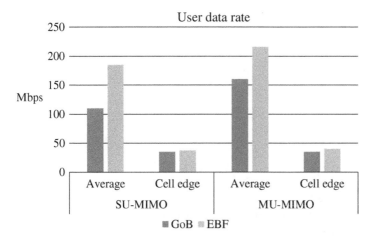

Figure 10.38 Downlink throughput with Grid of Beams (GoB) vs. user-specific beamforming (eigen beamforming (EBF)).

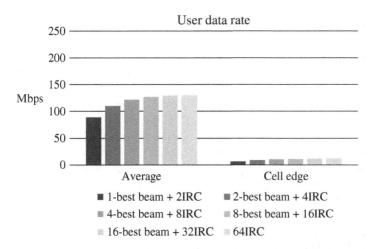

Figure 10.39 Uplink throughput with beam selection vs. Interference Rejection Combining (IRC).

10.7.2.2 Zero Forcing

The beamforming algorithm can be based on signal level maximization or signal-to-interference ratio maximization, which is also known as zero forcing or null-steering. The zero forcing algorithm requires that the interference information be known by the transmitter. The exact interference that affects the UE is known based on UE feedback, and the channel of the victim UE is known at the gNodeB based on the uplink SRS, which can then be used to zero-force toward that UE, that is, null the interference. This description assumes zero-forcing toward a victim UE in the same cell (MU-MIMO). To zero-force toward a victim UE in a neighbor cell, the cross-channel to the neighbor cell victim UE needs to be estimated by the gNodeB, and we would need also some coordination between the schedulers. The zero forcing algorithm can bring substantial improvements to the performance. Figure 10.40 illustrates the spectral

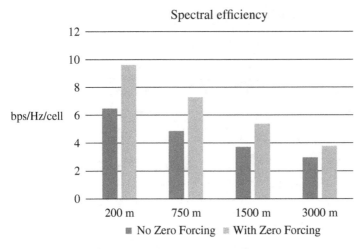

Figure 10.40 Downlink spectral efficiency with 32TRX with and without the zero forcing algorithm.

efficiency with and without zero forcing for the intra-cell coordination. The gain from zero forcing is 30–50% depending on the cell size.

10.7.2.3 Hybrid Beamforming

The number of TRXs in the beamforming antenna can be lower than the number of antenna elements. Part of the beamforming is done by digital processing in the baseband, and part of the beamforming by analog RF beamformers. This solution is called hybrid beamforming. Fully digital beamforming requires each antenna element to have a dedicated RF chain. The motivation for hybrid beamforming is to minimize the number of RF chains, which may lower the cost and complexity of the beamforming antenna. Hybrid beamforming is illustrated in Figure 10.41. This section illustrates the system performance of hybrid beamforming compared to fully digital beamforming at the 3.5 GHz band in urban microcells.

Hybrid beamforming simulations are run with 20 MHz bandwidth and with 10 UEs per cell, full buffer traffic, and single-user wideband scheduling. The user data rates would naturally be much higher with 100 MHz and fewer UEs with bursty traffic. The antenna is assumed to have 128 elements with eight rows, eight columns, and cross-polarization. The antenna element spacing is half wavelength. Four cases are

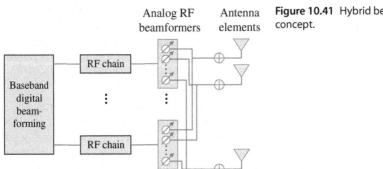

Figure 10.41 Hybrid beamforming concept.

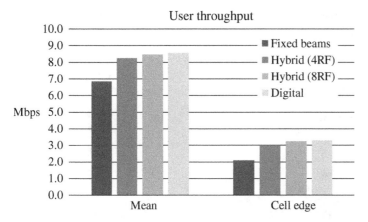

Figure 10.42 Downlink user throughputs with different beamforming solutions.

studied: fixed analog beams, fully digital beamforming with 128 RF chains, and hybrid solutions with four and eight RF chains. The results are illustrated in Figure 10.42. The simulations show that digital beamforming gives the best performance, while fixed beams have the lowest performance. The hybrid beamforming performance lies between the two other options. The difference between fully digital and hybrid beamforming is larger in the cell edge performance than in the average data rate. The average data rate with hybrid beamforming is 1–5% lower than with digital beamforming. The difference in the cell edge data rate is up to 10%. The gap between hybrid beamforming and full digital beamforming would become larger with frequency domain multiplexing of different UEs, and multi-user MIMO, given that the analog beamforming weights need to be applied jointly for all the scheduled UEs in the frequency and spatial domains. However, having twice the number of RF chains as the number of multiplexed streams in the spatial and frequency domains is expected to be enough to provide close-to-optimal performance [10]. Fixed beamforming provides a clearly lower data rate than digital beamforming: the difference is 20% in the average data rates and 35% in the cell edge data rates. The results show that it is feasible to consider hybrid beamforming from the performance point of view to simplify the practical implementation of the beamforming antenna.

10.7.3 Radio Network Architecture and Functionality Split

There are different options for splitting the baseband functionality in cloud RAN; see Figure 10.43. If more processing is located in the Radio Unit (RU) close to the antenna, the performance can be improved. If more processing is located in the baseband Distributed Unit (DU), the virtualization benefits are higher, and the complexity of the distributed unit is lower. This section compares the different uplink algorithms reflecting different functionality splits.

The benchmarking focuses on Options 7-2a and 7-3. Option 7-3 has more processing in the remote unit, while Option 7-2a has more processing in the centralized unit. The main differences between these two options are listed below:

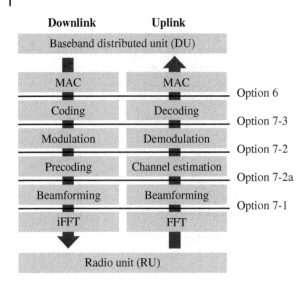

Downlink   **Uplink**

Baseband distributed unit (DU)

MAC	MAC

Option 6

Coding	Decoding

Option 7-3

Modulation	Demodulation

Option 7-2

Precoding	Channel estimation

Option 7-2a

Beamforming	Beamforming

Option 7-1

iFFT	FFT

Radio unit (RU)

Figure 10.43 Functionality split options between centralized and remote radio units.

Option 7-2a

- Best beams information for multi-user pairing and receiver processing is done either through wideband SRS (TDD) or from the UE feedback (TDD or FDD).
- Only beam space signals corresponding to the N best beams are made available (post digital beamforming) for channel estimation and further baseband processing.

Option 7-3

- N best beams for receiver processing are computed using reference signals over the allocated resource blocks for each UE.
- Best beams information for multiuser pairing is done either through wideband SRS (TDD) or from the UE feedback (TDD or FDD).
- Beamspace signals corresponding to all the beams are made available for channel estimation.
- Only the signals corresponding to the N beams are made available for further baseband processing.

Figure 10.44 shows uplink user throughputs with Options 7-2a and 7-3. The results are shown for 4RX IRC and 8RX IRC for both the 16TRX and 64TRX cases. 10 UEs per cell is assumed in the simulations with 10 MHz bandwidth. Option 7-3 uses SRS-based wideband beam selection for UE pairing and Demodulation Reference Signal (DMRS)-based beam selection for receiver processing. Option 7-2a uses SRS-based wideband beam selection for both UE pairing and receiver processing within fronthaul bandwidth limitations. Eight cross-polarized beams are assumed for Option 7-2a. The simulations indicate that Option 7-3 gives better performance than Option 7-2a in all cases, and the differences range from 1% to 4%. The smallest difference between the two options is with 4 RXs, 16 TRXs, and 3 km/h· while the largest difference is with 8 RXs, 64 TRXs, and 60 km/h. The benefits of Option 7-3 are greatest when a larger number of inputs to the IRC receiver are used, when a larger number of TRXs is available, and when the UE speed is high. On the other hand, Option 7-2a allows the use of more

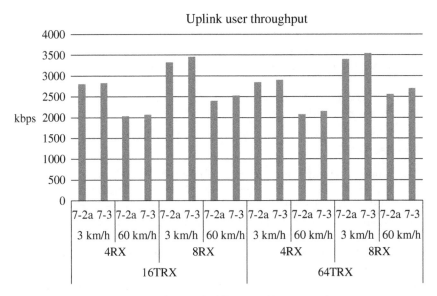

Figure 10.44 Uplink user throughput with different architecture options.

advanced receivers at the baseband module with higher processing capabilities. Option 7-2a also allows the realization of more gains from distributed MIMO when compared to 7-3 (Table 10.17).

10.7.4 RF Solution Benchmarking

Massive MIMO provides the most advanced radio performance, but it may not fit into all sites from the size or power consumption point of view. A multiband passive antenna with separate 4TRX or 8TRX RF heads may be an alternative solution. Table 10.18 illustrates high level benchmarking of 64TRX and 16TRX massive MIMO solutions and

Table 10.18 Benchmarking of different RF solutions for 3.5 GHz.

	Massive MIMO 64TRX	Massive MIMO 16TRX	Passive antenna 8TRX	Passive antenna 4TRX
Coverage gain in uplink vs. 2RX and 17 dBi antenna	9 dB[a]	9 dB	6 dB[b]	4 dB[c]
Capacity gain in downlink vs. 2TRX	4×	3×	1.8×	1.4×
EIRP	77 dBm with 200 W	74 dBm with 100 W	74 dBm with 8 × 30 W	71 dBm with 4 × 40 W
Multiband support	Hybrid antenna	Hybrid antenna	Yes	Yes

a) 7 dB antenna gain and 2 dB lower losses.
b) 4 dB diversity gain and 2 dB antenna gain (19 dBi).
c) 2 dB antenna gain and 2 dB antenna gain (19 dBi).

8TRX and 4TRX RF head solutions with a passive antenna at 3.5 GHz. The conclusion is that a high gain (19 dBi) passive antenna with an RF head can provide quite good coverage compared to massive MIMO, but the capacity will be lower than with massive MIMO beamforming. The RF head next to the passive antenna could be utilized if the site space is very limited. In general, we can note that there are more antenna and RF options available for 5G deployment compared to LTE. Most LTE networks use a passive antenna with 2×2 or 4×4 MIMO.

10.7.5 Distributed MIMO

Inter-cell interference is one of the main challenges in the mobile networks in the sub-6-GHz bands. One solution for improving the performance is distributed Multiple Input Multiple Output (dMIMO), which refers to coherent transmission and reception from multiple cells so that a set of devices is served simultaneously with many more dynamically controllable antenna elements than the number of devices. The concept is shown in Figure 10.45. The solution is known also as multi-TRP (TX/RX point) or Coordinated Multipoint (CoMP).

Figure 10.46 illustrates the distribution of the number of cells that can serve each user with 10 and 20 dB windows in a macrocell environment. The median number of cells with a 10 dB threshold is approximately 2 and with a 20 dB threshold is more than 5. The cell edge values (90%) are correspondingly 5 and 15 cells. It shows that there is a major potential to use multicell transmission and reception for improving the performance, especially for cell edge users.

Figure 10.47 shows the uplink user data rates in four different cases:

1. No distributed MIMO. Just single cell uplink reception.
2. Intra-site distributed MIMO. This approach is relatively simple since there are no transport requirements for multi-site coordination. Intra-site uplink CoMP is commercially deployed in most LTE networks today.
3. Inter-site distributed MIMO with 10 dB threshold.
4. Inter-site distributed MIMO with 20 dB threshold.

The results show that intra-site dMIMO can improve the performance by 40%, and it clearly makes sense to utilize that feature because cell edge UE transmission is likely

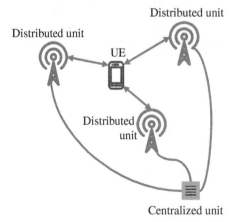

Distributed unit

Distributed unit

UE

Distributed unit

Centralized unit

Figure 10.45 Distributed MIMO uses multicell transmission and reception.

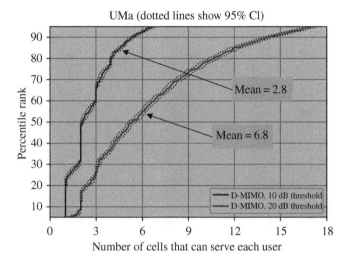

Figure 10.46 Number of cells that can serve the user in a macro environment.

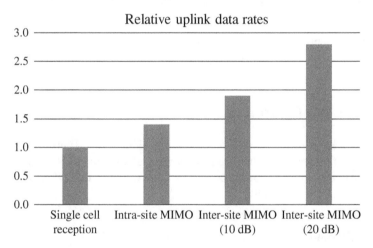

Figure 10.47 Relative uplink user throughputs with distributed MIMO options.

be received by multiple cells. Inter-site dMIMO can further improve the data rates by 40–100%. The relative gain in the cell edge data rates is even higher. It is an expected result that cell edge users benefit more from dMIMO than average users.

Figure 10.48 illustrates the potential downlink spectral efficiency gains with dMIMO. The studied case is small cells with 200 m distance, and each user is connected to a maximum of seven cells. The following cases are studied:

- Single cell transmission where each cell operates independently by using multi-user MIMO with four antennas.
- Seven site fixed disjoint clusters. Precoding coefficients are determined within each non-overlapping cluster.

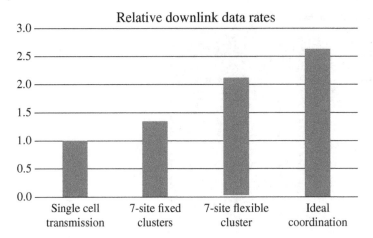

Figure 10.48 Relative downlink user throughputs with distributed MIMO options.

- Seven site clustering per user. Each user connects to sites within 26 dB of the strongest access, but to a maximum of seven sites. Precoding coefficients are determined centrally.

The results show that it is possible to even double the spectral efficiency by using per-user clustering compared to single cell transmissions.

10.8 Packet Scheduling Algorithms

3GPP does not define beamforming or packet scheduling algorithms. There are a number of options for configuring and optimizing scheduling algorithms in 5G radio. This section briefly illustrates two scheduling aspects which are different from LTE: low latency scheduling and mini-slot scheduling.

10.8.1 Low Latency Scheduling

The packet scheduler is responsible for enforcing Quality of Service (QoS) differentiation in the radio interface. A high-level overview of 5G scheduler functionality is depicted in Figure 10.49. The Medium Access Control (MAC) scheduler is the controlling entity for multi-user radio resource allocations, which is subject to several constraints but also has many options for efficiently serving the different devices. The enlarged number of options for the 5G scheduler offers performance improvements, but also presents the nontrivial problem of how to best utilize those degrees of freedom in an efficient manner. The 5G scheduler works by dynamically allocating radio transmission resources on a per-user basis for downlink and uplink transmissions. The scheduler aims at fulfilling the QoS service targets for all the bearers of the served users. The scheduler must support multi-cell connectivity mode, where devices are configured to be simultaneously served by multiple cells. There may also be other multi-cell coordination constraints including inter-cell interference coordination

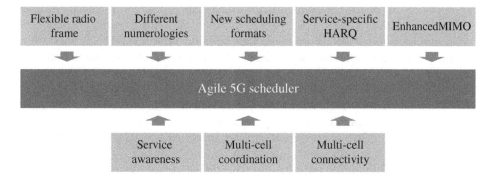

Figure 10.49 Agile 5G packet scheduler.

between neighboring cells where certain radio resources are dynamically muted, and hence not available for dynamic scheduling of users.

5G comes with a new flexible structure, consisting of 10 ms radio frames and 1 ms subframes. The subframes are constructed of slots that consist of 14 Orthogonal Frequency Domain Multiplexing (OFDM) symbols in the default setting. To support operation in different frequency bands, the physical layer numerology is configurable. The subcarrier spacing can scale from the base value by a factor 2 to the power of N from 15 to 120 kHz, or even larger values. The slot duration is 1 ms for 15 kHz, while it equals 0.5 ms for 30 kHz. Furthermore, mini-slots of two, four, or seven OFDM symbols are defined as well, with two-symbol mini-slots typically considered as the default. The smallest time-domain scheduling resolution for the MAC scheduler is mini-slot, but it is also possible to schedule users on slot resolution, or on the resolution of multiple slots. Thus, dynamic scheduling with different TTI sizes is supported. This enables the MAC scheduler to more efficiently match the radio resource allocations for different users in coherence with the radio condition, QoS requirements, and cell load conditions. The short transmission time is needed for Ultra-Reliable Low Latency Communication (URLLC) use cases, but is not restricted to such traffic. In the frequency domain, the minimum scheduling resolution is one physical resource block of 12 subcarriers, corresponding to 180 kHz for 15 kHz subcarrier spacing and 360 kHz for 30 kHz spacing. The MAC scheduler can freely decide how to schedule its different users on the carriers, and it is not visible to the RLC layer how this is done. But it is possible to enforce some restrictions via higher-layer control signaling to schedule data from certain bearers only on a given physical numerology, and a certain transmission size. Each scheduling allocation is sent to the UE via the downlink control channel carrying the scheduling grant. The downlink control channel is flexibly time-frequency-multiplexed with the other downlink physical channels, and can be mapped contiguously or non-contiguously in the frequency domain. This constitutes a design where the relative downlink control channel overhead can take values from below 1% values if, for example, scheduling few users with a long transmission time, and up to tens of percentages if scheduling a larger number of users with a very short transmission time. The design therefore overcomes the control channel blocking problems. These advantages are achieved by migrating toward a user-centric design, as compared to the predominantly cell-centric LTE design. Another advantage brought about by the more flexible 5G downlink control channel design is the support of UEs that operate only on a fraction of the carrier bandwidth, like narrowband IoT devices.

Efficient scheduling of LLC is rather challenging as such traffic is typically bursty and random, and requires immediate scheduling with a short transmission time to fulfill the corresponding latency budget. Instead of pre-reserving radio resources for LLC traffic bursts, it is proposed to use preemptive scheduling. Figures 10.50 and 10.51 illustrate the principle of preemptive scheduling for efficient transmission of LLC traffic. The basic principle is as follows: mobile broadband traffic is scheduled on all the available radio resource. Once an LLC packet arrives at the base station, the MAC scheduler immediately transmits it to the designated device by overwriting part of an ongoing scheduled transmission, using mini-slot transmission. This has the advantage that the LLC payload is transmitted immediately without waiting for ongoing scheduled transmissions to be completed, and without the need for pre-reserving radio resources for LLC traffic. The price of preemptive scheduling is paid by the user, parts of whose transmission are overwritten. To minimize the impact on the user that experiences the preemption, related recovery mechanisms are introduced. These include an indication to the victim device that part of its transmission has been preempted. This enables the terminal to take this effect into account when decoding the transmission; that is, it knows that part of the transmission is corrupted. Moreover, options for smart HARQ retransmission options are considered, where the damaged part of the preempted transmission is first retransmitted.

We show next that the flexible 5G design together with agile scheduler allows the tough performance requirements to be fulfilled. Results from system-level simulations are presented to illustrate the benefits of some of the 5G scheduling enhancements. We first show mobile broadband performance results for file download over the Transmission Control Protocol (TCP). Figures 10.52 and 10.53 show the performance for short (0.14 ms) and long (1 ms) TTIs, considering both the case with low offered traffic and high offered traffic. One of the reported performance metrics is the smoothed RTT of TCP packets. It is observed that the best performance is achieved for a short transmission time at a low offered load. This is due to the lower air interface latency that helps to quickly overcome the slow-start TCP phase. The higher control channel overhead from operating with a short transmission time is not a problem at the low offered load.

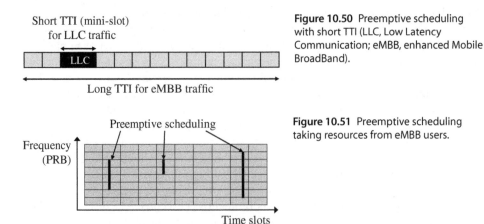

Short TTI (mini-slot) for LLC traffic

LLC

Long TTI for eMBB traffic

Figure 10.50 Preemptive scheduling with short TTI (LLC, Low Latency Communication; eMBB, enhanced Mobile BroadBand).

Preemptive scheduling

Frequency (PRB)

Time slots

Figure 10.51 Preemptive scheduling taking resources from eMBB users.

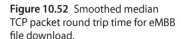

Figure 10.52 Smoothed median TCP packet round trip time for eMBB file download.

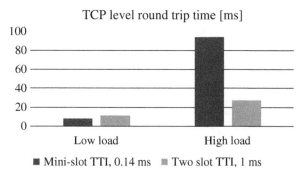

Figure 10.53 Median TCP level user throughput for eMBB file download.

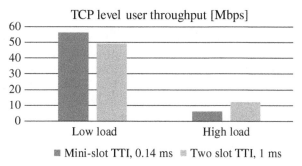

However, at the high offered load case, the best performance is observed for the case with a long transmission time. This is due to the fact that using a longer transmission time results in higher average spectral efficiency with lower control channel overhead. The results clearly show the benefit of being able to dynamically adjust the transmission time size.

Next, we present the downlink performance for a mixture of mobile broadband and LLC type of traffic. In this particular example, there are on average five active mobile broadband users per macrocell, performing a download of a 500 kB file size, using TCP. As soon as one of the mobile broadband users finishes its file download, the user is removed, and a new one is generated at a random location. In addition, there are on average 10 LLC users per cell, where small latency critical payloads of 50 bytes are sporadically generated according to a Poisson process. As this scenario corresponds to a fully loaded network, the mobile broadband users are scheduled with a transmission time of 1 ms, using all the available resources. Hence, no radio resources are reserved for potentially incoming LLC traffic. Instead, preemptive scheduling is applied whenever LLC payloads arrive. Those latency critical payloads are immediately scheduled on arrival with mini-slot resolution, overwriting part of the ongoing mobile broadband transmissions. Due to the urgency of the LLC traffic, we assume a block error rate of only 1% for such transmissions to avoid too many HARQ retransmissions. The average cell throughput is illustrated in Figure 10.54, where the performance is shown for cases with/without LLC traffic. For the cases with LLC traffic, the offered load is such that approximately 12% of radio resources is used for LLC. Two sets of results are shown for the case with LLC traffic: one where the full transport block is retransmitted for failed mobile broadband HARQ transmissions, and one where only the damaged part of the mobile broadband transmission that has been subject to preemption is retransmitted,

Average aggregated cell capacity [Mbps]

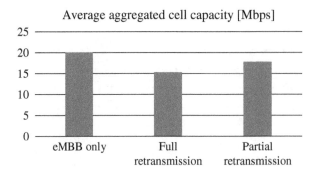

Figure 10.54 Average cell capacity for cases with LLC/eMBB and preemptive scheduling.

labeled as partial retransmission in Figure 10.54. The latter option is the most promising solution, as fewer radio resources for HARQ retransmissions of mobile broadband transmissions that have suffered from puncturing are used. However, the cost of using this approach is a slightly larger latency for the mobile broadband users, as the probability of triggering a second HARQ retransmission is higher, as compared to the case where the first HARQ retransmission includes the full transport block.

Figure 10.55 shows the complementary cumulative distribution function of the latency of LLC traffic. The latency is measured from the time when the LLC payload arrives at the base station until it is correctly received by the terminals, including the aggregated latency from scheduling, frame alignment, transmission time, and TX/RX processing delays. The ccdf shows that even under the considered full load conditions, the performance for the LLC traffic fulfills the challenging URLLC target of 1 ms latency with an

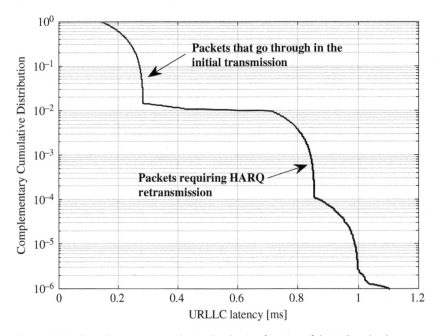

Figure 10.55 Complementary cumulative distribution function of the LLC packet latency.

outage of only 0.001% $(=10^{-5})$. Hence, the preemptive scheduling scheme fulfills its purpose: the ability to efficiently schedule the LLC traffic in line with its challenging latency and reliability constraints, while still having efficient scheduling of mobile broadband traffic without the need for pre-reservation of radio resources for sporadic LLC traffic.

10.8.2 Mini-Slot Scheduling

The mini-slot allows data transmission using just a part of the slot. Mini-slot allocations can improve the delay and throughput performance of users with small amounts of data. Using mini-slots for UEs with large amounts of data in their buffer will hurt system performance due to the additional control channel (PDCCH) overhead. Mini-slots can also result in fragmentation and inefficient use of resources, as all symbols may not be usable. The packet scheduler can decide how to share the resources and how to optimally utilize mini-slots. This section addresses these trade-offs and presents a scheme that can optimally use the flexibility of mini-slots to improve the performance of mobile broadband traffic. The objective is to dynamically determine the optimum mini-slot duration on a per-slot and per-user basis, which means determining the set of users which would benefit from mini-slot allocation and the optimum duration of the mini-slot to be used for each such user. The algorithm allocates users with smallest workloads first in order to minimize the slow truck effect, where users with a larger workload would delay others with a small workload. The algorithm also tries to minimize the time domain allocation and maximize the frequency domain allocation in order to minimize the latency.

A simple example is shown in Figure 10.56 with four UEs having data in the buffer. UE1 and UE2 have only small data amounts in the buffer, UE3 has slightly more data, and UE4 has a large amount of data. UE1 and UE2 are allocated two-symbol mini-slots to minimize their observed latency, UE3 is allocated the next four symbols, and UE4 is served last. All allocations are maximally using resources in the frequency domain as long as enough data is available in the buffer.

The performance of slot-based and mini-slot based resource allocations is compared, and the regime in which each of these approaches is optimal is determined. All users are scheduled on a slot (14-symbol duration) boundary. The reference case is slot-based resource allocation where each user allocation has a fixed duration of 14 symbols. Dynamic mini-slot allocation can utilize 2-, 4-, 7-, or 14-symbol allocation. A user may be scheduled multiple times within a slot under this scheme. Simulation results are shown in Figure 10.57 for an FTP traffic model with file sizes ranging from 1 kB to 10 kB, and the 20-MHz cell TDD cell loading ranging from 10 to 40 Mbps. The results show that flows with small file sizes and low loads provide the best gain with mini-slot-based scheduling. A gain as high as 130% is observed with 1 kb files. Large gains are due to

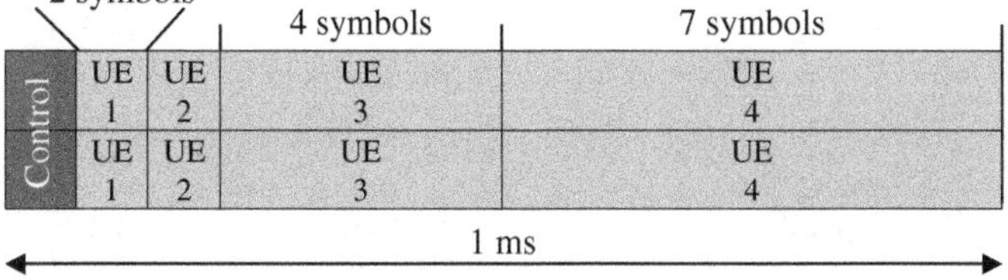

Figure 10.56 Example user multiplexing in time and frequency domains.

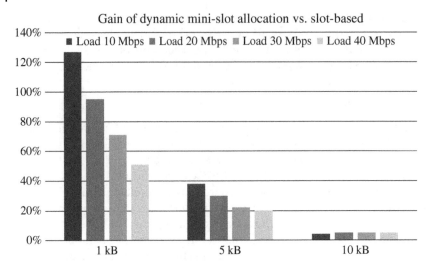

Figure 10.57 Gain of dynamic mini-slot allocation compared to static-slot-based allocation.

the use of mini-slots to transmit the packet, as opposed to the use of a slot. Gains of mini-slot based scheduling diminish rapidly both with increasing load and increasing file size. For a given load, as the file size increases, longer mini-slots would be needed to transmit the file, so the difference between the two schemes decreases. For a given file size, as the load increases, there is likely some queueing delay which degrades the performance gain from using mini-slots.

10.9 Spectral Efficiency and Capacity

10.9.1 Downlink Spectral Efficiency in 5G Compared to LTE

One of the 5G targets is to improve spectral efficiency compared to LTE. Since the single link performance is limited by the Shannon theorem, there are no simple solutions to improve the link efficiency. The efficiency improvements must come from more advanced usage of antennas and from system-level efficiency. Table 10.19 illustrates the

Table 10.19 5G features improving spectral efficiency compared to LTE.

5G features	Sub-1-GHz FDD 10 MHz 4 × 2 MIMO	Sub-6-GHz TDD 100 MHz Massive MIMO
Spectral usage	+4%	+8%
Lean carrier	+10%	+10%
Wideband carrier	—	+10%
MIMO optimization	+15%	+25%
Total	Approx. +30% (=1.04*1.10*1.15)	Approx. +60% (=1.08*1.10*1.10*1.25)

Table 10.20 Expected downlink spectral efficiency.

Spectrum	Bandwidth	Antennas	LTE (bps/Hz/cell)	5G (bps/Hz/cell)
Sub-1-GHz	10 MHz FDD	2 × 2 MIMO	1.7	2.2
2 GHz	20 MHz FDD	4 × 4 MIMO	2.5	3.3
3.5 GHz	100 MHz TDD	mMIMO	6.3	10.4

main technologies that can improve the efficiency in 5G radio compared to LTE: spectral usage, lean carrier, wideband carrier, and MIMO codebook. More efficient spectral usage in 5G yields a small improvement in efficiency. The LTE transmission bandwidth is 90% of the channel spacing. For example, 10 MHz LTE uses 9 MHz for transmission and 1 MHz for guard bands. 5G is designed to utilize the channel spacing more efficiently. 5G in 10 MHz with 15 kHz subcarrier spacing has 94% utilization, and 5G in 100 MHz with 30 kHz subcarrier spacing has 98% utilization.

Lean carrier means that there are no cell-specific reference signals in 5G. The relative impact of reference signals depends on LTE cell loading and transmission power. Reference signals make 5% of LTE cell transmission power at full load or 50% at low load. We assume here 10% gain from lean carrier by assuming 50% loading in adjacent cells. When the load is low, the relative gain from lean carrier is higher especially for the cell edge users.

Wideband carrier helps 5G performance when the spectrum is more than 20 MHz. LTE can utilize more spectrum than 20 MHz with multiple carriers, but there is a penalty in efficiency since each carrier must have common channels, and load balancing is needed between carriers. We assume here a modest 10% gain for the 100 MHz carrier compared to the 5 × 20 MHz case. The gain depends on the traffic type and network algorithms.

Multiantenna MIMO performance depends on channel measurements, signaling capability, and algorithms. We assume here a 15–25% gain in 5G compared to LTE because of the more advanced MIMO codebook and calibration in 5G. Also, common channels can take advantage of beamforming in 5G. The total gain of 5G compared to LTE is then expected to be +30% for low band FDD and +60% for sub-6-GHz TDD with massive MIMO.

In practice, the relative gain from 5G is higher with massive MIMO because many legacy LTE devices do not support massive MIMO.

The expected spectral efficiency values for LTE and for 5G for typical antenna configurations are shown in Table 10.20. We assume that the LTE 2 × 2 spectral efficiency is 1.7 bps/Hz/cell based on live network experience. We further assume that 5G provides a 30–65% gain over LTE. We also assume that 4 × 4 gives a 50% gain over 2 × 2, and massive MIMO gives a 150% gain for LTE. These results illustrate that there can be up to six times higher spectral efficiency with 5G, massive MIMO, 4RX UEs, and 100 MHz carrier compared to LTE with 2 × 2 MIMO.

10.9.2 Downlink Spectral Efficiency with Different Antenna Configurations

The 5G spectral efficiency depends on a number of assumptions: base station antenna solution, UE antenna solution, and also on the baseband capability in terms of the maximum number of parallel MIMO streams. The 5G downlink spectral efficiencies for different configurations are summarized in Figure 10.58. The efficiencies naturally depend also on the propagation environment, cell size, network optimization, and algorithms. We want to illustrate the expected efficiencies and the impact of the main design parameters. The target is to show the technology capability in optimized cases. The results are based on the number of simulations and field measurements. The following assumptions are used for obtaining the numbers:

- LTE 2×2 MIMO efficiency is 1.7 bps/Hz/cell.
- 5G gain over LTE is 30% at low bands.
- 4RX UE gives 30% gain over 2RX UE.
- 4TX and 8TX gains over 2TX in downlink are 40% and 80%.
- Massive MIMO with 64TX gives 250% gain over 2TX.
- 64TRX efficiency is 30% more than 16TRX efficiency.
- 64TRX efficiency is 10% more than 32TRX efficiency.
- Doubling the number of MIMO layers gives 5% more capacity.

10.9.3 Uplink Spectral Efficiency

The typical traffic asymmetry in LTE networks today is 10:1, which means that the downlink data volume is 10 times more than the uplink data volume. The downlink direction is the bottleneck in FDD deployments, but the importance of uplink efficiency grows in TDD, where a larger part of the frame is allocated for the downlink direction. The uplink efficiency in 5G is improved by using Orthogonal Frequency Division Multiple Access

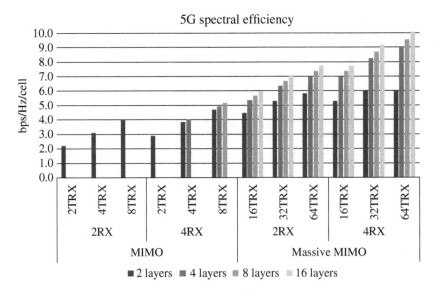

Figure 10.58 Summary of 5G downlink spectral efficiency.

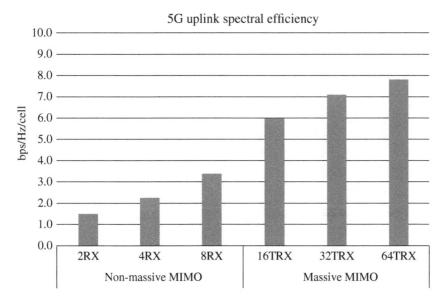

Figure 10.59 Uplink spectral efficiency with different antenna configurations.

(OFDMA) compared to a single carrier in LTE. The single carrier solution is fine for the coverage-optimized cases and for low spectral efficiency. OFDMA gives a clearly higher efficiency than the single carrier. A generic uplink spectral efficiency table is shown in Figure 10.59 using the following assumptions:

- LTE 2RX uplink is the reference point with a spectral efficiency of 1.0 bps/Hz/cell.
- 5G gain over LTE is 50%.
- 4RX gain over 2RX is 50%.
- 8RX gain over 4RX is 50%.
- Massive MIMO 16TRX efficiency is 6.0 bps/Hz/cell.
- 64TRX efficiency is 30% more than 16TRX efficiency.
- 64TRX efficiency is 10% more than 32TRX efficiency.

10.9.4 IMT-2020 Performance Evaluation

ITU has defined the minimum performance requirements that IMT-2020 technology must fulfill. One of the requirements was spectral efficiency in different environments in the downlink and in the uplink. Simulations were run to find out if 5G technology is able to meet the requirements. The downlink requirements and simulation results are shown in Figure 10.60, and the uplink requirements and simulation results in Figure 10.61. Three different environments are studied: indoor, dense urban, and rural cases. An indoor hot spot considers a very dense small cell deployment with 20 m inter-site distance. A dense urban environment has an inter-site distance of 200 m with 80% users indoors and 20% outdoors. A rural environment has an inter-site distance of 1732 m with 50% indoor and 50% in-car users. The carrier frequency was 4 GHz in all cases, 8RX UE was assumed, the zero forcing multiuser-MIMO algorithm, and non-ideal SRS-based beamforming. The cell edge spectral efficiency was simulated

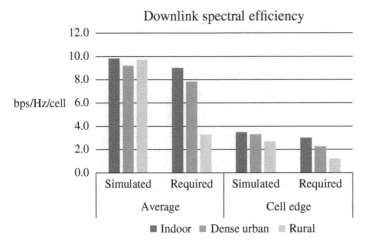

Figure 10.60 Downlink spectral efficiency evaluation for IMT-2020.

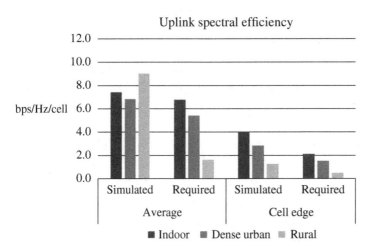

Figure 10.61 Uplink spectral efficiency evaluation for IMT-2020.

with 10 UEs per cell. We show here the cell edge UE throughput multiplied by 10 for compatibility with the average spectral efficiency.

The results show that 5G is able to meet the IMT-2020 requirements in all cases. The simulated downlink spectral efficiency is 9–10 bps/Hz/cell and the uplink efficiency 7–9 bps/Hz/cell. The cell edge efficiency is naturally lower than the average efficiency. The difference between average and cell edge data rates is approximately three times in the downlink. The difference in the indoor case and in the dense urban case is approximately two in the uplink, while the difference is seven in the large rural case, where the cell edge performance is heavily limited by the UE power.

For more details, see [11].

10.9.5 5G Capacity at Mid-Band

The combination of larger bandwidth and beamforming antenna results in a substantial increase in user data rates and capacity. Therefore, 5G deployment with massive MIMO using 100 MHz bandwidth at 2.5/3.5 GHz can boost mobile broadband performance compared to the LTE network. The capacity increase is so large that 5G can potentially also be used for fixed wireless access in addition to mobile broadband. The fixed wireless users tend to consume a lot more capacity than mobile users: mobile customers use typically 10–20 GB/month, while fixed wireless customers use 100–200 GB/month. 5G may provide enough capacity to satisfy even the needs of fixed wireless access. Figure 10.62 illustrates the typical user data rate and sector capacity for LTE with 2×2 MIMO and $20 + 20$ MHz spectrum compared to 5G with massive MIMO and 100 MHz spectrum. The 5G spectral efficiency is assumed to be 10 bps/Hz with 80% of the frame allocated for the downlink, which gives a sector capacity of 800 Mbps. A typical LTE network in urban area uses $20 + 20$ MHz aggregation. 5G can increase network capacity by a factor of 10 while reusing the same base station sites.

10.10 Network Energy Efficiency

There is a clear motivation for addressing energy efficiency in 5G networks: we want to keep the energy consumption of the mobile network on the current level, or even lower it, while data traffic keeps increasing and number of base stations keeps increasing. That means up to 1000 times more traffic with similar energy consumption. The network energy efficiency was not a target when 3G or 4G technologies were defined; the focus was only on the device power efficiency. The thinking was that the network energy efficiency is just an implementation topic. As the implementation technology becomes

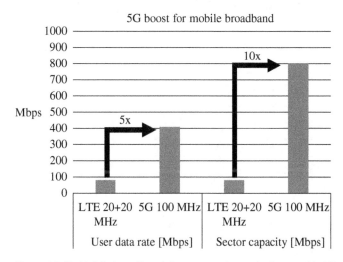

Figure 10.62 Mobile broadband data rate and capacity boost with 5G at 3.5 GHz.

more efficient, the energy efficiency is improved. But this evolution is too slow compared to the traffic increase. Therefore, 5G networks need to include more efficient system-level solutions to reduce the power consumption. The focus has been especially on the energy efficiency during low loading. The learning from LTE networks has been that the average network level utilization is typically 10–25% over 24/7 for the whole network. The relatively low utilization can be explained by the fact that the traffic is unequally distributed over a 24-hour period and unequally distributed over the geographical area. The loading is low in the network during the night time, and there are number of cells in the network that are needed for coverage and carry only low traffic volumes. All this means that base stations send nothing 75–90% of the time even in high-loaded networks. The LTE base station power consumption is relatively high also during the idle times because of the transmission of common reference symbols. 5G does not have common reference symbols, and 5G can better utilize power saving techniques in the base station. See Chapter 3 for more details on the technology components. If the low load power consumption can be minimized with 5G, there is a major potential for improving the network-level energy efficiency. Figure 10.63 shows the share of off-time in the base station transmission without any data transmission. The off-time share varies between 91% and 99% depending on the size of Synchronization Signal Block (SSB) transmission. It shows that the power consumption could theoretically be more than 90% lower for an idle base station compared to full power transmission.

Figure 10.64 shows the estimated gains of the power savings technology. The simple assumption is that the base station power consumption is the same for LTE and for 5G during 100% loading. When the loading becomes lower, the power consumption also decreases both in LTE and in 5G radios, but the impact of loading is clearly higher in 5G with more efficient power saving solutions. The estimation shows that 5G macro base station power consumption can be even below 50 W in the very best case. For more details, see [12].

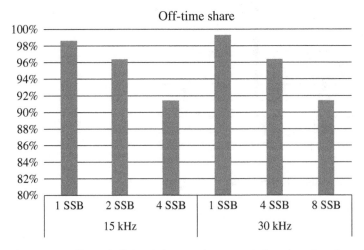

Figure 10.63 Share of off-time in base station transmission when there is no data transmission.

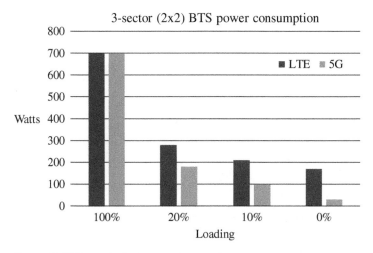

Figure 10.64 Example three-sector base station power consumption.

Radio power consumption in terms of kWh per GB of transmitted data is presented below. Three cases are assumed:

- 3G: Three-sector macro base station with 10 MHz of spectrum
- 4G: Three-sector macro base station with three carriers with a total of 50 MHz of spectrum
- 5G: Three-sector massive MIMO base station with 100 MHz of spectrum
- 5G: Small cell with mmWave and a total of 800 MHz of spectrum

The electricity price is assumed to be 20 cent/kWh. The detailed assumptions are shown in Table 10.21, and the results in Figure 10.65. The 3G macro base station delivers mobile data at 5 cent/GB and LTE 2 cent/GB from the electricity bill point of view. 5G massive MIMO and 100 MHz brings so much more capacity that the energy efficiency improves considerably by a factor of 10 compared to LTE. The 5G base station delivers

Table 10.21 Assumption for the energy efficiency calculations.

System parameter	3G macro	4G macro	5G macro	5G small
Efficiency (bps/Hz/cell)	1	2	10	6
Spectrum (MHz)	10	50	100	800
Sectors	3	3	3	1
Busy hour loading	80%	80%	80%	80%
Busy hour share	7%	7%	7%	7%
Sectors for 50% traffic	15%	15%	15%	10%
BTS power consumption (W)	500	2000	2000	50
Electricity cost (cent/kWh)	20	20	20	20

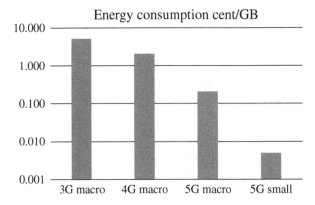

Figure 10.65 Energy consumption in eurocent/GB.

data at 0.2 cent/GB. The 5G small cell at mmWave can improve the efficiency further by a factor of 40 compared to the macrocell down to 0.005 cent/GB.

10.11 Traffic and Device Density

The very high traffic density up to 1 Tbps (1000 Gbps) per km^2 is one of the 5G targets. The traffic density can be increased by using more spectrum, more sites, and higher spectral efficiency. 5G brings together all these components for increasing capacity. Let us start first by estimating the highest traffic densities in the current LTE network. We exclude mass events from this calculation. We make the following assumptions for the traffic estimation: 100 000 population/ km^2, operator market share 40%, 5–10 GB/sub/month usage, and busy hour usage 7%. This simple calculation gives 1–2 Gbps/km^2. The typical LTE network in the busy area uses 50 MHz of spectrum and 30 base station sites per km^2. We further assume an LTE spectral efficiency of 2 bps/Hz/cell. We also take into account that the traffic is not equally distributed between the sites. The assumption is that the highest loaded sites carry 4× more traffic than the average site. These assumptions give a capacity of 2 Gbps/km^2, matching with the highest traffic densities (Table 10.22).

Table 10.22 Traffic density assumptions.

System parameter	LTE today	5G below 6 GHz	5G at 24–39 GHz	5G at >50 GHz
Spectrum	50 MHz	100 MHz	800 MHz	3000 MHz
Sites	30/km^2	50/km^2	150/km^2	300/km^2
Sectors per site	2.5	2.5	2.5	2.5
Spectral efficiency	2.0 bps/Hz	6.0 bps/Hz	6.0 bps/Hz	6.0 bps/Hz
Traffic distribution	4	6	15	15
Traffic density	1.9 Gbps/km^2	13 Gbps/km^2	120 Gbps/km^2	900 Gbps/km^2

The capacity can be increased by fully utilizing the existing frequencies below 6 GHz for 5G. We assume that up to 200 MHz spectrum per operator is available, including some unlicensed 5 GHz band. We also assume that the site density can be increased from 30 to 50/km^2 and the spectral efficiency can be increased by a factor of 3 compared to LTE-Advanced. We also assume that the traffic distribution becomes less equal with a higher site density. The result is a capacity up to 13 Gbps.

If the site density needs to be increased further, the cell range decreases to 50–100 m, and it makes sense to utilize mmWave. We assume 800 MHz of spectrum and 150 sites/km^2 with 24–39 GHz. This deployment gives more than 100 Gbps/km. Spectrum above 50 GHz gives access to even more spectrum and higher site density. We assume 3000 MHz and 300/km^2 corresponding to a cell range below 50 m. This deployment case gives approximately 1 Tbps/km^2 (Figure 10.66).

In addition to the high traffic density, 5G must also support high device density. The 5G target is to support 1 million devices per km^2. The high device density is needed for the massive number of IoT objects. There are no theoretical limits for the device density if the data volumes are low. We present here the required signaling capacity to support a large number of devices. The signaling (setup) capacity and the connectivity are likely the first bottlenecks in the LTE system. We assume 50 base stations per km^2 and a traffic distribution factor of 4.0 (see Table 10.23). The resulting signaling capacity depends on the frequency of the transmission. We consider here transmission frequencies from once per 10 s to once per hour. The results are shown in Figure 10.67. The required signaling capacity per base station is up to 12 000/s in the extreme case with a transmission frequency of once per 10 s. The corresponding signaling is 2000 setups per second per base station with once per minute and 33 setups per second per base station with once per hour activity. For reference, the existing LTE solutions can support more than 1000 setups per second per base station, which already allows a very high IoT density.

Table 10.23 Assumptions for the device density.

Number of devices	1 million
Sites	50/km^2

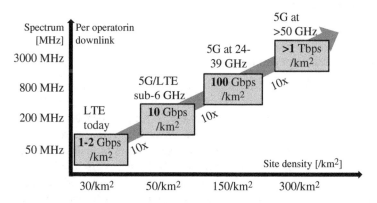

Figure 10.66 Maximum traffic density per km^2.

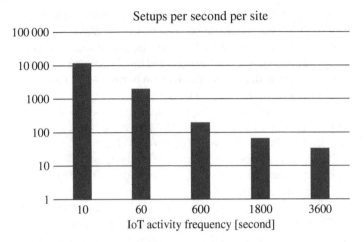

Figure 10.67 Signaling transactions per base station per second.

10.12 Ultra-Reliability for Mission-Critical Communication

The quality of the radio connection has a major impact on the overall system reliability and latency. SINR is used to measure the quality of the radio link. The higher the SINR, the lower the packet error probability, which results in higher reliability and lower latency communication. We will consider the technical solutions that can improve SINR values.

10.12.1 Antenna Diversity

The increased number of antennas results in better coverage. 2TX2RX is commonly used in today's LTE networks, with some operators also using the 4TX4RX base station both in FDD and in TDD, and 8TX8RX in TDD. Some LTE networks utilize also massive MIMO with 64TRX. For the 5G system, significant diversity enhancements using a very large number of antennas can be utilized.

10.12.2 Macro-Diversity and Multi-Connectivity

Mobility and handover are critical for ultra-reliability and low latency. The break-before-make method of LTE produces interruption times of a few 10 ms, which become an obstacle for reliable communication. Virtually zero interruption time must be the target for 5G, with make-before-break at every handover for always-available data connectivity. The main concept consists of multiple base stations transmitting synchronously the same information, which is then combined at the receiver. This helps combat shadowing effects, by means of diversity and redundancy, and increases the total received power of the desired signal. Multi-connectivity can also be obtained with Packet Data Convergence Protocol (PDCP) duplication.

10.12.3 Interference Cancelation

Mitigating the interference with either network-based or terminal-based techniques has been identified as a promising complementary solution to improve the SINR. The received interference from neighboring base stations or terminals can be reduced in order to improve the SINR.

10.12.4 HARQ (Hybrid Automatic Repeat Request) for High Reliability

The use of retransmissions is a way to increase reliability at the expense of latency. HARQ is an error correction mechanism based on retransmissions. The receiver produces either an ACK for the case of error-free transmission, or a NACK if errors are detected. Upon reception of a NACK message, the desired packet will be sent again.

The contributors to the HARQ RTT (Hybrid Automatic Repeat Request Round Trip delay Time) are: BTS and UE processing time, the propagation time, the backhaul delay, and the TTI size. In the context of a flexible design, the duration of the ACK/NACK is also flexible and configurable per UE. Such a scheme enables the HARQ RTT to be set not according to the "worst condition UE" in the cell, but optimized per link. The latency can be also reduced by exploiting early feedback based on a prediction of the detection success.

The practical system-level reliability is studied in system- and link-level simulations [13, 14]. First, system-level simulations are run to find out the SINR distribution. Second, the link-level performance as a function of the SINR is shown for the control and data channels. System-level radio reliability can be evaluated by combining these two simulations. The evaluation approach is shown in Figure 10.68.

The system simulations follow the urban macrocell environment. The studies show that the 5th percentile cell edge SINR values at 4 GHz are 14.5 dB in the downlink and 3.2 dB in the uplink. The corresponding values at 700 MHz are 9.4 and 8.0 dB. We note that the higher band (4 GHz) has a higher downlink SINR due to a more limited interference propagation, while the lower band (700 MHz) has a higher uplink SINR due to better signal propagation since the uplink performance is limited by the UE output power.

Figure 10.68 System-level reliability evaluation.

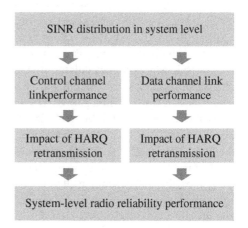

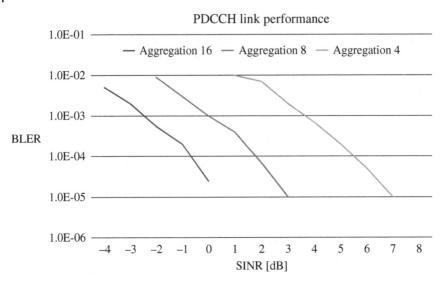

Figure 10.69 Physical Downlink Control Channel (PDCCH) performance.

The link performance of the PDCCH is shown in Figure 10.69, where the performance for the aggregation levels of 4, 8, and 16 are shown. The results show that the link-level reliability can be improved by higher aggregation levels. The simulations use a 39-bit payload on one OFDM symbol with precoder cycling at 4 GHz.

The link performance of the PDSCH/PUSCH was simulated for mini-slots with four and seven OFDM symbols per slot using a 30 kHz subcarrier spacing using a non-LOS channel (363 ns delay spread). The results are shown in Figure 10.70. The number of data bits per slot was 256 bits (32 bytes, cyclic redundancy check [CRC] overhead not considered) for four symbols per slot and 288 bits (36 bytes, with CRC overhead) for seven symbols per slot. The results show that the seven-symbol mini-slot with a longer transmission time has better performance than the four-symbol mini-slot. The 1 ms latency requirement with URLLC can be met with these four- and seven symbol mini-slots. From these results, we can see that an SINR of at least 5.4 dB is required to achieve a BLER of less than 10^{-5} with seven symbols/slot. With seven symbols/slot, an SINR of at least 7.6 dB is required. These values provide a lower bound for data channel reliability, as control channel reliability is also taken into account.

The total reliability is calculated by using the methodology where the effects of the control and data channel reliability are combined over one or more HARQ transmission attempts. A retransmission gain of 2 dB is assumed with soft combining of all transmissions and using four OFDM symbols per slot. For both the downlink and uplink, one front-loaded DMRS is used as overhead, and a full band (10 MHz) allocation is used for the PDSCH and PUSCH. An aggregation level of 16 is selected for the PDCCH, and the payload is 39 bits. The results show that s BLER of less than 10^{-5} can be obtained at the 5th percentile (95% of time) in all four cases at 4 GHz and 700 MHz both in the downlink and uplink. Other cases are in the non-LOS channel except for the 4 GHz uplink in the line of sight. The system-level reliability results are summarized in Table 10.24.

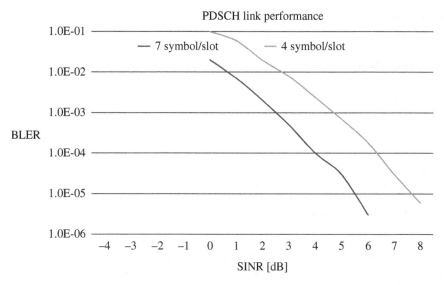

Figure 10.70 Physical Uplink/Downlink Shared Channel (PUSCH/PDSCH) performance with 30 kHz subcarrier spacing, 10 MHz bandwidth, and 3 km/h.

Table 10.24 Probability for BLER $>10^{-5}$.

Number of transmissions	4 GHz		700 MHz	
	Downlink (%)	Uplink (%)	Downlink (%)	Uplink (%)
1	0.6	4.2	2.5	4.5
2	< 0.1	4.0	0.9	2.3
3	< 0.1	4.0	< 0.1	1.4

10.13 Mobility and High-Speed Trains

The IMT-2020 requirement for mobility is 500 km/h, which is also a relevant case for high-speed trains like Shinkansen in Japan, Train à Grande Vitesse (TGV) in France, Inter-City Express (ICE) in Germany, Eurostar in France, or the Korean Train Express (KTX). Fast-moving UEs are challenging because the Doppler effect degrades the frequency tracking accuracy and channel estimation accuracy. This leads to low throughput caused by loss of orthogonality between subcarriers. The Doppler shift also affects detection of PRACH preambles. 3GPP has considered high-speed mobility in 5G physical layer design by including the following items:

- DMRS, whose frequency can be flexibly defined per UE depending on the mobility requirements. For low-speed scenarios, DMRS uses a low density in the time domain. For high-speed scenarios, the time density of DMRS is increased to track fast changes in the radio channel.
- PRACH design includes considerations for high-speed mobility. For short sequences, large subcarrier spacings (15 and 30 kHz) provide in-built protection against the

Doppler shift. For long sequences and low subcarrier spacings (1.25 and 5 kHz), restricted sets can be used. These are based on specific cyclic shifts that can be used for the Zadoff Chu sequence with high Doppler.

- Layer 3 (RRC based) mobility framework with flexible filtering and reporting capability.
- Beam management framework with configurable Channel State Information Reference Signal (CSI-RS) transmission and reporting.

When the UE is moving, it will experience a Doppler shift that changes the carrier frequency of the base station radio transmission. The amount of Doppler shift depends on the angle α between the direction of motion of the UE and the propagation vector of the radio wave. The Doppler shift is calculated as follows:

$$Dopper \; shift = \frac{v}{c} f_c \cos \alpha$$

where v is the UE speed in meters per second, c is the speed of light in meters per second, fc is the carrier frequency in Hz, and α is the angle between the velocity vector and LOS with the receiver. The Doppler shift is maximum if the UE moves along the radio propagation line, while it is zero if the UE moves perpendicularly against the radio propagation vector. The Doppler shift and the change of shift are illustrated in Figure 10.71 for the case where the UE passes the base station and the Doppler shift changes rapidly from positive to negative values. The practical Doppler shift experienced by the UE can be smaller since the UE synchronizes to the downlink signal. On other hand, the Doppler shift in the uplink can be correspondingly higher.

OFDM-based systems rely on the orthogonal spacing between the subcarriers, and receiver performance depends on their good localization in the frequency domain. Otherwise, OFDM receivers suffer from inter-carrier and inter-symbol interferences. Frequency error may also be caused by drifts in local oscillators in the UE and in the base station. 3GPP has defined the maximum frequency error in the UE as ±0.1 ppm and in the base station as ±0.05 ppm The worst-case Doppler shift is then the combination of the UE movement and the UE and base station frequency error. Table 10.25 shows the worst-case Doppler shift when the UE is moving at 500 km/h and when the UE and base station frequency error is included.

In order to avoid interference between OFDM subcarriers, as a rule of thumb, the subcarrier spacing should be 10–20 times higher than the maximum Doppler shift. Following this assumption, 3.5 GHz must utilize at least 30 kHz subcarrier spacing, and 28/39 GHz must utilize at least 240 kHz. If the Doppler effect is compensated by the

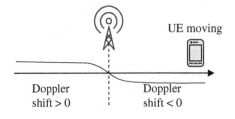

Figure 10.71 Change of Doppler shift when the UE passes a base station.

Table 10.25 Worst-case Doppler shift with UE moving at 500 km/h.

Doppler shift	3.5 GHz	4.5 GHz	28 GHz	39 GHz
Doppler due to UE mobility (kHz)	1.6	2.1	13.0	18.1
Doppler due to UE + BTS error (kHz)	0.5	2.7	4.2	5.9
Total Doppler (kHz)	2.1	2.8	17.2	23.9

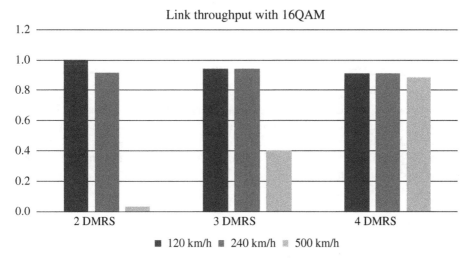

Figure 10.72 Relative link performance with different frequency of Demodulation Reference Signals (DMRSs) data from [14].

physical layer solutions, a lower subcarrier spacing should be possible. DMRS can be used for the channel estimation and for the Doppler compensation. For low UE speeds, single front-loaded DMRS may provide a sufficiently accurate channel estimate for the entire slot. For high velocities, additional DMRSs need to be included also later in the slot, as the channel estimate done based on the DMRS in the beginning may be already inaccurate due to the short coherence time. We can estimate the maximum UE speed with a single DMRS by assuming that the channel coherence time is $0.4 \times 1/\text{Doppler}$ shift. Using the 0.5 ms slot at 3.5 GHz enables 300 km/h, and using the 0.125 ms slot at 28 GHz enables 150 km/h. These calculations assume no frequency errors in UEs and base stations. If higher UE speeds are required, then multiple DRMSs per slot should be utilized. Figure 10.72 shows the link throughput with different number of DMRS and different UE speeds. 16-QAM modulation is used here. Two DMRSs are fine for 120 and 240 km/h, while four DMRSs are required for 500 km/h. The results also show that more DRMSs degrade throughput for 120 km/h because the DMRS takes resources away from data transmission. Therefore, the number of DMRS should be low for low-to-medium UE speeds, and more DMRS should only be used for very-high-speed UEs.

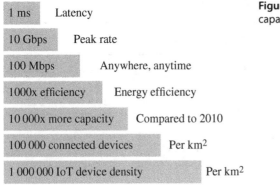

Figure 10.73 Summary of 5G radio target capabilities.

10.14 Summary

5G mobile networks will have huge capabilities in terms of data rate, capacity, latency, and reliability. The target 5G capabilities are summarized in Figure 10.73. Designing networks with such high targets will be different compared to LTE networks. This chapter illustrated 5G performance characteristics and the main factors affecting the performance and its optimization.

References

1 NGMN white paper about 5G requirements, 2015. https://www.ngmn.org/fileadmin/ngmn/content/downloads/Technical/2015/NGMN_5G_White_Paper_V1_0.pdf

2 ITU-R 5G requirements, 2015. http://www.itu.int/dms_pubrec/itu-r/rec/m/R-REC-M.2083-0-201509-I!!PDF-E.pdf

3 "Performance characteristics of 5G mm Wave wireless-to-the-home", Frederick W Vook; Eugene Visotsky; Timothy A. Thomas; Amitava Ghosh, 2016 50th Asilomar Conference on Signals, Systems and Computers, Year: 2016

4 Shannon, C.E. (1948). A mathematical theory of communication. *The Bell System Technical Journal* 27, pp. 379–423, 623–656.

5 3GPP Technical Specifications 38.214 "Physical layer procedures for data", 2019.

6 Khlass, A., Laselva. D. and Järvelä. R. "On the Flexible and Performance-Enhancing Radio Resource Control of 5G NR", IEEE International Symposium on Personal, Indoor and Mobile Radio Communications (PIMRC) 2019.

7 Rodriguez, I., Nguyen, H., Jørgensen, N., Sørensen, T., Elling, J., Gentsch, M. and Mogensen, P. "Path Loss Validation for Urban Micro Cell Scenarios at 3.5 GHz Compared to 1.9 GHz", Globecom, Wireless Communication Symposium, 2013.

8 3GPP Technical Specifications 38.133 "NR; Requirements for support of radio resource management", 2019.

9 3GPP Technical Specifications 38.101 "User Equipment (UE) radio transmission and reception", 2019.

10 Sohrabi, F. and Yu, W. (2016). Hybrid digital and Analog Beamforming Design for Large-Scale Antenna Arrays. *IEEE Journal of Selected Topics in Signal Processing* 10 (3): 501–513.

11 3GPP TSG-RAN WG1 Meeting #93, R1–1807284 Busan, Korea, 21st - 25th May 2018

12 "5G Network Energy Efficiency", Nokia White Paper, 2016

13 3GPP TSG-RAN WG1 Meeting #93, R1–1807285, Busan, Korea, 21st - 25th May 2018.

14 3GPP TSG RAN WG1 NR Ad-hoc #3, R1–1716509, Nagoya, Japan, 18th – 21st September 2017.

11

Measurements

Yoshihisa Kishiyama and Tetsuro Imai

NTT DOCOMO, Japan

11.1   Introduction

This chapter presents propagation measurements for above 6 GHz and field experiments on 5G radio technologies using sub-6-GHz and millimeter waves (mmW).

5G radio utilizes a wide range of frequency bands from below 6 to 100 GHz. In 4G and earlier generation systems, frequency bands of above 6 GHz have not been used due to difficulties involving mobile communication systems, for example, higher propagation loss. Use of frequency bands above 6 GHz is a new challenge for the development and operation of mobile communication systems. It is important to know the fundamental characteristics of above-6-GHz bands used in 5G radio. Propagation measurements and field experiments are essential to comprehend the fundamental characteristics of above-6-GHz bands.

Compared to above 6 GHz, below 6 GHz is valuable spectrum for mobile communication because of good penetration and robustness to obstacles. However, spectrum opportunities below 6 GHz are expected to be limited to a few hundreds of megahertz. To obtain a higher data rate from multi-Gbps to 10 Gbps with spectrum below 6 GHz, it is essential to employ higher rank Massive Multiple Input Multiple Output (MIMO), which is evaluated in experiments using sub-6-GHz bands.

On the other hand, propagation loss in mmW is significantly higher compared to that for the sub-6-GHz bands. Beamforming and beam tracking techniques, which are evaluated in the mmW experiments, are essential in such scenarios to achieve high availability of coverage and seamless mobility among cells. Beamforming can be thought of as a

5G Technology: 3GPP Evolution to 5G-Advanced, Second Edition.
Edited by Harri Holma, Antti Toskala, and Takehiro Nakamura.
© 2024 John Wiley & Sons Ltd. Published 2024 by John Wiley & Sons Ltd.

high-gain or narrow-beamwidth antenna pattern, while beam tracking refers to updating the orientation of that pattern over time to achieve some objective, for example, maximize signal strength.

These propagation measurements [1–15] and field experiments [16–25] were conducted in Japan. The field experiments on 5G radio technologies were conducted via the collaboration between Nokia and NTT DOCOMO during the period from 2014 to 2017.

11.2 Propagation Measurements Above 6 GHz

11.2.1 Fundamental Experiments

Fundamental experiments were conducted in the NTT DOCOMO R&D center in Yokosuka, Japan. In this section, the characteristics of path loss in open space, building corner diffraction loss, building penetration loss, scattering effect on rough surfaces, and human blockage effects are shown.

11.2.1.1 Path Loss in Open Space

Path loss measurements were carried out in a wide parking space. Figure 11.1 shows the measurement conditions. In the measurement site, there are few buildings that reflect waves strongly. A continuous wave (CW) is transmitted from a Tx antenna, and the received power level is recorded. The Tx and Rx antenna heights are set to 1.5 m. The antennas used for the Tx and Rx are sleeve antennas. The measured frequencies are 5.2 GHz and 26.4 GHz. In this case, it is expected that the direct wave and ground-reflected wave arrived at the receiver.

Figure 11.2 shows the path loss over the distance from the Tx to Rx. Fading caused by interference between the direct wave and the ground-reflected wave is observed. Note that these results are within the break point [26]. The path loss difference between frequencies is approximately identical to that in free space, that is, 14.1 dB (=20log(26.4/5.2)).

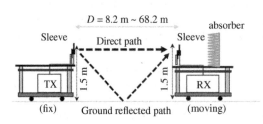
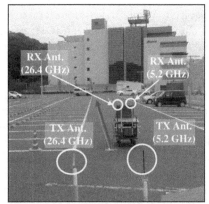

Figure 11.1 Measurement site and conditions.

Figure 11.2 Path loss in open space.

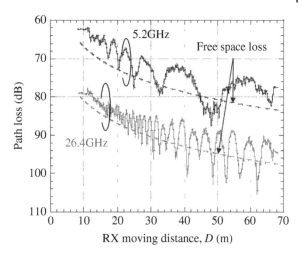

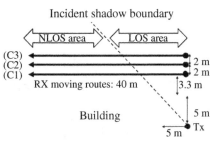

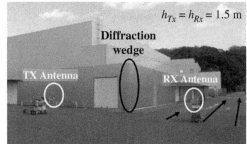

Figure 11.3 Measurement site and conditions.

11.2.1.2 Building Corner Diffraction Loss

Measurements were carried out for the building facing the parking space. Figure 11.3 shows the measurement site and conditions. In the measurement site, there are few buildings that reflect waves strongly. A CW is transmitted from a Tx antenna, and the received power level is recorded. The Tx and Rx antenna heights are set to 1.5 m. The antennas used for the Tx and Rx are sleeve antennas. The measured frequencies are 5.2 and 26.4 GHz. In this case, the difference in diffraction loss between the frequencies is theoretically 21.2 dB ($=30\log(26.4/5.2)$); note that the component of the free space path loss is included [27].

Figure 11.4 shows the measurement results. When Rx moves to the non-line-of-sight (NLOS) area, the path loss for 26.4 GHz increases more steeply than that for 5.4 GHz. This trend is consistent with theory. However, the average value of the difference between frequencies in the NLOS area is approximately 16 dB, which is less than above-mentioned theoretical value. It is thought that there are other propagation paths.

11.2.1.3 Building Penetration Loss

Measurements were carried out in a long corridor of the building facing the parking space. Figure 11.5 shows measurement site and conditions. In the measurement site, there are few buildings that reflect waves strongly. A CW is transmitted from a Tx antenna, and the received power level is recorded on the 3rd, 4th, 5th, 6th, and 7th

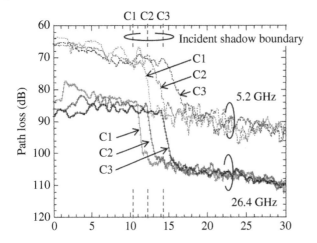

Figure 11.4 Path loss vs. moving distance.

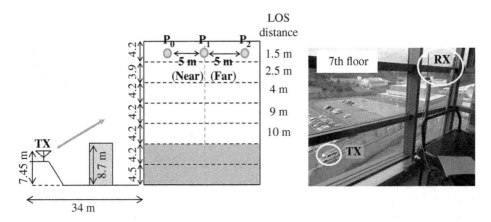

Figure 11.5 Measurement site and conditions.

floors. Note that the line-of-sight (LOS) distance in the figure means the distance from the window facing the Tx. The measured frequencies are 5.2 and 26.4 GHz. In this case, the difference in the building penetration loss between the frequencies is expected to be 14.1 dB ($=20\log(26.4/5.2)$) if the direct wave is dominant, or 21.2 dB ($=30\log(26.4/5.2)$) if the diffracted wave is dominant. Note that the frequency dependence of glass is not considered.

Figure 11.6 shows the measurement results. Here, the vertical axis represents the median value in the near or far region. The average value of the difference between frequencies is 16–22 dB.

11.2.1.4 Scattering Effect on Rough Surface

Measurements were carried out for the building facing the parking space [1]. Figure 11.7 shows the measurement site and conditions. In the surroundings, there is an aluminum wall with a regular unevenness on the order of millimeters. In front of the aluminum wall, antennas of the transmitters (Txs) and receivers (Rxs) of the channel sounders for

Figure 11.6 Building penetration loss on each floor.

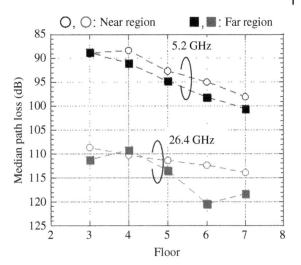

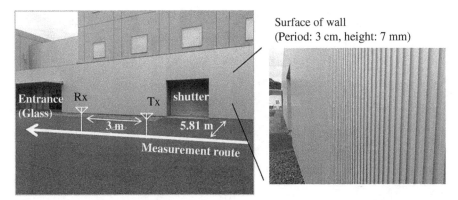

Figure 11.7 Measurement site and conditions.

the 2 GHz and 20 GHz bands were located. The distance between the Txs and the Rxs was fixed as 3 m. The distances between the Txs, the Rxs, and the wall were also fixed. In the measurements, the Txs and the Rxs were moved parallel to the wall within 50 cm in 30 s and recorded data 30 times at 1 s intervals. For each time of recoding, 10 delay profiles with 1 ms interval were recorded. From the delay profiles, the received powers due to scattering of the aluminum wall were obtained and normalized by the power which was calculated for the case of no roughness and no reflection loss of the wall.

Figure 11.8 shows the measurement results of the relative powers. In the wall part, the relative power at the 2 GHz band is not changed and is almost equal to 0 dB, whereas the relative power at the 20 GHz band fluctuates intensely around 5 dB. From these results, it is verified that the roughness of the wall strongly affects the scattering power at the high frequency band.

11.2.1.5 Human Blockage Effects

Measurements of the human body shadowing loss were conducted in an anechoic chamber [2, 3]. Figure 11.9 shows the measurement conditions. A CW is transmitted from a

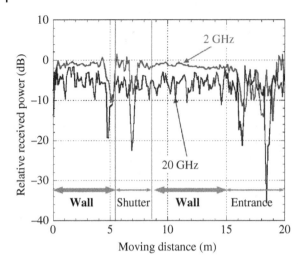

Figure 11.8 Measurement results.

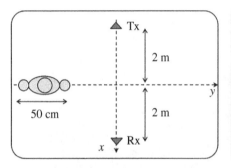

Figure 11.9 Measurement in anechoic chamber.

Tx antenna, and the received power level is recorded. Here, the antenna heights are 1.31 m, corresponding to the chest height of the human body. The distance from the human body to the Tx and Rx are the same at 2 m. The antennas used for the Tx and Rx are horn antennas with a gain of 20 dBi. The measured frequencies are 12, 15, and 18 GHz.

Figure 11.10a shows the relationship between the position of the human body and the shadowing loss. In this figure, the width of the area with a shadowing loss greater than 0 dB at 12 GHz is shown as the shadowing width. Figure 11.10b shows the frequency dependency of the shadowing width and the median value of the shadowing loss within the shadowing width area obtained from Figure 11.10a. In this figure, the calculated results based on the uniform geometrical theory of diffraction (UTD) and a simple human body model (lossy dielectric plates) are shown. The calculation values agree with the measurement relatively well. The results confirm that when the frequency increases (from 6 to 30 GHz), the first Fresnel radius decreases for the human body and the shadowing width decreases, whereas the shadowing loss increases (from 8 to 14 dB).

Measurements to confirm human body blockage effects were conducted in the meeting room [4]. Figure 11.11 shows the measurement conditions. The height of the meeting room is 3.7 m. In the measurements, a CW was transmitted from a Tx antenna, and the

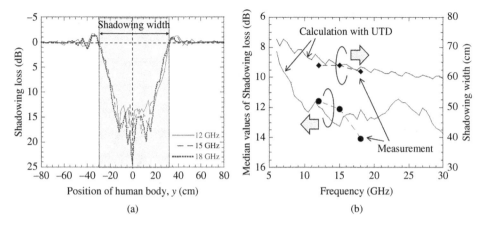

Figure 11.10 Human body shadowing loss: (a) relationship between position of human body and shadowing loss, (b) frequency dependency.

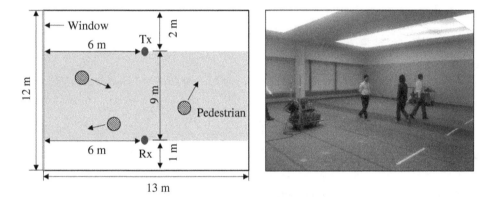

Figure 11.11 Measurement in meeting room.

received level was measured. The measured frequencies are 0.81, 2.2, and 19.85 GHz. Tx and Rx antennas are sleeve antenna, and the antennas were fixed during the measurement. Moreover, the measurement was performed at the Tx antenna height, h_{Tx}, of 1.3, 1.8, and 2.5 m to clarify the antenna height characteristics. The height of Rx, h_{Rx}, is 1.3 m in all measurements. The distance between the Tx and Rx antennas is 9 m, and the received levels were measured in 180 s when pedestrians walked randomly in the indoor environment. The measurement was performed when the pedestrian number, N, was 1, 3, and 5. Pedestrians walked only within the gray area between the Tx and Rx antennas in Figure 11.11.

Figure 11.12 shows the distributions of the received levels for each pedestrian number in 0.81, 2.2, and 19.85 GHz, where $h_{Tx} = h_{Rx} = 1.3$ m. Here, the received levels are normalized to a 99% value in each result. The variation range of the received levels becomes large with an increase in the pedestrian number N. The distributions of the received levels at 0.81 and 2.2 GHz change smoothly; however, the gradients of the distributions at 19.85 GHz change significantly, and the received levels tend to drop when the cumulative distribution function (CDF) value is relatively small. It is considered that these

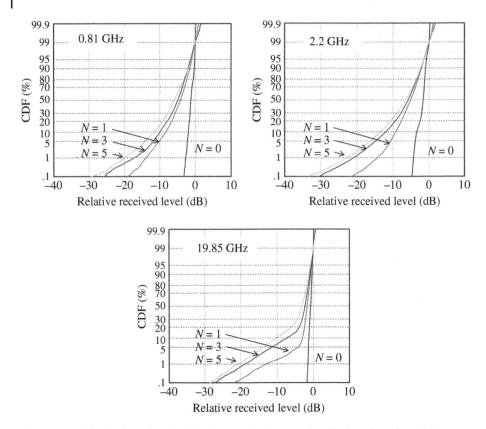

Figure 11.12 Distribution of received levels vs. pedestrian number, N, where $h_{Tx} = h_{Rx} = 1.3$ m.

phenomena occur due to the following reasons. In the cases of 0.81 and 2.2 GHz, the Fresnel zone formed between the transmission and reception antennas is relatively large, and therefore, the Fresnel zone is not blocked by pedestrians completely when they block the line between the Tx and Rx antennas, and significant drops in the received levels do not occur and the distributions of the received levels change smoothly. Conversely, there are cases in which most of the Fresnel zone is blocked by pedestrians in 19.85 GHz, and significant drops occur due to these phenomena. The literature [4] has confirmed that increasing the transmission antenna height is effective in reducing the influence of human body blockage. This can be explained by considering the Fresnel zone as well.

11.2.2 Urban Microcellular Scenario

11.2.2.1 Measurement of Path Loss
Path loss measurements were carried out in a dense urban area in Tokyo [2, 5, 6]. Figure 11.13 shows a picture of the measurement site and Tx antenna setting. In the measurement site, the average building height is 18 m, the average road width is 29 m, and buildings occupy 46% of the area. A CW is transmitted from a base station (BS) antenna, and the received power level is recorded at the MS while the MS is moving. The BS antenna height is set to 10 m. The MS antenna is established on a measurement

Figure 11.13 Measurement site and setting of Tx antenna.

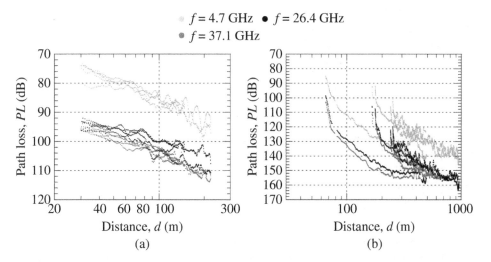

Figure 11.14 Path loss vs. distance: (a) LOS, (b) NLOS.

vehicle, and the height of the antenna is 2.5 m. The antennas used for the BS and MS are sleeve antennas. The measured frequencies are 0.81, 2.2, 4.7, 26.4, and 37.1 GHz. The distance between the BS and the MS was between 30 and 676 m in the LOS case, and between 56 and 959 m in the NLOS case. The path loss is obtained every meter by taking the median value of the received power levels within 10 m (before and after 5 m).

Figure 11.14 shows the path loss over the distance from the BS to the MS at 4.7, 26.4, and 37.1 GHz for the LOS and NLOS cases. Regression analysis is applied to the measurement data by using the CI and ABG models [7, 8]. A summary of these regression results is shown in Table 11.1.

The CI model is expressed as

$$PL(d) = 10n\log_{10}(d) + \text{FSPL}(f_C, 1) + \chi_\sigma \tag{11.1}$$

$$\text{FSPL}(f_C, d) = 32.44 + 20\log(f_C) + 20\log(d) \tag{11.2}$$

Table 11.1 Regression analysis.

		Parameters			
		n/α	*β* (dB)	*γ*	σ_{SF} (dB)
LOS	CI	2.09	N/A		3.1
NLOS	CI	3.21			6.2
	ABG	3.47	25.3	2.07	6.2

Table 11.2 Channel sounder specifications.

Center frequency	19.85 GHz
Bandwidth	50 MHz
Transmission signal	OFDM
Number of subcarriers	449
Transmission power	30 dBm
Tx antenna	Sleeve (2.4 dBi)
Rx antenna	Horn (19.1 dBi: 20 deg. HPBW)

where PL is the path loss in dB, d is the distance from the BS to the MS in meters, f_c is the frequency in GHz, and χ_σ is the shadow fading factor in dB, which is the log-normal distribution with the standard variation of σ_{SF}.

The ABG model is expressed as

$$PL(d) = 10\alpha\log_{10}(d) + \beta + 10\gamma\log_{10}(f_C) + \chi_\sigma \tag{11.3}$$

11.2.2.2 Measurement of Channel Model Parameters

Table 11.2 gives the specifications for the channel sounder for the 20 GHz band. On the transmitter (Tx) side, Orthogonal Frequency-Division Multiplexing (OFDM) signals with a 50 MHz bandwidth at the 20 GHz band are generated and transmitted from a 2.4 dBi sleeve antenna. The transmission power is 30 dBm. On the receiver (Rx) side, a 19.1 dBi horn antenna with a 20° half-power beamwidth (HPBW) is used. After down-converting the received radio frequency (RF) signals to an intermediate frequency (IF) band, the complex amplitudes of the subcarriers (transfer functions) are stored in a PC. The power delay profiles (PDPs) are obtained by converting the stored data into channel impulse responses.

Measurements were conducted in the campus of Niigata University, Japan [9]. Figure 11.15 shows the measurement environment. The Rx (emulating a BS) was set on the rooftop of building B0 at position BS0. The Tx (emulating a mobile station) was set on the rooftop of a car at points MS1–MS6. The condition between the Tx and the Rx was LOS. The antenna heights of the Rx and the Tx were 11.5 and 2.5 m, respectively. The measurement area is surrounded by some buildings. For the measurement, we rotated the horn antenna and measured the PDPs. The ranges of rotation varied from

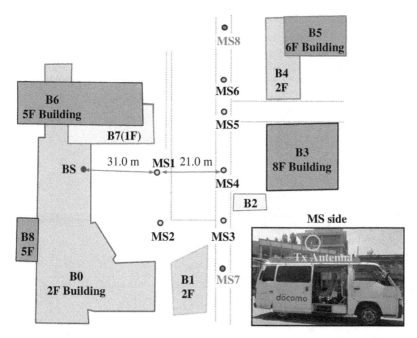

Figure 11.15 Measurement environment.

−180° to +180° in 1° steps for the azimuth angle, and from −50° to 0° in 10° steps for the elevation angle.

In the data analysis, all the PDPs were combined and paths were detected. The delay times, azimuth angles, elevation angles, and path gains of the paths were used to calculate the delay spreads (DSs), azimuth spreads (ASDs), and elevation spreads (ESDs). The K-Power-Mean method was used to estimate the number of clusters.

Table 11.3 shows the number of paths, DSs, ASDs, ESDs, and the number of clusters depending on the Tx's locations. From this table, it is verified that the differences of these values when the Tx location changes from MS1 to MS6 are not large. The average of the numbers of paths for all locations is about 30, and the average of the DSs, ASDs, and ESDs are 49.6 ns, 19.7°, and 2.0°, respectively. The average of the numbers of clusters is about five. This means that, for the case of using the high frequency band for an outdoor small cell environment like an open square, the numbers of paths and clusters tend to be small.

11.2.3 Indoor Hotspot Scenario

11.2.3.1 Measurement of Path Loss

Path loss measurements are carried out in an office of NTT DOCOMO R&D center in Yokosuka, Japan [6]. Figure 11.16 shows the measurement site and measurement route. The measurement was performed in only the LOS environment. In the measurements, a CW is transmitted from a BS antenna, and the received power level is recorded at the MS while the MS is moving. The BS antenna height is set to 2.6 m. The MS antenna is established on a hand truck with a height of 1.5 m. The antennas used for the BS and MS are sleeve antennas. The measurement frequency is 19.85 GHz. The distance between

Table 11.3 Measurement results for UMi scenario – open square.

Tx position	Number of paths	Delay spread (ns)	Azimuth spread (deg.)	Elevation spread (deg.)	Number of clusters
MS1	23	58.5	17.2	1.4	5
MS2	34	58.7	26.8	3.3	7
MS3	43	57.1	32.4	2.2	5
MS4	32	48.0	13.2	1.5	5
MS5	26	40.0	19.2	2.1	4
MS6	24	35.6	9.6	1.3	6
Average	**30.3**	**49.6**	**19.7**	**2.0**	**5.3**
Standard deviation	**7.6**	**10.1**	**8.5**	**0.7**	**1.0**

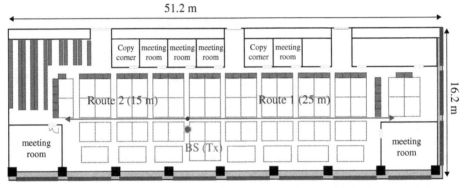

Figure 11.16 Measurement site and route.

the BS and MS varies from 3 to 25 m. The path loss is obtained every 0.1 m by taking the median value of the received power levels within 1 m (before and after 0.5 m).

Figure 11.17 shows the path loss over the distance from the BS to the MS. Regression analysis is applied to the measurement data by using CI, and the regression results are summarized in Table 11.4. Note that the CI model is expressed in Eq. (11.1).

11.2.3.2 Measurement of Channel Model Parameters

Figure 11.18 shows the channel sounder configurations, and Table 11.5 gives the specifications for the channel sounder for the 20 GHz band. On the transmitter (Tx) side, OFDM signals with a 50 MHz bandwidth at the 20 GHz band are generated and transmitted from a 2.4 dBi sleeve antenna. The transmission power is 30 dBm. On the receiver (Rx) side, a 19.1 dBi horn antenna with a 20° HPBW is used. After down-converting the received RF signals to an intermediate frequency (IF) band, the complex amplitudes of the subcarriers (transfer functions) are stored in a PC. The PDPs are obtained by converting the stored data into channel impulse responses.

Figure 11.17 Path loss vs. distance.

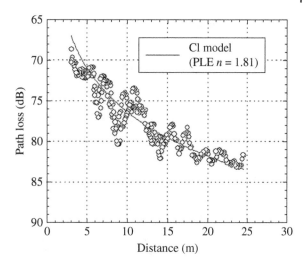

Table 11.4 Regression analysis.

		Parameters	
		n	σ_{SF} (dB)
LOS	CI	1.81	1.7

The channel sounding for indoor hotspot scenarios was performed in an office of NTT DOCOMO R&D center in Yokosuka, Japan [6, 10]. The office size is 51.2 × 16.2 × 2.7 m. There are wooden or metal desks and drawers, chairs, and lockers arranged in the office. There are also some meeting rooms and office equipment rooms that are surrounded by glass with metal frames and metal doors. The windows are of glass with aluminum frames. Near the windows, there are some concrete pillars.

In the measurements, the transmitter of the channel sounder was considered to be the mobile station and was located at points A1-A6, B1-B6, C1, C4, C5, D1, and D4 as shown in Figure 11.19. The height of the sleeve antenna (Tx) was 1.5 m. The receiver of the channel sounder was considered to be the indoor BS and was located at point P0 with an antenna height of 2.3 m. At each Tx location, we rotated the horn antenna in both azimuth and elevation and measured five PDPs at each angle combination on the Rx side. At A3, the azimuth angle of the horn antenna was changed from −180° to +180° with a 1° step, and the elevation angle was changed from −80° to +50° with a 1° step. For other Tx locations, the range and resolution of the azimuth angle was not changed, but the elevation angle was changed from −50° to +50° with a 10° step. There were no people moving in the office during the measurements. The antennas were kept within LOS. The data acquisition was performed automatically.

Table 11.6 shows the measurement results for indoor hotspot scenarios or offices. Here, the K-Power-Mean method was used to cluster multi-paths. The delay times,

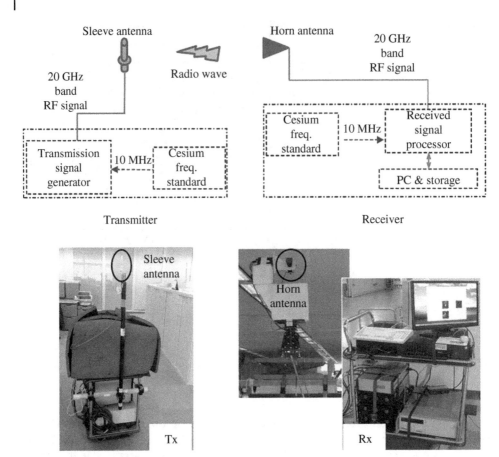

Figure 11.18 The channel sounder configurations.

Table 11.5 Channel sounder specifications.

Center frequency	19.85 GHz
Bandwidth	50 MHz
Transmission signal	OFDM
Number of subcarriers	449
Transmission power	30 dBm
Tx antenna	Sleeve (2.4 dBi)
Rx antenna	Horn (19.1 dBi: 20 deg. HPBW)

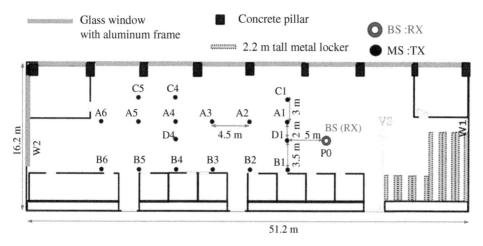

Figure 11.19 Channel sounding in office.

azimuth angles, and elevation angles of cluster centroids were used to obtain the DS, ASD, and ZSD at each position.

11.2.4 Outdoor-to-Indoor Scenario

11.2.4.1 Measurement of Path Loss

Table 11.7 shows the specifications of the measurement equipment. The transmitter sends a CW, and the received power is recorded at the receiver.

The measurement was conducted in the campus of Niigata University, Japan [6, 11, 12]. The measurement environment is illustrated in Figure 11.20. The frequencies used for the measurement were 0.81, 2.2, 4.7, 8.45, 26.4, and 37.1 GHz. Six sleeve antennas for each frequency were installed on a car roof (at B1–B6) or on the roof of a building (at A1–A3) as the BS with simultaneous transmission of the CW signal. The antenna heights were 2.5 or 11.5 m. The measurement was conducted with two receiver units of hand trucks. Each one had three different sleeve antennas installed on it with an interval of 1.5 m between the floor and the antennas. The measurement was repeated on the 1st, 2nd, 4th, 6th, and 8th floors (namely, 1F, 2F, 4F 6F, and 8F). Note that vertically polarized waves were transmitted and received in the measurement.

The measured data of received power were post-processed in the following manner: (1) reference points were established every meter. (2) The data at the reference point and other data acquired within 0.5 m before/after the reference point were considered as a data set belonging to the reference point. (3) The median of the data set was determined as the received power at the reference point. (4) Finally, the power was converted to the path loss.

In the conventional model [13], the O2I path loss in dB can be basically modeled by

$$PL = PL_b + PL_{tw} + PL_{in}. \tag{11.4}$$

Here, PL_b is the "basic path loss," which represents the loss in the outdoor scenario, PL_{tw} is the penetration loss into the building, and PL_{in} is the loss inside the building.

Table 11.6 Measurement results for indoor hotspot scenarios – office.

TX positions	Tx-Rx distances	DS (ns)	ASD (deg.)	ZSD (deg.)	# of paths	# of clusters	#of rays per cluster	Cluster DS (ns)	Cluster ASD (deg.)	Cluster ZSD (deg.)
D1	5	57.3	58.7	1.2	61	8	7.6	42.5	16.2	6.1
A1	5.39	59.4	64.2	5.3	70	8	8.8	41.1	14.2	5.7
B1	6.1	42.5	75.2	7.6	72	5	14.4	51.0	25.1	10.1
C1	7.07	36.9	64.7	0.0	65	7	9.3	43.0	35.7	0.0
A2	9.71	46.4	52.9	1.3	52	9	5.8	29.8	15.3	5.4
B2	10.12	49.0	95.8	1.9	80	7	11.4	56.1	19.1	7.2
A3	14.14	57.5	46.2	4.3	54	8	6.8	36.7	23.6	2.6
B3	14.43	46.7	67.4	1.6	67	6	11.2	41.2	28.7	1.8
D4	18.5	36.1	62.9	0.4	82	5	16.4	60.6	26.5	2.1
A4	18.61	58.0	58.9	1.2	77	6	12.8	62.0	17.7	3.3
B4	18.83	27.3	59.0	0.3	70	5	14.0	60.2	30.1	3.8
C4	19.16	45.0	61.8	1.0	82	7	11.7	47.6	14.5	2.3
A5	23.09	39.8	41.4	0.1	72	6	12.0	48.7	22.6	2.2
B5	23.26	55.5	89.0	0.4	86	7	12.3	33.6	17.5	2.7
C5	23.54	40.2	53.2	0.3	80	7	11.4	43.9	16.7	1.6
A6	27.57	61.6	69.2	0.5	87	9	9.7	46.7	19.0	2.4
B6	27.72	61.2	82.9	1.3	92	9	10.2	53.6	16.0	2.8
Mea. Std.		**48.3**	**64.9**	**1.7**	**73.5**	**7.0**	**10.9**	**47.0**	**21.1**	**3.6**
		10.0	13.9	2.0	11.0	1.3	2.7	9.2	6.1	2.4
log 10	μ	−7.33	1.8	0	—	—	—	—	—	—
	σ	0.1	0.09	0.48	—	—	—	—	—	—

Cross-correlations

Cross-correlations			correlations	
ASD vs. DS	0.26		ZSD vs. SF	0.02
ASD vs. SF	−0.68		ZSD vs. DS	0.24
DS vs. SF	−0.33		ZSD vs. ASD	0.1

Table 11.7 Specifications of measurement equipment and parameters.

Parameters		Values					
Tx (BS)	Frequency (GHz)	0.81	2.2	4.7	8.45	26.4	37.1
	Power (dBm)	43	41	41	38	39	33
	Signal	CW					
	Antenna	Sleeve antenna					
	Antenna height	2.5 or 11.5 m					
Rx (MS)	Antenna	Sleeve antenna					
	Antenna height	1.5 m					
	Floor number of measurement	1, 2, 4, 6, 8					

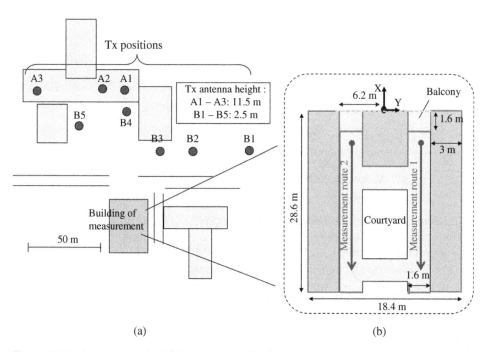

Figure 11.20 Measurement site: (a) Tx positions and building of measurement, (b) layout of building and measurement routes.

Here, PL_b is assumed to be the free space loss:

$$PL_b = 20\log(d_{3D-out} + d_{3D-in}) + 20\log f_{GHz} + 32.4, \tag{11.5}$$

and PL_{in} is assumed to be

$$PL_{in} = 0.5\, d_{2D-in}. \tag{11.6}$$

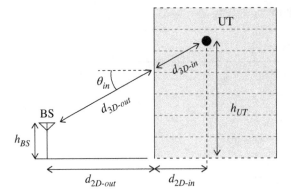

Figure 11.21 Definitions of parameters.

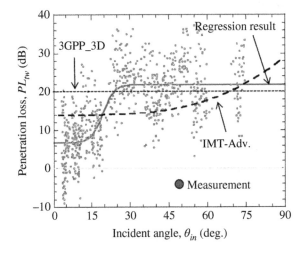

Figure 11.22 Penetration loss characteristics.

Equations (11.5, 11.6) are same as those in the 3GPP 3D channel model [13], f_{GHz} is the frequency in GHz, and the definitions of other parameters are shown in Figure 11.21. The normalized measured data obtained by calculated PL_b and PL_{in} are regarded as PL_{tw}.

Figure 11.22 shows the penetration loss characteristic; all measured data are plotted here. In this figure, the horizontal axis represents the incident angle, θ_{in}, and the vertical axis represents the penetration loss, PL_{tw}. Note that the penetration losses in the 3GPP _3D channel model [13] and in the IMT-Advance channel model [14] are shown as a reference. In Figure 11.22, PL_{tw} increases when θ_{in} becomes large, and there are upper and lower limits. This means that the characteristic can be modeled as a sigmoidal function. The solid line is a regression result based on a sigmoidal function, and it is expressed by

$$PL_{tw} = 6.8 + \frac{21.9 - 6.8}{1 + \exp\{-0.453(\theta_{in} - 19.4)\}}. \tag{11.7}$$

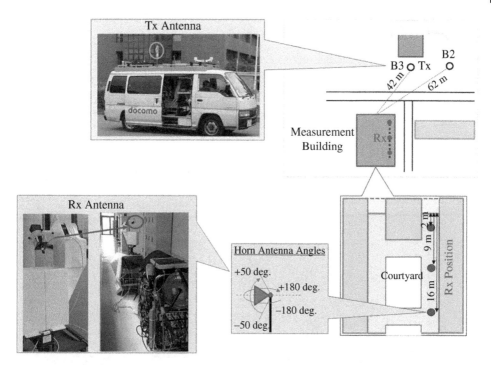

Figure 11.23 Channel sounding setups for outdoor-to-indoor scenario.

Here, the values of 21.9 and 6.8 represent the upper limit and lower limits, respectively. The RMS value of the residual error is 6.8 dB. Note that it has been confirmed that the frequency dependence of PL_{tw}, is very small in this measurement.

11.2.4.2 Measurement of Channel Model Parameters

The channel sounding for the outdoor-to-indoor (O2I) scenario was performed in the campus of Niigata University, Niigata, Japan, as shown in Figure 11.23 [6, 15]. In the measurement, the Tx of the channel sounder was located at points B2, B3 with an antenna height of 2.5 m. On the other hand, the Rx of the channel sounder was located in the building. Measurement data were acquired in the corridors on the 1st, 2nd, 4th, 6th, and 8th floors (namely, 1F, 2F, 4F 6F, and 8F) with positions 2, 9, and 16 m from the glass windows. The height of the horn antenna was 1.5 m. Here, at each Rx position, the horn antenna was rotated, and the PDPs were measured. The ranges of rotation varied from $-180°$ to $+180°$ with a $1°$ step for the azimuth angle, and from $-50°$ to $+50°$ in $10°$ steps for the elevation angle.

Table 11.8 shows the measurement results for the outdoor-to-indoor (O2I) scenario. Here, the K-Power-Mean method was used to cluster multipath channels. The delay times, azimuth angles, and elevation angles of the cluster centroids were used to obtain the DS, ASA, and ZSA at each position.

Table 11.8 Measurement results for outdoor-to-indoor scenarios.

Floor	TX positions	RX positions (m)	DS (ns)	ASA (deg.)	ZSA (deg.)	# of paths	# of clusters	# of rays per cluster	Cluster DS (ns)	Cluster ASA (deg.)	Cluster ZSA (deg.)
1F	A1	2	45.7	96.6	3.6	82.0	11.0	7.5	39.9	20.3	3.3
		9	40.2	30.7	0.7	58.0	5.0	11.6	47.6	36.5	2.5
		16	43.7	33.6	2.7	36.0	12.0	3.0	22.0	8.8	2.8
	A2	2	56.4	52.2	1.7	61.0	4.0	15.3	53.8	37.4	3.9
		9	51.7	21.6	1.1	60.0	3.0	20.0	61.7	69.2	3.5
		16	52.7	52.7	1.9	68.0	6.0	11.3	23.5	55.2	4.0
2F	A1	2	35.0	45.3	1.8	47.0	3.0	15.7	32.6	102.0	6.7
		9	52.3	87.9	2.7	68.0	9.0	7.6	36.8	19.9	4.2
		16	11.0	26.0	1.1	19.0	8.0	2.4	10.3	10.7	1.0
	A2	2	82.2	64.1	2.7	75.0	9.0	8.3	35.6	16.3	3.4
		9	64.2	25.6	0.8	64.0	3.0	21.3	71.7	63.5	4.7
		16	62.4	26.2	2.0	54.0	3.0	18.0	62.3	56.9	6.2
4F	A1	2	47.0	7.3	12.8	18.0	3.0	6.0	19.9	55.6	4.4
		9	55.4	46.3	2.2	82.0	3.0	27.3	61.1	66.3	7.0
		16	77.1	25.7	0.4	72.0	4.0	18.0	57.3	56.3	6.1
	A2	2	3.5	50.2	6.7	43.0	3.0	14.3	50.1	40.4	10.5
		9	12.9	64.8	3.4	71.0	3.0	23.7	91.5	45.9	5.1
		16	104.8	32.6	1.2	50.0	3.0	16.7	40.1	50.2	6.2

		n									
6F	A1	2	32.6	43.9	16.1	43.0	8.0	5.4	36.3	15.3	4.6
		9	53.0	32.8	4.4	45.0	5.0	9.0	65.9	37.1	7.6
		16	151.6	70.9	8.5	37.0	12.0	3.1	27.6	7.7	3.8
	A2	2	77.9	43.0	18.0	45.0	10.0	4.5	49.1	12.2	6.5
		9	77.8	20.0	5.5	16.0	8.0	2.0	5.5	6.7	3.7
8F	A1	2	47.4	50.7	25.2	47.0	4.0	11.8	43.2	38.4	9.3
		9	98.1	39.1	3.4	33.0	8.0	4.1	32.1	18.1	6.4
		16	129.0	67.0	2.3	16.0	9.0	1.8	10.4	2.3	1.2
	A2	2	123.6	81.0	21.6	58.0	8.0	7.3	58.6	29.9	8.0
		9	98.3	63.0	3.1	17.0	5.0	3.4	18.7	17.1	6.0
		Mea	**63.8**	**46.5**	**5.6**	**49.5**	**6.1**	**10.7**	**41.6**	**35.6**	**5.1**
		Std	**34.7**	**21.5**	**6.6**	**19.8**	**3.0**	**7.1**	**20.2**	**23.8**	**2.2**
	log10	**μ**	**−7.29**	**1.62**	**0.5**	—	—	—	—	—	—
		σ	**0.34**	**0.24**	**0.46**	—	—	—	—	—	—

Cross-correlations

ASA vs. DS	0.25
ZSA vs. DS	0.12
ZSA vs. ASA	0.15

11.3 Field Experiments with Sub-6-GHz 5G Radio

11.3.1 Experimental System with Higher Rank MIMO

Figure 11.24 shows the sub-6-GHz experimental system. The system uses time division duplexing (TDD) with Cyclic Prefix OFDM (CP-OFDM) modulation in both the uplink and downlink. To enable shorter scheduling latency, the system uses a bidirectional control structure for the TDD frame, enabling transmitting and receiving of scheduling information in every subframe [28]. Adopting CP-OFDM is also beneficial in such frequent link direction switching because CP-OFDM waveforms are localized in the time domain compared to filter-bank multi-carrier (FBMC)-based waveforms, which are localized in the frequency domain but have a pre- and post-tail in the time domain, resulting in the need for a longer guard period (GP) for link direction switching [29]. The subframe structure is symmetric in the downlink and uplink so that each subframe can be used either for downlink data or for uplink data statically or dynamically. One subframe is formed by 11 OFDM symbols for either downlink or uplink data, followed by one OFDM symbol for downlink control, one OFDM symbol for uplink control, and one OFDM symbol for the demodulation reference signal (DM-RS) with the same link direction as the OFDM symbol for data. This is illustrated in Figure 11.25.

The major radio link parameters are given in Table 11.9. The carrier bandwidth is 100 MHz, and up to four-carrier carrier-aggregation is supported as the system bandwidth. Eight sets of modulation and coding schemes (MCSs) are supported and can be either automatically configured based on the average error rate or configured manually. In this trial, modulation and coding are manually configured. Single-user MIMO is supported up to Rank-8. Up to two transport blocks are generated and are mapped to 1,

BTS (Base Transceiver Station)   UE (User Equipment)

Figure 11.24 Appearance of sub-6-GHz experimental system.

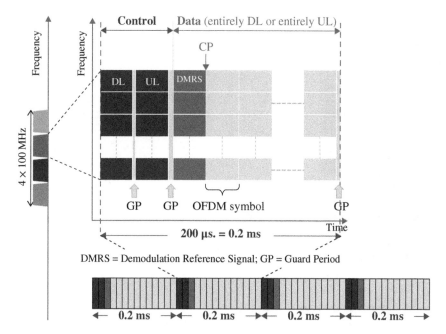

Figure 11.25 Subframe structure.

Table 11.9 Major radio link parameters.

Center frequency	4.65 GHz
System bandwidth	400 MHz (4 × 100 MHz)
Duplex	TDD
Waveform	CP-OFDM both uplink and downlink
TTI length	0.2 ms
Subcarrier spacing	75 kHz
CP and GP length	0.75 μs and 0.95 μs
Number of subcarriers	1280 subcarriers per carrier
Modulation	QPSK (R = 1/6, 1/3, 0.5, 0.8), 16-QAM (R = 0.5, 0.8), 64-QAM (R = 0.5, 0.8)
MIMO	8-stream MIMO
Number of RF branches	8 (14 dBm at each branch, 23 dBm in total)

2, 4, or 8 layers depending on the Rank. At the receiver side, MIMO is decoded based on channel estimation using the DM-RS symbol, which is pre-coded with the same precoding matrix as the data symbols, and based on MMSE-based channel equalization. Error detection is based on cyclic redundancy check (CRC) bits attached at each transport block. One precoding matrix is supported for each rank. The same MCS has been used for each MIMO stream. The theoretical maximum throughput of the system is greater than a total of 10 Gbps throughout the downlink and uplink by using rate 0.8 coded 64-QAM and the 8 × 8 MIMO scheme. In this trial, the TDD ratio is set to 1 : 9 as

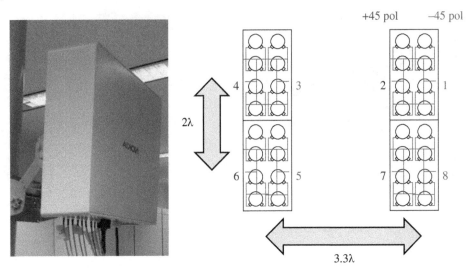

+45 pol   −45 pol

2λ

4   3   2   1

6   5   7   8

3.3λ

The distance between antenna sub-array

Figure 11.26 BTS antenna configuration.

uplink: downlink; hence, the theoretical throughput is 1.07 and 9.67 Gbps in the uplink and downlink, respectively.

The RF system has eight branches and is configured to have 14 dBm per TX, 23 dBm in total from the system. The base transceiver station (BTS) and user equipment (UE) use the same RF system but with different antenna configurations. The BTS uses a patch array antenna with ±45° polarization. The distance between the antenna sub-arrays is 3.3λ in the horizontal direction and 2λ in the vertical direction. Each sub-array has eight antenna elements and is connected to one RF branch. This is illustrated in Figure 11.26. For the UE side, we used two different antenna configurations: (1) same antenna configuration as the BTS and (2) omnidirectional (OMNI) antenna configuration. In the OMNI antenna configuration, eight rod antennas are used, with four rod antennas being attached at 90° angles for polarization.

The baseband part has been implemented on a baseband unit commercially used for Long Term Evolution (LTE), sized 3U (U = Unit) of a 19-inch rack, with the new baseband card supporting 100 MHz, eight-layer MIMO, and 2.5 Gbps throughput. In this trial, four baseband cards have been used. The baseband system is connected to the RF system via sixteen 10.1 Gbps Common Public Radio Interfaces (CPRIs) (four QSFP) and is also connected to a generic IA server hosting Layer 2 and application generating traffic via four 10G Ethernet links.

11.3.2 Field Experiments

The field experiments were conducted in a shopping mall environment in the Roppongi Hills high-rise complex in Tokyo and a long corridor environment in the NTT DOCOMO R&D center in Yokosuka, Japan. The shopping mall has complex reflection paths within its NLOS areas. The long corridor environment exhibits LOS propagation with many reflection paths.

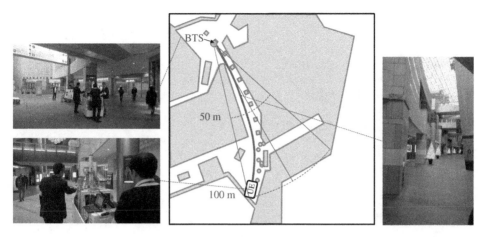

Figure 11.27 Measurement course (shopping mall).

11.3.2.1 Field Experiment in a Shopping Mall Environment

Figure 11.27 shows the measurement course and environment for the field experiments. The environment is a shopping mall and has complex reflection paths including NLOS propagation conditions. The length of the course is approximately 108 m. The propagation is LOS from 8 to 40 m from the BTS and NLOS from 40 to 112 m. The antenna height and downtilt of the BTS are 2.0 m and 0°, respectively. The BTS is located at the side of a walkway and directed toward the UE. The UE moves on the course along the walkway. The antenna height and the average speed of the UE are 1.7 m and 2.4 km/h, respectively. The speed is assuming walking speed in the shopping mall. The UE antenna direction is tangential to the course in the direction of the BTS.

Figure 11.28 shows the CDF of the downlink throughput. The throughput is averaged over a time of 500 ms. A patch antenna is used in both the BTS and UE. The throughput is more than 4.8 Gbps in the downlink in 20% of the environment when the

Figure 11.28 CDF of downlink throughput with the same Tx/Rx antenna configuration.

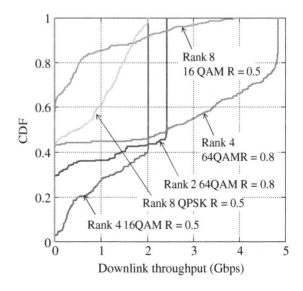

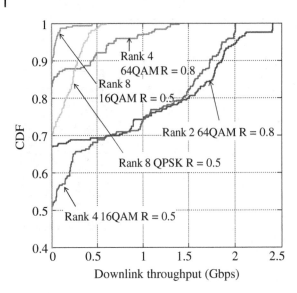

Figure 11.29 CDF of downlink throughput with OMNI Rx antenna configuration.

MCS is configured as Rank-4 64-QAM with a coding rate of R = 0.8. However, Rank-8 transmission is not effective in this environment, since it achieved only up to 4.5 Gbps instantaneously. The spatial correlation of BTS and UE should be low enough to support higher order MIMO.

In the first step, to make the spatial correlation low on the UE side, the UE antenna is changed to an OMNI antenna. Figure 11.29 shows the CDF of the downlink throughput in the OMNI antenna configuration in the UE. However, the result is like achieving only 2.4 Gbps instantaneously and degraded compare to the result of Figure 11.28. This is because the spatial correlation of the BTS side is still high even when the correlation of the UE side is changed. In addition, the antenna gain is degraded by −11 dB due to the change in the UE antenna configuration. However, this antenna gain degradation is unavoidable when the OMNI antenna is used.

In the second step, to make the spatial correlation low on both the BTS and UE side, the UE keeps using the OMNI antenna, and the space of the BTS antenna branch is changed to 10λ in the horizontal direction and 6.6λ in the vertical direction. Figure 11.30 shows the CDF of the downlink throughput in the configuration. The correlation is low enough, achieving more than 4.8 Gbps even in Rank-8 of 64-QAM R = 0.8, as well as Rank-4 64-QAM R = 0.8. The characteristics of both Rank-4 and Rank-8 are improved compared to Figure 11.29 when the low correlation antenna configuration is used on the BTS side.

11.3.2.2 Field Experiment in a Long Corridor Environment

Figure 11.31 shows the measurement area and environment of the long corridor environment. The propagation is LOS with many reflection paths. The antenna height and down tilt of the BTS are 2.0 m and 0°, respectively. The BTS is installed at one side of the corridor and directed through a passage to another side of the corridor. The UE is fixed at each point in 5 m steps from 5 to 60 m away from the BTS. The antenna height of the UE is 1.7 m. A patch antenna is used in both the BTS and the UE.

Figure 11.30 CDF of downlink throughput with distributed Tx antenna configuration.

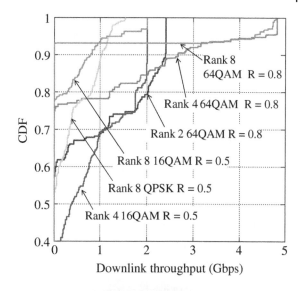

Figure 11.30 CDF of downlink throughput with distributed Tx antenna configuration.

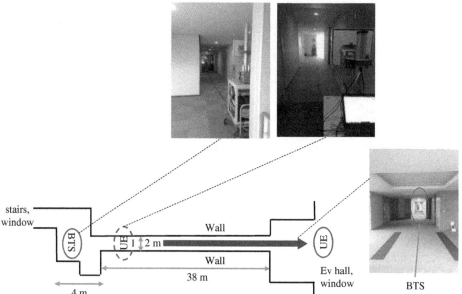

Figure 11.31 Measurement environment (long corridor).

Figure 11.32 shows the average value of the block error rate (BLER) for 10 s at each point according to the distance from the BTS in each MCS. From 10 to 25 m away from the BTS, the BLER remains 0 in most of the MCS except Rank-8 of 64-QAM R = 0.8. MCS is sensitive to the position of the UE and spatial correlation affects when 10–20 m away from the BTS. Figure 11.33 shows the CDF of the downlink throughput 25 m away from the BTS in Rank-8 of 64-QAM R = 0.8 keeping the same position for 20 s. More than 9.6 Gbps in the downlink was achieved at the 25 m point, which is the peak through-out this trial. Rank-8 transmission can be realized provided the environment does not

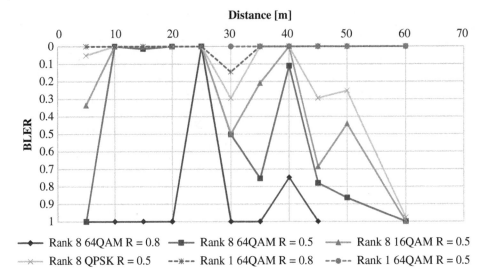

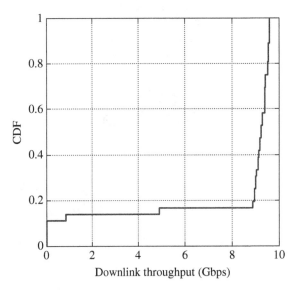

Figure 11.32 Distance vs. BLER.

Figure 11.33 CDF of downlink throughput at 25 m from BTS (Rank-8 of 64-QAM R = 0.8).

depend on the distance from the BTS. By optimizing the UE position, Rank-8 transmission can be achieved in the long corridor environment exhibiting LOS propagation with many reflection paths.

11.4 Field Experiments of Millimeter Wave 5G Radio

11.4.1 Experimental System with Beamforming and Beam Tracking

The system uses TDD with Null Cyclic Prefix Single Carrier (NCP-SC) modulation. The NCP-SC is a form of single-carrier modulation proposed for mmW broadband radio access in [30]. The PoC system consists of an access point (AP) and a user device (UD).

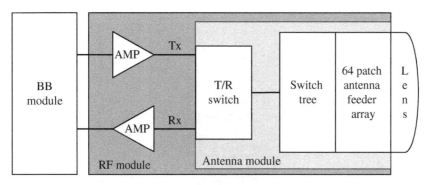

(a) **AP**

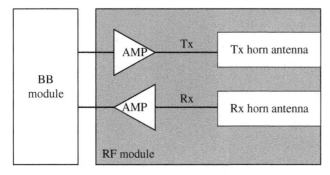

(b) **UD**

Figure 11.34 Configuration of PoC transceiver: (a) AP, (b) UD.

Figure 11.34 shows the configuration of the PoC transceiver. The beamforming and beam tracking are implemented in the AP. The AP and UD have identical baseband units and identical RF units except for the antenna. The AP uses a 28-dB-gain dielectric lens antenna with a switched antenna feeder configuration. The switching is accomplished in less than 1 μs and is driven by the baseband processing unit synchronously with the time division multiplexing slot structure. The HPBW is approximately 3° in both azimuth and elevation. The beam can be steered approximately ±4° in elevation and ±17.5° in azimuth by selecting one of 64 patch antennas organized in a 4-row by 16-column rectangular array. Figure 11.35 shows each beam's theoretical coverage area for an AP height and downtilt of 2.8 m and 10°, respectively. The UD uses a 10-dBi gain horn antenna with a HPBW of 48°. The system achieves a peak data rate greater than 2 Gbps using rate 7/8 coded 16-QAM modulation.

The major radio link parameters for the transceiver are given in Table 11.10. The system bandwidth is 1 GHz, and the center frequency is 73.5 GHz. The total transmit power of the power amplifier is +23.6 dBm (0.229 W). Adaptive modulation and coding (AMC) using four MCSs is implemented. AMC is performed at the transmitter based on the average error rate reported by the receiver. In the transmitter, the information data sequence is turbo-encoded with the coding rate of R. The available MCSs are BPSK (R = 1/5), QPSK (R = 1/2), and 16-QAM (R = 1/2, 7/8). As shown in Figure 11.36, in the trials, a 102.4 μs transmission time interval (TTI) frame consists of 150 NCP-SC blocks, and each NCP-SC block consists of 480 single-carrier-modulated data symbols and 32

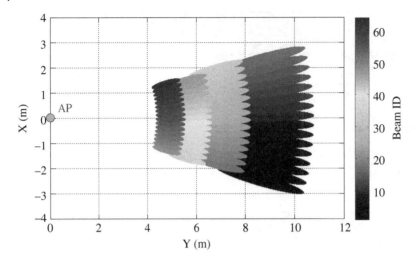

Figure 11.35 Example of coverage area of beams and definition of the beam ID.

Table 11.10 Major radio link parameters.

Center frequency	73.5 GHz
System bandwidth	1 GHz
Duplex	TDD (Downlink 188, Uplink 12 slots)
TTI length	102.4 µs
Access system	NCP-SC (Null Cyclic Prefix Single Carrier)
Modulation	NCP-SC BPSK (R=1/5), QPSK (R=1/2), 16-QAM (R=1/2, 7/8)
Antenna configuration (rank-1 is used in these measurements)	AP: 1Tx/Rx Lens (28 dBi) UD: 1Tx/Rx Horn (10 dBi)
AP feeder configuration	64 (Horizontal: 16, Vertical: 4)
AP scan range	± 17.5 deg. (Horizontal) ± 4.0 deg. (Vertical)

null symbols that act as a cyclic extension to facilitate frequency domain equalization at the receiver.

On the transmit side, after conversion to baseband in-phase (I) and quadrature (Q) components using 1.25 GS/s 14-bit digital-to-analog converters, quadrature modulation is performed. The intermediate frequency (IF)-modulated signal is up-converted to a RF signal and amplified by the power amplifier. In the receiver, the down-converted RF signal is first linearly amplified by an automatic gain control amplifier. The I and Q signals from the quadrature detector are converted into a digital format using 1.5 GS/s 8-bit analog-to-digital converters. Sixteen parallel instances of the LTE turbo decoder and encoder, NCP-SC modulation, NCP-SC demodulation, and lower MAC processing are implemented in field-programmable gate arrays. A CPU implements the upper MAC and radio resource control.

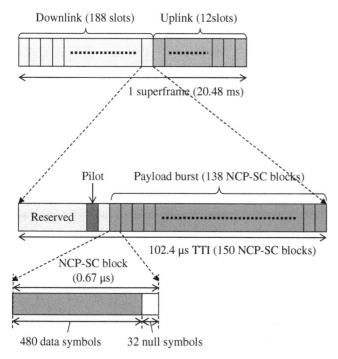

Figure 11.36 NCP-SC frame structure.

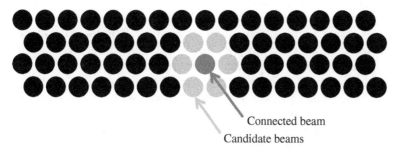

Connected beam
Candidate beams

Figure 11.37 Image of the current candidate beams while connected.

Beam tracking is achieved with two phases of beam selection: an initial beam searching phase and a beam tracking phase. Figure 11.37 shows an image which indicates the current candidate beams while connected. The beam is selected by using the 4 row by 16 column array of patch antennas that feeds the lens.

In the initial beam searching phase, the AP samples all available beams by transmitting a TTI-length pilot signal and listening for a TTI-length interval on each candidate beam. Each pilot transmission contains information indicating when the AP will attempt reception on each beam. Specifically, the AP will utilize 64 TTIs to transmit the pilot signal on each beam and will also utilize an additional 64 TTIs as it sequentially listens on all 64 candidate beams in the initial beam searching phase. The UD detects a subset of the candidate beams, synchronizes to the frame, measures their signal-to-noise ratios (SNRs) and sends a measurement report with this information back to the AP at the time that corresponds to the beam with the best SNR. Upon reception of this report, the AP will transition to the tracking phase by updating the candidate beam list to contain the best beam and all surrounding beams. The candidate beam set is transmitted every superframe or 20.48 ms, where a superframe consists of 200 TTIs, each exactly 102.4 μs. The AP averages over five superframes before making the beam selection decision (i.e. the beam used for the payload data). Thus, the beam update interval is 102.4 ms.

After establishing the connection, the phase changes to the beam tracking phase, and the pilot signal is transmitted from seven candidate beams: the selected beam used for the payload and its six neighbor beams.

The present dielectric lens antenna with its 3° HPBW corresponds to massive MIMO using approximately 1000 antenna elements. Using a dielectric lens has a size advantage compared to other equivalent massive MIMO implementations.

11.4.2 Field Experiments

The field experiments were conducted in a courtyard environment in the NTT DOCOMO R&D center in Yokosuka, Japan, and a shopping mall environment in a Roppongi Hills high-rise complex in Tokyo and in a street canyon environment in the Minato Mirai area in Yokohama, Japan. These crowded environments are representative of typical 5G scenarios. The defined course in the shopping mall environment consists of 69% NLOS conditions and numerous reflectors providing opportunities for multiple-bounce delay paths.

11.4.2.1 Field Experiment in a Courtyard Environment

Figure 11.38 shows the measurement course for the field experiments, while Figure 11.39 includes two pictures that show the measurement environment and the AP and UD installation. The environment is a courtyard with mainly LOS conditions. The length of the course is approximately 10 m, and the course is approximately 23 m from the AP. The antenna height and downtilt of the AP are 5.8 m and 14°, respectively. The UD moves along the 10 m course; its antenna height and average speed are 1.7 m and 4.0 km/h, respectively. The speed is consistent with walking speed.

Figure 11.40 shows the SNR at the UD versus the elapsed time. The SNR is greater than 15 dB except for a few sparse measurement points. Figure 11.41 shows the downlink throughput versus the elapsed time. The throughput is consistently greater than

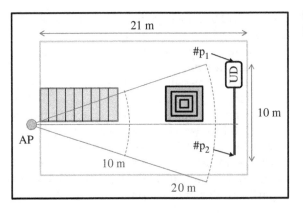

Figure 11.38 Measurement course (courtyard environment).

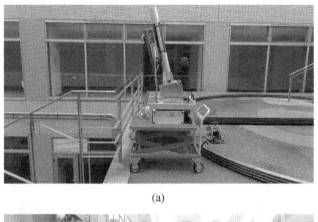

(a)

(b)

Figure 11.39 Measurement areas for field experiments (courtyard environment): (a) AP, (b) UD (view from point p2 to the direction for AP).

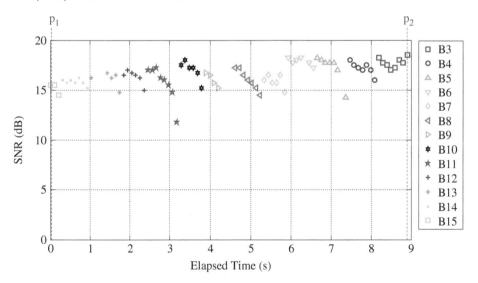

Figure 11.40 Downlink SNR.

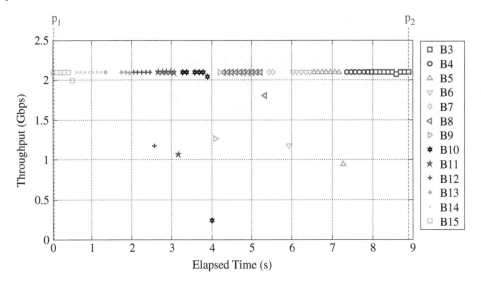

Figure 11.41 Downlink throughput.

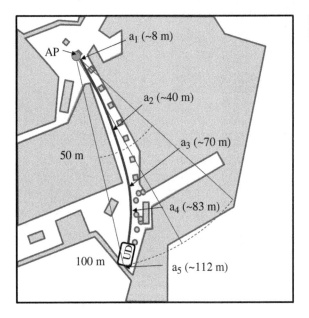

Figure 11.42 Measurement course (shopping mall environment).

2 Gbps except for a few sparse measurement points. These sparse points are at the beam switching times where SNR degradations are also observed. However, the results also show that the beam is tracking the UD, thus showing that beam tracking worked properly in this environment.

11.4.2.2 Field Experiment in a Shopping Mall Environment

Figure 11.42 shows the measurement course for the field experiments, while Figure 11.43 includes two pictures that show the measurement environment and the AP and UD

(a)

(b)

Figure 11.43 Measurement areas for field experiments (shopping mall environment): (a) AP, (b) UD (view from point a3 to the direction for AP).

installation. The environment is a shopping mall with mainly NLOS conditions and complex reflections. The length of the course is approximately 108 m. The propagation is LOS from a1 to a2 in Figure 11.42 (8–40 m from the AP) and NLOS from a2 to a5 (40–112 m from the AP). In the shopping mall environment, 69% of the defined course exhibits NLOS conditions. The antenna height and downtilt of the AP are 2.0 m and 2.5°, respectively. The AP is installed at a side of a street of the shopping mall and directed through a passage to another side of the street. The UD moves on the course across the street. The antenna height and the average speed of the UD are 1.7 m and 1.7 km/h, respectively. The speed is consistent with the average walking speed in a shopping mall. The UD antenna is pointed toward the AP, tangential to the UD's path.

Figure 11.44 shows the SNR at the UD versus its distance from the AP. The SNR shown is the maximum value measured in each time frame from the candidate beam set. The marker shows the selected beam ID. The SNR is maintained at more than 15 dB in the range from 8 to 45 m from the AP and decreases from 45 to 49 m as the distance increases. The SNR increased rapidly around 58 m and peaks at around 71 and 87 m. Beyond ~30 m (slightly before point a2), the AP mainly utilizes beam #8. At the NLOS locations (e.g. 71 and 87 m), the propagation path is completed via a square pillar located between points a2 and a3. The peak SNR is 18.5 dB at 14 and 19 m, 17.3 dB at 71 m, and 15.3 dB at 87 m.

Figure 11.45 shows the downlink throughput versus the UD distance from the AP. The throughput, averaged over 102.4 ms, is more than 2 Gbps from 8 to 45 m (except for a few measurements around 10 and 31 m). The throughput peaks at 1.2 Gbps around 71 and 87 m with a momentary jump to 2 Gbps at the former distance.

Figure 11.46 shows the relative delay profile at point a3. The figure shows the main signal path and some additional delay paths. The delay paths are approximately 2.0, 3.3, and 18.0 ns after the main path, corresponding to 0.6, 1, and 5.4 m of the additional path length, respectively. The result suggests that some delay paths are multiple reflections from the ground and the wall.

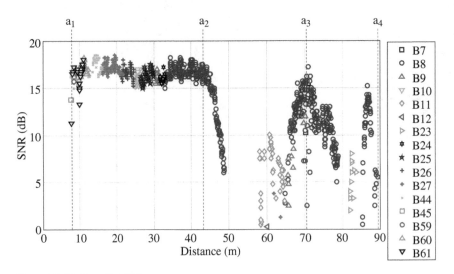

Figure 11.44 Downlink SNR.

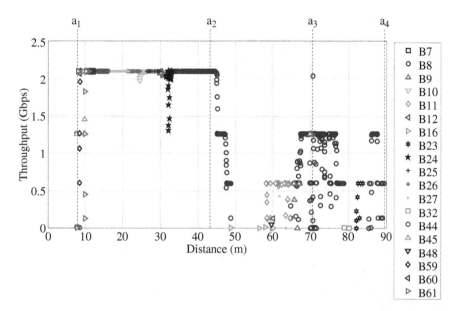

Figure 11.45 Downlink throughput.

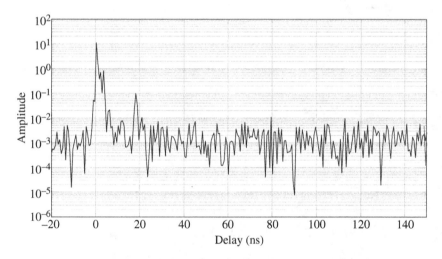

Figure 11.46 Power delay profile at point a3.

11.4.2.3 Field Experiment in a Street Canyon Environment

Figure 11.47 shows the measurement course, while Figure 11.48 includes two pictures that show the measurement environment and the AP and UD installation. The environment is a street canyon with entirely LOS conditions. The course the UD traversed is from 6 to 172 m from the AP (b1 to b3 in Figure 11.47). The AP antenna height and downtilt are 2.8 and 2.7 m, respectively, and the antenna is installed on the far side of a cross street. The UD moves on the course across the street. The antenna height and the average speed of the UD are 1.7 m and 4 km/h, respectively. The speed is consistent with walking speed. The UD antenna points toward the AP.

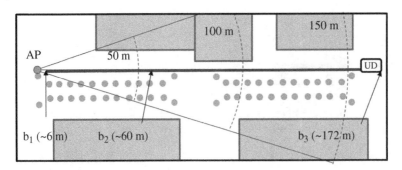

Figure 11.47 Measurement course (street canyon environment).

(a)

(b)

Figure 11.48 Measurement areas for field experiments (street canyon environment): (a) AP, (b) UD.

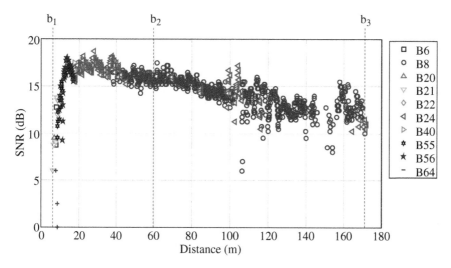

Figure 11.49 Downlink SNR.

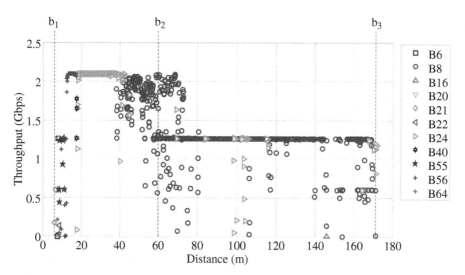

Figure 11.50 Downlink throughput.

Figure 11.49 shows the SNR at the UD versus its distance from the AP. From 10 to 80 m, the SNR is greater than 15 dB and generally decreases with increasing distance. Figure 11.50 shows the downlink throughput versus distance. The throughput is greater than 2 Gbps from 12 to 40 m and fluctuates from 40 to 72 m. Even during this fluctuation, a throughput of 2 Gbps is frequently achieved. The cause of the variation may be reflected paths from the side building windows and its effects on beam selection and receiver performance. Even with these variations, the throughput is greater than 1 Gbps from 72 to 170 m except for a few sparse measurement points.

Figure 11.51 shows the CDF of the downlink throughput in the courtyard, shopping mall, and street canyon environment. The maximum of 2 Gbps is achieved in the

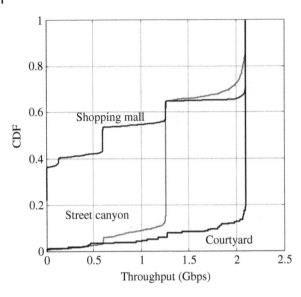

Figure 11.51 Downlink throughput CDF.

downlink for 87%, 34%, and 28% of each of the respective environments. 1 Gbps is achieved in the downlink for 95%, 45%, and 91% of each of the respective environments. The shopping mall environment, which is 69% NLOS, shows an outage of 35%. The beam tracking achieves high availability of coverage and seamless mobility not only in LOS environments but also under NLOS conditions through the reflected paths. It has been commonly believed that mmW communications is possible only in LOS propagation environments and thus is not suitable for mobile applications. The results here show the possibility of using these high frequencies for mobile communications by employing beamforming and beam tracking techniques to facilitate connections through reflected paths.

11.5 Summary

In this chapter, Section 11.2 presented the propagation characteristics of frequency bands above 6 GHz based on measurements. First, fundamental experimental results for the 20 (or 26) GHz band were shown, where the characteristics of the path loss in open space, building corner diffraction loss, building penetration loss, scattering effects on rough surface, and human blockage effects were clarified. These are useful for understanding the basic characteristics of the above 6 GHz bands. Next, the measurement results of frequency bands above 6 GHz (mainly the 20 GHz band) were shown, where the path loss characteristics and channel model parameters in the UMi, InH, and O2I scenarios were clarified. Note that in the UMi and O2I scenarios, frequency-dependent characteristics of the path loss were also discussed based on the measurement results from the 0.8 to 37 GHz band. These are helpful for understanding the characteristics in the environment where 5G will actually be deployed.

Section 11.3 presented field experiments on the downlink throughput performance of a 5G centimeter wave radio access system at 4.5 GHz with 400 MHz bandwidth employing higher rank MIMO in a shopping mall environment and a long corridor environment. The shopping mall has complex reflection paths within its NLOS areas. The long corridor environment exhibits LOS propagation with many reflection paths. The results show that a maximum throughput of more than 4.8 Gbps in the downlink is achieved in 20% of the course in the shopping mall environment using patch antenna in both the BTS and the UE when configuring the MCS as Rank-4 64QAM R = 0.8. However, Rank-8 transmission is not effective in this environment, since it achieved only up to 4.5 Gbps instantaneously because of the spatial correlation. The correlation is low enough, achieving more than 4.8 Gbps even in Rank-8 of 64-QAM R = 0.8 when the UE side is using an OMNI antenna and the space of the BTS antenna branch is changed to 10λ in the horizontal direction and 6.6λ in the vertical direction. Moreover, we also showed that Rank-8 transmission of more than 9.6 Gbps in the downlink can be achieved in the long corridor environment exhibiting LOS propagation with many reflection paths.

Section 11.4 presented field experiments on the downlink throughput and SNR performance of a beam tracking PoC system targeting 5G millimeter wave radio access at 73 GHz with 1 GHz bandwidth in courtyard, shopping mall, and street canyon environments. The majority of the area in the courtyard and street canyon environment is LOS, while the shopping mall environment is 69% NLOS. The results show that a maximum throughput of over 2 Gbps using rate 7/8 coded 16-QAM modulation is achieved for 87%, 34%, and 28% of each of the respective environments. 1 Gbps is achieved in the downlink for 95%, 45%, and 91% of each of the respective environments. The beam tracking achieves high availability of coverage and seamless mobility not only in LOS environments but also NLOS conditions through reflected paths.

References

1 N. Tran, T. Imai, K. Kitao and Y. Okumura, "A Study on Wall Scattering Characteristics Based on ER Model with Point Cloud Data," ICCEM2017, pp. 252–253, March 2017.

2 T. Imai, K. Kitao, N. Tran, N. Omaki, Y. Okumura, M. Sasaki, W. Yamada, "Development of High Frequency Band over 6 GHz for 5G Mobile Communication Systems," EuCAP2015, April 2015.

3 Tran, N., Imai, T., and Okumura, Y. (2015). Model for estimating effects of human body shadowing in high frequency bands. *IEICE Transactions on Communications* E98-B (5): 773–782.

4 K. Kitao, N. Toran, T. Imai, Y. Okumura, "Frequency Characteristics of Changes in Received Levels by Human Body Blockage in Indoor Environment," URSI AP-RASC 2016, Aug. 2016.

5 K. Kitao, T. Imai, N. Tran, N. Omaki, Y. Okumura, M. Inomata, M. Sasaki and W. Yamada, "Path Loss Prediction Model for 800 MHz to 37 GHz in NLOS Microcell Environment," PIMRC2015, pp. 516–520, 2015.

6 Workshop in conjunction with IEEE Globecom'15, White paper on 5G Channel Model for bands up to100 GHz, Dec. 2015. (http://www.5gworkshops.com/5GCM .html)

7 3GPP TR38.901 V14.0.0, "Study on channel model for frequencies from 0.5 to 100 GHz (Release 14)," March 2017.

8 Report ITU-R M. 2412-0, "Guidelines for evaluation of radio interface technologies for IMT-2020," Oct. 2017.

9 N. Tran, T. Imai, K. Kitao, Y. Okumura, "Study on Propagation Channel Characteristics at 20-GHz Band for Outdoor Small Cell," ISAP2017, 2017.

10 N. Tran, T. Imai, and Y. Okumura, "Measurement of Indoor Channel Characteristics At 20 GHz Band," ISAP2015, 2015.

11 T. Imai, K. Kitao, N. Omaki, Y. Okumura, "Study on Extension to Higher Frequency Band of 3GPP Outdoor-to-Indoor Path Loss Model," ISAP2015, 2015.

12 T. Imai, K. Kitao, N. Tran, N. Omaki, Y. Okumura, K. Nishimori, "Outdoor-to-Indoor Path Loss Modeling for 0.8 to 37 GHz Band," EuCAP2016, 2016.

13 3GPP TR 36.873 (V1.2.0), "Study on 3D channel model for LTE," Sep. 2013.

14 ITU-R, Report ITU-R M.2135–1, Guidelines for evaluation of radio interface technologies for IMT-Advanced, 2009.

15 N. Tran, T. Imai and Y. Okumura, "Outdoor-to-Indoor Channel Characteristics at 20 GHz," ISAP2016, pp. 612–613, 2016.

16 Y. Kishiyama, T. Nakamura, A. Ghosh, and M. Cudak, "Concept of mmW Experimental Trial for 5G Radio Access," IEICE Society Conference, B-5-58, 2014.

17 Y. Inoue, Y. Kishiyama, Y. Okumura, J. Kepler, and M. Cudak, "Experimental Evaluation of Downlink Transmission and Beam Tracking Performance for 5G mmW Radio Access in Indoor Shielded Environment," IEEE PIMRC, Sept. 2015.

18 Y. Inoue, Y. Kishiyama, S. Suyama, J. Kepler, M. Cudak, and Y. Okumura, "Field Experiments on 5G mmW Radio Access with Beam Tracking in Small Cell Environments," IEEE Globecom Workshops, Dec. 2015.

19 P. Weitkemper, J. Koppenborgy, J. Bazzi, R. Rheinschmitty, K. Kusume, D. Samardzijaz, R. Fuchsy, and A. Benjebbour, "Hardware Experiments on Multi-Carrier Waveforms for 5G," IEEE WCNC, Apr. 2016.

20 S. Yoshioka, Y. Inoue, S. Suyama, Y. Kishiyama, Y. Okumura, James Kepler, and Mark Cudak, "Field Experimental Evaluation of Beamtracking and Latency Performance for 5G mmWave Radio Access in Outdoor Mobile Environment," IEEE PIMRC Workshops, Sept. 2016.

21 M. Cudak, T. Kovarik, T. A. Thomas, A. Ghosh, Y. Kishiyama, and T. Nakamura, "Experimental mmWave 5G Cellular System," IEEE Globecom Workshops, Dec. 2014.

22 Y. Inoue, S. Yoshioka, Y. Kishiyama, S. Suyama, Y. Okumura, James Kepler, and Mark Cudak,"Field Experimental Trials for 5G Mobile Communication System Using 70 GHz-Band," IEEE WCNC Workshops, Mar. 2017.

23 Inoue, Y., Yoshioka, S., Kishiyama, Y. et al. (2017). Field experimental evaluation on 5G millimeter wave radio access for Mobile communications. *IEICE Transactions on Communications* E100-B (8).

24 Y. Inoue, S. Yoshioka, Y. Kishiyama, S. Suyama, Y. Okumura, T. Haruna, T. Tanaka, A. Splett, and H. Liljeström, "Field Experimental Evaluation of Low SHF 5G Radio Access System Employing Higher Rank MIMO," IEEE VTC 2017 Spring Workshops, June 2017.

25 Y. Takahashi, Y. Inoue, S Yoshioka, Y. Kishiyama, S. Suyama, J. Mashino, A. Splett, and H. Liljeström, "Field Experimental Evaluation of Higher Rank MIMO in Quad-directional UE Antenna Configuration for 5G Radio Access System," IEEE WPMC 2017, 2017.

26 Goldsmith, A. (2007). *Wireless Communications*. Maruzen & Cambridge.

27 Ikegami, F., Yoshida, S., Takeuchi, T., and Umehira, M. (1984). Propagation factors controlling mean field strength on urban streets. *IEEE Transactions on Antennas and Propagation* AP-32 (8): 822–829.

28 E. Lähetkangas, K. Parjukoski, E. Tiirola, G. Berardinelli, I. Harjula, and J. Vihriälä, "On the TDD Subframe Structure for Beyond 4G Radio Access Network", Future Network & Mobile Summit, 2013.

29 Eeva Lähetkangas Kari Pajukoski, Jaakko Vihriälä, Gilberto Berardinelli, Mads Lauridsen, Esa Tiirola, and Preben Mogensen, "Achieving low latency and energy consumption by 5G TDD mode optimization," ICC'14 Workshop on 5G Technologies. Sydney, Australia 2014.

30 S.G. Larew, T.A. Thomas, M. Cudak, and A. Ghosh, "Air interface design and ray tracing study for 5G millimeter wave communications," 2013 IEEE Globecom Workshops (GC Wkshps), pp. 117, 122, 9–13, 2013.

12

5G RF Design Challenges

Petri Vasenkari[1], Dominique Brunel[2], and Laurent Noël[3]

[1] *Nokia Bell Labs, Finland*
[2] *Skyworks, France*
[3] *Skyworks, Canada*

CHAPTER MENU

12.1 Introduction

5G brings changes compared to Long-Term Evolution (LTE) for the user equipment (UE) radio frequency (RF) design aspects in many ways. There is larger bandwidth with higher bandwidth occupancy, and there is dual connectivity with two uplink (UL) transmissions as baseline feature as well as operation with higher frequency bands than LTE could achieve. This chapter covers the impacts on the UE RF design from both standalone 5G as well as from the dual connectivity perspective, considering the large number of frequency bands that must be supported. Also, the impacts originating from mmW support, including the need for Over the Air Testing (OTA), is covered. This chapter concludes with an analysis of the multi-band as well as mmW UE implementation issues, including mmW UE antenna placement considerations.

5G Technology: 3GPP Evolution to 5G-Advanced, Second Edition.
Edited by Harri Holma, Antti Toskala, and Takehiro Nakamura.
© 2024 John Wiley & Sons Ltd. Published 2024 by John Wiley & Sons Ltd.

12.2 Impact of New Physical Layer on RF Performance

12.2.1 New Uplink Waveforms

In 5G New Radio (NR), a number of innovations have targeted improvements in the uplink (UL) and downlink (DL) waveforms:

- Variable subcarrier spacing (SCS)
- Waveform filtering/windowing for better spectrum confinement
- Use of Cycle Prefix-Orthogonal Frequency-Division Multiplexing (CP-OFDM) and Direct Fourier Transform-spread Orthogonal Frequency-Division Multiplexing (DFT-s-OFDM) in the UL
- Low peak to average power ratio (PAPR) waveforms with spectral shaping

LTE waveforms use only 15 kHz SCS (with the exception of the 3.75 kHz case in NB-IoT); in NR, variable SCS is introduced. SCS is in the form of $15 \times (2)^{\mu}$ kHz, with $\mu = 0\text{--}5$, covering SCS from 15 to 480 kHz. This was required to enable the following:

- Covering new frequencies of operation well beyond the current LTE maximum frequency of 6 GHz and especially mmWave frequencies, where the increased phase noise requires a higher SCS, but also the shorter channel coherence time requires shorter frames.
- At a given frequency, higher SCS enables shorter symbols, thus enabling lower latencies.
- Addressing higher Doppler shifts due to higher mobility (up to 500 km h^{-1}) and higher frequencies like mmWave.
- Support of larger channel bandwidths with a reasonable Fast Fourier Transform (FFT) size.

Waveform filtering (filtering in the frequency domain) or windowing (filtering in the time domain) has been introduced, which enables the spectrum to be better concentrated around the allocated resource blocks (RBs). This improves both in-channel in-band emissions (IBE) but also out-of-channel emissions with a higher number of RBs resulting in better spectrum utilization (SU). In the study phase of NR, a number of filtering or windowing techniques were in competition, but as shown in Figure 12.1, once power amplifier (PA) nonlinearity occurs, the output spectrum (bottom) is dominated by the PA spectral regrowth regardless of the input waveform filtering or windowing technique used (top). Based on this, it was decided that the spectrum confinement technique would not be standardized but indirectly specified in RAN4 via IBE and SU requirements.

Compared to LTE, where CP-OFDM was already used in the downlink, the Single Carrier Frequency-Division Multiple Access (SC-FDMA) waveform was used in the uplink to reduce the PAPR and achieve higher output power capability from the UE power amplifier. In NR, it is mandatory for the UL to support both CP-OFDM and DFT-s-OFDM:

- Quadrature Phase Shift Keying (QPSK)-modulated DFT-s-OFDM offers a low PAPR (as does LTE SC-FDMA) but is restrained to single stream operation (cannot be used for Uplink Multiple Input Multiple Output (UL MIMO)) and has limitations in

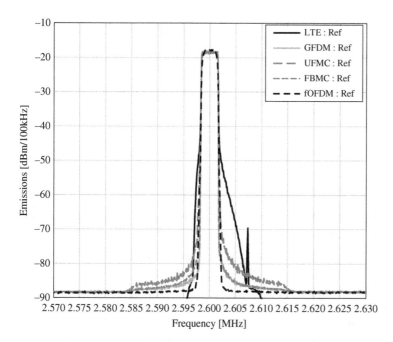

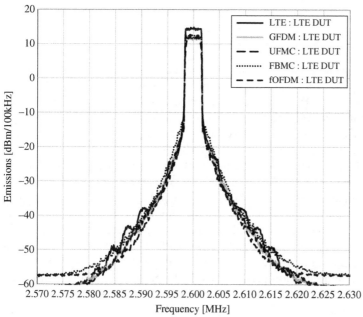

Figure 12.1 Spectrum using with different filtering/windowing techniques at (top) PA input and (bottom) output.

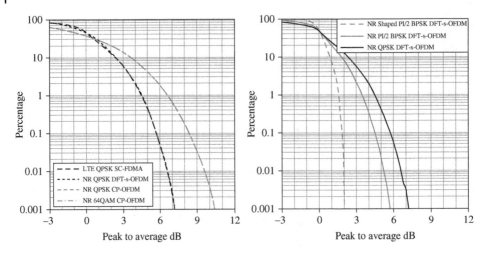

Figure 12.2 PAPR for (left) single-carrier LTE QPSK vs. 5G DFT-S-OFDM and CP-OFDM waveforms, (right) 5G shaped and un-shaped Pi/2 BPSK vs 5G QPSK DFT-S-OFDM.

possible RB allocations that must follow the rule $N_{RB} = 2^x \times 3^y \times 5^z$. Also, the PAPR increases with higher order modulations.

- CP-OFDM has an approximately 2.5 dB higher PAPR but is not subject to limitations in RB allocation, and thus it is used for multiple stream links. Although it has a higher PAPR, it does not vary for higher order modulations compared to QPSK (Figure 12.2 left), and thus it is more efficient than DFT-s-OFDM for higher order modulations.

While CP-OFDM waveforms increase the PAPR and thus the linearity requirements for the UE power amplifier increases, NR also introduces lower PAPR UL modulation for DFT-s-OFDM waveforms that are dedicated to challenging link conditions. Beyond QPSK, 16 state Quadrature Amplitude Modulation (16-QAM) and 64-QAM that are mandatory to support, and 256-QAM that has optional support, Pi/2 Binary Phase Shift Keying (BPSK) modulation has been added for DFT-s-OFDM. Since the gain in the PAPR with straight Pi/2 BPSK is barely more than 1 dB, spectrum shaping is also allowed, which can further improve the PAPR by close to 5 dB compared to the QPSK case. Pi/2 BPSK support is optional in FR1 (below 7 GHz) and mandatory in FR2 (mmWave) as the link conditions are particularly difficult, but also the power amplifier efficiency is much lower than in FR1. It has been decided that the exact spectral shaping filter coefficients would not be specified, but rather that RAN4 would have requirements in terms of higher power capability (due to the lower PAPR) and restrictions in spectrum flatness to still guarantee demodulation performance at the base station (BS). Figure 12.3 shows un-shaped and shaped Pi/2 BPSK spectrums, where the higher power capability and spectrum flatness aspects are visible.

12.2.2 New Frequency Range Definition

The applicable frequency range for 5G is extremely wide as it spans from ultra-high frequency (UHF) to millimeter waves. Therefore, 3GPP Release 15 has divided this range into two parts:

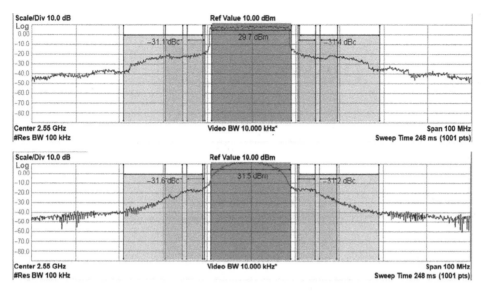

Figure 12.3 Spectrum plots with maximum output power at linearity limit for FR1 for (top) unshaped and (bottom) shaped Pi/2 BPSK.

- Frequency range 1 (FR1) = 410–7125 MHz
- Frequency range 2 (FR2) = 24.250–52.600 GHz

The requirements for FR1 operating bands are captured in TS 38.101-1 and are intended to be verified by conducting measurements similar to LTE and has been recently extended to 7.125 GHz to encompass newly available spectrum above 5.925 GHz. The requirements for FR2 operating bands are captured in TS 38.101-2 and can only be verified using radiated measurements, which is a big difference compared to LTE. However, RF requirements are not directly defined for frequency ranges; instead, RF requirements are still associated with operating bands and channel bandwidths, and therefore in the next paragraph we discuss how the new 5G operating bands are numbered.

12.2.2.1 5G Operating Band Numbering Scheme

3G Wideband Code-Division Multiple Access (WCDMA) band numbers were using Roman numerals, that is, capital letters, band I being the most widespread WCDMA band at 2 GHz. Then, in 4G, LTE band numbers used standard Arabic numerals, and where LTE and WCDMA have the same frequency arrangement, then the band numbers match. For 5G, it was decided to use Arabic numerals similar to LTE but with the n-prefix as it was deemed beneficial to keep an easy linkage between 5G, LTE, and WCDMA band numbering. Therefore, if an LTE and 5G band have the same frequency arrangement, then LTE and 5G have the same band number, but 5G uses the n-prefix. As an example, WCDMA band I corresponds to LTE band 1 and 5G band n1.

From Release 15 onward, 3GPP decided to allocate operating band numbers in such a way that 5G FR1 bands will have band numbers between n65 and n256, and 5G FR2 bands will have band numbers starting from n257. New LTE Frequency Division Duplex (FDD) bands will share band numbers with 5G FR1 bands, and band numbers

Table 12.1 5G vs. LTE band number scheme.

	LTE		5G FR1/FR2	
	Band #	Duplex	Band #	Duplex
Legacy LTE FDD band numbers	1	FDD	n1	FDD
	32		n32	
Legacy LTE TDD band numbers	33	TDD	n33	TDD
	64		n64	
New LTE/5G FR1 band numbers	65	All duplex modes	n65	All duplex modes
	256		n256	
New 5G FR2 band numbers	N/A	N/A	n257 and above	TDD

are assigned on a first-come, first-served basis. As an exception, new LTE Time Division Duplex (TDD) bands are allocated band numbers from the 33–64 range as there are still unused band numbers left. This scheme is summarized in Table 12.1.

12.2.3 Impact of NSA Operation on the 5G UE RF Front-End

For Non-Standalone Architecture (NSA), Evolved-Universal Terrestrial Radio Access (E-UTRA) and NR dual connectivity (EN-DC) are introduced. EN-DC/NSA uplink coverage benefits and principles are covered in Chapter 13. EN-DC dual uplink operation is not a novelty since UL-CA is already part of LTE; however, while this feature remained optional for LTE UEs and had limited deployment, it becomes mandatory for the 5G EN-DC UE and constitutes one of the toughest RF front-end challenges. Through power amplifier nonlinearity, Figure 12.4 shows that several inter-modulation distortion (IMD) products are generated, some of which may overlap the UE's own primary downlink channels and/or secondary downlink channels. IMD3 and IMD5 are the two strongest products which may create huge receiver desensitization and may violate emission requirements. For EN-DC combinations where such conditions occur, Section 12.5.4 discusses how both maximum transmit power and minimum receiver sensitivity are impacted, that is, potentially degrading both the UL and DL link budget.

Figure 12.4 (left) illustrates the case of a UL MIMO-capable UE where two separate PA-filter-antenna switch modules (PAM) are available. Through limited antenna-to-antenna isolation, each radio access technology (RAT) signal reaches the other RAT PA output port, and IMDs are generated via reverse mixing. UEs with dual-PA architectures would typically generate lower IMD levels than those with the single PA architecture because the reverse interferer is attenuated by antenna isolation, while IMDs in single PA solutions are due to forward mixing (cf. Figure 12.4 (right)). 3GPP assumes 10 dB antenna isolation for MB/HB operation, while for low bands, dual antenna architecture is considered impractical due to the difficulty of creating sufficient physical separation between antennas in a smartphone at low frequencies, and therefore only single PA architectures are assumed for LB. When these emissions become too difficult to handle, the UE is allowed to request Single Uplink Operation (SUO), whose concept is illustrated in Figure 12.5.

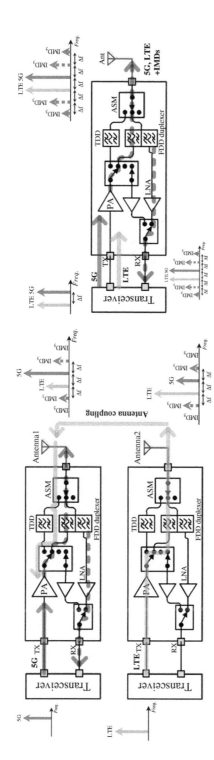

Figure 12.4 Dual uplink inter-modulation distortion (IMD) products for dual-PA architecture for (left) inter-band and some intra-band cases, (right) single PA UE architecture for intra-band EN-DC.

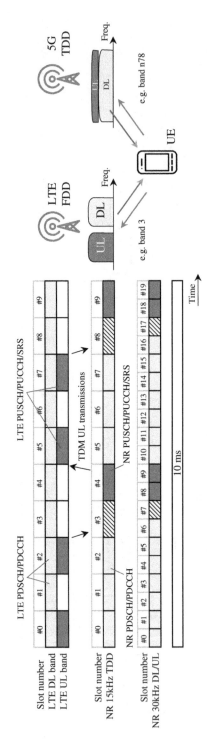

Figure 12.5 Solving dual uplink transmission IMD issues using Single Uplink Operation: example of an FDD-TDD EN-DC combination with 15 and 30 kHz SCS for 5G. Receive slots: light gray, transmit slots: dark gray, special slots (TDD): dashed.

SUO uses Time Division Multiplexing (TDM) to avoid 5G and LTE UL simultaneous transmissions and therefore eliminates any two carrier related IMDs. It is also used for UEs which are not capable of performing dynamic power sharing (DPS), as explained in Section 12.5.

To optionally request SUO, the UE needs to fulfill two conditions:

1. The UE's own LTE or NR primary downlink channels are victim of either an IMD2 or IMD3 product.
2. IMD3 products are only considered when LTE and 5G UL transmissions operate either both in the "low frequency range" (i.e. 410–960 MHz), or when both operate in the mid-band range (i.e. 1427–2690 MHz).

This is justified by the fact that interferers cannot be suppressed easily due to their close proximity in the frequency domain. SUO may also be allowed when emission requirements are difficult to meet, such as intra-band operation in band 41. In all other cases, the UE is mandated to support simultaneous dual uplink transmissions.

It is worth noting that even if an EN-DC combination is SUO eligible, SUO is only allowed for operating carrier frequencies and channel BW for which a direct hit occurs. As a result, operators who plan to deploy 5G in a "difficult" EN-DC band combination may not necessarily have to pay the SUO penalty. Operators need to perform a detailed IMD analysis based on their spectrum holdings to determine if direct hit applies. Take, for example, the case of DC_5_n66 shown in Figure 12.6 for which a direct IMD 2 hit occurs to band 5 DL carrier 886 MHz when n66 is modulated at 1727 MHz and band 5 at 841 MHz; that is, SUO is allowed (IMD2 = 1727 − 841 = 886 MHz). However, if band 5 is deployed at 830 (UL)/875 (DL) MHz, the IMD2 does not overlap band 5 DL, and therefore the UE must support dual uplink transmissions.

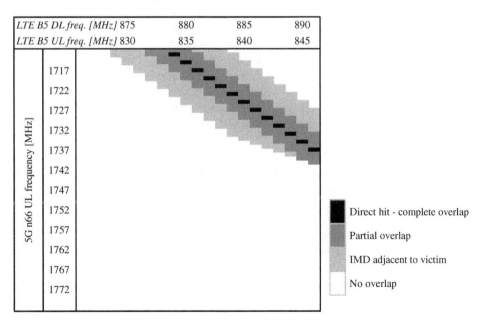

Figure 12.6 Example of detailed IMD2 analysis for DC_5_n66 or DC_66_n5. Direct hits have been derived assuming 5 MHz CBW for each RAT and 1 MHz raster.

Dual connectivity in 5G is also specified for NE-DC, where NR is the Master Cell Group (MCG) and LTE the Secondary Cell Group (SCG), and NN-DC, which combines both 5G CA (for example, with 2 FR1 bands with n2-n5) and 5G dual uplink connectivity between FR1 and FR2 bands, for example, n1-n257.

12.2.4 New Features Impacting UE RF Front-End

12.2.4.1 Impact of Beam Forming in FR2

While beamforming can be implemented with a large number of antenna elements at the gNb, Figure 12.7 shows that the limited volume of modern smartphones and shrinking area for additional antennas creates new trade-off challenges to deliver sufficient spherical coverage with a limited number of mmWave antenna panels.

For beam correspondence (BC), Release 15 assumes dual-polarized signal paths and a minimum of two antenna panels and four antenna elements per panel. However, the number of antenna panels required to achieve spherical coverage requirements and to mitigate blockage issues due to user interaction, such as head and hand proximity, is purely an original equipment manufacturer (OEM) implementation choice. In 3GPP RAN4, OEMs have presented simulation and measurement data across a large range of UE architectures, with two to four antenna panels, and four to eight antenna elements per panel, some using combinations of dipoles and antenna arrays [1–5]. Figure 12.8 (left) illustrates the two most common architectures used at 3GPP, using two and three antenna panels.

For the FR1 subsystem, remote antennas such as D1, D2, and D3 (Figure 12.7) can be routed back to the RF transceiver either via printed circuit board (PCB) tracks or coaxial cables. In FR2, mmWave routing must be kept as short as possible as routing mmWave through the UE PCB material (e.g. FR4) would cause detrimental insertion losses. First-generation UEs therefore use the superheterodyne architecture illustrated in Figure 12.8 (right). FR2 waveforms are modulated at an intermediate frequency (IF) where PCB trace or coaxial cables losses are tolerable (10.56 GHz [6], 6.5 GHz [7]). The remote mmWave RF transceiver needs to be tightly connected to the antenna array and performs final up-conversion, phase shifting, and amplification for transmission and for reception. This is a radical change to FR1 legacy architectures, and Section 12.4.3 discusses some challenges associated with mmWave.

12.2.4.2 Impact of UL MIMO Operation

UL MIMO is already part of the LTE specification, but in 5G the adoption of CP-OFDM for UL MIMO transmissions brings several additional benefits. In addition to removing RB length restrictions inherent to DFT-S-OFDM, CP-OFDM also offers a 1.6–3.9 dB uplink signal-to-noise ratio (SNR) gain for QPSK and an up to 5 dB gain for 64-QAM over DFT-S-OFDM (cf. Chapter 3, Table 3.5), thereby offsetting the penalty in the PAPR. For example, Section 12.3.4.2 shows that up to 15% PA power consumption savings may be achieved by using QPSK CP-OFDM vs. QPSK DFT-S-OFDM.

UL MIMO is therefore bound to play an important role to meet the high cell-edge spectral-efficient targets of 5G. From a UE RF front-end perspective, the price to pay for this feature is to double the number of transmitter paths, that is, two power amplifiers and two separate antennas (the Rx diversity antenna can be reused for UL diversity). Due to the limited space for additional antennas (Figure 12.7), and

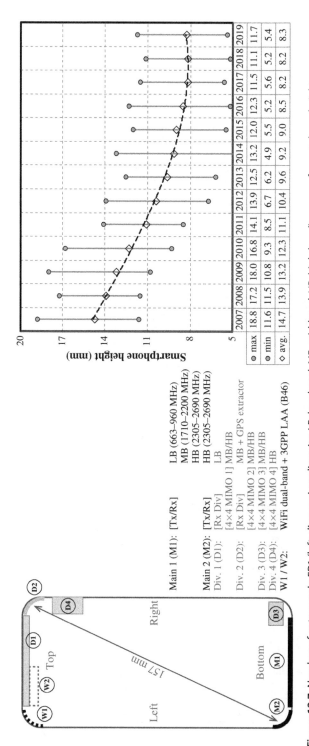

Figure 12.7 Number of antennas in FR1 (left, div: receiver diversity, LB: low-band, MB: mid-band, HB: high-band), survey of smartphone height.

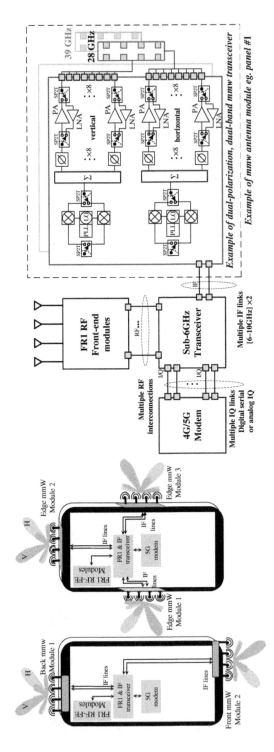

Figure 12.8 (Left) Impact of mmWave support on a 5G smartphone high-level architecture, (right) example of mmWave antenna module using eight antenna elements per band of operation and RF phase shifting.

since good MIMO performance requires low antenna correlation [8], this feature can only be practically supported in mid- and high-frequency bands (MB/HB). Dual-PA architectures (Figure 12.4) may also help deliver better uplink coverage due to (1) reduced IMD levels thanks to antenna-to-antenna isolation and (2) transmit diversity gain at the cell edge.

12.2.4.3 Impact of Sounding Reference Signal (SRS) Switching as Enhancement to Downlink MIMO

In FR1, the UE is mandated to support 4 receive (RX) antennas in high-frequency bands (n7, n38, n41, n77, n78, and n79). In the TDD bands, the 4×4 DL MIMO system performance is suboptimal as channel reciprocity can only be accurately exploited using the port on which SRS symbols are transmitted, that is, on the primary antenna port. At best, for UL MIMO-capable UEs, SRS sounding can be performed on two out of four antennas, thereby partially improving the 4×4 matrix channel estimation.

To further enhance channel estimation and to better exploit channel reciprocity, 5G supports "SRS antenna switching," which requires the UE to perform uplink sounding on all RX physical antenna ports. This feature complicates the RF front-end multiplexing /de-multiplexing architecture as several additional RF switches are needed to connect each RX antenna port to the PA module(s). This feature is defined for one transmitter (1T) two receivers (1T/2R), 1T/4R, 2T/4R, and an equal number of Tx/Rx. The concept is illustrated in Figure 12.9 for 1T/4R (left) and 2T/4R (right). To account for additional insertion losses due to new RF switches and due to the large routing distance separating the PA from some of the remote Rx diversity antennas, a 3 dB transmit power reduction is allowed for SRS antenna switching in the Pcmax equations.

One of the challenges for UL MIMO-capable UEs is to provide an RF-FE (RF-front end) architecture which allows the UE to reconfigure from 2T4R to 1T4R at little or no additional insertion losses. At the time of writing, there are still many of those aspects that need completion, such as support for intra-band EN-DC operation, or support for FR2, where SRS switching could be performed across panels, or between the two polarizations of a given panel when transmissions occur on a single polarization.

12.2.5 RAN4 Technical Specification (TS) Survival Guide

Compared to LTE where everything is included in TS 36.101, for NR, 3GPP has decided to split 5G UE RF and demodulation performance requirements into four different specifications in order to keep the content more compact and focused:

- TS 38.101-1 NR; User Equipment (UE) radio transmission and reception; Part 1: Range 1 Standalone
- TS 38.101-2 NR; User Equipment (UE) radio transmission and reception; Part 2: Range 2 Standalone
- TS 38.101-3 NR; User Equipment (UE) radio transmission and reception; Part 3: Range 1 and Range 2 Interworking operation with other radios
- TS 38.101-4 NR; User Equipment (UE) radio transmission and reception; Part 4: Performance requirements

TS 38.101-1 includes FR1 UE transmitter and receiver RF requirements for single carrier, 5G carrier aggregation, 5G NR-NR dual connectivity, supplementary uplink, and

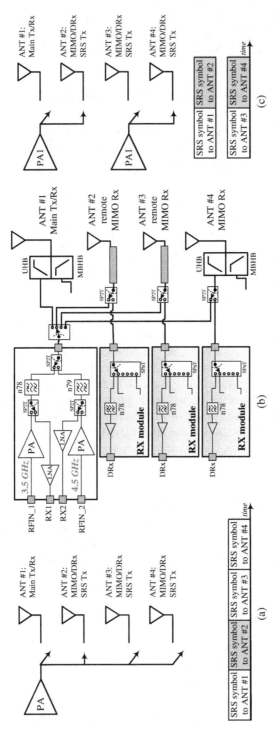

Figure 12.9 SRS antenna switching: (a) 1T4R concept, (b) example relying on external switches, (c) 2T4R concept.

uplink MIMO operation. All requirements are intended to be verified with conducted tests.

TS 38.101-2 includes FR2 UE transmitter and receiver RF requirements for single carrier, 5G carrier aggregation, and uplink MIMO operation. All requirements are intended to be verified with radiated tests.

TS 38.101-3 include UE transmitter and receiver RF requirements for 5G CA between FR1 and FR2, EN-DC (dual connectivity between LTE and 5G), and NR-DC (5G dual connectivity between FR1 and FR2). In most of the cases, TS 38.101-3 does not state the actual requirements; instead, there is a reference to the applicable LTE (TS 36.101), 5G FR1 (TS 38.101-1), or 5G FR2 (TS 38.101-2) specification. A big part of TS 38.101-3 is just listing what are the specified 5G CA, EN-DC, NE-DC, or NR-DC configurations that can be deployed.

TS 38.101-4 include UE requirements for demodulation performance and channel state information (CSI) reporting for both FR1 and FR2.

12.3 5G Standalone Performance Aspects in Frequency Range 1

12.3.1 New Channel Bandwidths and Improved SU

As discussed in Section 12.2.1, NR has introduced a new SCS which enables larger channel bandwidths beyond the 20 MHz maximum channel bandwidth of LTE. This is important for the new spectrum available above 2.5 GHz in FR1 with bands offering more than 100 MHz total bandwidth. Instead of having to use intra-band contiguous carrier aggregation of 20 MHz channels, NR offers channel bandwidths up to 100 MHz.

As can be seen from the header of Table 12.2, a large choice of channel bandwidths from 5 to 100 MHz is provided with good granularity, and the maximum number of RBs that can be allocated per channel is provided for the different SCSs valid for FR1.

In FR1, valid SCSs are 15, 30, and 60 kHz. 15 kHz SCS is compatible with LTE and thus allows an easy refarming of LTE bands with NR. Higher SCS values not only enable shorter frames and improved Doppler immunity, but also higher channel bandwidths.

The minimum and maximum channel bandwidths realizable for a given SCS are dictated by two aspects:

- The maximum FFT size of 4096 bins sets the maximum number of RBs and thus the maximum channel bandwidth. For 15 kHz SCS, the maximum bandwidth would be $4096 \times 15\,\text{kHz} = 61.44\,\text{MHz}$, but accounting for the margin needed for the cycle prefix, it is actually limited to 50 MHz channel bandwidth.
- The number of subcarriers (107) for the synchronization channel also sets the minimum channel bandwidth that can be accessed by the UE. For 60 kHz SCS, the synchronization channel needs to be larger than $107 \times 60\,\text{kHz} = 6.42\,\text{MHz}$; thus, 5 MHz channel bandwidth is not achievable.

It is to be noted that the 60 KHz SCS would allow channel bandwidths up to 200 MHz, but in the time frame of Release 15, there was no operator holding that would require such bandwidth. Therefore, it was decided to limit channel bandwidth at 100 MHz, mandate UEs to support 15 and 30 kHz SCS and all channel bandwidths, while 60 kHz SCS

Table 12.2 Supported channel bandwidths in FR1 and associated spectrum utilization.

Parameters	SCS (kHz)	FR1 Channel BW (MHz)											
		5 N_{RB}	10 N_{RB}	15 N_{RB}	20 N_{RB}	25 N_{RB}	30 N_{RB}	40 N_{RB}	50 N_{RB}	60 N_{RB}	80 N_{RB}	90 N_{RB}	100 N_{RB}
CP-OFDM SU [#RB]	15	25	52	79	106	133	160	216	270	NA	NA	NA	NA
	30	11	24	38	51	65	78	106	133	162	217	245	273
	60	NA	11	18	24	31	38	51	65	79	107	121	135
DFT-s-OFDM SU [max#RB]	15	25	50	75	100	128	160	216	270	NA	NA	NA	NA
	30	10	24	36	50	64	75	100	128	162	216	243	270
	60	NA	10	18	24	30	36	50	64	75	100	120	135
CP-OFDM SU [max%]	15	90%	94%	95%	95%	96%	96%	97%	97%	NA	NA	NA	NA
	30	79%	86%	91%	92%	94%	94%	95%	96%	97%	98%	98%	98%
	60	NA	79%	86%	86%	89%	91%	92%	94%	95%	96%	97%	97%
DFT-s-OFDM SU (max%)	15	90%	90%	90%	90%	92%	96%	97%	97%	NA	NA	NA	NA
	30	72%	86%	86%	90%	92%	90%	90%	92%	97%	97%	97%	97%
	60	NA	72%	86%	86%	86%	86%	90%	92%	90%	90%	96%	97%
Min guard band (kHz)	15	242.5	312.5	382.5	452.5	522.5	592.5	552.5	692.5	NA	NA	NA	NA
	30	505	665	645	805	785	945	905	1045	825	925	885	845
	60	NA	1010	990	1330	1310	1290	1610	1570	1530	1450	1410	1370

is optional. 60 kHz SCS is only supported for bands >1 GHz as the lower bands do not require wide channel bandwidths.

Filtering or windowing techniques have been introduced to obtain better spectrum confinement as already discussed in Section 12.2.1. One of the goals was to obtain better SU (the ratio of allocated RB bandwidth versus channel bandwidth) than for LTE, which is limited to 90% (for example, a 20 MHz channel has the maximum number of RBs of 100 which is 18 MHz). In Table 12.3, the number of maximum allocated RBs is provided for CP-OFDM versus SCS in the first set of rows; this was derived by simulating the Adjacent Channel Leakage Ratio (ACLR), SEM, and the required guard band for both the UE and the BS. Given that different companies have used different spectrum confinement techniques, the end result is not always consistent within a few RBs. In the second set of rows, SU is derived for DFT-s-OFDM, which cannot always achieve the same SU as CP-OFDM because the number of RBs must meet the $N_{RB} = 2^x \times 3^y \times 5^z$ rule.

In the next two set of rows, the SU in % is reported for CP-OFDM and DFT-s-OFDM. If for small channel bandwidths the result is similar, and for high SCS, sometimes lower than the 90% of LTE, for channel bandwidths above 40 MHz close to 98% SU is achieved.

One particular aspect of NR waveforms is that the subcarriers are shifted by half a subcarrier such that the DC spur falls between two subcarriers and thus can be easily cancelled at the receiver without affecting the data associated with the adjacent subcarriers. In LTE, this shift does not exist, and the DC spur falls on a subcarrier that does not carry data. This shift results in a small asymmetry, and together with the variable SU, results in a variable minimum guard band as shown in the last set of rows. This one is defined using the half subcarrier shift on the lowest valid SCS, and higher SCSs are

Table 12.3 Maximum power reduction (MPR) for power class 2 and 3.

Modulation		Edge RB allocations (PC2 only)	Outer RB allocations	Inner RB allocations
DFT-s-OFDM	PI/2 BPSK	≤3.5	0.5	0
	QPSK	≤3.5	≤1	0
	16-QAM	≤3.5	≤2	≤1
	64-QAM	≤3.5	≤ 2.5	
	256-QAM		≤4.5	
CP-OFDM	QPSK	≤3.5	≤3	≤1.5
	16-QAM	≤3.5	≤3	≤2
	64-QAM		≤3.5	
	256-QAM		≤6.5	

aligned such that the first subcarrier of all RBs is aligned within all the valid SCSs. This alignment is needed to ensure that multiple signals using different SCSs can be received within the same FFT process.

12.3.2 Impact of Large Channel Bandwidths on PA Efficiency Enhancement Techniques

In LTE, a number of power amplifier efficiency enhancement techniques have been developed to increase the phone battery life. There are mainly two techniques that are implemented which both use a DC/DC converter to provide the supply to the power amplifier:

- The DC/DC converter voltage follows the power level requested at the power amplifier output. This technique is called adaptive power tracking (APT). The DC/DC converter only needs to dynamically follow the Transmit Power Control (TPC) requests and thus use relatively low switching frequencies and achieve efficiencies >90%. The power amplifier still relies on its intrinsic linearity, but its supply is adjusted to the minimum voltage needed.
- The DC/DC converter output voltage follows the envelope of the RF signal fed at the power amplifier output. This technique is called envelope tracking (ET). In this case, the DC/DC converter needs to support a bandwidth that is approximately twice the baseband bandwidth (equivalent to the channel bandwidth) and thus use high switching frequencies, which impacts its efficiency. In this case, the power amplifier operates closer to saturation and has a higher efficiency. The overall efficiency obtained for ET is usually better than for APT, but depends on the bandwidth to be supported.

The state-of-the-art ET implementations in LTE support a 40 MHz bandwidth with carrier aggregation of two contiguous 20 MHz channels; some support up to 60 MHz.

In order to support up to 100 MHz channel bandwidth, ET technology will require a significant step in performance with much higher switching frequencies resulting in lower DC/DC efficiency. The improved performance compared to APT, which has no bandwidth limitation, may be nullified until new ET techniques are developed that have improved bandwidth with good DC/DC conversion efficiency.

Release 16 will probably increase Tx bandwidths to 200 MHz. Also, ET is not suitable for FR2 with 400 MHz bandwidth and the multiple PA supplies to be supported in a phase array system.

12.3.3 FR1 Frequency Bands

In Release 15, 5G is specified in 30 frequency bands, out of which only 3 bands are specific to 5G operation, the remaining other bands being legacy LTE bands used for NR TDD/FDD or Supplemental Uplink (SUL) operation. In this respect, the added cost in RF-FE for supporting 5G in FR1 is limited to the features introduced in Section 12.2.4 and to the support of the new bands n77/78 and n79. In some solutions, these bands are supported in a dedicated PA module. It should be noted, however, that n79 is a completely new band, whereas n77/n78 can be considered as an extension of LTE Bands 42 and 43. Figure 12.10 shows the key differences compared to LTE:

- In LTE refarmed bands, extended channel bandwidth (CBW) support has been agreed upon for the bands n3/n80, n5, n8/n81, n39, n40, n41, n51, n66/n86, and n70 for downlink mode operation only. In bands with a short duplex distance/duplex gap such as bands 5, 8, and 12, this leads to additional maximum sensitivity degradation (MSD) issues as discussed in Section 12.3.3.1. Further band extensions are requested for Release 16, with, for example, the request to support 50 MHz CBW in n7, thereby creating additional MSD challenges for the RF front-end.
- In bands n3, n7, n38, n41, n77, n78, and n79, it is mandatory for the UE to support four receive antennas (cf. note 1 in Figure 12.10). The only allowed exception to this rule has been granted to vehicular UEs in order to account for cost/implementation trade-offs. 3GPP analyses have shown that thanks to the higher performance of rooftop mounted antennas, the link budget for 2RX vehicular UEs is similar to that of smartphones with for four low profile, built-in antennas [8].
- n40, n41, n77, n78, and n79 are the most attractive bands since 100 MHz CBW is supported at SCS 30 and 60 kHz, thereby paving the way to exploit fully the promises of the 5G physical layer.

12.3.3.1 Impact of Extended Channel Bandwidth on MSD in Refarmed Bands

Similar to previous studies done, for example, in the LTE band 12/17, in bands with a short duplex distance and short gaps, increasing the transmitter CBW places the PA Tx noise leakage in too close proximity to the UE's own downlink carrier to prevent receiver self-desensitization. The measurements in Figure 12.11 have been performed taking into account only the counter-intermodulation order 3 (counter-IM3) products at the PA input (Figure 12.11 left). The resulting MSD is 0.8 and 1 dB for 20 MHz cell bandwidth operation in bands 8 and 5, respectively [9]. Recent 3GPP contributions have requested to increase these MSD levels to account for the presence of additional counter-IM5 products that may be present at the PA input port. Depending on the counter-IM5 rejection ratio, an increase in the transmitter excess noise level can be expected due to (1) the short Tx-Rx frequency separation inherent to bands n8 and n5 and (2) the fact that the counter-IM5 PSD extends at the 45 MHz duplex distance into the receive channel (Figure 12.11 left).

NR Band	UL/DL frequencies [MHz]	SCS [kHz]	UE Channel bandwidth [MHz] 5	10	15	20	25	30	40	50	60	80	90	100
n1 FDD	LTE refarm	15	X	X	X	X								
	UL: 1920 – 1980	30		X	X	X								
	DL: 2110 – 2170	60		X	X	X								
n2 FDD	LTE refarm	15	X	X	X	X								
	UL: 1850 – 1910	30		X	X	X								
	DL: 1930 – 1990	60		X	X	X								
n3 FDD	5G BW extension	15	X	X	X	X	X	X						
	UL: 1710 – 1785	30		X	X	X	X	X						
	DL: 1805 – 1880	60		X	X	X	X	X						
n5 FDD	5G BW extension	15	X	X	X	X								
	UL: 824 – 849	30		X	X	X								
	DL: 869 – 894	60												
n7[1] FDD	LTE refarm	15	X	X	X	X								
	UL: 2500 – 2570	30		X	X	X								
	DL: 2620 – 2690	60		X	X	X								
n8 FDD	5G BW extension	15	X	X	X	X								
	UL: 880 – 915	30		X	X	X								
	DL: 925 – 960	60												
n12 FDD	5G BW extension	15	X	X	X									
	UL: 699 – 716	30		X	X									
	DL: 729 – 746	60												
n20 FDD	LTE refarm	15	X	X	X	X								
	UL: 832 - 862	30		X	X	X								
	DL: 791 – 821	60												
n25 FDD	LTE refarm	15	X	X	X	X								
	UL: 1850 – 1915	30		X	X	X								
	DL: 1930 – 1995	60		X	X	X								
n28 FDD	LTE refarm	15	X	X	X	X								
	UL: 703 – 748	30		X	X	X								
	DL: 758 – 803	60												
n34 TDD	LTE refarm	15	X	X	X									
	UL/DL: 2010-2025	30		X	X									
		60		X	X									
n38[1] TDD	LTE refarm	15	X	X	X	X								
	UL/DL: 2570-2620	30		X	X	X								
		60		X	X	X								
n39 TDD	5G BW extension	15	X	X	X	X	X	X	X					
	UL/DL : 1880-1920	30		X	X	X	X	X	X					
		60		X	X	X	X	X	X					
n40 TDD	5G BW extension	15	X	X	X	X	X	X	X	X				
	UL/DL : 2300-2400	30		X	X	X	X	X	X	X	X	X		
		60		X	X	X	X	X	X	X	X	X		
n41[1] TDD	5G BW extension	15		X	X	X			X	X				
	UL/DL : 2496-2690	30		X	X	X			X	X	X	X	X[2]	X
		60		X	X	X			X	X	X	X	X[2]	X
n50 TDD	5G BW extension	15	X	X	X	X			X	X				
	UL/DL : 1432-1517	30		X	X	X			X	X	X	DL		
		60		X	X	X			X	X	X	DL		

NR Band	UL/DL frequencies [MHz]	SCS [kHz]	UE Channel bandwidth [MHz] 5	10	15	20	25	30	40	50	60	80	90	100
n51 TDD	LTE refarm	15	X											
	UL/DL : 1427-1432	30												
		60												
n66 FDD	5G BW extension	15	X	X	X	X			X					
	UL: 1710 - 1780	30		X	X	X			X					
	DL: 2110 - 2200	60		X	X	X			X					
n70 FDD	LTE refarm	15	X	X	X	DL	DL							
	UL: 1695 - 1710	30		X	X	DL	DL							
	DL: 1995 - 2020	60		X	X	DL	DL							
n71 FDD	LTE refarm	15	X	X	X	X								
	UL: 663 - 698	30		X	X	X								
	DL: 617 - 652	60												
n74 FDD	LTE refarm	15	X	X	X	X								
	UL: 1427 – 1470	30		X	X	X								
	DL: 1475 - 1518	60		X	X	X								
n75 SDL	LTE refarm	15	DL	DL	DL	DL								
	UL: N/A	30		DL	DL	DL								
	DL: 1432 - 1517	60		DL	DL	DL								
n76 SDL	LTE refarm	15	DL											
	UL: N/A	30												
	DL: 1427 - 1432	60												
n77[1] TDD	5G specific	15		X	X	X			X	X				
	UL/DL: 3300-4200	30		X	X	X			X	X	X	X	X[2]	X
		60		X	X	X			X	X	X	X	X[2]	X
n78[1] TDD	5G specific	15		X	X	X			X	X				
	UL/DL: 3300-3800	30		X	X	X			X	X	X	X	X[2]	X
		60		X	X	X			X	X	X	X	X[2]	X
n79[1] TDD	5G specific	15							X	X				
	UL/DL: 4400-5000	30							X	X	X	X		X
		60							X	X	X	X		X
n80 SUL	B3 UL band	15	X	X	X	X	X	X						
	UL: 1710 -1785	30		X	X	X	X	X						
	DL: N/A	60		X	X	X	X	X						
n81 SUL	B8 UL band	15	X	X	X	X								
	UL: 880 - 915	30		X	X	X								
	DL: N/A	60												
n82 SUL	B20 UL band	15	X	X	X	X								
	UL: 832 - 862	30		X	X	X								
	DL: N/A	60												
n83 SUL	B28 UL band	15	X	X	X	X								
	UL: 703 - 748	30		X	X	X								
	DL: N/A	60												
n84 SUL	B1 UL band	15	X	X	X	X								
	UL: 1920 – 1980	30		X	X	X								
	DL: N/A	60		X	X	X								
n86 SUL	B66 UL band	15	X	X	X	X			X					
	UL: 1710 -1780	30		X	X	X			X					
	DL: N/A	60		X	X	X			X					

NOTE 1:Four Rx antenna ports shall be the baseline for this operating band except for two Rx vehicular UE.

Figure 12.10 Release-15 5G bands for FR1. White: LTE refarmed bands, light gray: 5G BW extension, dark-gray: 5G dedicated bands. (Note 1: Four Rx antenna ports shall be the baseline for this operating band except for two Rx vehicular UEs. Note 2: 90 MHz channel bandwidth is optional for Release 15.)

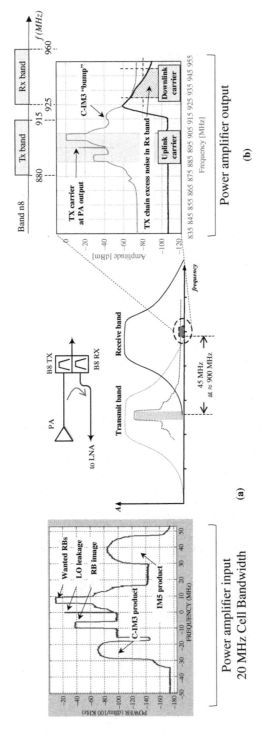

Figure 12.11 Measurement example of PA Tx excess noise falling in receive Band n8 for NR operation at 20 MHz Cell BW and 25RB allocated closest to the downlink band.

12.3.4 Transmitter Chain Aspects

12.3.4.1 Maximum Power Reduction and Inner/Outer Allocation Concept

The maximum output power reduction (MPR) concept is adopted in 3GPP to enable the usage of optimized power amplifiers that are dimensioned to deliver the best efficiency at the maximum output power. This is achieved by allowing the UE to transmit signals which are more challenging in terms of PA linearity and power consumption with a lower output power than a certain reference signal, which is a DFT-s-OFDM QPSK-modulated inner allocation waveform. This reference signal defines the 0 dB MPR point, which means that the UE shall meet all the necessary unwanted emission requirements such as ACLR and spectrum emissions mask with nominal output power according to UE power class. In the next paragraph, the specifics of inner, outer, and edge allocations are discussed.

12.3.4.1.1 Inner versus Outer Allocation Concept Figure 12.12 (left) presents MPR simulation results of required power reduction for all possible allocations within a 5 MHz channel bandwidth using a CP-OFDM signal which is QPSK-modulated with 15 kHz SCS. The simulation was performed using a power class 3 amplifier model. In this plot, the x-axis is RB_{start} and the y-axis is L_{CRB}. RB_{start} is the lowest RB index of the transmitted RBs within the channel, and L_{CRB} is the transmission bandwidth expressed as the length of contiguous RBs. As an example, the top of the triangle is $RB_{start} = 0$ and $L_{CRB} = 25$, and this represents the fully allocated 5 MHz channel bandwidth signal with a 15 kHz SCS.

When examining Figure 12.12 (left), we see that the required MPR follows certain patterns; for example, when L_{CRB} increases, then the required MPR also increases. This is caused by the fact that wider signals cause more unwanted emissions, and in order for the UE not to exceed unwanted emission limits, it may need to reduce the output power. Another noticeable pattern is that if the allocation is in the middle of the channel, then the need for MPR is lower, and in Figure 12.12 (left), this region is the white-colored triangle.

After exhaustive simulations and measurements work, 3GPP RAN4 decided to specify the 5G MPR based on outer, inner, and edge allocations scheme, which is presented in Figure 12.12 (right). It is to be noted that all allocation sizes L_{CRB} are possible for CP-OFDM waveforms, but the DFT-s-OFDM waveforms have restrictions on the allocation sizes which, by RAN1 design, must have a length of the form $2^x \times 3^y \times 5^z$, with x, y, and z being integers; these are denoted by a cross in Figure 12.12 (right).

This definition of different regions for allocations allowed the optimization of MPR values. The outer allocations are on the edge of the channel, and thus spectral regrowth due to PA nonlinearity extends almost entirely outside the channel edge, and the UE may fail out of band emission requirements like ACLR and SEM. The inner allocations are in the middle of the channel and in such a way that most of the spectral regrowth is within the channel and thus might only fail in band emissions or error vector magnitude (EVM), which, for low modulation orders, have a relaxed linearity requirement compared to ACLR or SEM. This inner-outer concept also benefits from the fact that 3G/UTRA (WCDMA)-related ACLR requirements are not generic since bands where both 3G and 5G will be deployed are an exception, and only for these bands is additional MPR allowed to meet UTRA ACLRs.

The allowed MPR for power class 2 (PC2: 26 dBm) is the same as for PC3 (23 dBm), with the exception that it introduces the concept of edge allocation in addition to inner

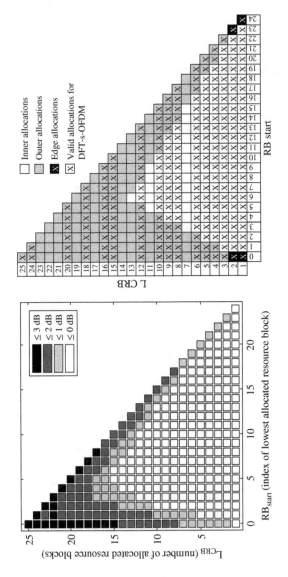

Figure 12.12 Required power back-off for (left) all 5 MHz NR channel possible allocations, (right) inner and outer allocation for a 5 MHz 15 kHz SCS NR channel.

and outer allocations. An edge allocation is located at the lowermost or uppermost edge of the channel with $L_{CRB} \leq 2$ RB's, in which case the allowed MPR is 3.5 dB. The reason for allowing more MPR for PC2 edge allocations is that these edge allocations are limited by the spectrum emission mask, which is at an absolute power level, and as PC2 has 3 dB more output power compared to PC3, it requires this additional relaxation, especially as in NR, because it has higher SU, these allocations are closer to the channel edge than for the LTE case.

The 5G MPR for power class 2 and 3 operation is presented in Table 12.3.

12.3.4.2 Impact on Power Amplifier Power Consumption

Compared to LTE, the advantage of the outer and inner MPR concept in 5G is that there are more allocations that benefit from a low MPR, as in LTE the MPR is determined only based on L_{CRB} and therefore the benefits of the inner allocation concept were not utilized. In particular, it enables CP-OFDM waveforms to be transmitted closer to the UE maximum output power. From a PA power consumption perspective, the UL SNR gains of CP-OFDM (cf. Chapter 5) are not sufficient to completely offset the difference in MPR with DFT-S-OFDM at large outer RB allocations. However, with the inner allocation concept, CP-OFDM transmissions may benefit either from equal or lower power consumption than DFT-S-OFDM. The example in Figure 12.13 shows that a gain in PA power consumption up to 15% may be achieved using QPSK CP-OFDM vs. DFT-S-OFDM assuming a 1.6 dB SNR gain, a simplistic PA power efficiency model, and 4 dB post PA loss.

12.3.4.3 MPR for Almost Contiguous Allocations and PI/2 BPSK Power Boosting

The basic MPR concept was enhanced already during Release 15 by introducing power boosting and MPR for almost contiguous allocations. Power boosting is a UE capability, and if the UE supports it and the network allows it, the UE can increase the DFT-S-OFDM PI/2 BPSK modulation output power beyond nominal PC3 limits. The allowed increase in output power is 3 dB compared to the nominal PC3 output power of +23 dBm. UEs operated in power boost mode do not have to increase the output power by a full 3 dB; instead, it is sufficient if the output power increase is 2.8 dB for inner allocations, 1.8 dB for outer allocations, and 0 dB for edge allocations compared to a nominal PC3 output power of +23 dBm. Power boosting is not allowed for any other modulation than DFT-S-OFDM PI/2 BPSK, and it is initially limited to the TDD mode in bands n40, n41, n77, n78, and n79 under specific duty cycle constrains.

Uplink transmissions that are DFT-S-OFDM modulated must always be contiguous, so that all of the RBs are transmitted also in the middle of the allocation; this is due to the RAN1 physical layer specifications. However, CP-OFDM uplink transmissions can be fully non-contiguous according to RAN1, but the RAN4 basic MPR definitions only cover contiguous transmissions. To allow more flexibility for network operation, the concept of almost contiguous allocation was introduced in 3GPP RAN4, enabling CP-OFDM uplink transmission to be non-contiguous with some restrictions. An uplink transmission is considered to be an almost contiguous allocation if it satisfies the following constraints: $N_{RB\_gap}/(N_{RB\_alloc} + N_{RB\_gap}) \leq 0.25$ and $N_{RB\_alloc} + N_{RB\_gap}$ is larger than 106, 51, or 24 RBs for 15, 30, or 60 kHz SCS, respectively, where N_{RB_gap} is the total number of unallocated RBs inside the allocated RBs, and N_{RB_alloc} is the total

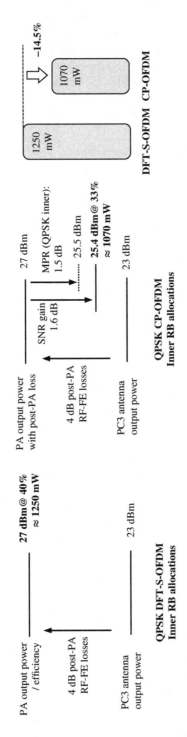

Figure 12.13 Combined impact of reduced MPR for CP-OFDM QPSK inner RB allocations and SNR UL gain on PA power consumption in a power class 3 UE.

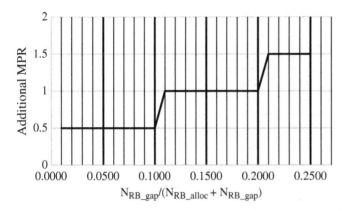

Figure 12.14 A-MPR Additional MPR for almost contiguous allocations.

number of allocated RBs. For these signals, the MPR can be increased by CEIL (10 $\log_{10}(1 + N_{RB\_gap}/N_{RB\_alloc})$, 0.5) dB; see Figure 12.14.

12.4   5G Standalone Performance Aspects in mmWave Frequency Range 2

12.4.1   Channel Bandwidths and SU

The new spectrum available in FR2 with bands offering more than 1 GHz total bandwidth requires channel bandwidths up to 400 MHz. As can be seen from the header of Table 12.5, a few channel bandwidths from 100 to 400 MHz are specified. In order to accommodate that, valid SCSs are 60 and 120 kHz.

Similar to FR1, the minimum and maximum channel bandwidth realizable for a given SCS are dictated by the maximum FFT size and synchronization channel bandwidth:

- The maximum FFT size of 4096 bins results in the maximum channel BW of 200 MHz for 60 kHz SCS.
- The number of subcarriers (107) for the synchronization channel would set the minimum channel bandwidth to 10 MHz for 60 kHz SCS, but the current channelization in FR2 bands only requires 50 MHz minimum channel bandwidth.

Given the smaller frequency range in FR2 (about one octave) compared to FR1 (more than a decade), only two SCSs are useful.

Table 12.4 reports the same elements as for FR1 in Section 12.3.1 in terms of the maximum RBs for CP-OFDM and DFT-s-OFDM, related SU in %, and associated minimum guard bands for the 60 and 120 kHz SCSs. It is to be noted that only up to 95% SU is achieved. Refer to Section 12.3.1 for a detailed explanation of the sub-carrier and RB alignment rules which are common to all frequency ranges and the resulting half SCS shift and SU.

12.4.2   FR2 Bands

In the Release 15 specifications, there are a very limited number of FR2 bands, and most of them are concentrated in the 24–30 GHz region with some overlap (Table 12.5). The

Table 12.4 Supported channel bandwidths in FR2 and associated spectrum utilization.

Parameters	SCS (kHz)	FR2 Channel BW (MHz)			
		50 N_{RB}	100 N_{RB}	200 N_{RB}	400 N_{RB}
DFT-s-OFDM and CP-OFDM SU [#RB]	60	66	132	264	NA
	120	32	66	132	264
DFT-s-OFDM and CP-OFDM SU [max%]	60	95%	95%	95%	NA
	120	92%	95%	95%	95%
Min. guard band (MHz)	60	1.21	2.45	4.93	NA
	120	1.9	2.42	4.9	9.86

Table 12.5 Release 15 FR2 bands.

Band	Frequency range (MHz)	Regions
n257	26 500–29 500	Korea, Japan
n258	24 250–27 500	Europe, China
n260	37 000–40 000	United States
n261	27 500–28 350	United States

reason for a separate band in that sub-range is linked to the regional footprint of bands, while implementing the full sub-range would be challenging due to the 20% relative bandwidth. All bands use TDD duplex mode, which is a prerequisite for beamforming operation in the current specification.

12.4.3 FR2 Key RF Parameters

Unlike for the FR1 operation, UE RF conformance tests at FR2 cannot be performed at the antenna port using coaxial cables. Consequently, all of the 38.101-2 requirements are measured OTA, and in the beamforming operation. In FR1, OTA measurements consist of total radiated power (TRP), total isotropic sensitivity (TIS), and LTE MIMO/SISO throughputs, and these tests do not constitute a 3GPP pass/fail criteria: they are, for example, part of the PTCRB conformance tests based on the CTIA OTA test plans, giving operators a useful insight into UE performance [8]. Translating all core requirements into the OTA test cases has created a lot of discussion and test methodology challenges, so much so that many test cases are not yet completely finalized at the time of writing (cf. Section 12.4.6). For the transmitter chain, two key performance indicators are defined in addition to TRP:

- The peak effective isotropic radiated power (EIRP) is a measure of the UE maximum transmitted power with the UE Tx beam locked in the beam peak direction. The UE beam peak direction is found with a 3D EIRP scan separately for each polarization, for which three measurements techniques are considered: Direct Far Field (DFF), Indirect Far Field (IFF), and Near-Field to Far-Field transform (NFTF). The average UE Tx power is measured with a spectrum analyzer, and is then converted into the EIRP using several correcting OTA factors, such as path loss due to distance, measurement antenna factor, and so on.

- The EIRP spherical coverage (SC) is key to assessing the UE ability to steer its numerous beam patterns and the performance of UE beam management algorithms. The uplink SC is evaluated by the cumulative distribution function (CDF) of the EIRP, where CDF = 100% corresponds to the peak EIRP. For the FR2 PC3 UE, the uplink SC has been agreed upon after numerous contributions at CDF = 50% as a compromise between the UE model using only one antenna panel and the UE model using two panels [10]. Table 12.6 below summarizes the PC3 transmit power KPIs and shows that at 28 GHz, the difference between SC and the peak EIRP is 10.9 dB. An ideal UE with perfect isotropic radiation pattern would exhibit a 0 dB difference; on the contrary, a UE with a very large difference would indicate limited beam steering capabilities as a result of, for example, a limited number of panels, beam granularity, constraints on the panel location, and constraints on the steering angles for each panel.
- BC is the ability of the UE to select a suitable beam for UL transmission based on DL measurements with or without relying on UL beam sweeping. BC ensures that every Tx beam has an identical corresponding Rx beam. The core requirements for BC are particularly difficult to define as many uncertainties in the UE mmWave antenna panel may introduce errors depending on the mmWave transceiver architecture. For example, the impact of phase shifter errors and PA/LNA gain errors are some examples the numerous ongoing discussions [11],

For the receiver chain effective isotropic sensitivity (EIS), SC was added to ensure that sensitivity performance can be measured using all directions surrounding the UE, not just one panel at a time. For the PC3 UE, the EIS SC is defined similarly to the EIRP SC, as the 50%-tile CCDF of the EIS measured over the full sphere around the UE.

All requirements are applicable to each antenna panel, and are verified with the UE beam locked into the beam peak direction. They must be tested in extreme temperature conditions (ETCs), that is, versus temperature, battery voltage, and across all channel bandwidths (50, 100, 200, and 400 MHz), some at 60 kHz SCS, others at 120 kHz, and for

Table 12.6 Power class 3 UE maximum transmit power OTA KPIs summary.

Operating band	Min EIRP at 50%-tile CDF (dBm), i.e. spherical coverage	Min peak EIRP (dBm), i.e. at 100%-tile CDF
n257	11.5	22.4
n258	11.5	22.4
n260	8	20.6
n261	11.5	22.4

each supported band, the UE must be tested at low–mid–high channels. One can easily picture that adding OTA uncertainties and limitations would only add to the overall legacy "conducted" test complexity.

12.4.4 Transmitter Aspects

12.4.4.1 Power and Device Classes

As discussed in Section 12.4.3, in FR2, most UE KPIs are defined with radiated performances. For this reason, the UE power classes are tightly coupled to the beamforming aspects, which are directly constrained by the UE form factor and use case. It is thus different for a smartphone, a customer premises equipment (CPE) used in fixed wireless access, or a vehicle; the main parameters are as follows:

- The EIRP, which depends on both the power capability (battery capacity) and the practical phase array size
- SC, which depends on the need to cover the entire sphere

With regard to these two criteria:

- A smartphone has to cover all directions as it is both mobile and is manipulated, has limited room available for arrays, and limited battery power: thus, full SC is needed and a moderate EIRP can be achieved.
- A CPE has mostly to cover the outside and can be a few cm square: thus, half SC is sufficient, and a higher EIRP is feasible.
- A vehicle mostly has to cover the horizon and above it and has less stringent size and power supply limitations than a smartphone: thus, less than half SC and a higher EIRP is feasible.
- Finally, for relays and repeaters, space and supply are not critical limitations, and they can be physically oriented: thus, half SC and a high EIRP are targeted.

These different UE characteristics have resulted in four UE power classes being defined for FR2 as summarized in Table 12.7, for which requirements are defined in terms of the minimum EIRP, maximum TRP, and SC CCDF. The maximum EIRP corresponds to the regulatory limits for different types of UE.

12.4.4.2 ACLR vs. FR1

One of the early (2017) 5G NR studies in FR2 was to assess adjacent channel coexistence performance accounting for the beamforming aspects. Coexistence requirements result

Table 12.7 FR2 UE power classes for band n257.

UE Power class	UE type	Min EIRP (dBm)	Max EIRP (dBm)	Max TRP (dBm)	Min EIRP CDF (percentile/dBm)
1	Fixed wireless access (FWA) UE	40.0	55	35	85%/32.0
2	Vehicular UE	29.0	43	23	60%/18.0
3	Handheld UE	22.4	43	23	50%/11.0
4	High power non-handheld UE	34.0	43	23	20%/25.0

in two KPIs: the ACLR for the transmit side and the ACS (Adjacent Channel Selectivity) for the receive side. Since with beamforming, the probability of blockers coming from the same direction as the BS or of transmitting toward other UEs is much lower, both the ACLR and ACS can be relaxed. Furthermore, it is common practice that the BS performance requirement is more stringent than the UE performance requirement, and a higher ACS on the BS side further relaxes the ACLR on the UE side. Coexistence simulation results based on a 5% throughput loss together with the consensus on the BS ACS vs. UE ACLR balance resulted in an ACLR requirement for FR2 of 17 dBc instead of 30 dBc in FR1.

12.4.4.3 MPR and A-MPR

In FR1 and the QPSK full allocation case, the difference between the out-of-channel (like the ACLR of 30 dBc) and in-channel requirements (like an EVM of 17.5%, which correspond to 15 dBc SNR) is large, and thus the linearity requirement in the PA is dictated by the ACLR. In FR2, however, both are very similar (17 dB ACLR and 15 dBc EVM), and IM3-related spectral regrowth being dominant in-channel, the PA linearity is nominally dictated by the EVM and IBE.

If these were the only requirements to be fulfilled, the maximum power as defined by the power class could be obtained for a fully allocated waveform. However, another parameter that is part of regulation in some regions needs to be met: an occupied bandwidth (OBW) of 99%. This means that 99% of the emitted power need to fit within the channel bandwidth; thus, that only 1% should lie outside. Considering the power on adjacent channels on both side of the channels, this means that the ACLR cannot be higher than 23 dBc (and that is assuming there is only negligible power further away from the allocated channel). With 17 dBc ACLR on each side, at least 4% of the power is outside the channel.

In order to keep the system parameters aligned and thus enable best UE efficiency and maximum power, it was decided that the maximum output power as defined by the power class would only be achievable for partial allocations in the middle of the channel.

The waveform defined by 100 MHz channel bandwidth, 120 kHz SCS, DFT-S-OFDM QPSK, and 20RB23 allocation occupies the center third of the channel, and thus its third order inter-modulation regrowth range is contained within the channel, and the associated 4% power equivalent to the 17 dBc ACLR linearity level being within the channel, the 99% OBW criterion is met. For this reason, this particular waveform is defined as the 0 dB MPR reference waveform. Given this reference, the power class 3 MPR table below could be defined. For higher order modulations than QPSK, the EVM requirement dominates, and the MPR level is similar for inner and outer allocations. As for FR1, CP-OFDM has a higher MPR due to its higher PAPR.

In Table 12.8, inner allocations are within the center third of the channel, or $RB_{start} \geq Ceil(1/3 N_{RB})$ AND $RB_{end} \leq Ceil(2/3 N_{RB})$; and outer allocations are located in the two other thirds at the edges of the channel, or $RB_{start} < Ceil(1/3 N_{RB})$ OR $RB_{end} > Ceil(2/3 N_{RB})$.

12.4.4.4 Impact on FR2 Relaxed ACLR Requirements on MPR Gating Factor

The impact of ACLR relaxation for low order modulation schemes is illustrated in Figure 12.15, where the OBW and ACLR are evaluated versus the PA transmit power at 24 GHz using a fully allocated DFT-S-OFDM QPSK waveform in 100 MHz CBW

Table 12.8 MPR table for PC3 channels ≤200 MHz.

Waveform type	Modulation order	Inner allocations	Outer allocations
DFT-s-OFDM	Pi/2 BPSK	0.0	≤2.0
	QPSK	0.0	≤2.0
	16-QAM	≤3.0	≤3.5
	64-QAM	≤5.0	≤5.5
CP-OFDM	QPSK	≤3.5	≤4.0
	16-QAM	≤5.0	≤5.0
	64-QAM	≤7.5	≤7.5

[12]. Figure 12.15b shows that at 0.7 dB back-off from the 0 dB MPR reference level, the OBW is just met but the ACLR is far from reaching the minimum requirements of 17 dBc. In Figure 12.15c, 99% of the total power is no longer constrained within the channel bandwidth and starts "splashing" into the adjacent channels. Yet the ACLR requirements are still met. The exhaustive simulations in [13] lead to the same conclusions.

12.4.5 Multi-Band Support and Carrier Aggregation

The beamforming operation in FR2 essentially depends on the antenna array design, and especially the physical distance between the elements in the array is related to the half-wavelength at the frequency of operation. Thus, unlike for FR1, the frequency range of operation of an antenna array is limited to 20–30% relative bandwidth, and multiple band support is not possible other than for adjacent or overlapping bands. This is actually the cases for bands n257, n258, and n261, but as they cover 20% of the relative bandwidth and designs can be optimized for a given region, a relaxation of the minimum EIRP and SC is allowed. Table 12.9 reports that the allowed relaxations depend on the band combinations. It is to be noted that relaxation is also allowed for combinations with band n260, although this one ideally requires a different array due to the significantly smaller wavelength compared to the 28 GHz bands. The two arrays can belong to the same module, and thus the array performance is influenced by the proximity of the other array, especially at large steering angles.

For Release 15, only intra-band carrier aggregation is specified; with large bandwidths and fixed wireless access use cases, a large number of intra-band contiguous bandwidth classes have been defined when compared to the LTE legacy. These can be found in Table 12.10, which covers fallback groups based on 100 MHz, 200 MHz, and 400 MHz channel bandwidths and up to 8 carriers in the 100 MHz channel bandwidth case.

12.4.6 OTA Conformance Test Challenges

OTA testing brings a whole set of measurement uncertainties (MUs) in addition to what formed the baseline of FR1 3GPP conducted conformance tests. For example, even if measurements could be performed conducted using "V" type coaxial connectors

Figure 12.15 Example of OBW vs. ACLR as MPR gating factor for 100 MHz SCS120 kHz QPSK DFT-S-OFDM 64RB0 waveforms at 24.3 GHz. PA output power back-off relative to 0 dB MPR waveform: (a) 1.4 dB, (b) 0.7 dB, (c) 0.1 dB.

Table 12.9 EIRP additional relaxation versus multi-band support.

Supported bands (2 Bands)	EIRP relaxation	Supported bands (>2 Bands)	EIRP relaxation
n257, n258	≤1.3	n257, n258, n260	≤1.7
n257, n260	≤1.0	n257, n258, n261	≤1.7
n258, n260	≤1.0	n257, n260, n261	≤0.5
n258, n261	≤1.0	n258, n260, n261	≤1.5
n260, n261	0.0	n257, n258, n260, n261	≤1.7

Table 12.10 NR FR2 intra-band CA bandwidth classes.

Bandwidth class	Aggregated channel bandwidth	Number of contiguous CC	Fallback group/ Maximum CC BW
A	$BW_{Channel} \leq 400\,MHz$	1	1, 2, 3, 4
B	$400\,MHz < BW_{Channel\_CA} \leq 800\,MHz$	2	1/400 MHz
C	$800\,MHz < BW_{Channel\_CA} \leq 1200\,MHz$	3	
D	$200\,MHz < BW_{Channel\_CA} \leq 400\,MHz$	2	
E	$400\,MHz < BW_{Channel\_CA} \leq 600\,MHz$	3	2/200 MHz
F	$600\,MHz < BW_{Channel\_CA} \leq 800\,MHz$	4	
G	$100\,MHz < BW_{Channel\_CA} \leq 200\,MHz$	2	
H	$200\,MHz < BW_{Channel\_CA} \leq 300\,MHz$	3	
I	$300\,MHz < BW_{Channel\_CA} \leq 400\,MHz$	4	
J	$400\,MHz < BW_{Channel\_CA} \leq 500\,MHz$	5	3/100 MHz
K	$500\,MHz < BW_{Channel\_CA} \leq 600\,MHz$	6	
L	$600\,MHz < BW_{Channel\_CA} \leq 700\,MHz$	7	
M	$700\,MHz < BW_{Channel\_CA} \leq 800\,MHz$	8	
O	$100\,MHz \leq BW_{Channel\_CA} \leq 200\,MHz$	2	
P	$150\,MHz \leq BW_{Channel\_CA} \leq 300\,MHz$	3	4/100 MHz
Q	$200\,MHz \leq BW_{Channel\_CA} \leq 400\,MHz$	4	

designed to operate up to 60 GHz, the MU can vary from one test bench to another due to factors such as cable quality, connector aging, and torque spanning. To gain a better insight, note that for LTE, the conducted MU is set to ±0.7 dB while the LTE OTA MU is already ±2.6 dB. To add to these, because of the limited test equipment (TE) noise floor, the fact that a minimum of 10 dB SNR is required to obtain better than 0.4 dB accuracy, and the large path loss due to the distance separating the UE from the TE (e.g. 68 dB free space loss 1.5 m at 43 GHz), the dynamic range in OTA is limited. The impact on the conformance test is that a minimum transmit power has been agreed upon as −13 dBm, while both the ACS test case 2 and the UE maximum input power requirement of −25 dBm at the UE antenna ports are considered impractical today due to high OTA losses, making the TE transmit power requirement too high. For the transmitter

OFF power, the requirement of −35 dBm would need to be relaxed by 24 dB to become "test-able," bringing the "Off" Tx power at higher level than the minimum transmit power requirement. With the OFF power not being finalized, the dynamic On/Off transient time is also problematic.

Another major challenge in OTA conformance is the related test cases that must be tested under ETC conditions, where the temperature and battery voltage must be varied. For temperature control, it has been proposed to enclose the UE in "air bubbles" to meet the −10 + 55 °C range. For the battery, the UE must be modified to accommodate a dummy battery that is connected to an external power supply. With these constraints, any test that requires device rotation becomes impractical to measure at ETC. For example, there has been a recent agreement that TRP, EIS, and EIRP SC are not tested at ETC, but the question of peak searches is still being debated as it is considered not practical to perform device rotation over an hemisphere or even a full sphere within an "air bubble" [14]. Considering that OTA testing is intrinsically more time consuming than conducted testing, test cases that are selected for ETC should also consider the impact of the test time and therefore the product development cost.

Other TE limitations such as lack of UL MIMO decoding capability limit uplink testing to be performed one polarization at a time.

12.5 Dual Uplink Performance Challenges for NSA Operation

12.5.1 From Single UL to Dual UL Operation

With many bands and features common to LTE, the support of 5G for standalone operation in FR1 constitutes an evolution from legacy solutions. As such, 5G radio solutions inherit from the successive challenges imposed as we moved from single-carrier operation, to downlink carrier aggregation to dual uplink operation. Figure 12.16 summarizes the most important challenges to complexity resulting from supporting multiple carriers simultaneously. The following subsections give a few examples of key challenges for EN-DC operation.

Single-carrier operation (1UL-1DL) key challenges are inherent to all FDD systems in which the UE's own receiver, due to the close proximity of its own Tx carrier in the frequency domain, may be subject to MSD due to the following:

1. Tx noise not sufficiently rejected by duplexer bandpass filters as seen in the example of n8 bandwidth extension (Section 12.3.3.1)
2. Tx signal leakage, which, through receiver mixer nonlinearity and self-mixing, can generate an in-band IMD2 product

Other challenges are deliberately omitted in this section. In downlink CA (1UL- xDL), intra-band non-contiguous CA is a special case where MSD can be aggravated due to an even shorter gap separating either the PCC or SCC from the UE's own Tx chain. In the case of inter-band CA, additional MSD may occur if one of the transmit carrier's harmonics directly hit one of the SCCs. One well-known example is that of LTE-CA between Band 12 and Band 4/66, where the third harmonic of Band 12 may completely overlap the Band 4 receive channel.

Any components in the transmit path are potential sources of harmonics, starting with the transceiver driver amplifier and limited transceiver pin to pin isolation, filters,

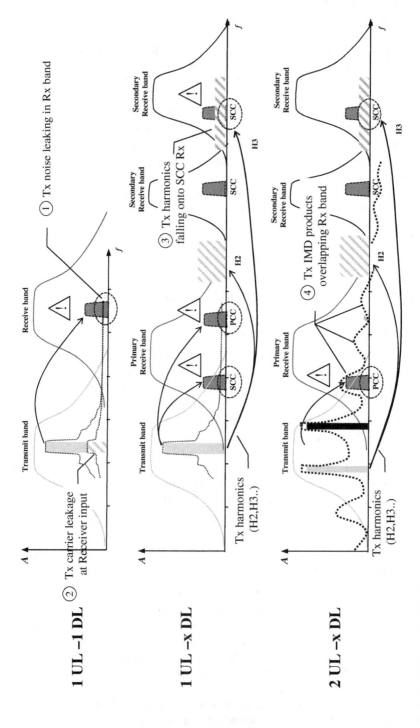

Figure 12.16 From single UL/single DL, to dual UL intra-band multiple DL operation.

antenna switches, antenna tuners, with the PA being the largest contributor. It is no surprise that nearly 10 years after its introduction, only a few smartphone models support CA_4_12 with band 12 as Pcell. Moving on to 2UL-xDL, the UE now has to face additional sources of self-interference from IMD products due to the simultaneous uplink operation. IMDs may impact the UE's own Pcell as well as several Scells, as explained in the next section.

12.5.2 EN-DC: Explosion of LTE-CA Combinations as Baseline to 5G

With LTE-CA commercial deployments being the baseline of further 5G EN-DC combinations, Figure 12.17 shows that 5G starts with an impressive number of 400+ distinct combinations. In comparison, it took LTE several years to reach 667 combinations. It is worth noting that FR2 plays an important role in the number of combinations, and the number of bands is likely to expand further. The pace at which new combinations are introduced is so high that these numbers will probably be obsolete by the time of printing. This obviously further fuels complexity and challenges RF-FE solutions to deliver the best cost/number of supported combinations trade-off.

12.5.3 FR1 UE Types and Power Sharing in EN-DC

There are two types of UEs in respect of uplink power sharing between LTE and 5G, namely, (1) UEs that support DPS between LTE and 5G and (2) UEs that do not support DPS. The UE informs the network which category it belongs to with the UE capability parameter d*ynamicPowerSharing*. Support for DPS will become mandatory after initial deployments.

UEs that do not support DPS cannot coordinate LTE and 5G transmission powers, and both modems independently set their own transmission powers without knowledge of other RAT transmissions. Therefore, these UEs rely on the network to guarantee by scheduling that $P\_LTE + P\_NR < Pcmax$, in other words, that the combined output power of LTE and 5G does not exceed the UE's output power capabilities. In an event that $P\_LTE + P\_NR$ may exceed Pcmax, the network needs to configure the UE to use *tdm-Pattern* for *single UL-transmission* in which case only either LTE or 5G carrier is transmitted in a time-multiplexed way. UEs that do not support DPS have to support *tdm-Pattern* for *single UL-transmission* capability, which is also known as SUO.

In the case where the UE supports DPS, the 5G modem is aware of the LTE transmission power and allocation properties, and it can ensure that the UE's maximum output power capability, that is, Pcmax, is not exceeded. The LTE modem is not required to be aware of the 5G transmission power or allocation properties, and thus it will set its output power regardless of the 5G transmission. If $P\_LTE + P\_NR$ exceeds Pcmax, then the 5G modem scales the 5G transmission power down to satisfy $P\_LTE + P\_NR = Pcmax$. If the amount of necessary scaling is more the network signaled value *xScale*, then the UE is allowed to drop the 5G transmission, that is, not to transmit the 5G uplink at all for that subframe.

12.5.4 Dual Uplink Challenges for EN-DC Operation in FR1

The root cause of challenges for DPS-capable EN-DC UE operation resides in the basic assumption that for non-standalone (NSA) initial deployments, there is no scheduler

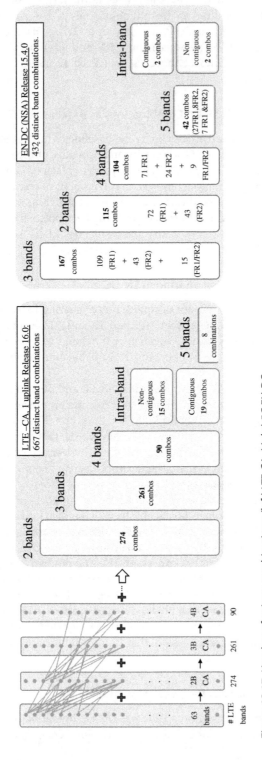

Figure 12.17 Number of unique combinations: (left) LTE-CA, (right) 5G EN-DC.

coordination between the eNb and the gNb, thereby making the UE responsible for deciding when and how to perform power sharing between LTE and 5G. In addition to this, for the special case of intra-band EN-DC operation, the UE must also be operated in conditions which can lead to huge receiver self-desensitization for FDD bands, a much larger MSD than that witnessed in LTE-CA. In the TDD and FDD bands, a large back-off may also be required to meet unwanted emissions resulting from the inter-modulation products of the two UL signals.

12.5.4.1 Intra-Band Challenges

In the Release 15 specifications, uplink intra-band EN-DC was specified for two bands: 41/n41 and 71/n71. Both bands are used in the United States and are subject to special emission requirements from FCC. Furthermore, uplink intra-band EN-DC is very challenging from an emissions point of view as LTE and NR are transmitted simultaneously on one band; therefore, special MPR solutions were developed for these bands. Note that intra-band EN-DC was also specified for the band 3/n3, but the uplink operation was restricted to single switched uplink only, and therefore there was no need for a special MPR solution as both LTE and NR followed single-carrier MPR definitions.

The intra-band EN-DC configurations that needed special MPR were DC_(n)41AA, DC_(n)71AA, and DC_41A-n41A. The first two are contiguous intra-band EN-DC configurations, and the third one is a non-contiguous intra-band EN-DC configuration. The term "contiguous" means that the LTE and NR carriers are adjacent and non-contiguous, meaning that there is a frequency gap between them. Special MPR due to specific regional requirements (FCC in these cases) is called A-MPR in the 3GPP specifications; thus, we will be using it from now on.

How intra-band EN-DC UE calculates A-MPR depends first of all on whether the UE supports DPS or not. If the UE supports DPS, then the LTE modem sets its output power as defined in the LTE specification 36.101. The NR modem calculates A-MPR from formulas developed by simulations and measurements to take into account joint LTE and NR transmissions. After calculating the A-MPR, the NR modem takes also into account concurrent LTE transmissions and decides whether it needs to further scale the NR transmission power down as explained in the previous chapter.

If the UE does not support DPS, both LTE and NR modems make a worst-case assumption in terms of the allocation of the other modem and then decide the A-MPR based on a formula defined in TS 38.101-3. As mentioned in Section 12.5.3, the network is responsible for ensuring that the non-DPS UE does not exceed Pcmax. Hence, in addition to the A-MPR that the UE calculates, the network can configure single switched uplink operation or cap the LTE and NR transmission powers to a certain limit, for example, to +20 dBm by signaling p-MaxEUTRA and p-NR-FR1, respectively.

12.5.4.2 Inter-Band Challenges

Inter-band EN-DC is far more popular in the Release 15 specifications than intra-band EN-DC, and all kinds of EN-DC band combinations have been specified. The most complex combination comprises five LTE bands and one FR1 5G band. Furthermore, band combinations involving LTE together with FR2 5G or LTE together with FR1 and FR2 5G are specified.

How the UE treats MPR/A-MPR is a bit simpler for inter-band EN-DC compared to intra-band EN-DC. No matter whether the UE supports or does not support DPS, both LTE and 5G modems follow single-carrier MPR/A-MPR definitions as defined in 36.101 and 38.101-1 for LTE and 5G, respectively. If the UE does support DPS, then the network is ultimately responsible for ensuring that the UE does not exceed PCmax. If the UE supports DPS, then the NR modem ensures that PCmax is not exceeded, as explained in the previous paragraph.

Intra- and inter-band EN-DC configurations are illustrated in Figure 12.18 using the following examples, where MSD results from the presence of IMD products due to simultaneous uplink transmissions:

- Large MSDs may occur for FDD intra-band contiguous configurations in bands with short distance duplex as shown in Figure 12.18a. For example, in rel-15 DC_(n)71AA NR PCC, MSD due to IMD 7 can be as high as 30 dB and LTE MSD due to IMD9 as high as 17 dB (cf. Section 12.5.4.3). Note that despite these large MSDs, this configuration is not eligible for SUO operation.
- Large MSDs may also occur in the case of inter-band configurations such as DC_1A-n77A (Figure 12.18b), where an IMD2 product may lead to approximately 30 dB LTE MSD. This configuration is eligible for SUO, and if the UE decides to request this mode, no MSD would occur. This group is currently specified for 2, 3, 4, and 5 bands (e.g. DC_1A-3A-5A-7A_n78A), but nothing precludes adding more bands.
- Figure 12.18c illustrates the case of FDD intra-band non-contiguous operation, such as that of DC_3A_n3A, where depending on the gap width, an IMD3 may overlap with either LTE or NR PCC and lead to a large desense. In Rel-15, only SUO is supported, but in Rel-16, the proposed MSD levels beat all 3GPP records with 44 dB MSD using 1 PA UE architecture and approximately 33 dB using 2 PA architecture. Note that this can apply equally to LTE or NR Pcell as their relative positions can be swapped, that is, LTE below NR or NR below LTE.
- Inter-band combinations can also operate with a third transmission in FR2, as shown in Figure 12.18d with the example of DC_1A_n77A-n257A. The addition of n257 does not contribute to further desense since mmWave modules are physically separated (separate antenna module, Tx chains) and isolated by a large frequency distance. This group is currently specified for 2, 3, 4, and 5 bands (e.g. DC_1A-3A-5A-7A-7A_n257A).

Although not shown in Figure 12.18, if one of several bands is added to a "parent 2-band," the EN-DC configuration may lead to new sources of IMD. RAN4 has decided to specify only the additional MSD cases. For example, DC_1A-3A_n77A is a three band extension of case (B) DC_1A_n77A, where

- The two band "parent" band 1 MSD due IMD2 between n77 and Band 1 remains at 30 dB and is not specified to avoid overloading the tables.
- However, band 1 as an SCELL may also undergo 31 dB MSD due to an IMD2 between Band 3 and n77.
- Band 3 as an SCELL may also undergo 31 and 8.5 dB MSD due to an IMD2 and an IMD4, respectively, between Band 1 and n77 transmissions.

Other EN-DC variants include FR1 and FR2 and are currently specified up to six bands (e.g. DC_1A-3A-5A-7A_n78A-n257A) and NR-DC between FR1 and FR2 (e.g. DC_n77A-n257A). Other sources of MSD due to harmonic mixing and due to

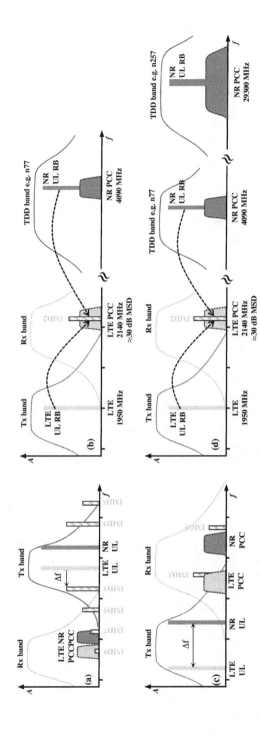

Figure 12.18 Examples of intra- and inter-band EN-DC configurations: (a) FDD intra-band contiguous, (b) inter-band within FR1 (two bands), (c) intra-band non-contiguous, (d) inter-band including FR1 and FR2 (three bands).

cross-band isolation are not covered in this section but are part of the REFSENS exceptions in the standard. Overall, it is easy to understand that EN-DC induces a large expansion of conformance test cases and test time since every MSD test point must be tested.

12.5.4.3 Example of MSD/A-MPR/MPR Challenge with DC_(n)71AA

The example of FDD intra-band contiguous EN-DC operation in Band 71, denoted DC_(n)71AA, best illustrates both MSD and A-MPR challenges. Due to the uncoordinated operation of the LTE and 5G schedulers, the UE may be operated across all possible RB offset positions and RB lengths. Figure 12.19 shows spectral measurements collected at the power amplifier output port for two RB allocations: on the left, the LTE and RB allocations are nearly contiguous, the LTE RB being transmitted at their highest offset position, and the NR RB at their lowest. In this case, the modulated waveforms exhibit near-single-carrier behavior, where the LTE anchor point suffers from low self-desensitization as the transmitter interference is dominated by its $1/f$ noise and not the IMD products, resulting in less than 2 dB MSD. In Figure 12.19 (right), the RB allocations are such that a gap is created by using the lowest offset position for the LTE RBs while keeping the 5G RB position identical to Figure 12.19 (left). The LTE and NR receive channels are now victims of IMD 9 and IMD7 products that can be clearly distinguished. The power spectral density of these IMD products is approximately 10 to 20 dB higher than in the previous case, resulting in a moderate MSD increase. In Release 15, 4 MSD TPs have been adopted to ensure good understanding of UE behavior when operated at or near to its maximum output power: the first three TPs correspond to Figure 12.19 (left) with the MSD ranging from 1.6 to 1.8 dB, while the fourth TP ensures that the MSD is measured with the largest possible frequency gap: i.e. LTE RBs and 5G RBs are positioned at their lowest and highest frequency offset, respectively. In this case, the LTE and 5G downlink carriers are impacted by IMD 5 and IMD 7 products, leading to 17 dB and 29 dB desense for LTE and 5G, respectively. It is worth noting that for these allocations, an IMD5 product falls in the band 12 downlink frequency band for which the UE is allowed to use an additional maximum power reduction to meet band 12 protection limits.

For LTE/5G power sharing and A-MPR evaluation, the situation is nearly as severe, with the added complexity that both analyses are more complicated. A-MPR evaluation consists in sweeping the 5G carrier power, denoted by "P_{NR}," for a given LTE uplink transmit power, denoted by "P_{LTE}." For each P_{NR}/P_{LTE} pair, the resulting IMD product must be evaluated through simulation or measurements to ensure that all the emission requirements for this band are met, that is, ACLR, SEM, general spurious emissions, and Band 12 and Band 29 neighbor band protection. The power sweep is stopped as soon as one of these gating factors is violated. This work needs to be repeated for all combinations of LTE and 5G RB allocations (size and offset position), each CBW combination, and each waveform type, that is, creating a multimillion point search space.

Figure 12.20 (left) shows the results of both simulation [15] and measurements data collected for NR 15 MHz/LTE 5 MHz CBW and plots the 5G transmit power back-off (A-MPR) to pass emissions requirements vs. P_{LTE} vs. the sum of LTE + NR numbers of allocated RBs, denoted by RB_{tot}. At large RB_{tot} values, the waveform properties are similar to those of Figure 12.19 (left); that is, spectral regrowth leads to A-MPR being ACLR dominated. On the other hand, as the number of RBs is reduced, the IMD products are

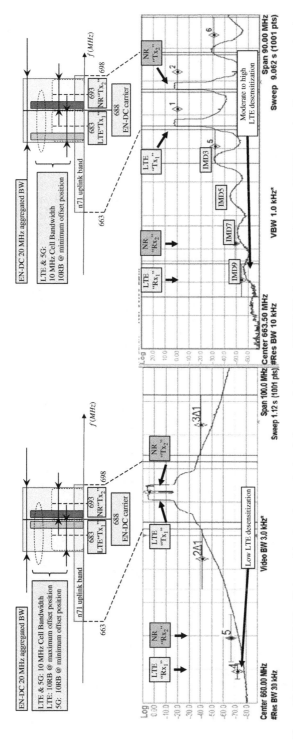

Figure 12.19 5G EN-DC UE self-desensitization due to dual uplink transmissions vs. LTE/5G RB allocations. Left: LTE and 5G RBs nearly adjacent to each other. Right: LTE and 5G RBs positioned at their respective minimum offset position.

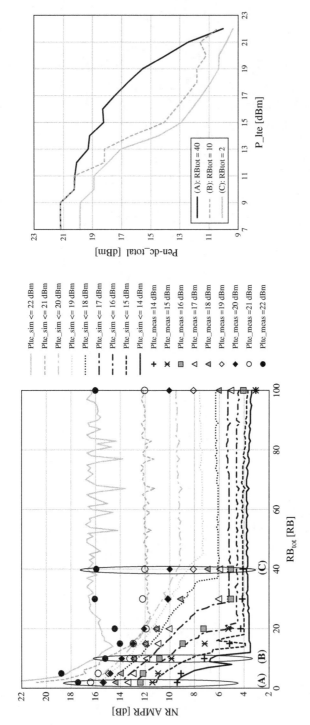

Figure 12.20 Left: Simulated (lines) vs. measured (symbols or dots) NR A-MPR as function of RB_{tot} for NR15_L5, NR QPSK configuration. A, B, and C slices correspond to RB_{tot} = 2, 10, 40. Right: Measured maximum EN-DC total transmission power vs. P_{LTE} for slices A, B, and C.

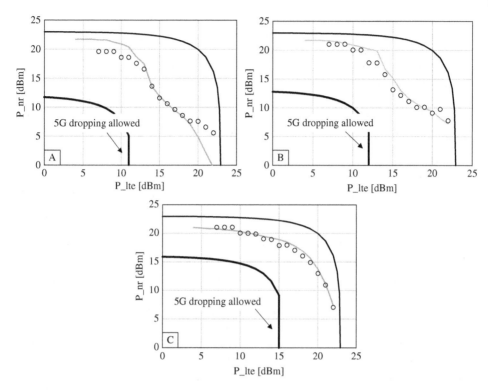

Figure 12.21 5G A-MPR (maximum allowed P_{NR}) vs. P_{LTE} for slices A (2 RB_{tot}), B (10RB_{tot}), and C (40 RB_{tot}). Plain dark: power sharing for PC3 23 dBm antenna power; dashed black: allowed 3GPP 5G A-MPR; gray: simulated maximum allowed 5G transmit power; dots: measured maximum allowed 5G transmit power.

more discrete, the PSD in each IMD increases, and 5G A-MPR becomes either SEM or band protection dominated. Figure 12.20 (right) plots for RB_{tot} values of 2, 10, and 40 the measured EN-DC total transmit power that could be delivered by this power amplifier.

Further details of the impact of dual uplink IMDs on DPS are shown in Figure 12.21 for RB_{tot} slices A, B, and C. It can be seen that for a PC3 UE, when P_{LTE} is transmitted at 22 dBm, the 5G minimum back-off is 16 dB ($RB_{tot} > 40$), and in the worst case may reach up to 22 dB for $RB_{tot} = 2$, meaning that 5G could, at best, be transmitted only at +1 dBm. This is a huge link budget penalty. 3GPP A-MPR plotted in dotted lines are the results of a trade-off between taking enough margins to ensure products are feasible, while trying to minimize the impact on system performance. When the NR transmit power is too low, 3GPP allows the UE to drop NR transmissions.

DC_(n)71AA and DC_3A_n3A have been great learning tools in Rel-15 and will form the framework for additional intra-band combinations that are already under study for Release 16, such as the no less challenging DC_2_n2, DC_5_n5, and DC_(n)5AA.

12.5.5 Dual Uplink Challenges for EN-DC and NN-DC Operation in FR2

With NR operation in FR2, SA operation is limited to fixed wireless access. In mobility conditions, however, it is not foreseen that the UE would be able to perform handovers, registrations, and initial access using an FR2 link. To enable all legacy services like 911, voice, or messaging and also roaming and operation in low-coverage areas, an FR1 link is needed to provide link robustness and ubiquity of the baseline services. This will first be provided by using an LTE anchor in the EN-DC configuration, but once refarmed to NR, using an FR1 NR band in an NN-DC combination.

As for EN-DC in FR1, this will imply concurrent UL operation in FR1 and FR2. At the early stages of NR specifications, in 3GPP, in-device (UE) coexistence between FR1 and FR2 was discussed to understand if some requirements or exceptions like MSD and MPR/A-MPR would be required.

With FR1 finishing at 7.125 GHz (in fact, with the current specified bands at 5 GHz) and FR2 starting at 24.25 GHz, there is not only a large gap in frequency but also the antennas are separate and use very different technologies and radiation patterns, with FR1 being omnidirectional and FR2 being beamformed. This guarantees that there is large antenna large isolation and that the filter, matchings, and DC coupling implemented in both frequency range paths also provide significant attenuation for interfering signals transmitted by the other side. For this reason:

- IMDs resulting from both ULs are negligible as the signal coming from the one frequency range is largely attenuated before it mixes with the other side; even so, the generated IMD products are also at a very large offset and will see further attenuation before they reach any victim band.
- Harmonics from FR1 overlapping with FR2 are above 3rd order (5th order for band n79, which is the highest band today) and will experience significant attenuation and large antenna isolation before it reaches any FR2 receiver.
- Similarly, harmonic mixing is of 5th order (only odd orders are of concern), and the FR2 UL signal is very significantly attenuated before it reaches any FR1 receiver.

For all these reasons, it was decided that no specific coexistence study will be conducted for combinations of FR1 and FR2 bands, and thus no MSD or MPR/A-MPR is needed in 38.101-3 for this case.

It is to be noted, though, that in Release 16 there are studies underway for the frequency range 7.125–24.25 GHz which will probably result in single and dual UL issues if operated concurrently with either FR1 or FR2.

12.6 Examples of UE Implementation Challenges

12.6.1 More Antennas, More Bands to Multiplex, and More Concurrency

The latest LTE phones already represented a significant challenge in terms of RF front-end architecture with a large number of bands and band combinations, and 4×4 DL MIMO supported in a number of MB and HB bands. However, bands were limited to up to 2.7 GHz in the expectation of some implementation supporting Band 42 in Japan, and dual UL-CA was not implemented. With the advent of 5G, even without considering the addition of mmWave, the RF front-end sees yet another step up in complexity with the addition of the 3.5 and 4.5 GHz bands, which have mandatory support for 4×4 DL MIMO but will most probably support 2×2 UL MIMO. 5G NSA operation adds further complexity as additional transmitter paths and improved filter performance are required to handle the challenging EN-DC out-of-band emissions.

Table 12.11 FR1 RF front-end filter count in high-end smartphone – antenna multiplexing and band.

	2019 4G phone (34 cellular bands)				2020 5G phone (35 cellular bands)			
Nplexing	Primary		Diversity/MIMO		Primary		Diversity/MIMO	
Filters	7		9		9		11	
Total		16				20		
Band Group	FDD Tx	FDD Rx	TDD	Diversity/MIMO	FDD Tx	FDD Rx	TDD	Diversity/MIMO
LB	10	10		10	10	10		10
LMB	2	2		3	2	2		3
MB	4	4	2	18	4	4	2	18
HB	2	2	4	12	2	2	4	12
UHB			1	3			4	4
n79							4	4
WiFi			4				6	
GPS				4				4
Total	18	18	11	50	18	18	20	55
Grand Total			97				111	

Table 12.11 only illustrates the increase in complexity in terms of the number of filters, limited to FR1, needed to implement a 2019 high-end 4G smartphone on the right and a 2020 high-end 5G smartphone on the right.

First, the multiplexing of the different subbands, bands, and modes to antennas increases the overall number of paths from 16 to 20, thus allowing up to 20 paths to operate concurrently. This is essentially due to the addition of the 3.5 and 4.5 GHz NR bands, but also due to the potential addition of the 6 GHz unlicensed band extension of the current 5.15–5.925 GHz band (5.925–7.125 GHz in the United States and 5.925–6.425 GHz in Europe). This 25% increase in potential concurrent RF paths also results is a 15% increase in required band filters.

Although Table 12.11 already provides a good picture of the size and routing complexity of the RF portion of a 5G phone, Figure 12.22 gives a better sense of the need to implement the RF front-end in sub-modules:

- To manage the routing complexity and manage the interconnect losses from the RF transceivers to the sub-modules and from the sub-modules to the antennas.
- To manage the isolation needed between functions that will operate concurrently.
- To enable flexibility in supporting different level of features, band support, or regional skews.

This is especially the case in FR2 where multiple band, SC, and blockage require multiple arrays to be supported, as is illustrated on the right side of the block diagram. Here, unlike for FR1, it is not possible to multiplex non-adjacent bands to the same antenna system, and the mmWave transceiver needs to be as close as possible to the antenna as the losses associated with connection traces at these frequencies are impossible to recover. This is why the transceiver and antenna array constitute a single module in FR2: a module may have more than one array to support 28 GHz and 39 GHz bands, for example (as illustrated on the top-right corner of the block diagram).

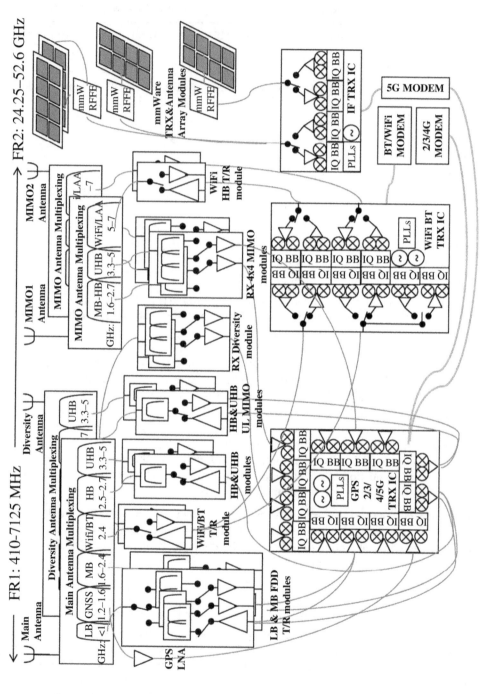

Figure 12.22 High-level block diagram of the RF portion of a 5G UE supporting LTE and NR in FR1, NR in FR2, and all required wireless connectivity.

The block diagram here is not necessarily how a real implementation will work, and the interconnect is extremely simplified to show what is connected to what and, for example, does not show all the switches needed to support feature like SRS antenna switching and the different Tx/Rx paths modes. Also, it shows an implementation using only four antennas to support all the multiplexing and concurrent modes, but in most implementations, extra antennas will be added (especially for the higher FR1 bands as they are relatively small) to simplify the interconnect and multiplexing and provide further isolation between concurrent modes.

In terms of concurrency, this block diagram illustrates one of the most complex sets from the new 5G NR features:

- In UL: Triple UL FR1 + FR2 EN-DC with concurrent operation of 2.4 GHz WiFi/Bluetooth and 5 GHz WiFi (up to five simultaneous transmissions, including 2×2 UL MIMO in HB/UHB NR and 5 GHz WiFi)
- In DL: 4×4 DL MIMO in 2 FR1 bands + 3 Band 2×2 DL MIMO + FR2 polarization diversity + 2.4 GHz WiFi/Bluetooth and 5 GHz WiFi + Global Navigation Satellite System (GNSS) (multiple band simultaneously)
- All these concurrent modes have to be supported with complex multimode cellular and connectivity RF transceivers with the addition of the IF transceiver needed for FR2 to provide the parallel BB signals to/from their respective modems, which further increases the number of RF connections.

12.6.2   FR2 Antenna Integration and Smartphone Design

If the integration and performance of antennas have already been a challenge in smartphones until now, the problem will be further exacerbated with the integration of the mmWave antenna array panels in 5G. Given that the phone aesthetics and design is a prime concern for a high-end smartphone, the material used (glass, metal casing) and dimensions (thickness, screen size) are not favorable for the efficiency of mmWave antennas, and additional power losses on the order of 2–3 dB are often discussed. This is a big challenge to the success of mmWave in the phone as it adds to all the mobility challenges that are big and brand new for the network.

In FR1, SAR considerations are the main reason for placing transmit antenna ports at the bottom of the UE, so as to minimize head exposure during voice calls in the case of hand and head test cases. Proximity sensors are also sometimes used to back-off transmission power in case SAR might be violated, a scheme that is also known as Power Management UE Maximum Power Reduction (P-MPR). Unless the user is deliberately trying to cover all primary Tx/Rx and diversity Rx antennas simultaneously, it is difficult under FR1 normal cell coverage conditions to entirely block signal reception.

On the other hand, hand blockage at mmWave can easily cause a 30 dB signal attenuation, which, considering the difficult link budget, can make mmWave more susceptible to dropping the link. Steering the beam peak away from blockage, for example, to attempt to maintain the link using a multipath reflection may not be always successful as absorption on walls/buildings materials might be as detrimental as hand blockage. Even though 3GPP assumptions for SC are a compromise to accommodate single-antenna panel UEs, one common mitigation technique to cope with blockage consists in equipping the UE with a minimum of two panels, so that it can benefit from panel switching/diversity. To ensure maximum SC, the antenna placement strategy adopted in FR1 is no longer an option; that is, the TX antenna port can no longer be constrained to the bottom part of the smartphone. For example, the two panel placements shown in Figure 12.7 (left; Section 12.2.4.1), uses one panel located at the top-left corner, on the UE back side, while the second panel is located at bottom-right corner on the front side. With the top panel being located right in front of the ear, placing the panel on the back side, and ensuring that beams are directed away from the head is the best strategy to minimize body exposure. In case panel switching is not an option, P-MPR and restrictions on the maximum uplink duty cycle are additional schemes considered to comply with maximum permissible exposure (MPE) safety requirements. This, in turn, may imply dedicating a proximity sensor for each antenna panel.

The number of antenna module also heavily depends on the UE form factor and the choice of cover materials. The impact of materials on the antenna performance is exacerbated by the fact that the trend in UE thickness is pushing for ever thinner devices, thereby leaving little gap between the display glass and the back cover material. For example, 3D simulations in [5] show how the LCD screen may impact the radiated beam pattern. Data in [1, 16] show that the when panel is in a "sandwich" between two glass layers (LCD screen and glass back cover), not only are implementation losses 1–2 dB higher with a glass than with a plastic cover, but glass also tends to scatter radiation patterns, providing amplification in certain directions and attenuations in others. Low-end devices with a plastic back cover might therefore be able to deliver good performance with only two panels, while for high-end devices, where both the glass back cover and metallic bezels are used, it might be necessary to use three or even four modules, with a combination of face/back and edge modules. The higher the number, the better the chance of mitigating MPE and blockage as well as meeting SC, but this comes at the expense of signal routing complexity since two intermediate frequency lines are required per panel, and at the expense of cost.

Details of the implementation of the first-generation "high-end" mmWave-capable smartphones are available in the public domain and are examples of OEMs choosing three [17] or four antenna modules [18].

12.7 Summary

From many viewpoints, 5G can be considered as an evolution of LTE-A with minimum impact on existing UE architectures in general. This is the case of frequency bands in FR1, which, with the exception of n79 and n77, are extensions of bands that are already covered in legacy platforms. On the other hand, operation at mmWave, mandatory support of simultaneous dual uplink transmissions, and UL MIMO are features that bring major challenges to the RF front-end. The huge physical layer flexibility offers unique opportunities to address a wide range of use cases. The price to pay, however, resides in increasing RF-FE multiplexing/demultiplexing and test time complexity. Even though it is too early to say if all features will fulfill their promises, 5G has certainly created a lot of opportunities for innovations, fueled by several thousand of contributions yearly for 3GPP. Who would have thought that within a one-and-a-half years of 5G standardization work, smartphones would be able to surpass LTE-A complexity and deliver mmWave transmission capability using active phase arrays?

References

1 Sony, RAN4#86. R4-1802868 UE Spherical coverage at mmWave 28GHz. March 2018.
2 NTT DOCOMO Inc., Sharp, RAN4 NR ad-hoc-3. R4-1709394 EIRP CDF for mmWave UE. September 2017.
3 Qualcomm Incorporated, RAN4#85. R4-1712381 Spherical coverage of realistic design. November 2017.
4 LG Electronics, RAN4 #87. R4-1806676 Measurement EIRP levels for spherical coverage of NR UE at FR2. May 2018.
5 Samsung, Apple, Intel, RAN4#84. R4-1711036 Consideration of EIRP spherical coverage requirement. October 2017.
6 Benjamin Jann et al. (2019). A 5G Sub-6GHz Zero-IF and mm-Wave IF Transceiver with MIMO and Carrier Aggregation, IEEE International Solid- State Circuits Conference – (ISSCC). 17–21 Feb. 2019.
7 J. D. Dunworth et al. (2018). A 28GHz Bulk-CMOS dual-polarization phased-array transceiver with 24 channels for 5G user and basestation equipment. IEEE International Solid - State Circuits Conference – (ISSCC). 11–15 Feb. 2018.
8 Holma, H., Toskala, A., and Tapia, P. (eds.) (2014). *HSPA+ Evolution to Release 12: Performance and Optimization*", Chapter 14. Wiley.
9 Skyworks Solutions, Inc. RAN4#86. R4-1802203 n5 and n8 MSD Measurements. March 2018.
10 Apple Inc., Intel Corporation, vivo, Xiaomi, OPPO, SGS Wireless, TCL, Spreadtrum, Asus, Kyocera, Hisense, Hytera, Meizu, IITH, IITM, CEWiT, RAN4#87. R4-1808173 Proposals for concluding the spherical coverage requirement for FR2 handheld UEs. May 2018.
11 Samsung, RAN4#90Bis .R4-1905186 Ad-Hoc Meeting Minutes for Beam Correspondence. April 2019.
12 Skyworks Solutions Inc, RAN4#91. R4-1907163 FR2 PC3 MPR PA Measurements for DFT-S-OFDM and CP-OFDM for Rel-16. May 2019.
13 Nokia, Nokia Shanghai Bell, RAN4#90bis. R4-1903524 FR2 MPR due to OBW. April 2019.
14 Keysight Technologies, RAN4#90bis. R4-1904191 On ETC FR2 EIRP and EIS Testing. April 2019.
15 Nokia, Nokia Shanghai Bell, RAN4#89. R4-1814973 On DC_(n)71AA new A-MPR approach and intraband contiguous EN-DC Pcmax. November 2018.
16 Sony, Ericsson, RAN4-AH-1801. R4-1800888 UE Spherical coverage at mmWave 28GHz. January 2018.
17 Samsung Electronics Co Ltd. SMG977U (Galaxy S10 variant) Portable Handset RF Exposure Info Near Field Power Density Simulation Report. https://fccid.io/A3LSMG977U/RF-Exposure-Info/Near-Field-Power-Density-Simulation-Report-4235748#download
18 Motorola Mobility LLC. Mobile 5G MOD T56XL1 FCC Power Density Test Reports. https://fccid.io/IHDT56XL1

13

5G Modem Design Challenges

YihShen Chen[1], Jiann-Ching Guey[1], Chienhwa Hwang[1], PeiKai Liao[1], Guillaume Sébire[2], Weide Wu[1], and Weidong Yang[3]

[1] *MediaTek, Taiwan*
[2] *MediaTek, Finland*
[3] *MediaTek, United States*

CHAPTER MENU

13.1   Introduction

5G NR – the new radio access technology at the heart of the 5G system – is designed to push the boundaries of mobile communications to unprecedented levels in terms of data rates, latency, reliability, and connectivity. Purposely versatile, 5G NR is engineered to address the needs, known and future, of many industries, to enable new services and experiences, through a high degree of in-built flexibility and configurability.

Thus, an engineering challenge in itself, 5G NR is also a major engineering challenge for mobile device implementation, especially the modem. Indeed, the promise of 5G NR to deliver multi-Gbps data rates, ultra-reliability, and low latency is only viable if the overall complexity of the modem and its power consumption are all but fully mitigated for mobile devices. This chapter provides expert insights into how this difficult equation can be resolved so that 5G NR can quickly become a reality for consumers. It assumes a working knowledge of 5G NR, as described in other chapters.

To shed light on these challenges, it is necessary to understand the basic structure of a modem and what constitutes its complexity. Figure 13.1 shows a generic block diagram inside a 5G modem System on Chip (SoC). It typically consists of a baseband processing unit, a mixed signal circuits unit, and a configurable RF circuits unit. Outside of the

5G Technology: 3GPP Evolution to 5G-Advanced, Second Edition.
Edited by Harri Holma, Antti Toskala, and Takehiro Nakamura.

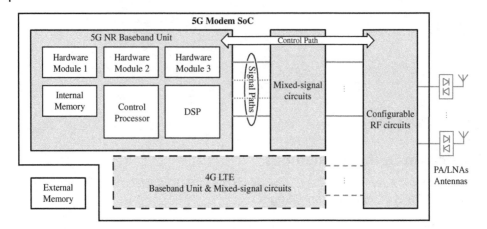

Figure 13.1 Generic 5G modem architecture.

5G modem SoC lie the external RF components such as the power amplifier/low-noise amplifier (PA/LNA) and antenna units, as well as an external (and less expensive) memory unit that can be shared by other integrated circuits and the Application Processor (AP). This generic architecture may vary slightly depending on the operating radio frequency of the modem.

An optional 4G baseband and mixed signal unit (dotted lines in Figure 13.1) may also be present in a multi-mode device. Multi-mode support is, as for all of the past generations, essential to ensure a smooth migration to the newer generation and thereby provide continuity of service in areas where the newer generation is yet to be deployed. In the early phase of the migration, this 4G unit is likely to share part of the RF unit with the 5G modem. For cost reduction, it is necessary that the overall *modem architecture* also continues evolving at later stages with a higher degree of integration and hardware sharing among all supported radio access technologies, whether cellular or other wireless connectivity.

The baseband processing unit comprises a number of hardware modules, one or more general-purpose Digital Signal Processing (DSP) cores, an internal high speed memory unit, and a control processor. The hardware modules are bespoke functions for computationally intensive tasks such as channel coding, cell search, and Radio Resource Management (RRM) measurement. Together with the DSP core that performs general-purpose computation, they account for the bulk of the modem's *computational complexity* that delivers NR's multi-Gbps data rates. The control processor is responsible for scheduling the various modules and coordinating their communications between each other through the internal memory for each procedure in the NR protocol stack. This *control flow complexity* is, compared to Long Term Evolution (LTE), much higher, due to 5G NR's in-built flexibility and configurability as well as its much more stringent latency requirement.

The cumulative computational, control, and architectural complexity naturally lead to yet another great implementation challenge; that of mitigating *power consumption*. NR introduces a number of features to keep these issues in check. Most notable of all is the BandWidth Part (BWP) adaptation, an effective mechanism to reduce the modem's power consumption resulting from wide bandwidth usage in 5G NR.

Table 13.1 Impact of 5G NR features on modem design.

	NR Features			
Modem impacts	High data rate and system flexibility	Low latency and flexible timing	Multi-RAT coexistence	Wide bandwidth
Computational complexity	★★★	★★☆	★☆☆	★★☆
Control flow complexity	★★☆	★★★	★☆☆	★☆☆
Modem architecture	★☆☆	★☆☆	★★★	★☆☆
Power consumption	★★☆	★☆☆	★☆☆	★★★

Table 13.1 illustrates how the key 5G NR features contribute to impacting the modem. For example, multi-(Radio Access Technology) RAT coexistence primarily impacts the modem architecture.

This chapter exposes the many challenges facing the 5G NR modem design and the solutions implemented to address those. Each part in this chapter focuses on one key NR feature (column in Table 13.1) and its associated modem impacts (rows in Table 13.1) with a particular emphasis on the main challenge to overcome. First, high data rates and system flexibility are discussed, focusing on tackling the associated computational complexity issue. Then, low latency and flexible frame structure are addressed from the main standpoint of control flow complexity. Thereafter, multi-RAT coexistence and the challenge it poses on modem architecture are debated. Finally, NR wide bandwidth operation is analyzed in relation to power consumption.

13.2 High Data Rate, System Flexibility, and Computational Complexity

The IMT-2020 requirements for 5G specify a target peak data rate of 20 Gbps for downlink (network to mobile device) [1], a 20-fold increase compared to the IMT-Advanced requirement (1 Gbps). In addition, IMT-2020 requirements target a threefold increase in spectrum efficiency compared to IMT-Advanced.

To meet the above requirements, higher-layer Multiple Input Multiple Output (MIMO) schemes as well as flexible reference signal configurations to minimize the system overhead are cornerstones of the 5G NR design.

This part elaborates on how the complexity in the device scales and can be managed to achieve a 20-fold increase in peak downlink throughput while ensuring the rich system flexibility for 5G NR.

13.2.1 Channel Coding Aspects Versus UE Complexity

To achieve a 20-fold increase in data rate, the error correcting decoder is the critical modem component. For capacity-achieving performance, turbo code and advanced iterative decoding algorithms are utilized in 3G and 4G systems. However, not only is the corresponding arithmetic complexity very high, the dedicated accelerator is also typically one of the largest hardware modules in the modem. With the same coding scheme,

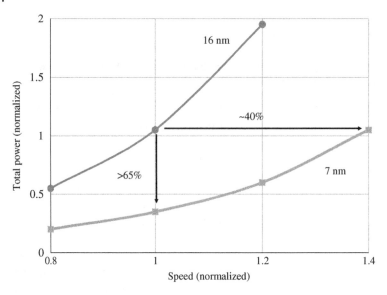

Figure 13.2 Processing speed characteristics.

a 20-fold increase in data rate will fundamentally imply a prohibitive 20-fold increase of the decoder die area.

An advanced foundry process node[1] is necessary but not sufficient. Considering a common LTE category supporting two component carriers, the baseline process node is 16 nm. While 7 nm technology is two generation more advanced than 16 nm, it can only provide ~0.5× area reduction and up to ~1.4× faster speed than the 16 nm process node; see Figure 13.2. In other words, advancing the process node by two generations can only improve the throughput by a factor of ~2.8×. A fundamental change in data channel coding is therefore necessary to close the remaining ~7.2× throughput gap.

Low Density Parity Check (LDPC) code is considered for NR due to its superior area efficiency compared to turbo code used in LTE, particularly at higher code rates. In [2], the area efficiencies, in terms of Gbps/mm, of an LTE turbo decoder and an LDPC decoder (extended from the 802.11ac decoder design) are compared. As can be observed in Table 13.2, the LDPC decoder can provide up to ~7.5× better area efficiency at high code rates compared to the turbo decoder. It should be noted that the turbo decoding process *always* operates at a mother code rate of 1/3 – that is, there is no complexity reduction when the code rate is higher and the redundancy is less. Furthermore, the sequential processing for the trellis-based algorithm and the interleaving/de-interleaving also cause long processing latency. In contrast, the LDPC decoding complexity is code rate dependent. Therefore, the complexity decreases as the code rate increases (or as the redundancy decreases); the decoding operates over the parity check matrix, and as there are fewer check equations with higher code rates (and

---

1 Manufacturing process for semiconductors – an Xnm terminology does not bear any direct relation with the *actual* size of a transistor but refers instead to a particular generation of such a process. The same X does not imply the same transistor size between foundries. However, the relative size reduction from one generation to the next is typically similar between foundries. The smaller the X, the more advanced the process, the smaller, faster, and more power efficient the transistor.

Table 13.2 Area efficiency comparison.

	Turbo decoder (mother code rate = 1/3)	LDPC decoder (Code rate = 2/3)	LDPC decoder (Code rate = 0.88)
Area efficiency (Gbps/mm^2)	3.66	12.48	27.72

Table 13.3 LDPC decoding complexity.

Code rate	Edge count	Relative complexity
8/9	79	*1.0 (reference)*
3/4	122	1.54
2/3	144	1.82
4/9	240	3.04
1/3	316	4.0

fewer redundancy bits), the decoding is as a result simpler. This allows for a significant data rate boost at reasonable costs for the adopted foundry process, and hence makes the targeted peak data rate of 20 Gbps for NR feasible from that standpoint.

5G NR LDPC is a rate-compatible design in order to support HARQ. Retransmissions can add redundancy bits for error correction; however, the decoding complexity per codeword is increased. Table 13.3 shows the complexity increment with respect to a high code rate of 8/9. When the code rate is reduced to 1/3, the decoding complexity is increased by a factor of 4. To confine the worst-case complexity, the HARQ buffer size is limited to accommodate code rates down to 2/3 according to the maximal available data volume among all configured BWP settings. This is also an important NR design to ensure reasonable decoder complexity.

In addition to computation complexity, the following two issues also pose challenges to area-efficient NR UEs:

- *HARQ buffer size.* The total HARQ buffer size can be determined by the product of the peak data rate, transmission time interval (TTI) time duration, and HARQ process number divided by the limited HARQ code rate, 2/3. To deliver multiple Gbps data rates and accommodate up to 16 HARQ processes, the required total buffer size is huge. This calls for sophisticated designs both to compress the HARQ content and to manage the HARQ processes in order to balance the HARQ performance and buffer complexity.
- *Decoder multi-mode sharing.* In Universal Mobile Telecommunications System (UMTS) and LTE, turbo coding is used. A multi-mode (i.e. UMTS and LTE) turbo decoder can be developed to minimize the decoder cost. However, NR uses LDPC to achieve the 20-fold increase in peak data rate at a reasonable chip cost. Because of the fundamental difference in the decoding algorithm between the turbo decoder and LDPC decoder, the realization of an efficient multi-mode (UMTS, LTE, NR) decoder requires a sophisticated design in both the decoder algorithm and architecture.

In summary, this subsection describes how a 20-fold peak data rate can be realized for NR with a reasonable die area. Introducing LDPC codes subject to a limited HARQ code rate for the maximum data volume is the key enabler. The implementation challenges of accommodating huge HARQ buffer size and realizing a multi-mode data decoder are also highlighted. A lesson learned from NR coding design is that the joint consideration of both the power efficiency (in terms of the required power per bit) and the area efficiency (in terms of the required die area per bit) can fundamentally reduce the foundry requirement and realize the 20-fold peak data rate in the near future.

13.2.2 MIMO and Network Flexibility Versus UE Complexity

MIMO technology has been the key driver to achieve ever-higher spectrum efficiency in mobile communications. The network can use analog beamforming, digital beamforming, or hybrid beamforming for Physical Downlink Control Channel (PDCCH) and/or Physical Downlink Shared Channel (PDSCH) transmissions; and different beam widths can be used for the PDCCH and PDSCH, for example, a wide beam for the PDCCH to achieve good coverage, and narrow beams for the PDSCHs to allow MU-MIMO operations. However, improving spectrum efficiency alone is not sufficient to deliver the targeted user and system throughputs in NR; a large channel bandwidth or large aggregated channel bandwidths is/are needed, which are more readily available at higher frequencies, for example, mmW. Hence, the support for mmW is an important feature of NR.

Lessons learned in LTE have driven many design aspects of 5G NR. In LTE, the Cell-specific Reference Signal Common Reference Signals (CRS) serves multiple purposes. Considering those relevant for PDSCH reception, they include: *timing-frequency synchronization* for idle state UEs, *fine-timing frequency synchronization* for connected mode UEs, *channel estimation* for PDCCH demodulation, for the PDSCH, and for Channel State Information (CSI) acquisition.[2] With so many functionalities depending on CRS, it is therefore always ON at the prescribed occasions, *no matter* what the loading situation in a cell may be. In 5G NR, those functionalities are handled not by one signal as in LTE but instead by different signals: PDCCH Demodulation Reference Signal (DMRS), PDSCH DMRS, CSI-RS for beam management, CSI-RS for CSI acquisition, and CSI-RS for tracking. Consequently, the density/frequency of each signal can be configured separately in a flexible way, for example, by considering interference to other cells, base station/UE power consumption, and latency.

While good for the 5G NR radio efficiency itself, such flexibility comes at the cost of additional UE implementation complexity. For example, the PDCCH with a Downlink Control Information (DCI) scheduling PDSCH can be sent with beam 1, and the scheduled PDSCH can be transmitted with another beam, say, beam 2. As a UE may need to apply different receiver parameters for different beams (e.g. beams 1 and 2), a UE can either be indicated with the selected beam with the "TCI state" in the DCI or assume a default beam under some conditions. The characteristics of PDSCH such as numerology, time domain resource allocation, frequency domain resource allocation, DMRS positions, DMRS type, and DMRS waveform can all be indicated *dynamically*. Note also that in contrast to CRS in LTE, the UE cannot deduce all such characteristics of the demodulation signal for the PDSCH until the necessary DCI fields are parsed.

While new spectrum has been identified for 5G NR, allowing 5G NR to operate on spectrum available to LTE is also fundamentally important (e.g. for wide area coverage).

---

2 For transmission modes 1–6.

Hence, coexistence with LTE is also a key consideration for the design of 5G NR. To avoid generating interference to LTE CRS signals, rate matching of the PDSCH over LTE CRS is supported in NR. Also, as the COntrol REsource SET (CORESET) configuration is flexible, and a relevant search space may be located on any symbol in a slot, resource sharing among the NR PDCCH and NR PDSCH is also key. Rate matching in NR provide a unified framework to support all the features.

In contrast to LTE, in 5G NR there is no link between the transmission scheme and the CSI feedback scheme configured for the device. While to a large degree the basic CSI acquisition schemes developed in LTE are retained, the CSI acquisition framework with 5G NR is more flexible, and importantly, beam management reference signals and procedures are introduced. At a high level, however, CSI acquisition can still be divided into downlink-signal-based and uplink-signal-based.

For the downlink-signal-based CSI acquisition, the SSB (Synchronization Signal Block), CSI-RS, and SSB plus CSI-RS can be used for beam management; CSI-RS with 2–32 antenna ports can be used for codebook-based transmission. Compared with LTE (e)FD-MIMO, the codebook design in 5G NR is simplified to some extent. Roughly two types of codebooks are specified in 5G NR: Type I for SU-MIMO and Type II for MU-MIMO. For Type I, the single panel codebook supports uniform spacing among antenna elements at a base station, and the multiple panel codebook supports uniform spacing among antenna elements within a panel, but the spacing between antenna elements from different panels can be different. From a device implementation point of view, separate processing procedures for Type I and Type II may be needed to achieve optimized spectrum efficiency. For both Type I and Type II, a number of sub-modes are defined to achieve trade-offs between feedback accuracy and feedback overhead depending on the device's capability and the network's configuration, and hence device implementations need to be optimized for those modes as well. 5G NR supports a fast CSI report which is for up to four ports and is subject to a number of restrictions and normal CSI reports. Hence, the network can decide which CSI report to trigger depending on channel conditions, deployment scenario, or operating spectrum (e.g. over unlicensed spectrum a fast CSI report may deliver CSI feedback with less chance of being stopped due to listen-before-talk), and a device's reported processing capabilities. Due to all the provided flexibility, many derived configurations from Radio Resource Control (RRC) signaling need to be accessed by a UE at a fast pace depending on the dynamic signaling from the network; for that reason, they often have to be stored in internal memory, which is typically limited in its capacity and also expensive. Consequently, the size of RRC signaling itself, especially that of *CSI-MeasConfig* information element [3], which provides all the configurations for CSI acquisition at a UE, has been a key consideration in 5G NR design to mitigate signaling overhead in the system and the memory need on the device side. With the maximum allowable values, the *CSI-MeasConfig* information element alone can take almost 150 kB per carrier. In contrast, the size of an LTE RRC message is usually small, for example, less than 1 kB. Consequently in Rel-15 NR, the size of an RRC message is limited to no more than 9 kB [4].

An understanding of NR's flexibility and of the challenges in UE implementation can be obtained by examining a few of the parameters in *CSI-MeasConfig*. Figure 13.3 shows that a UE can be configured with up to 128 trigger states for aperiodic CSI reporting at a given component carrier. Among them, 63 trigger states can be selected as the active trigger states, whereby each active trigger state can be indicated dynamically by the

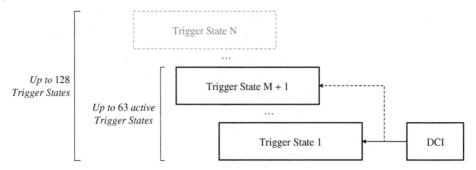

Figure 13.3 Trigger state configuration and indication for aperiodic CSI reporting.

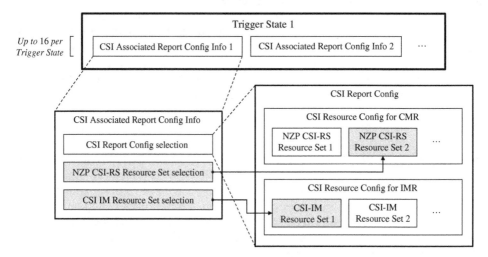

Figure 13.4 Link between trigger state for aperiodic CSI reporting and CSI report configuration.

network to the UE. As shown in in Figure 13.4, in NR, up to 16 CSI reports can be requested simultaneously with a single trigger state. Further for each CSI report, the measurement resources for the desired channel and interference can be selected.

As for uplink data transmission, NR supports both codebook-based transmission and non-codebook-based transmission. To address the uplink coverage shortage in some deployment scenarios, a device may also need to support the so-called Supplemental Uplink (SUL) transmission. In that case, a device transmits on the SUL frequency, which is at a lower carrier frequency than the normal UL frequency, or the normal frequency depending on dynamic signaling from the network. The PUSCH transmission characteristics over SUL and UL can be very different, for example, subcarrier spacing, and hence different baseband and RF processing are needed for SUL and UL.

13.3 Low Latency, Flexible Timing, and Modem Control Flow Complexity

This section addresses the UE modem design challenges for supporting low latency and high system flexibility. Low latency is an important design characteristic of 5G NR, reflected not only in the intrinsic details of its radio interface but also in the entities processing the related radio signals. The processing time requirements on the device side have been considerably tightened compared to LTE. In addition, with the goal of addressing the diverse requirements of a plethora of services, 5G NR features a very high degree of flexibility. By means of flexible configurations, the network can adapt its many parameters to meet the required capabilities of any particular service of various use cases and usage scenarios. High flexibility coupled with low latency pose various challenges for modem processing.

In Section 13.3.1, the impacts of 5G NR low latency on the modem processing capability are investigated. That is, the physical layer techniques used to produce low latency, including shorter slot duration, mini-slot transmission, multiple PDCCH monitoring occasions per slot, and so on, are addressed in terms of the challenges they pose to the modem. For each technique, four main aspects are discussed (when applicable):

- Identifying the new NR features and requirements for low latency that increase modem design complexity
- The modem design challenges to support these features
- The methods introduced in 3GPP specifications to mitigate the impact on the modem
- What can be expected in future releases of 5G NR.

Section 13.3.2 addresses system flexibility, namely, flexible slot format indication and flexible scheduling, alongside the challenges they mean for the design of the modem, following a similar analysis as for low latency.

13.3.1 Low Latency Aspects Versus Modem Processing Capability

To improve performance and enable new use cases, low latency is one of the key performance indicators for the design of the 5G system. In a cellular system, the latency of a packet transmission results from its transit through the various system entities (each with its own processing delay) and interfaces on the data path and varies as a function of the system's topology on that path. This part focuses on the latency caused by the physical layer at the access network, which includes the following:

(1) Queueing delay due to scheduling
(2) Time to a transmit instance (e.g. the start of a slot)
(3) Duration of the packet transmission
(4) Processing delays in the base station (gNB) and in the device
(5) Latency due to scheduling request, retransmission, device feedback, and so on.

To achieve low latency, 5G NR features various enhancements compared to LTE, as summarized in Table 13.4. These include shorter slot duration, mini-slot transmission, multiple PDCCH monitoring occasions per slot, shorter PDSCH/PUSCH processing time, preemption indication, front-loaded DMRS, and Orthogonal Frequency Domain Multiplexing (OFDM) symbol-based Physical Uplink Control Channel (PUCCH).

In the following, the details of these enhancements and the associated impact on the modem processing capability, if any, are presented.

Table 13.4 Overview of NR physical layer techniques for low latency.

Enhancement	Approach	Description
Shorter slot duration	Reduction of (2), (3)	Compared with LTE, larger SCS is supported in NR. Durations of an OFDM symbol and of a slot are reduced.
Mini-slot transmission	Reduction of (2), (3)	Transmission over a fraction of a slot is allowed.
Multiple PDCCH monitoring occasions per slot	Reduction of (2)	PDCCH monitoring can take place at any OFDM symbol of a slot.
Shorter PDSCH/PUSCH processing time	Reduction of (4), (5)	The values of "the time between PDSCH and HARQ-ACK" and "the time between uplink grant and PUSCH" in NR can be shorter than in LTE.
Preemption indication	Reduction of (1)	A scheduled resource can be preempted and used for latency-critical service.
Front-loaded DMRS	Reduction of (4)	DMRS is located at the beginning of a data transmission.
OFDM symbol-based PUCCH	Reduction of (5)	The duration of PUCCH can be as short as one OFDM symbol.

13.3.1.1   Shorter Slot Duration

The subcarrier spacings supported by NR are $2^\mu \times 15\,\text{kHz}$ with $0 \leq \mu \leq 4$. For $\mu \geq 1$, a larger subcarrier spacing Δf than in LTE means a shorter OFDM symbol duration of $1/\Delta f$. This reduction can contribute to low latency.

On the other hand, it requires a change of the common symbol duration and the frame structure. A shorter symbol duration directly leads to a faster baseline system. Specifically, for a shorter slot duration, the clock rate should be increased to complete the tasks scheduled in a slot. This increases the power consumption. In order to keep the clock rate at a sustainable level (for the purpose of reducing the energy demand at the device) under a tighter processing requirement, tasks can be distributed into more parallel pipes. However, such a design consumes a larger chip area and cost, and more sophisticated algorithm design is needed to handle the interaction among the pipes. Besides, function-specific hardware modules (see Figure 13.1) can be used to improve the processing speed. However, this again increases the device cost.

To make the processing burden at similar levels for all subcarrier spacings, the processing time requirement is subcarrier spacing specific. For a task, more OFDM symbols are allowed for the modem to process a certain task at a larger subcarrier spacing.

13.3.1.2   Mini-Slot Transmission

NR allows packet transmission over a fractional slot duration of 2–14 OFDM symbols, sometimes referred to as "mini-slot." Combined with a shorter slot duration, the duration of an NR packet can be much shorter than that of an LTE packet. For example, the duration of a 2-OFDM-symbol data packet in a 120 kHz subcarrier spacing is only 1/56th of the duration in LTE, where the length of a transmission (i.e. a TTI) is 14 OFDM

Table 13.5 Maximum number of monitored PDCCH candidates.

Subcarrier spacing (kHz)	Slot duration (ms)	Maximum number of PDCCH candidates per slot and per serving cell
15	1	44
30	1/2	36
60	1/4	22
120	1/8	20

symbols of 15 kHz subcarrier spacing. With a short transmission time, a small packet can be transmitted with a much lower latency in NR than in LTE.

To reduce the modem processing burden, the processing time requirement is generally defined starting from the end of a reception.

13.3.1.3 Multiple PDCCH Monitoring Occasions Per Slot

In LTE, the PDCCH is transmitted during the first three OFDM symbols of a slot. To enable mini-slot transmission to take effect under low latency, NR allows the transmission of the PDCCH over any of the OFDM symbols in a slot. In so doing, the time to wait for a transmit instance can be greatly reduced.

To avoid the modem implementation burden due to more frequent PDCCH monitoring occasions, NR specifies an upper limit on the number of blind decoding candidates to be monitored as well as the number of control channel elements in which the channel estimation is to be performed within a slot for various PDCCH monitoring cases. These requirements have similar levels as in LTE for the case of 15 kHz subcarrier spacing. For other subcarrier spacings, the capability of handling PDCCH within 1 ms is larger in NR than in LTE. See Table 13.5 for the maximum number of monitored PDCCH candidates per slot and per serving cell. The tasks involved in handling PDCCH include channel estimation, noise estimation, detection/decoding, pruning of blind decoding results, DCI parsing, and so on.

The impact of mini-slot transmission and multiple PDCCH monitoring occasions per slot can be considered together. First, the power consumption will be increased. In LTE, the arrival patterns of PDCCH and PDSCH can be anticipated: the PDCCH occurs at the first three OFDM symbols of a subframe, followed by the associated PDSCH until the end of the subframe. Thus, the activities of a modem can be carefully designed to save the consumed power. For instance, when no PDCCH is detected at the beginning of a subframe, the modem can turn off some modules until the next subframe. In NR, when there are multiple PDCCH monitoring occasions in a slot, the OFF time becomes shorter. Considering the required transition time for ON/OFF, some components may not be turned off. The strategy for deciding ON/OFF states of a module becomes more complicated when the mini-slot transmission is used. In this case, the time gap between the end of a PDSCH and the start of the coming PDCCH monitoring occasion is flexible.

Second, multiple PDCCH monitoring occasions per slot and potentially more associated PDSCH instances incur more jobs for control and data channel processing, leading to a more complicated design for baseband pipeline processing. More communication overheads are needed for the interaction among memory, hardware, software, and firmware.

13.3.1.4 Shorter PDSCH/PUSCH Processing Time

The requirements of a UE in the processing capability for the PDSCH and PUSCH are much tighter in NR compared to LTE. For example, in LTE, the time between the reception of a PDSCH and the transmission of the associated HARQ-ACK is typically 4 ms. In NR, this duration can be as short as 0.2 ms (i.e. 20 OFDM symbols with 120 kHz subcarrier spacing). Similarly, the time from the reception of an uplink grant to the PUSCH is in a similar range.

A short processing time comes at a higher modem cost. This implies that the processing time cannot be made *arbitrarily* short. Accordingly in NR, a UE can report its capability regarding the minimum PDSCH/PUSCH processing time. Two device capabilities are defined, capability 1 and capability 2. The former corresponds to a baseline mode for the capability of the minimum PDSCH/PUSCH processing time, while the latter is an aggressive mode which requires a tighter processing schedule. All UEs are required to fulfill at least the baseline mode. A UE may also report that it is capable of a more aggressive processing timeline.

The impact of the more aggressive processing capability is similar to the impact of a shorter slot duration. The design involves a balance among the clock rate, the power consumption, the parallel processing capability, the chip area/cost, and so on. Either a higher energy demand, or a higher device cost for components of hardware, memory, and processors is needed, or more sophisticated software programming is needed to handle the scheduling of concurrent jobs.

13.3.1.5 Preemption Indication

Assume, under high traffic load, that all downlink time-frequency resources are occupied as a result of which the network is unable to accommodate an incoming time-critical service. NR supports the functionality of re-allocating the resource originally scheduled to a first UE1 to another UE2 with a tight latency requirement. To do so, a preemption indication is sent to UE1 notifying some of its resources have been preempted and used for other purposes. In this way, for UE2, the latency due to scheduling delay can be saved.

The network and UE1 should take further action to compensate for the resource preemption. In the example above, UE1 has to perform some soft buffer handling. For instance, the part of the soft buffer which corresponds to the preempted resource should be flushed because the data intended to UE2 carried in the preempted resource have corrupted the information stored in the soft buffer. This soft buffer handling should be completed before the arrival of the associated retransmission of the preemption indication; otherwise, the retransmission is wasted. For PDSCH with tight HARQ timing, this is difficult for the device.

Other alternatives for assigning resource for time-critical service are available. The simplest method is reserving a dedicated resource for such service. When the traffic load is high and a dedicated resource is not desirable, the signal intended for UE2 can be superposed with the signal for UE1, instead of preempting the latter. The signals for the two UEs can be separated at the receiving sides if the favorable beam directions of the two users are disjoint. When signal superposition is used, it is also possible to perform power boosting for the signal with a higher reliability demand.

13.3.1.6 Front-Loaded DMRS

By locating the reference signals at the beginning of a PDSCH transmission, a UE can start performing channel estimation early on. Once the channel estimate is obtained, the receiver can process the received symbols immediately without having to buffer the received signal prior to data processing. This design can reserve some more processing time for the modem.

Besides the front-loaded DMRS, NR supports additional DMRS in later OFDM symbols within the resource allocated to the PDSCH. The channel estimation at front-loaded and additional DMRS can be interpolated to improve the accuracy of channel estimates.

When there is no additional DMRS for the PDSCH, the design for channel estimation becomes more complicated. In this case, extrapolation of channel estimation is needed. This either degrades the performance of channel estimation or requires a more sophisticated algorithm to maintain a comparable performance.

13.3.1.7 OFDM Symbol-Based PUCCH

When low latency is required, the delay due to a scheduling request, UE feedback, retransmission, and so on, should be taken into account. All these activities are relevant to the transmission of the PUCCH. Unlike in LTE where a PUCCH takes one subframe duration, the duration of a PUCCH in NR can be as short as one OFDM symbol.

The above features provide the basic support for low latency in the first release of NR. In the next release (Release 16), a service called Ultra-Reliable Low Latency Communications (URLLCs) is studied as further use cases with tighter requirements have been identified as an important area for NR evolution; also, Release 15 use cases need to be enhanced. The potential physical layer improvements for further latency reduction include the following:

- *PDCCH enhancements*: Compact format of DCI, increased PDCCH monitoring capability
- *Uplink control information enhancements*: Enhanced HARQ feedback methods such as increased number of HARQ transmission possibilities within a slot
- Enhancements to scheduling/HARQ/CSI processing timeline

13.3.2 System Flexibility Versus Modem Control Timing

ITU-R defined a set of capabilities needed for an IMT-2020 technology to support various use cases and usage scenarios. This set includes the peak data rate, user experienced data rate, area traffic capacity, spectrum efficiency, network energy efficiency, mobility, latency, and connected density. These capabilities need not be met simultaneously – various combinations of these capabilities allow the provision of different types of services. For instance, a mobile broadband service with a very high peak data rate and user experienced data rate need not feature very low latency. For this reason, one of the design philosophies of NR is flexibility. The network, by means of flexible configurations, can adapt its many parameters to meet the required capabilities of any particular service.

Such flexibility leads to challenges in the design and implementation of a device. Job multitasking becomes more complicated with flexibility in configurations. In a modem, multiple operations are executed pipelined and in parallel. They include the support of layer 1 control, data channel processing, HARQ-related signaling, physical layer procedures, measurements, and so forth. These tasks are executed at device components such as DSPs, hardware, memory, and processors as shown in Figure 13.1.

Consider as an example the reception of the PDCCH with flexible configurations. A PDCCH candidate is first blindly decoded at the hardware. The decoding result is passed to other modules for the cyclic redundancy check and parsing of the DCI to judge whether the decoding result is valid. If it is considered valid, the DCI is passed to the software to proceed to the next actions. However, due to flexible configuration, the software is generally not waiting for the result of the PDCCH decoding but rather is carrying out some other tasks. In this case, the software will be interrupted to handle the PDCCH decoding result as well as the associated upcoming tasks and goes back to the original work after that. Complex system scheduling and data traffic management are essential to realize low device cost, small latency, and power saving. This causes software programming burden.

In the following, two NR features that enable system flexibility are identified, as well as the challenges they pose to the design of the modem.

13.3.2.1 Flexible Slot Format Indication

In NR, the link direction of an OFDM symbol in a slot can be classified as "downlink," "flexible," or "uplink." One of the key features of NR is the support for very flexible configurations for link directions. Four different signaling mechanisms can provide information to the UE on the link directions:

(1) Semi-static higher-layer signaling of slot format indication: three possible states "downlink," "flexible," or "uplink" for the link direction
(2) UE-specific RRC configuration from which a link direction "downlink" or "uplink" can be inferred: for instance, the link direction of an OFDM symbol can be assumed as "uplink" if a UE is configured to perform periodic Sounding Reference Signal (SRS) transmission at the symbol
(3) Dynamic data scheduling signaling from which a link direction "downlink" or "uplink" can be inferred: for instance, the link direction of an OFDM symbol can be assumed as "downlink" if a UE is scheduled with a PDSCH at the symbol
(4) Dynamic slot format indication: three possible states of "downlink," "flexible," or "uplink"

The purpose of dynamic link direction indication is to allow for low latency and flexibility. For example, a resource configured as "flexible" by semi-static higher-layer signaling and used for CSI-RS measurement at UE can be dynamically reconfigured as "uplink" when an urgent uplink traffic demand is present.

A set of overwriting rules are specified in NR to handle the situation when the link directions signaled by the above mechanisms conflict each other. In general, the status levels are

"downlink"and"uplink"of (1) > (3) > (4) > (2) > "flexible"of (1)

The information of link direction provided by a signaling with higher status level can overwrite the previous information given by a lower status level signaling.

To provide guidance to avoid overdesign at the UE, NR specifies the processing time requirements for the link direction overwriting behaviors of "a downlink reception is cancelled" and "an uplink transmission is cancelled."

When the link direction is overwritten, the action includes cancellation within a given processing time for measurement, reception, or transmission. The change of link direction at the radio frequency components is also needed; however, this is not a serious issue as long as the switching time is sufficient. In addition, the set of scheduled tasks in the data pipeline may also need to be rescheduled. To deal with such flexible direction configurations, the software at the UE should constantly track the latest information. The reaction time for software is generally very short, because some demanding action had better not be performed in advance to save extra efforts once the link direction is not the same as originally expected. Further, for non-full-duplex UE operating in the carrier aggregation (CA) scenario, the overwriting rules of link direction should also be applied to different cells. This increases the complexity of software decision making significantly in comparison with LTE, where event timing is fairly static.

13.3.2.2 Flexible Scheduling

The flexibility of NR scheduling is realized by configuring the following:

- Time domain resource allocation
- Time gap from PDCCH reception to PDSCH/PUSCH
- Timing for HARQ-ACK transmission, and so on

In LTE, the corresponding configurations for the above assignments are static. The PDCCH is located at the beginning of a subframe, and the associated PDSCH/PUSCH has a fixed starting OFDM symbol until the end of the subframe. The HARQ-ACK feedback is then provided with a constant timing, for example, 4 ms later for Frequency Division Duplex (FDD). In NR, the arrival time of these tasks can occur at any OFDM symbol in a slot and may change in a per-slot basis.

The flexibility in scheduling requires a UE to tolerate different levels of timing requirements and the associated UE behaviors configured by the network. At the same time, the UE should distribute its computing resources and timing to various tasks and maintain its power consumption as low as possible. Moreover, planning for the resource such as the amounts of memory, hardware, and so on, becomes difficult because the usage scenarios to be supported by the device are unclear at the planning stage. Furthermore, the testing activities become involved because abundant test cases are needed to cover all target scenarios.

13.4 Multi-RAT Coexistence and Modem Architecture

The deployment of a new mobile communication system generation can take several years, whether as a new greenfield deployment or as a complement to an existing, older, generation. The latter scenario requires a smooth migration from the older to the newer generation, which implies both graceful coexistence and continuity of service between the two systems. The definition of a new 5G radio technology, that is, NR, and of a new 5G System was made with full consideration of the widespread use of LTE systems worldwide and of the various migration plans mobile operators may have. To cater for the above points, five system architectural options[3] addressing different deployment scenarios over various transitional phases were defined. From the modem's perspectives, the main challenges in supporting these architectural variants lie in the sharing, and eventually the integration, of LTE and NR functionalities. In this section, some key considerations are highlighted for balancing performance and cost-effectiveness when implementing both LTE and NR modems in an SoC.

13.4.1 Dual Connectivity and Modem Architecture

Option 3, also known as EN-DC, of the five system architectures is a very popular option for mobile operators because it leverages existing LTE networks to provide full coverage mobility and the newly deployed NR network to provide super-fast data transmission. Not requiring the deployment of a new core network, this option can accelerate not only the availability of 5G services but also their adoption. However, from the UE viewpoint, simultaneous operation of two radio technologies is required, which brings new challenges to the UE design, especially the modem architecture.

One option is for a UE to be equipped with completely separated yet coordinated LTE and NR hardware modules to handle data transmission/reception with LTE and NR base stations independently. However, the resulting hardware cost would be prohibitive (e.g. the die area).

The other option is hardware sharing. Hardware sharing uses a resource pooling mechanism to allocate hardware/computing capabilities dynamically based on the need for LTE/NR at any given time. Configuration messages from the network allow the dynamic allocation to operate. The capability of a UE to report its hardware capability to the network on a per LTE/NR band combination basis was introduced into the 5G specifications. For example, a UE can be designed in a way such that the lower MIMO capability is declared for low frequency bands, due to antenna placement considerations. Hardware sharing is an effective means to reduce hardware cost, but it comes at the price of a considerably higher design complexity, which itself pushes the hardware cost higher. Thus, a design trade-off between hardware sharing and design complexity is needed.

In an EN-DC network, the LTE and NR base stations make independent scheduling decisions. Should the two base stations grant simultaneous uplink transmissions, uplink power sharing becomes an issue because the total UE transmit (Tx) power is bound by a single power class. To tackle this, one approach is to use a fixed power sharing concept. This method configures a maximum Tx power for each of the LTE and NR links independently such that the resulting aggregated Tx power never exceeds the upper Tx power bound of the UE. The drawback of this scheme, however, is that of coverage impact for cell-edge UEs due to lower maximal Tx power. To deal with the power sharing issues, two options have been defined:

a. Single-uplink-operation (SUO) based Time Division Multiplexing (TDM) sharing
b. Dynamic power sharing (DPS).

Figure 13.5 illustrates the basic concept of these schemes. For the non-cell-edge case, there is no difference between the two options because the aggregated Tx power never

---

3 Option 2: NR Standalone in 5GS; Option 3: LTE + NR dual connectivity in EPS; Option 4: NR + LTE dual connectivity in 5GS; Option 5: LTE Standalone in 5GS; Option 7: LTE + NR dual connectivity in 5GS.

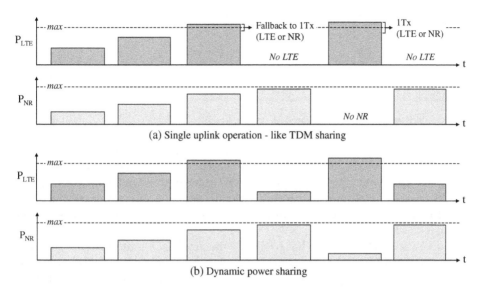

Figure 13.5 LTE/NR Power sharing schemes: (a) SUO-like TDM sharing and (b) dynamic power sharing.

exceeds the power class of the UE. However, as a UE moves toward the edge of the cell, the SUO-based scheme falls back to single-uplink mode; that is, a TDM pattern is applied for EN-DC uplink transmission. On the other hand, the DPS scheme can use dual uplink transmissions as long as residual power is available. As a result, to maximize the overall system throughput, the DPS scheme is a better option.

From a modem implementation viewpoint, DPS requires the modem to coordinate the power allocation between the LTE and NR links. Coordinating the power allocation for different timings is challenging. In LTE, there is a fixed 4 ms timing relationship between an uplink grant and the corresponding uplink transmission. However, in NR, the timing relationship is dynamic and configured by the network (see Section 13.3.2). Figure 13.6 shows one example of DPS operation. In this example, NR uplink is granted two slots ahead of the NR transmission, while LTE uplink is granted four subframes ahead of LTE transmission. Both transmissions happen to occur at subframe $N + 5$. The internal processing time for power allocation between the two links becomes stringent,

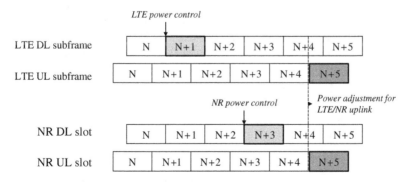

Figure 13.6 Illustrative example of DPS operation.

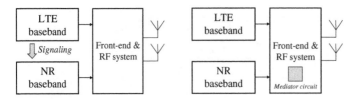

Figure 13.7 Illustrative examples of shared RF system.

compared to the LTE-only design. In the example, the 5G NR module is unaware of NR power control information until subframe $N + 3$ arrives. Therefore, the available time to calculate the appropriate power allocation between the two transmissions is effectively two subframes less the timing advance and the decoding of NR's power control command. The situation is even more critical in the mixed TTI scenarios (e.g. with LTE in FR1 and NR in FR2).

Figure 13.7 shows two approaches to solving the problem. On the left is an architecture in which the LTE baseband module signals its power control parameters to the NR baseband modules, which then calculates the adjustment NR transmission needs to take to accommodate the LTE module's power usage. Alternatively, the architecture on the right adds a mediator circuit in the front-end and RF system module. The circuit takes in power control parameters from both baseband modules and can thus dynamically balance between the two to achieve the optimal usage of the power.

13.4.2 Impact of LTE/NR Coexistence on Modem Design

For FR1 deployment, NR could operate in the new NR band (e.g. C-band[4]) or a lower legacy LTE band. The design challenges for uplink transmission in these two scenarios are described in the following.

13.4.2.1 Operating in the New NR Band
The C-band is one of the most important NR sub-6-GHz bands; sitting between the low bands for wide area coverage and the high bands for high capacity, the C-band provides a sound compromise between coverage and capacity which can help build NR coverage at reasonable costs while meeting the expected demand in data capacity. The C-band is likely to operate in Time Division Duplex (TDD) mode, for which a link budget issue was observed [2]: the uplink coverage is much smaller than the downlink coverage.

To bridge this coverage gap, the first option is a stronger gNB[5] antenna capability under the assumption that the gNB has better reception capability than the UE. A second option is to transmit uplink signals in the lower band for reducing the propagation loss. For this latter option, two mechanisms are standardized: Supplementary Uplink (SUL) and CA.

13.4.2.2 Supplementary Uplink
The UE is configured with an *additional* uplink (i.e. SUL) set in a different frequency band than the one used for the downlink and the normal uplink (NUL). One example

---

4 C-band: 3300–4200 MHz and 4400–5000 MHz (the exact availability depends on the regional regulations).
5 gNB: NR base station.

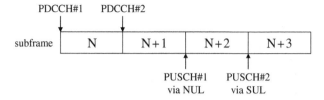

Figure 13.8 Example of switching between NUL and SUL.

is to configure an SUL in a wide area coverage band such as 900[6] or 1800 MHz[7], while the downlink and the NUL remain at 3.5 GHz.[8] It is up to network scheduling decision whether to allow uplink transmission in the SUL or in NUL – the corresponding decision is informed to the UE via DCI. The associated design challenge for the UE is therefore to switch the uplink transmission path *on the fly* by DCI. Figure 13.8 shows one example in which the UE alternates between SUL and NUL in consecutive subframes.

13.4.2.3 Carrier Aggregation
Asymmetric CA configuration (e.g. 2DL + 1UL) is a popular setting in LTE, reflecting the typical downlink bias of data traffic. This concept can be exploited to address the uplink coverage issue. The basic idea is that the original NR carrier is reconfigured as an SCell[9] (i.e. it is used for downlink-only transmission), while the PCell[10] is reconfigured to a lower wide area coverage band (e.g. 900 MHz), as shown in Figure 13.9.

From a UE implementation viewpoint, it is analogous to an intra-gNB handover procedure with the additional challenge of supporting mixed-numerology HARQ timing. The low band is likely to operate with a 15 kHz SCS to coexist with the incumbent

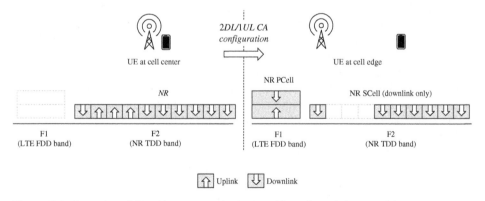

Figure 13.9 Illustration of CA with spectrum sharing to address the uplink coverage issue.

---

6 LTE Band 8 (B8).
7 LTE Band 3 (B3).
8 NR Band 77 (n77).
9 SCell: secondary cell in a CA configuration.
10 PCell: primary cell in a CA configuration (~equivalent to the serving cell in a non-CA configuration).

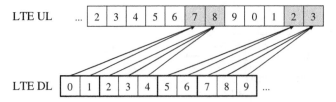

Figure 13.10 Illustrative SUO figure for EN-DC operation.

LTE system, but the new NR band is likely to operate with a 30 kHz SCS. The mixing numerology operation complicates the UE internal timing design.

13.4.2.4 Operating in the Legacy LTE Band

The new C-band might not be available in some regions (e.g. where the C-band has already been occupied by other applications). One alternative is a soft refarming-like approach [5] which deploys an NR system in a legacy LTE band. For this LTE/NR coexistence scenario, special handling is required to avoid any negative impact on the legacy UE in the field. For example, CRS is present in every subframe for the LTE system, so the NR PDSCH needs to perform CRS rate matching accordingly. As for uplink transmission, the UE will face new challenges for intra-band EN-DC scenarios (e.g. LTE B41 + NR n41[11] or LTE B71 + NR n71[12]) due to transmissions of LTE and NR signals in the same band. Since the NR PUCCH may only occupy a partial subframe, the gain control module in the UE has to handle the non-equal Tx power scenario within a transmission subframe (i.e. some OFDM symbols require a higher Tx power than others).

13.4.3 Uplink Transmission Design for Minimizing Intermodulation Effect

For certain EN-DC band combinations (LTE B3[13] + NR n77), it is observed that the intermodulation problem can degrade the LTE receiving sensitivity [6]. SUO, introduced above (see Section 13.4.1 point a), is adopted to solve the issue. In SUO, a predefined TDM pattern alternating between LTE and NR uplink transmissions is configured by the network. Figure 13.10 show a SUO example. In this example, it can be seen that LTE uplink transmissions are aggregated to some subframes (#7, #8, #2, #3) while NR uplink transmissions are over other subframes. For UE implementation, a specific controller is used to coordinate the EN-DC uplink transmissions, which impacts both LTE and NR modems.

As discussed in this section, NR uplink design deals with the challenges incurred by new 5G deployment scenarios, including uplink power sharing, uplink coverage gap, and intermodulation. Uplink design also needs to face the challenges posed by flexible NR configurations. Flexible hardware sharing and higher processing power are relevant UE implementation options to handle complicated scenarios (e.g. mixed TTI). Therefore, the trade-off between system flexibility and UE device complexity/cost is critical to the success of 5G.

---

11 LTE B41 and NR n41: TDD 2.5 GHz (2496–2690 MHz).
12 LTE B71 and NR n71: FDD 600 MHz (Uplink: 663–698 MHz; Downlink: 617–652 MHz).
13 LTE B3: FDD 1800 MHz (Uplink: 1710–1785 MHz; Downlink: 1805–1880 MHz).

13.5 Wider Bandwidth Operation and Modem Power Consumption

In order to achieve higher peak data rates than achievable with 4G LTE, a wide bandwidth (e.g. 100 MHz) is necessary in 5G NR. However, although necessary to yield higher data rates, the wide bandwidth introduces a power consumption challenge for mobile devices.

The power consumption of a mobile device is a very important criterion for it directly impacts the device's battery lifetime and in turn the user experience; the introduction of a new radio technology with greater capabilities should not be made at the expense of battery lifetime. While batteries with a higher power capacity can help address this challenge, it is not sufficient and also not always possible, for example, due to size constraints, and hence the necessity to systematically engineer a mobile device to be as energy efficient as possible. With the power consumption typically scaling with any increase in data rate, such an engineering challenge is inevitably one of modem design. Furthermore, other related issues such as heat dissipation and overheating are fundamental challenges for mobile devices that are not only size-constrained, but also often handheld.

This section focuses on 5G NR modem impacts for supporting wider bandwidth operation from a power consumption perspective. Section 13.5.1 analyzes the power consumption of a typical LTE smartphone usage and then identifies the key challenges that have to be overcome in 5G NR to maintain the same level of user experience while delivering a 20-fold increase in data rate. Section 13.5.2 focuses on bandwidth adaptation, which is one of the primary features introduced in NR to address power consumption issues. Finally, the impacts of implementing these features on modem design are given in Section 13.5.3.

13.5.1 Modem Power Consumption in Daily Use

The power consumption of a modem results both from the design of the modem itself and from the semiconductor manufacturing process used to physically implement it. Advances in semiconductor manufacturing processes have played an important role in improving the power efficiency of modems. Undoubtedly these have contributed to the phenomenal success of 4G LTE. However, as the physical limitations of the process node are being reached, so are the limitations of the corresponding power efficiency gains. The 20-fold increase in data rate presents a major challenge for modem design. In the following, this issue is introduced from the perspective of existing 4G LTE design based on which 5G NR conclusions are then drawn.

Based on extensive field logging, it is found that the modem power consumption accounts for less than 20% of the total day of use (DoU) power consumption of an LTE smartphone. The remaining power consumption comes from the device's screen, graphics processing unit (GPU), AP, and so on, Focusing on the modem only, Figure 13.11 shows the use time and power consumption of a typical LTE modem throughout a DoU. In these pie charts:

- *4G operation,* including *Data Service* and *Voice Service,* refers to the modem state in which there is data traffic for the modem to process.

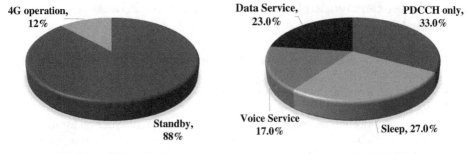

Use Time Distribution   Power Consumption Distribution

Figure 13.11 Example DoU for 4G LTE modem using 20 MHz.

- *Standby,* including *PDCCH-only* and *Sleep,* refers to the modem state in which there is no data traffic but data scheduling monitoring, synchronization tracking, and measurements are needed.

In the *PDCCH-only* state, the modem performs PDCCH blind decoding to monitor DL/UL data scheduling, but performs neither DL/UL data decoding nor encoding. The *Sleep* state comprises two major power consumption contributors: leakage current on one hand, and paging monitoring and RRM measurement on the other hand. However, only the latter can be improved through better system design and modem implementation.

Since there are no field data for a 5G device's DoU power consumption, only the phone's power consumption in a downlink data session (for sub-6-GHz) running at peak rate can be analyzed, as shown in Figure 13.12. The figure shows that the proportion of the modem's power consumption ranges from 70% to 80% of the total estimated power consumption of the device, depending on the number of component carriers. This number is not expected to go down much in a typical DoU and will likely be significantly greater than 20% for an LTE phone. Unless special measures are taken in the system design to lower the power consumption when the modem is not actually processing its data, the user experience in 5G may be worse than 4G. The reasons are mainly twofold. First, the 5G NR modem is very efficient at high peak rates since it consumes much less energy per bit compared to LTE due to LDPC code, as mentioned in Section 13.2.1. Therefore, operation at non-peak rates does not lower the peak power consumption much. Second, the primary source of power consumption in a modem is its RF circuits' current drain, which scales with the bandwidth. With 5G NR carriers typically having a much wider bandwidth (100 MHz for sub-6-GHz) than LTE, the standby state power consumption in the modem would be much higher than in LTE if the RF circuits were always tuned to the widest bandwidth. The situation would be even more severe in the mmW spectrum where the carrier bandwidth is upward of 400 MHz.

Based on Figures 13.11 and 13.12 and the above analysis, it is clear that efficiently and dynamically adapting the modem's operating bandwidth is the key to reducing the 5G modem's power consumption. To help pinpoint the main sources of current drain, the modem power consumption model captured in the 3GPP study on UE power saving in 5G NR [7][14] for operation in the sub-6-GHz spectrum is shown in Table 13.6. The data

---

14 The study is underway at the time of writing.

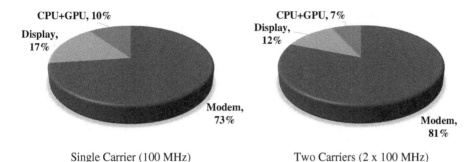

Figure 13.12 Estimated power consumption distribution for a 5G NR smartphone under peak data rate for DL in sub-6-GHz spectrum.

in Table 13.6(a) are normalized such that the smallest modem power consumption (deep sleep) corresponds to one power consumption unit. Deep sleep, light sleep, and micro sleep shown in Table 13.6(a) are defined as follows.

- *Deep sleep.* All components within the modem are turned off, and it requires longer time and larger additional energy for the modem to wake up, compared to light sleep.
- *Light sleep.* Almost all components within the modem turned off, and it requires the longer time and larger additional energy for the modem to wake up, compared to micro sleep.
- *Micro sleep.* Only RF components are turned off, and it requires the shortest time and almost no additional energy for the modem to wake up among the three states.

In Table 13.6(b), modem power consumption scaling factors are introduced that reflect the modem power consumption scaling when the scheduling is changed from same-slot scheduling to cross-slot scheduling and the reception bandwidth and carrier number of a modem increases – references are provided in *italic*.

From Table 13.6, the following candidate techniques can be derived that can help reducing the modem power consumption for *PDCCH-only* and *Voice Service* modem states:

- Bandwidth adaptation for connected mode
  - Narrow bandwidth for *PDCCH-only* and *Voice Service* modem states
  - Wide bandwidth for other states
- Adaptation of cross-/same-slot scheduling for connected mode
  - Cross-slot scheduling for *PDCCH-only* and *Voice Service* modem states
  - Same-slot scheduling for other states
- Adaptation of PDCCH monitoring periodicity for connected mode
  - Long periodicity for *PDCCH-only* and *Voice Service* modem states
  - Short periodicity for other states

During the *Sleep* modem state, as described above, it is of interest to reduce the power consumption due to paging monitoring and RRM measurement. Intuitively, wide bandwidth for paging monitoring and RRM measurement as well as frequent transitions between deep sleep and paging monitoring and RRM measurement are not power efficient. Therefore, the following candidate techniques are helpful to reduce the modem power consumption.

Table 13.6 Modem power consumption model data from [7].

(a) Relative power consumption level for each modem state

	Deep sleep	Light sleep	Micro sleep	PDCCH-only	SSB or CSI-RS Proc.	PDCCH + PDSCH	UL
Relative power level	1	20	45	100	100	300	250 @ 0 dBm 700 @ 23 dBm
Additional energy for state transition	450	100	0	0	0	0	N/A

(b) Scaling factor

Time domain

DL	Cross-slot scheduling: $0.7 \times PDCCH\text{-}only$
UL	N/A

Frequency domain

DL	Scaling for Y MHz $= 0.4 + 0.6 \times (Y - 20)/80$ (*Reference: 100 MHz*)
	2 carriers: 1.7×1 *carrier*
	4 carriers: 3.4×1 *carrier*
UL	2 carriers $= 1.7 \times 1$ *carrier* @ *0 dBm*;
	2 carriers $= 1.2 \times 1$ *carrier* @ *23 dBm*

- Narrow bandwidth for idle mode.
- Time window aggregation for paging monitoring and RRM measurement in DRX[15] operation.

13.5.2 Reducing Modem Power Consumption by Bandwidth Adaptation

There are two ways to support a wide bandwidth: using a single wideband carrier (e.g. 100 MHz) or carrier aggregation (e.g. five carriers of 20 MHz each). Compared to a single wideband carrier, carrier aggregation introduces not only a higher PDCCH overhead but also a higher mobile device processing requirement to process more PDCCH blind decoding candidates. A single wideband carrier is therefore a more efficient approach than carrier aggregation with contiguous spectrum, from both complexity and spectrum efficiency standpoints. However, a wideband carrier introduces a power consumption challenge for mobile devices. To mitigate this, 5G NR features BWP operation in both the DL and UL, which allows one to adapt the bandwidth used according to the need, thereby mitigating the modem power consumption. BWP supports the following techniques to reduce the modem power consumption:

- Bandwidth adaptation for connected mode

---

15 DRX: discontinuous reception whereby the UE only wakes up at set intervals (i.e. DRX cycle) to monitor paging.

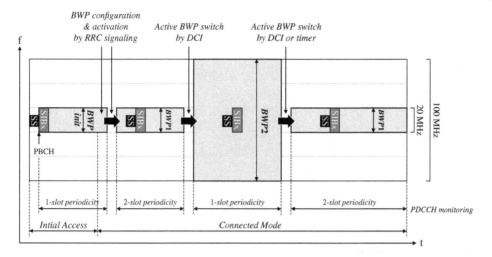

Figure 13.13 Example illustration of DL BWP operation in single wideband carrier.

- Adaptation of PDCCH monitoring periodicity for connected mode
- Narrow bandwidth for idle mode

Figure 13.13 illustrates an example how BWP operation can be used to support the above power saving techniques in the DL. When a mobile device performs an initial access or switches from idle mode to connected mode, it starts with an initial narrow DL BWP with a bandwidth of 20 MHz. Once in connected mode, dedicated DL BWP configurations (i.e. narrow DL BWP #1 of 20 MHz and wide DL BWP #2 of 100 MHz) are signaled to the mobile device by the network. The network simultaneously activates DL BWP #1. When there are data packets to be transmitted, the network switches the modem's active DL BWP from DL BWP #1 to DL BWP #2 via DCI for low latency. When there are no data packets to be transmitted to the device, the modem's active DL BWP is switched back from DL BWP #2 to DL BWP #1. When the mobile device switches to idle mode from connected mode, it reverts back to using the initial DL BWP. In Rel-15 5G NR, three ways of changing the active BWP are supported: higher-layer signaling (i.e. RRC signaling), layer-1 signaling (DCI indication), and timer. Compared to RRC signaling, DCI indication offers better latency for active BWP change ($2 \sim 3$ ms instead of $10 \sim 15$ ms) and less signaling overhead at the cost of higher modem complexity.

To illustrate how the modem's operation bandwidth affects its power consumption, the mixed signal circuits block in Figure 13.1 is expanded and shown in Figure 13.14. When the operation bandwidth is scaled down, the function blocks of A/D (analog-to-digital) converter, D/A (digital-to-analog) converter and baseband unit can use a lower voltage and/or a lower clock rate due to less data sample for processing. When the PDCCH monitoring periodicity is scaled up (i.e. longer periods, and hence slower), RF circuits can be turned off until the next PDCCH monitoring occasion and other function blocks can either be turned off or use a lower voltage and clock rate. To improve the modem power efficiency by BWP operation, the 5G NR modem should be designed to enable dynamic voltage/clock rate adjustment (because the power consumed by a semiconductor chip is proportional to the applied clock rate and the

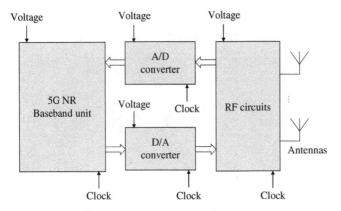

Figure 13.14 Function blocks in 5G NR modem.

Table 13.7 System-level evaluation assumptions.

Parameters	Assumptions
Deployment scenario	Dense urban (clustered 21 macro cells, ISD = 200 m)
Users per cell	7
Carrier frequency	4 GHz
FTP traffic model	4 packets per second, packet size = 1.75 MB
Video traffic model	H.264 (3840p × 2160p, 30 fps)
	Buffered video length: 1 s
DRX configuration	On duration: 8 ms
	Inactivity timer: 100 ms
	Retransmission timer: 4 ms
	Long cycle: 160 ms
	Short cycle: 40 ms
	Short cycle timer: 0 ms

ISD: Inter-site distance.

square of the applied voltage) and reduce additional energy required for turning on/off of function blocks for more frequent on/off transitions.

To illustrate how bandwidth adaptation improves the power efficiency of a 5G NR modem, the following three scenarios are benchmarked by system-level evaluation:

- 4G LTE with constant 20 MHz bandwidth
- 5G NR with constant 100 MHz bandwidth
- 5G NR with bandwidth adaptation between 20 MHz and 100 MHz

Table 13.7 shows the evaluation assumptions for the evaluation, and Figure 13.15 shows the evaluation results for File Transfer Protocol (FTP) and video traffic types.

In Figure 13.15, the average modem power consumption (i.e. the average energy unit per time unit) is equal to the total consumed energy over a period of time divided by the period of time. From Figure 13.15, the following three observations can be drawn:

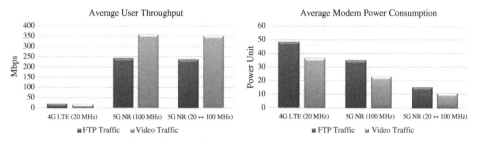

Figure 13.15 System-level evaluation results for FTP and video traffic.

- Higher user throughput is beneficial to modem power consumption reduction for data services requiring large data transfers and hence fast mobile broadband. This enables a shorter time for data transmission and reception and longer sleep time between data bursts.
- Bandwidth adaptation is an efficient means of reducing the power consumption of the mode, at the cost of a slight degradation in user throughput.
- Even with DRX operation, over 50% modem power consumption reduction in connected mode is observed for both FTP and video traffic types with 5G NR.

Based on the above observations, bandwidth adaptation in 5G NR can provide higher energy efficiency and lower power consumption than 4G LTE for data services requiring fast mobile broadband (services with large packet size) and may provide comparable modem power consumption to 4G LTE for data services with smaller packet size (e.g. tens of kilobytes) and voice service. For services that are not sensitive to latency, the technique of PDCCH monitoring periodicity adaptation can be applied at the same time for further reduction of the modem power consumption. BWP is a key feature in 5G NR to resolve the modem power consumption issue for single wideband carrier operation.

So far this section has mainly focused on the bandwidth adaptation in a single wideband carrier. As of Release 15, in addition to single wideband carrier, 5G NR retains the LTE carrier aggregation for wide bandwidth operation. The BWP adaptation can therefore operate in each of the individual component carriers. Since a mobile device supporting a wideband carrier requires more advanced implementation of RF components, not every mobile device may have such capability, especially for early 5G NR deployments. Carrier aggregation is a well-known technique introduced in 4G LTE that can be exploited in some 5G NR device implementations to cover the same bandwidth as a wideband device but over multiple narrowband carriers, for example, aggregating five 20 MHz carriers for a total bandwidth of 100 MHz.

Though component carrier activation/deactivation can be utilized to enable bandwidth adaptation for modem power consumption reduction, secondary carriers are rarely deactivated in real-life 4G LTE network deployments, due to the long activation time necessary to re-activate the secondary carrier (e.g. up to 34 ms in 4G LTE). In Rel-15 5G NR, the secondary carrier activation time is even slower than in 4G LTE (e.g. up to 90 ms) due to the sparser density of common pilots for modem automatic gain control (AGC) and synchronization.

Take video streaming, for example; the data typically arrive in large packets intermittently, and therefore a second component carrier is often activated by the network to quickly deliver these packets. Most of the time in between these large packets,

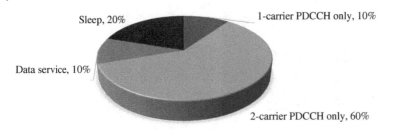

Figure 13.16 Power consumption distribution for 4G LTE modem with two-carrier carrier aggregation in video streaming application.

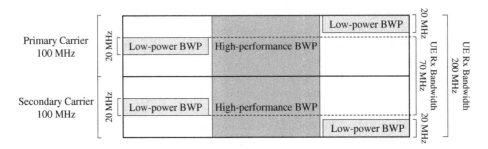

Figure 13.17 Example illustration of BWP operation in contiguous carrier aggregation.

however, there is little traffic activity while the modem continues performing PDCCH monitoring for both carriers. Figure 13.16 shows the power consumption breakdown for a 4G LTE modem with two-carrier carrier aggregation in a video streaming scenario. In this example, the secondary carrier is activated only for PDCCH monitoring in most of the use time. In 5G NR, the power consumption issue with carrier aggregation can be even more severe because the bandwidth of each component carrier can be larger than that of 4G LTE.

For non-contiguous carrier aggregation in 5G NR, different RF chains are applied for each carrier so the techniques of bandwidth adaptation and PDCCH monitoring periodicity adaptation in each carrier are beneficial for reducing the modem power consumption. However, for contiguous carrier aggregation, as shown in Figure 13.17, a single RF chain is applied to both carriers, and only the technique of PDCCH monitoring periodicity adaptation is useful for modem power consumption reduction. Therefore, even for carrier aggregation, BWP is also a key feature in Rel-15 5G NR to resolve the modem power consumption issue for carrier aggregation. However, as mentioned earlier, the 5G NR modem should be designed to reduce the additional energy required for turning on/off of function blocks for more frequent on/off transitions.

13.5.3 Impacts on Modem Design

Though the modem power consumption issue can be resolved by the support of BWP operation in 5G NR, it does not come for free. BWP operation introduces at least the following two modem impacts:

- High modem processing capability to support 2–3 ms switching delay for DCI-based active BWP switching

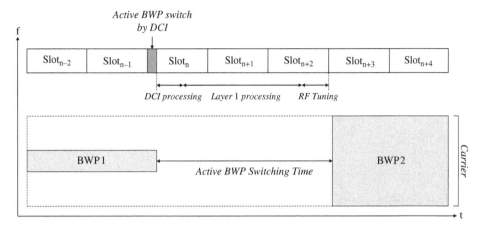

Figure 13.18 Modem processing timeline of DCI-based active BWP switching for 15 kHz subcarrier spacing.

Figure 13.18 illustrates the modem processing timeline of DCI-based active BWP switching for 15 KHz subcarrier spacing. From Figure 13.18, the overall active BWP switching delay consists of three major parts: (1) DCI processing, (2) Layer 1 processing, and (3) RF tuning. Among the three parts, layer 1 processing accounts for the largest portion of the overall active BWP switching delay because it involves the calculation of many detailed baseband and RF parameters due to BWP configuration switching, and there is a trade-off between the processing time and modem hardware cost. Though a modem's radio interrupts during the active BWP switching delay in Rel-15 5G NR, only RF tuning actually requires radio interruption. Since a longer radio interruption time introduces a larger user throughput degradation, it may prevent frequent use of active BWP switching for better modem power efficiency if the modem radio interruption time is equal to the active BWP switching delay. This is another aspect for further enhancements in future 5G NR releases.

In addition to modem power consumption, carrier aggregation also introduces other modem impacts, including at least the following:

- High modem processing capability for PDCCH blind decoding to support a large number of carriers
- Great difficulty for a modem to support short secondary carrier activation time with sparse common pilots (e.g. SS/PBCH block) for AGC settling and synchronization

In Rel-15 5G NR, up to 16 carriers can be supported in carrier aggregation. The greater the number of carriers supported in carrier aggregation, the more overall PDCCH blind decoding candidates are introduced and the higher the modem processing capability needed to meet the timeline for HARQ operation and UL scheduling supported in R15 5G NR. Though there are several techniques supported in Rel-15 5G NR to prevent an excessive number of overall PDCCH blind decoding candidates in carrier aggregation, higher modem processing capability is required for shorter service latency, compared to 4G LTE.

Figure 13.19 shows the timeline for a modem to complete secondary carrier activation in carrier aggregation. The time for cell search, AGC tuning, and synchronization

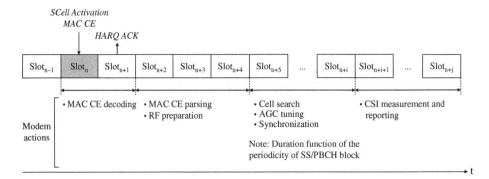

Figure 13.19 Secondary carrier activation timeline for 15 kHz subcarrier spacing.

account for the largest portion of the secondary carrier activation time. Due to sparse common pilots (e.g. SS/PBCH block) in 5G NR, it takes a longer time for a modem to get ready for signal reception, compared to 4G LTE. For example, it takes around 30 ms for a modem to activate the secondary carrier if the periodicity of the SS/PBCH block is 20 ms. Unless the modem performs CSI measurement periodically over a secondary carrier before receiving the secondary carrier activation command, it is very difficult for it to further reduce the secondary carrier activation time. However, it consumes more modem power in this way.

13.6   Summary

Compared to 4G LTE, 5G NR offers much lower latency, greater reliability, higher data rates operating in a wider bandwidth, and above all, an unprecedented level of flexibility in mobile communication systems. As a result, 5G modem designers are facing mounting challenges in keeping computational and control complexity under control, optimizing the modem architecture to save costs, and lowering the power consumption and mitigating overheating to improve the user experience.

This chapter provides a systematic analysis of the interconnection between the NR advancements and their impacts on modem design for mobile devices. One aspect that is examined is the new NR features and their requirements that raise the modem design complexity. For example, the 20-fold increase in peak data rate and a suite of new CSI feedback signals both increase computational complexity tremendously. Another aspect that is highlighted is the effort put into NR specification development to alleviate these impacts, for example, rate-compatible LDPC and BWP adaptation. Yet another aspect is hardware sharing, which, among other facets of modem architecture design, is one of those challenges that cannot be addressed by specification of a new radio technology but that are left to modem engineers to overcome. Finally, insights into the upcoming NR enhancement in UE power saving are presented. In particular, the introduction of a UE energy efficiency metric could enable the network to adapt its behavior dynamically, enabling it to provide the 5G user experience to customers without degradation of the device's battery lifetime compared to 4G.

5G NR has been engineered to address the needs, known and future, of many industries, to enable new services and experiences, through a high degree of in-built flexibility and configurability. Such flexibility allows network operators to source and deploy networks tailored to their needs. However good such flexibility is in providing a wide range of services and experiences, it is extremely difficult to accommodate it in a device's modem in a cost-effective way, for example, to properly dimension the various modules inside a modem. Yet, the final price point is critical in ensuring the wide and rapid adoption of 5G NR, especially in mass-market products such as the smartphone.

In conclusion, for 5G NR to be introduced smoothly into the market, all actors in the ecosystem, including operators, network and device vendors, and especially the modem suppliers, must work together to ensure the availability of a wide range of affordable devices served by stable and robust networks. Only with such a collaborative model can 5G really thrive and bring forth the productivity and convenience it promises to the end users.

References

1 3GPP TR 38.913 (2018). Study on scenarios and requirements for next generation access technologies. 3GPP 2018.

2 3GPP TSG-RAN R1-167531 (2016). Area, power and latency comparison for NR high throughput decoder. WG1 Meeting #86, Gothenburg, Sweden, 22–26 August 2016.

3 3GPP TS 38.331 v15.4.0 (2019). "NR; Radio Resource Control (RRC) protocol specification", 3GPP, January 2019.

4 3GPP R2-1813462 (2018). LS on the RRC message size restriction, Gothenburg, Sweden, 20–24 August 2018.

5 GTI (2018). 5G Sub-6GHZ Spectrum and Re-farming White Paper.

6 3GPP TR 37.863-01-01 v15.2.0 (2019). E-UTRA (Evolved Universal Terrestrial Radio Access) – NR Dual Connectivity (EN-DC) of LTE 1 Down Link (DL) / 1 Up Link (UL) and 1 NR band. 3GPP, January 2019.

7 3GPP TR 38.840 v1.0.0 (2019). Study on UE Power Saving; NR. 3GPP, March 2019.

14

Internet of Things Optimization

Harri Holma[1], Rapeepat Ratasuk[2], and Mads Lauridsen[3]

[1] *Nokia, Finland*
[2] *Nokia Bell Labs, United States of America*
[3] *Nokia, Denmark*

CHAPTER MENU

14.1 Introduction

Internet of Things (IoT) refers to the interconnection and autonomous exchange of data between devices which are machines or parts of machines, also called sensors and controls. IoT can also be described as a network of physical objects that are connected to the Internet. The future connected world is expected to have tens of billions of IoT devices. This chapter focuses on the cellular IoT technologies. IoT enables a huge number of use cases in the areas of homes and consumers, industries, utilities and environment, logistics and connected cars. Example use cases are illustrated in Figure 14.1: smart metering for the collection of electricity or gas meter readings, smart grid for better utilization of power networks, traffic telematics for improved road efficiency and safety, industry applications for productivity improvements, smart homes for handy control of lighting and heating, health applications for collecting data from the body, environmental measurements, smart cities for higher efficiency and safety, object tracking and agricultural applications, and preventive maintenance and smart monitoring for safety.

5G Technology: 3GPP Evolution to 5G-Advanced, Second Edition.
Edited by Harri Holma, Antti Toskala, and Takehiro Nakamura.
© 2024 John Wiley & Sons Ltd. Published 2024 by John Wiley & Sons Ltd.

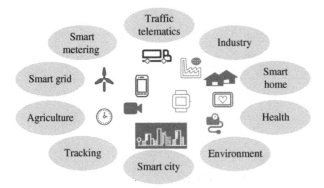

Figure 14.1 Example cellular IoT use cases.

The target of IoT optimization is to create IoT modules that can be integrated into billions of objects and provide wireless connectivity to the Internet. IoT optimization targets can be summarized as follows:

- Long battery life: Smartphones need daily charging of batteries, but many IoT devices must operate for very long times, often years, without charging. A good example is a fire alarm sensor sending data directly to a fire department. The battery change interval in such a device is a very important cost factor. The battery life should preferably be as long as the device life time since changing battery can be difficult in hard-to-access locations. A long battery life would also enable completely new connected device applications. Many objects around us currently do not have a cord, but are battery operated or even work without a battery. These devices can also be brought into the network. The target of the cellular IoT connectivity is 10 years of battery operation for simple daily connectivity.
- Low device cost: IoT connectivity will mostly serve users with a 10-fold lower revenue compared to mobile broadband subscriptions. To enable a positive business case for cellular IoT, the total cost of ownership including the device must be extremely low. The current industry target is for an IoT module cost of less than 5 USD.
- Low deployment cost: The network cost of IoT connectivity, including the initial Capital Expenditure (CAPEX) and annual Operating Expenditure (OPEX), must also be kept to a minimum. Deploying IoT connectivity on top of existing cellular networks can be accomplished by a simple, centrally pushed software upgrade, thus avoiding any new hardware, site visits, and keeping CAPEX and OPEX to a minimum.
- Simple subscription management: The mEbedded Subscriber Identity Module (eSIM) can be used to enable remote SIM provisioning of any mobile device. Remote provisioning and eSIM are expected to be important for small IoT devices.
- Full coverage: Enhanced coverage is important in many IoT applications. Simple examples are smart meters, which are often in basements of buildings behind concrete walls. Industrial applications such as elevators or conveyor belts can also be located deep indoors. This requirement has driven the IoT community to look for methods to increase coverage by tolerating lower signal strength and longer latency than is required for other devices. The target for the IoT link budget is an enhancement of 15–20 dB compared to Global System for Mobile communications

(GSM). The coverage enhancement would typically be equivalent to wall or floor penetration, enabling deeper indoor coverage.

- Support for a massive number of devices: IoT connectivity is growing significantly faster than normal mobile broadband connections, and there will be billions of connected devices over cellular IoT networks. The density of connected devices may not be uniform, leading to some cells having a very high numbers of devices connected. Therefore, IoT connectivity needs to be able to handle an extremely large number of simultaneous connected devices.

In order to realize the optimization targets, 3GPP defined new features for an enhanced IoT connection in Releases 13 and beyond. The two solutions are called LTE-M, known also as Category-M or eMTC (enhanced Machine Type Communication), and NB-IoT (Narrowband Internet of Things). Both solutions are fully integrated into LTE networks, which allows the existing LTE network equipment and LTE spectrum to be utilized. This chapter explains the IoT optimization technologies in 3GPP. The general IoT evolution in 3GPP is explained in Section 14.2. LTE-M is described in Section 14.3 and NB-IoT in Section 14. 4. Core network optimization is illustrated in Section 14.5. The link budget and coverage are presented in Section 14.6 and network capacity in Section 14.7. Power consumption minimization is discussed in Section 14.8 followed by power consumption measurements in Section 14.9. There are also alternative technologies for low-cost IoT connectivity, for example, LoRa and Sigfox. They are deployed on unlicensed spectrum and need a dedicated network which is not integrated into mobile networks. The solution benchmarking is illustrated in Section 14.10. 5G enhances IoT connectivity in particular for industrial, vehicular, and other use cases where low latency and ultra-high reliability are important. 5G IoT is presented in Section 14.11, and the chapter is summarized in Section 14.12.

14.2 IoT Optimization in LTE Radio

LTE evolution enhanced IoT capabilities by improving coverage, device power consumption, device and deployment cost, and connectivity. IoT operation on top of LTE networks was kicked off with Release 8 using Category 1 devices. LTE Release 8 baseline was not optimized for low-cost IoT, and there was a clear need to make enhancements in LTE specifications to improve IoT support. IoT optimization targets were substantially higher for coverage, power consumption, and cost. The targets are summarized in Table 14.1. The maximum path loss target is up to 20 dB higher than in LTE or LTE-Advanced. Operating time with two AA batteries should be 10 years, and the IoT device cost should be 80% lower from the chip set perspective and allow an IoT device cost of clearly less than 5 USD. The target is to keep the transaction latency below 10 s.

IoT optimization started in Release 12 with a new device category called Category 0 (Cat-0). That category was never implemented in the live networks because major IoT improvements were included in Release 13 with LTE-M and NB-IoT. Release 12 also included Power Saving Mode (PSM) which allows minimization of IoT power consumption with low activity frequency. PSM is utilized in commercial IoT devices. Release 13 included enhanced Discontinuous Reception (eDRX) for power consumption minimization, which practically refers to longer DRX cycles. Another

Table 14.1 IoT optimization targets.

	LTE-advanced	IoT optimization
Max. path loss	140–145 dB	164 dB
Operating time with two AA batteries	Less than 1 yr	10 yr
Device cost	More than 10 USD	Less than 5 USD
Maximum transaction latency	Typically <2 s	10 s

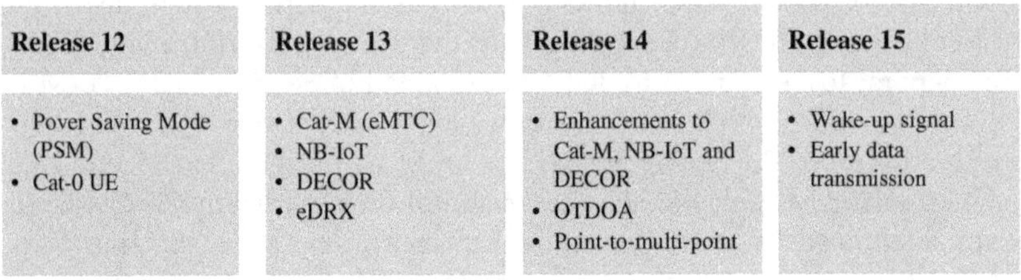

Release 12	Release 13	Release 14	Release 15
• Power Saving Mode (PSM) • Cat-0 UE	• Cat-M (eMTC) • NB-IoT • DECOR • eDRX	• Enhancements to Cat-M, NB-IoT and DECOR • OTDOA • Point-to-multi-point	• Wake-up signal • Early data transmission

Figure 14.2 Main IoT optimizations in 3GPP Releases 12–15.

improvement in 3GPP specifications was Dedicated Core network (DECOR) in Release 13. It enables dedicated core networks with specific characteristics to be assigned to certain users or devices. All these IoT enhancements turned out to be important for the industry, and therefore, further enhancements were included in Releases 14–15. One important feature in Release 14 is the Observed Time Difference of Arrival (OTDOA) positioning method, since many IoT applications need location information. Another feature is point-to-multipoint transmission, which may be used, for example, to update the firmware of multiple IoT devices simultaneously. Release 15 brought the wake-up signal for minimized power consumption during idle mode paging, and early data transmission for data transmission during the random access procedure for lower latency for small packet transmission. IoT evolution is summarized in Figure 14.2. The items are described in more detailed later in this chapter.

IoT operation on top of LTE networks started from Release 8 using Category 1 devices that were available from 2015. Cat-1 is similar to Cat-4, but the peak data rate is limited to 10 Mbps. The initial smartphones had Cat-4 modems with 150 Mbps rates, and the latest smartphones are already Cat-20 with 2 Gbps capability. The lower peak rate in Cat-1 allows some costs to be saved in the baseband. Cat-1 was simple from the network point of view because it was compatible with Release 8 and could be deployed without any network upgrades.

IoT device cost and power consumption were optimized in Release 12 with Cat-0 with a lower data rate of 1 Mbps, single-antenna reception, and half duplex RF. All earlier LTE devices were mandated to have two-antenna reception, while single-antenna reception can simplify the RF section of IoT devices. Another RF simplification was half duplex, which means that there is no simultaneous transmission and reception for the IoT device. The network has normal FDD operation with simultaneous uplink and downlink, but transmission and reception for a single IoT device are scheduled so that they do not occur at the same time. The advantage is that a half-duplex device does not need to have a duplex filter at all, reducing the cost and power consumption. Cat-0 was never implemented in practice because there were already more substantial IoT improvements in the pipeline in Release 13.

Cat-M in Release 13 brought narrowband RF with 1.4 MHz bandwidth. All earlier LTE devices had to support 20 MHz bandwidth. NB-IoT took it even further with 200 kHz bandwidth. The benefit of narrowband RF is lower cost for the implementation, reduced power consumption, and higher power spectral density for extended coverage. The 1.4 MHz bandwidth was relatively simple for LTE specifications because the smallest LTE bandwidth was 1.4 MHz already in Release 8. The 200 kHz bandwidth required more changes in the specifications because new channel definitions were included for the narrowband control and data channels. Cat-M supports a peak rate of 1 Mbps with full duplex and 375 kbps in half duplex. NB-IoT with 200 kHz bandwidth limits the data rates to lower values: 26 kbps in downlink and 62 kbps in uplink. Therefore, NB-IoT can only support those uses cases where the transmitted data volumes are relatively low. Cat-M and NB-IoT also allow a lower device output power: 20 dBm for Cat-M and later even 14 dBm for NB-IoT, while LTE smartphone categories use 23 dBm. One additional difference between the IoT UE categories is VoLTE capability: Cat-M can support VoLTE, while NB-IoT does not support VoLTE. Another difference is mobility capability: LTE-M supports handovers, while NB-IoT relies on cell reselections. IoT UE categories are summarized in Table 14.2.

Cat-M and NB-IoT are designed to be multiplexed within the LTE carrier with other LTE traffic. Both Cat-M and NB-IoT solutions were implemented and have been widely deployed in LTE networks starting 2017. IoT uses mostly sub-1-GHz bands in order to benefit from better propagation and larger coverage areas. Cat-M was typically a software upgrade to the network, while NB-IoT required new base station hardware in some cases. Cat-M usage started mainly in the United States, while NB-IoT started in Korea.

Table 14.2 IoT UE categories in Releases 8–14.

	Release 8		Release 12	Release 13		Release 14	
	Cat-4	Cat-1	Cat-0	Cat-M	Cat-NB1 (NB-IoT)	Cat-M2	Cat-NB2 (NB-IoT)
Downlink peak rate	150 Mbps	10 Mbps	1 Mbps	1 Mbps (375 kbps)	26 kbps	4 Mbps	127 kbps
Uplink peak rate	50 Mbps	5 Mbps	1 Mbps	1 Mbps (375 kbps)	62 kbps	7 Mbps	159 kbps
Number of antennas	2	2	1	1	1	1	1
Duplex mode	Full duplex	Full duplex	Half duplex	Half duplex	Half duplex	Half duplex	Half duplex
UE bandwidth	20 MHz	20 MHz	20 MHz	1.4 MHz	0.2 MHz	5 MHz	0.2 MHz
Transmit power	23 dBm	23 dBm	23 dBm	20/23 dBm	14/20/23 dBm	20/23 dBm	14/20/23 dBm
VoLTE	Yes	Yes	Yes	Yes	No	Yes	No

Cat-M deployment was slowed down in some markets because a number of legacy LTE devices turned out to be incompatible with Cat-M signaling in the master information blocks (MIBs), and all those legacy devices had to be software upgraded before Cat-M could be launched.

14.3 LTE-M

3GPP Releases 8–12 for LTE assumed that every device must support 20 MHz RF bandwidth even if the system can be deployed on smaller bandwidths down to 1.4 MHz Cat-M is the first UE category with narrowband RF. The actual transmission bandwidth is a maximum of six physical resource blocks (PRBs) or 1.08 MHz. The bandwidth of 1.4 MHz includes also guard bands. Cat-M users can be multiplexed within a normal wideband LTE carrier. A Cat-M UE can receive synchronization signals and broadcast channels from the wideband LTE carrier because those common channels are transmitted using the middle 1.4 MHz. The legacy Physical Downlink Control Channel (PDCCH) uses full-band transmission and cannot be received by Cat-M UEs. Therefore, Cat-M uses a new control channel, the MPDCCH (LTE-M PDCCH), which has 1.4 MHz bandwidth and can be located outside the middle 1.4 MHz. Cat-M UE retunes to that new frequency within the LTE carrier. The functionality is illustrated in Figure 14.3. The new control channel MPDCCH uses the resources that are normally used for the Physical Downlink Shared Channel (PDSCH). Note also that a Cat-M UE cannot receive legacy Physical Control Format Indicator Channel (PCFICH) and Physical Hybrid ARQ Indicator Channel (PHICH) that use full-band transmission. There is no need for PCFICH for Cat-M because the MPDCCH size is fixed, and there is no need for PHICH because HARQ feedback after uplink transmission is not provided at all. Instead, eNodeB informs the UE about the need for retransmission via a new data indicator. The new downlink physical channels for LTE-M are listed in Table 14.3.

The MPDCCH and PDSCH can be transmitted in different subframes. The reason is that a Cat-M UE uses only six PRBs, and there would not be much space in one subframe if some of the PDSCH area is used for the MPDCCH. Also, by allocating separate subframe after the MDCCH, the UE can have enough time to decode the PDCCH to prepare for the reception of the following PDSCH. This is called "cross-subframe scheduling."

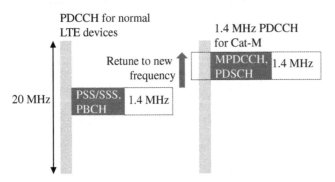

Figure 14.3 Cat-M multiplexing with legacy LTE. PSS = Primary Synchronization Signal; SSS = Secondary Synchronization Signal; PBCH = Physical Broadcast Channel.

Table 14.3 Downlink physical channels for LTE-M.

Physical channels	Channel usage with LTE-M
Primary Synchronization Signal (PSS)/ Secondary Synchronization Signal (SSS)	Fully reused by Cat-M UEs.
Cell Reference Signals (CRSs)	Fully reused by Cat-M UEs.
Physical Broadcast Channel (PBCH)	Fully reused by Cat-M UEs. Cat-M allows to use additional repetition of PBCH to improve signal-to-noise ratio.
Physical Control Format Indicator Channel (PCFICH)	Not used by Cat-M. PCFICH indicates number of PDCCH symbols for normal LTE UEs. The size of MPDCCH is given by RRC parameter in SIB1, so it is semi-static and not dynamic.
Physical Hybrid ARQ Indicator Channel (PHICH)	Not used by Cat-M. PHICH indicates retransmission information for uplink transmission. Not needed in Cat-M because the new transmission is indicated on MPDCCH.
Physical Downlink Control Channel (PDCCH/MPDCCH)	PDCCH cannot be used by Cat-M because it is wideband. Therefore, narrowband MPDCCH is needed.
Physical Downlink Shared Channel (PDSCH)	Used for Cat-M data transmission with appropriate bandwidth.
Physical Random Access Channel (PRACH)	Fully reused by Cat-M. The only difference is in the timing: it is possible to limit PRACH locations.
Physical Uplink Control Channel (PUCCH)	PUCCH for Cat-M only occupies 2 PRBs.
Physical Uplink Shared Channel (PUSCH)	Used for Cat-M data transmission with appropriate bandwidth.

However, it is still possible to allocate the MPDCCH and PDSCH in the same subframe but to different UEs. In this case, two PRBs are allocated for the MPDCCH, and the remaining four RBs are allocated for the PDSCH.

The possible locations of Cat-M within the 10 MHz LTE carrier are illustrated in Figure 14.4. There are a total of eight possible locations, each of which has six PRBs, called narrowband (NB) locations. The first and the last resource blocks cannot be used by Cat-M allocations. A Cat-M UE is instructed as to which NB location is to be used in the SIB. The information indicates the index of the narrowband MPDCCH location. A similar structure is also defined for the other LTE bandwidths: 20 MHz LTE has correspondingly 16 NB locations, and 5 MHz has 4 locations.

Cat-M includes two modes of coverage enhancements:

- Mode A. A small amount of repetition or no repetition, full mobility support, and VoLTE support. Operation in coverage enhancement Mode A would have an equivalent coverage as that of a Category 1 UE. The difference in coverage between LTE-M and a Category 1 UE lies in the fact that LTE-M uses only 1 Rx, 6 PRB narrowband transmission, and reduced uplink transmit power. The reduced uplink power in LTE-M is compensated by utilizing repetition coding.

- Mode B. This mode uses large repetitions and has limited mobility support. There are up to 15 dB coverage enhancements. The transmission power for both the Physical Uplink Control Channel (PUCCH) and Physical Uplink Shared Channel (PUSCH) is set to be MAX. Since the power does not change (i.e. no power control is performed) DCI formats 6-0B and 6-1B do not carry a transmission power control field. This mode is designed to be used in very poor propagation conditions.

Repetition is the main solution for coverage enhancements. The target of the repetition is to transmit more energy: a repetition of 2 theoretically gives a better coverage of 3 dB, and a repetition of 10 gives 10 dB. The maximum allowed repetition depends on the mode and channel type. Mode A allows a maximum of 32 repetitions for the PDSCH and PUSCH, while Mode B allows up to 2048 repetitions for those data channels. The selection between Mode A and B is controlled by eNodeB according to the UE measurements. In general, a UE is kept in Mode A unless it is experiencing very poor coverage.

Other Cat-M solutions for improving coverage are frequency hopping, multi-subframe channel estimation, and dynamic frequency domain scheduling. Frequency hopping means that the UE can hop from one narrowband to another for frequency diversity gain. Multi-subframe channel estimation refers to the case where the UE stays in one narrowband for a fixed amount of time prior to hopping. Dynamic frequency domain scheduling means that a UE can be scheduled in any narrowband location.

The instantaneous peak rate with Cat-M is 1 Mbps, while the sustained peak rate is lower. A full-duplex UE can receive in eight subframes out of 10 ms since the maximum number of HARQ processes is 8. That gives a peak rate of 800 kbps. The half-duplex UE data rate is lower because simultaneous uplink and downlink is not supported. The half-duplex peak rate in the downlink is 300 kbps. The corresponding peak rates in the uplink are 1 Mbps with full duplex and 375 kbps with half duplex. The downlink transmission timing is shown in Figure 14.5 and the peak rates in Table 14.4.

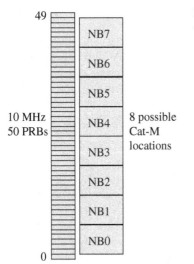

Figure 14.4 Possible Cat-M locations within 10 MHz LTE carrier.

Full duplex

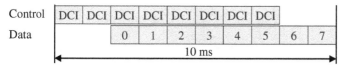

Half duplex

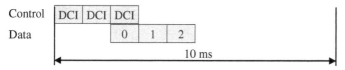

Figure 14.5 Full-duplex and half-duplex transmissions (DCI = Downlink Control Information).

Table 14.4 Cat-M peak data rates.

	Downlink	Uplink
Instantaneous peak	1 Mbps	1 Mbps
Sustained peak rate with full duplex	800 kbps	1 Mbps
Sustained peak rate with half duplex	300 kbps	375 kbps

3GPP Release 14 introduced enhancements to Cat-M including a new Cat-M2 UE category supporting higher data rates, multi-cast support using single-cell point-to-multipoint transmission, and enhancements for positioning.

14.4 Narrowband-IoT

NB-IoT takes the cost optimization even further than Cat-M. It is also known as Cat-NB. While LTE-M supports 1.4 MHz, NB-IoT only needs to support 200 kHz, which leads to a simpler implementation, even better coverage, and a lower data rate. NB-IoT is still designed to be deployed together with LTE. There are three options regarding how NB-IoT can be used with LTE. The in-band option has the NB-IoT carrier located within the LTE carrier. Most of the first NB-IoT deployments are based on the in-band option. Guard band options place the NB-IoT carrier just next to the LTE carrier within the guard band. Note that 10% of the LTE channel spacing is reserved for guard bands. 10 MHz LTE has a total of 1 MHz allocated for guard bands. The guard band option is the target deployment for many NB-IoT cases because it gives NB-IoT capability without sacrificing any existing LTE spectrum. The standalone option is used for those cases where NB-IoT is deployed together with GSM or WCDMA without any LTE carrier on the same spectrum. NB-IoT can replace one GSM carrier because both have 200 kHz bandwidth. NB-IoT can be deployed at the edge of the WCDMA carrier because the WCDMA transmission bandwidth is 3.84 MHz, while the typical channel spacing is 5 MHz. The deployment options are shown in Figure 14.6.

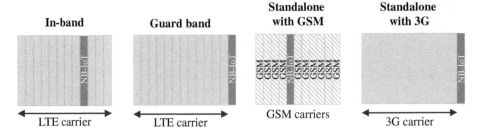

Figure 14.6 NB-IoT deployment options.

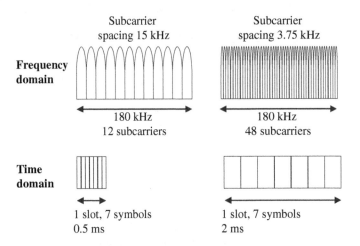

Figure 14.7 Uplink subcarrier spacings (SCSs) and frame structure.

NB-IoT allows the normal LTE subcarrier spacing of 15 kHz to be used or another option with 3.75 kHz subcarrier spacing in the uplink. The target of 3.75 kHz option is to further enhance the coverage. NB-IoT uplink has two further options:

- Single tone transmission, which is mandatory for NB-IoT UE. A single subcarrier of 15 or 3.75 kHz is used. The slot duration is 0.5 or 2 ms. See Figure 14.7.
- Multi-tone transmission, which is optional for NB-IoT UE. The number of subcarriers is 3, 6, or 12, and the subcarrier spacing is 15 kHz. The slot duration is 0.5 ms.

Common channels on an LTE carrier have a transmission bandwidth of 1.4 MHz. Those channels cannot be utilized by NB-IoT with only 200 kHz RF bandwidth. Therefore, new physical channels will be required by NB-IoT, including Narrowband Primary and Secondary Synchronization Signals (NPSS/NSSS), Narrowband Reference Signals (NRSs), and Narrowband Physical Broadcast Channel (NPBCH). The new physical signals and channels are listed in Figure 14.8, and the common control locations within 1 PRB are shown in Figure 14.9. The common control channels take 5 subframes out of 20. The NPBCH is based on the LTE Physical Broadcast Channel (PBCH) with a transmission every 10 ms in subframe #0. The NPBCH consists of eight independently decodable blocks spanning 80 ms. The NPDCCH is based on the LTE PDCCH with a maximum aggregation level of 2 and repetitions for enhanced coverage. Three new DCI formats are

Figure 14.8 New physical signals and channels for NB-IoT.

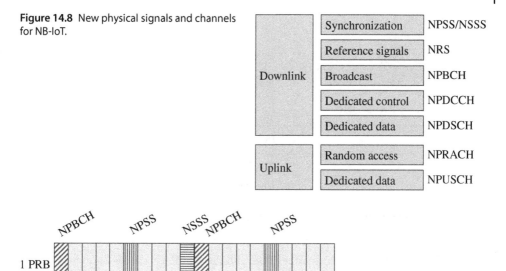

Figure 14.9 Physical channel locations.

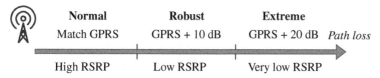

Figure 14.10 Coverage extension levels.

defined to support NB-IoT. The NPDSCH is based on the LTE PDSCH with one transport block that may be mapped to 1–10 subframes. The maximum NPDSCH transport block size is 680 bits. Single-process HARQ for the PDSCH is realized by adaptive and asynchronous timing transmission.

Three different coverage levels are signaled via the System Information Block Narrow Band (SIB-NB): normal, robust, and extreme. The coverage level is defined by Narrowband Reference Signal Received Power (NRSRP) thresholds. The thresholds depend on the deployment environment and on the system configuration. The number of repetitions and network parameters can be selected for each class. The target of the normal coverage level is to match General Radio Packet Service (GPRS) coverage, while extreme coverage will exceed GPRS coverage by 20 dB (Figure 14.10).

The Narrowband Physical Random Access Channel (NPRACH) is based on single tone transmission with frequency hopping. It uses a subcarrier spacing of 3.75 kHz, and the cyclic prefix length is designed for cell sizes up to 40 km. The NPUSCH has one resource unit schedulable for data transmission. The length with 15 kHz is 1 ms for 12 tones, 2 ms for 6 tones, 4 ms for 3 tones, and 8 ms for a single tone, and with 3.75 kHz 32 ms for a single tone. One transport block may be mapped to 1–10 resource units. The maximum TBS size for the NPUSCH is 1000 bits, and 1 HARQ process is used.

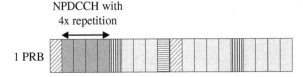

Figure 14.11 NPDCCH repetition.

Table 14.5 NB-IoT peak data rates.

	Downlink	Uplink
Instantaneous peak	170 kbps	250 kbps
Sustained peak rate without considering common channels	26.2 kbps	62.5 kbps
Sustained peak when common channel overhead is considered	19.6 kbps	56.3 kbps

All NB-IoT channels can utilize repetition, and the repetition can be configured individually per channel. Each repetition is self-decodable, and one ACK is used for all repetitions. An example repetition of four for the NPDCCH is illustrated in Figure 14.11. The repetition factor can be much higher: the NPDCCH allows repetitions up to 128, and the NPDSCH even allows 2048 repetitions. Modulation can only be QPSK or BPSK for the single tone case. No higher order modulation is supported.

Multiple NB-IoT carriers can be supported for higher IoT capacity. One carrier will be defined as the anchor carrier. This carrier contains NPSS/NSSS/NRS/NPBCH and SIB transmissions. Additional PRBs are configured by RRC signaling. An idle UE camps on the NB-IoT carrier on which the UE has received NPSS/NSSS, NPBCH, and SIB transmissions. An RRC-connected UE can be configured, via UE-specific RRC signaling, to another carrier, for all unicast transmissions. Both anchor and non-anchor PRBs shall be located outside the innermost PRB zone (in-band case).

The instantaneous peak rate with the NB-IoT downlink is 170 kbps, while the sustained peak rate is 26.2 kbps without considering the NPBCH/NPSS/NSSS overhead, and 19.6 kbps when 25% overhead is taken into account. The uplink instantaneous peak data is 250 kbps with 12 tones. The sustained peak rate is approximately 62.5 kbps without considering NPRACH overhead, and 56.3 kbps when 10% overhead is taken into account (Table 14.5).

3GPP Release 13 brought the RRC suspend and resume feature, which is similar to RRC Inactive in 5G radio. Release 14 introduced enhancements to NB-IoT including a new Cat-NB2 UE category supporting higher data rates, supporting two HARQ processes, multi-cast support using single-cell point-to-multipoint transmission, support for positioning including Enhanced Cell Identity (E-CID) and OTDOA, and support for paging and random access on non-anchor carriers.

14.5 IoT Optimization in LTE Core Network

IoT optimization in 3GPP is not limited to radio networks, but also there are optimizations included for the core network. The core network improvements can be motivated with the following reasons:

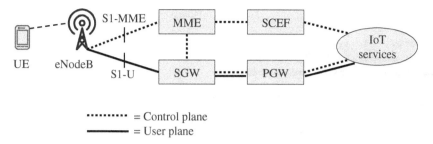

Figure 14.12 Network architecture for IoT optimization.

- IoT traffic may be very latency tolerant, such as utility reading that can tolerate a latency of minutes or even hours. Mobile broadband traffic is latency critical and needs to be scheduled immediately.
- IoT traffic may be very bursty. Monitoring of certain events or sensors may be correlated. If there is a major power outage, many IoT sensors may report the event simultaneously.
- An interface exposing the IoT data to an application platform needs to be provided.
- A dedicated MME may be required for subscription management as the pricing and changing models for IoT may be very different compared to voice and data.
- There may be large number of similar subscriptions, for example, millions of utility meters. Common subscription and signaling optimization may be useful for those subscribers.
- Group-based paging may be needed to optimize signaling in the network.

IoT optimization is done in the control and user planes. In the control plane optimization, uplink data can be transferred from eNodeB to MME. Data are further transferred either via the Serving Gateway (SGW) to the Packet Data Network Gateway (PGW), or to the Service Capability Exposure Function (SCEF), which, however, is only possible for non-IP data packets. The uplink data is then forwarded to IoT application services. The downlink data transmission uses the same delivery paths in the opposite direction. The IoT control plane solution does not need a data bearer setup, because data packets are sent on the signaling bearer. This control plane solution is most appropriate for the transmission of infrequent and small data packets. SCEF is a new node designed especially for machine-type data. It is used for delivery of non-IP data over the control plane and provides an abstract interface for the network services including authentication and authorization.

On the user plane optimization, data are transferred in the same way as the normal mobile broadband data traffic using data radio bearers via the SGW and PGW to the application server. There is some overhead caused by the connection setup. The relative signaling overhead is higher if the data volume is very low. User plane transmission supports both IP and non-IP data delivery. The system architecture is illustrated in Figure 14.12.

14.6 Coverage

Coverage optimization is one of the targets in LTE IoT design. Coverage can be measured as the maximum allowed path loss between the device and base station

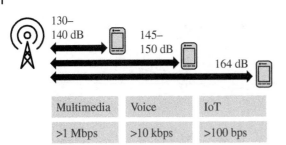

Figure 14.13 NB-IoT coverage target.

antennas. A good-quality data connection can be maintained with path loss values up to 130–140 dB and low-data-rate voice up to 145–150 dB. The target for IoT connectivity is a path loss up to 164 dB for a very-low-data-rate connection of only 100 bps = 0.1 kbps. The coverage extension is illustrated in Figure 14.13. In theory, the path loss can be improved by 20 dB when the data rate is reduced by 20 dB, that is, by a factor of 100. The practical solutions for coverage extension are narrowband transmissions with power boosting, repetition, and retransmission.

Link budgets are needed to estimate the maximum allowed path loss. The path loss value can then be converted into the maximum cell radius based on the propagation models. We use the following assumptions for the link budget calculations:

- The base station transmission power is assumed as 32 dBm per 200 kHz for LTE-M corresponding to 40 W for 10 MHz, which is a typical case for low band LTE FDD. 35 dBm per 200 kHz is assumed for NB-IoT including power boosting.
- The UE transmission power is assumed as 23 dBm.
- The base station noise figure is assumed as 3 dB and for the UE 5 dB.

  The IoT data rate is relatively low when using extreme coverage. Figure 14.14 illustrates the IoT data rate as a function of the path loss. LTE-M can support up to 161 dB path loss in the uplink, which is 5 dB more than the 3GPP target of 156 dB. NB-IoT can support even up to 167 dB path loss. We note that the data rate becomes very low at high path loss values. We also note that the uplink coverage is still the limiting factor. The downlink data rate at 156 dB is 20 kbps with LTE-M and 5 kbps with NB-IoT, while the uplink data rate is just 0.8 kbps. The data rate at 164 dB is 0.8 kbps downlink and 0.3 kbps uplink with NB-IoT [1]. The data rate capability needs to be considered from the application development point of view: a low data rate allows sending of only low amounts of data with IoT modules.

14.7 Delay and Capacity

High capacity is one of the targets in IoT optimization. The technology should be designed so that a very large number of IoT modules can be connected to the network. This section illustrates the maximum number of IoT modules that can be supported by the LTE-based IoT solutions. The simulations assume low band LTE deployment at 800 MHz with 10 MHz bandwidth. LTE-M capacity allocation is assumed to be six PRBs and NB-IoT one PRB. A macrocell network is assumed with three different levels of indoor penetration loss: 10, 20, and 30 dB corresponding to close to a window,

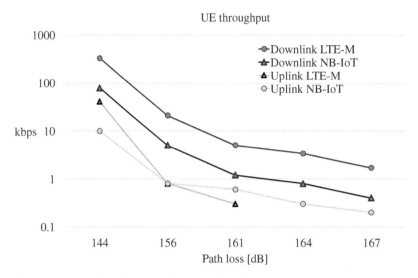

Figure 14.14 IoT throughput as a function of path loss.

average, and deep indoors like a basement. The macrocell network is based on a real operator's network configuration in terms of site location, antenna tilt, and direction. IoT activity is assumed to be one transaction per day. The capacity results can be scaled down if the transmission frequency is higher. Two applications are considered in the evaluation: application 1 with 4 uplink and 4 downlink transmissions and application 2 with 1 uplink transmission and acknowledgment in the downlink. The transmission payload is 128 bytes.

First, we estimate the session delay, which is a relevant factor in the system capacity evaluation. Figure 14.15 shows the estimated session delay for the different user

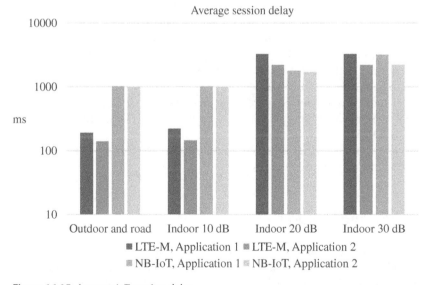

Figure 14.15 Average IoT session delay.

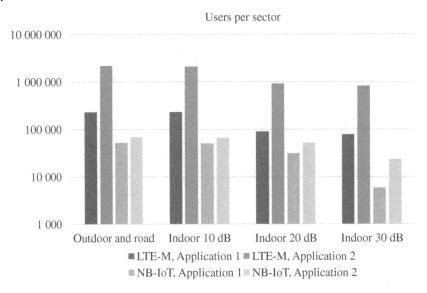

Figure 14.16 Number of supported IoT devices per sector.

groups applying the two applications. The session delay includes payload transfer, synchronization time overhead, and overhead due to RRC signaling. We note the delay is lower in a good signal than in a deep indoor environment due to the higher data rate. LTE-M shows a lower delay than NB-IoT because of the higher data rate. The delay for LTE-M is less than 300 ms and for NB-IoT approximately 1 s in a good signal environment. The delay with LTE-M and NB-IoT is similar with 20–30 dB indoor penetration loss, and the value is 1–4 s. Application 2 shows slightly lower delays than application 1 because of less data transmission. For more details, see [2].

Next, the number of supported IoT users per sector is calculated, based on the path loss values and on the estimated delays. LTE-M supports 5–8 times more users than NB-IoT, which is expected because six times more PRBs are allocated for LTE-M. The capacity in good signal areas is more than 200 000 users per sector with LTE-M for application 1 and even more than 1 million users per sector for application 2. That is an impressive capacity. NB-IoT can support more than 50 000 users per sector. Dense networks tend to have more than 30–50 sectors per km^2, which indicates that NB-IoT can support even 1 million IoT devices per km^2. If all users experience 20–30 dB indoor penetration loss, the capacity is lower because each transaction occupies more time and resources (Figure 14.16).

LTE IoT capacity can be further enhanced by allocation of more resources for IoT transmission with more PRBs and more frequencies.

14.8 Power Saving Features

In order to achieve a 10-year battery life for IoT modules, the power saving sleep mode must be efficiently utilized, which means the UE should wake up only very seldom to receive paging messages. Release 12 provides PSM, where the UE wakes up to perform

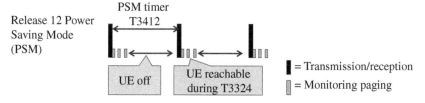

Figure 14.17 Power Saving Mode (PSM) in Release 12.

tracking area updates (TAUs) periodically. After performing the TAU, the UE will then remain in idle mode and monitor paging for some time, specified by T3324, which allows the network to send data to the UE. If no transmission need be sent to the UE, the UE will go to sleep mode until the next TAU. The sleep time is the difference between the PSM timer T3412 and the idle mode timer T3324. During the sleep time, no data transmission from the network to the UE is possible. It is possible to set T3324 = 0 such that the UE enters PSM directly. The PSM solution is illustrated in Figure 14.17. In comparison with powering the UE completely OFF, the UE using PSM avoids the detach procedure, and it may potentially also store system information, which makes the connection reestablishment faster.

Release 13 allows the use of the longer Discontinuous Reception (DRX) to minimize the device power consumption. The feature is called extended DRX (eDRX). The maximum DRX cycle in Release 12 was 2.56 s both in idle and in connected modes. The cycle was increased to 10.24 s in the connected mode and even 44–175 min in the idle mode. The benefit of the Release 13 eDRX solution is that the network can more frequently send data to the UE compared to PSM. It provides a lower application layer latency at the cost of higher energy consumption. The concept is described in Figure 14.18, and the maximum DRX cycle lengths are listed in Table 14.6.

Network Access Stratum (NAS) and RRC states with PSM are shown in Figure 14.19. The UE remains RRC idle in the eDRX state, while it does not have any RRC state in PSM.

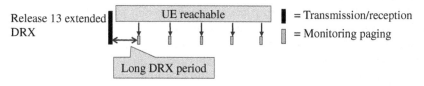

Figure 14.18 Long DRX period in Release 13.

Table 14.6 Maximum DRX cycle lengths.

	Release 8–12 DRX	Release 13 eDRX	Release 12 PSM
Connected mode	2.56 s	10.24 s	N/A
Idle model	2.56 s	43.69 min (2621.44 s) for Cat-M 174.76 min (10 485.76 s) for NB-IoT	413 days (9920 hr) applicable to both Cat-M and NB-IoT (timer T3412)

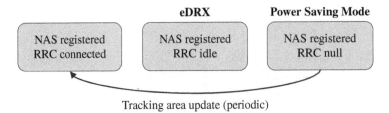

Figure 14.19 NAS and RRC states with eDRX and power saving mode (PSM).

14.9 NB-IoT Power Consumption Measurements

In 2018 and 2019, multiple commercial NB-IoT networks and UEs were launched. Therefore, the current state of the art in terms of NB-IoT UE power consumption can be measured, and using the live NB-IoT network configuration, the battery lifetime can be estimated.

The power consumption of three commercial NB-IoT UEs (Release 13) is listed in Table 14.7. The power consumption was measured in four power states using a lab test bed consisting of a Keysight UXM NB-IoT base station emulator and a power analyzer [3]. Active transmission at maximum output power is the most power consuming state, while reception requires three to six times less power. Clearly, it is important to use Connected Mode DRX during the data transfer states since it further reduces the power consumption to one tenth of the reception power state. Finally, the PSM power state has proved to be a valuable addition to the LTE IoT technologies as it is roughly 1000 times less power consuming than Connected Mode DRX.

The battery lifetime estimation is based on lab measurements, but to ensure a realistic setup of the base station emulator, the network configuration of a live NB-IoT cell in Denmark was observed and used as baseline configuration. Specifically, the network configuration was collected as a function of the coverage level. This was achieved by placing a UE in a shielded box and connecting it to an outside antenna using a step attenuator. By varying the signal attenuation, the configuration for each of the three coverage enhancement levels (ECLs) was captured in [3].

The estimated battery lifetime is illustrated in Figure 14.20 as a function of the payload size, packet interarrival time, and ECL. The estimate is based on device A of Table 14.7 and a battery capacity of 5 W, in accordance with 3GPP assumptions. In good coverage conditions (ECL0), a battery lifetime target of 10 years is achieved for both payloads with up to 4–5 updates per day. In average coverage conditions (ECL1), the 20 B payload can be sent every 10 h and still fulfill the battery lifetime target, while the larger packet

Table 14.7 Measured power consumption for three commercial NB-IoT devices data from [3].

Power state	Device A	Device B	Device C
Transmitting at 23 dBm	765 mW	731 mW	1030 mW
Receiving	242 mW	215 mW	168 mW
Connected DRX	29.1 mW	17.8 mW	17.7 mW
Power saving mode (PSM)	11.1 µW	14.1 µW	24.3 µW

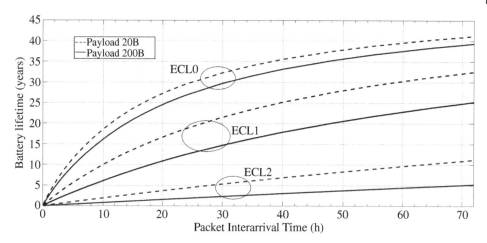

Figure 14.20 Estimated battery lifetime for device A as a function of packet interarrival time and coverage level based on data from [3].

may only be sent once every 18 h. In the most challenging coverage (ECL2), the battery lifetime target is very challenging due to the use of repetitions across all radio channels. If the 20 B payload is only updated every third day, it is achievable, and this may be a feasible solution, for example, for smart meters. Other event-driven use cases may also cope with this signaling limitation.

The results show that LTE IoT technology has great battery life as long as the transmission frequency remains relatively low and the coverage is good. If coverage extension is needed for NB-IoT, the battery life is reduced considerably. The further enhancements in Releases 14–15, for example, Early Data Transmission, will improve the battery lifetime, but for now it is important to match the traffic profile and the offered radio coverage with battery and latency targets, as major trade-offs are observed.

Compared with Cat-M measurements [4], it is observed that PSM, C-DRX, and transmit power consumption levels are similar for Cat-M and NB-IoT for multimode devices, while NB-IoT dedicated devices consume up to 50% less for cDRX and reception. The wider bandwidth of Cat-M seems costly in terms of instantaneous power consumption; however, the times to synchronize and actively receive are correspondingly shorter for Cat-M. Thus, it is difficult to determine which technology consumes the least energy – it depends on the application layer payload, latency requirement, and coverage level.

14.10 IoT Solution Benchmarking

IoT connectivity is the baseline for a number of new applications in the future connected world including wireless factories, smart cities, and homes. Some applications require reliable connectivity, low latency, flexible scalability, and high security. On the other hand, other use cases need low-cost devices with low power consumption. Wi-Fi networks at 2.4 and 5 GHz have been traditionally used for the local areas. There are also other options besides Wi-Fi for IoT connectivity: LTE Cat-1, Cat-M, NB-IoT, 5G, Bluetooth, and Zigbee. The radio technology options are shown in Figure 14.21.

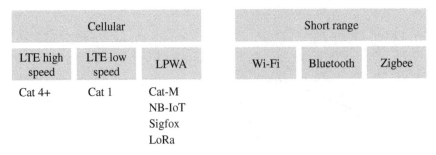

Figure 14.21 Main radio technology options for IoT connectivity.

The technologies differ in terms of data rate, latency, coverage, capacity, interference tolerance, mobility, cost, power consumption, and deployment options. Wi-Fi, Bluetooth, and Zigbee are limited to local area cases with short distances. This section focuses on Low Power Wide Area (LPWA) technologies, which are mainly LTE-based standardized solutions, LoRa and Sigfox.

LTE has been successfully utilized by mobile operators for smartphones on the licensed spectrum. LTE technology was enhanced for IoT optimization with Cat-M and NB-IoT for low-cost and low power consumption devices. Both solutions have been commercially deployed during 2018 and 2019 with more than 100 commercial networks by 2019.

LoRa and Sigfox operate in the unlicensed Industrial, Scientific, and Medical (ISM) band which is 868 MHz in Europe and at 915 MHz in the United States. Since the frequency is similar to that of low LTE bands, the propagation is similar, while the ISM band has limitations in terms of transmission power and activity factor. Also, the mode of operation is different compared to LTE. Sigfox is a global operator, while LoRa can be self-owned or a hybrid solution.

Figure 14.22 illustrates the technology positions in terms of range and data rate, and Figure 14.23 in terms of price and power consumption. Wi-Fi- and LTE-based solutions offer the highest radio capability, but also highest cost and power consumption. Wi-Fi-based solutions have a high data rate and low latency combined with wide availability and lower cost compared to LTE-based solutions. Bluetooth is optimized

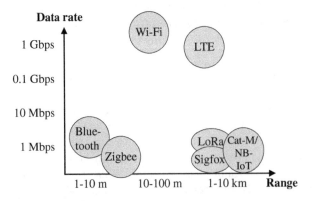

Figure 14.22 Range and data rate of IoT technologies.

Figure 14.23 Power consumption and price of IoT technologies.

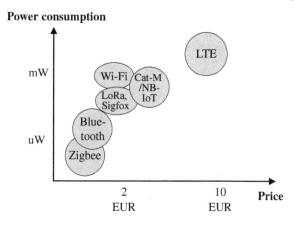

for lower cost and lower power while providing lower data rates than Wi-Fi and LTE. Zigbee goes even further into lower power consumption, lower cost, and lower data rate. LTE-based NB-IoT solutions can already compete with Wi-Fi solutions in terms of size and power consumption, and also in terms of cost. LoRa and Sigfox have similar data rate capabilities as NB-IoT, while the range is shorter due to the ISM band limitations. LoRa and Sigfox power consumption is on the same level as in NB-IoT, while the price is lower than that of NB-IoT. Note that the power consumption naturally depends heavily on the applications and configurations, and the price depends on the volumes.

14.11 IoT Optimizations in 5G

5G New Radio (NR) was standardized in 3GPP in Release 15. 5G was designed to support diverse requirements. Three important 5G use cases are enhanced mobile broadband (eMBB), ultra-reliable low latency communications (URLLCs), and massive machine type communications (mMTCs). Enhanced mobile broadband addresses high-data-rate and low latency services. URLLC addresses services requiring ultra-low latency and ultra-high reliability, while mMTC addresses delay-tolerant and low-data-rate IoT services. This allows for network deployment with scalability, configurability, ease of deployment, and lower cost. The various IoT verticals can be served using the appropriate 5G solutions according to their respective service requirements. Some examples of 5G IoT verticals and use cases are shown in Table 14.8. For instance, in a smart factory, delay-tolerant IoT devices such as sensors and asset trackers can be supported using mMTC, while mission-critical IoT components such as robotic control or augmented reality display require URLLC. Applications that require large bandwidth and low latency such as remote diagnostic or video camera can be supported using eMBB.

In term of URLLC requirements, 5G will support a user plane latency of 1 ms at high reliability. The intended use cases are for applications requiring very low latency and high reliability such as telemedicine, robotics, and industrial sensors. For mMTC, 5G will support the same targets as for eMTC and NB-IoT. They are ultra-low-complexity and low-cost IoT devices and networks with a maximum coupling loss (MCL) of 164 dB for a data rate of 160 bps at the application layer, connection density of 1 million devices

Table 14.8 Examples of IoT use cases.

	Automotive	Smart Home	Smart City	eHealth	Smart Factory
eMBB	Assisted driving, infotainment	Gaming, remote computing	Video surveillance	Remote diagnostics	Remote diagnostics, video
mMTC	Remote diagnostics, traffic management, insurance sensors	Smart meters, home security, connected appliances	Remote sensors, metering, fleet management, asset tracking, traffic management	Patient tracking, wearables, asset tracking	Remote sensors, security, asset tracking
URLLC	Automated driving, public safety	Virtual reality gaming, remote office	Drone, tactile Internet	Telemedicine, remote surgery	Industrial IoT, time-sensitive network

per km^2 in an urban environment, battery life in extreme coverage beyond 10 years, and latency of 10 s or less on the uplink to deliver a 20-byte application layer packet measured at 164 dB MCL.

The flexible 5G frame structure allows different services to be offered using the same carrier. For instance, mMTC can be offered at the same time as URLLC by supporting different slot sizes. Massive MTC services can use the 1 ms slot size, which minimizes overhead and maximizes coverage, while URLLC services can use a small slot size (as low as 0.125 ms) to minimize latency. In addition, 5G offers a lean carrier design with no cell-specific reference signal and flexible cell-specific signaling such as synchronization signals and broadcast channel. This enables flexible segmentation of the 5G carrier to simultaneously support different services.

In addition to physical layer optimization for IoT, 5G also supports network layer optimization such as edge cloud and network slicing. The edge cloud distributes computing capabilities across the network to where the traffic is. This allows cloud implementation to run at the edge of the network, which greatly benefits latency and efficiency. In addition, the edge cloud allows processing of a large amount of data at the edge without having to transport them to the centralized servers. For example, in an industrial IoT private factory network, having an edge cloud at the factory can allow for fast video analytics of the production process. Network slicing allows multiple logical networks with different end-to-end performance characteristics to be supported using the same physical infrastructure. Network slices can be independently configured and managed. Some of the performance attributes that can be configured per network slice include latency, throughput, reliability, capacity, mobility, security, analytic, and cost profile. This makes it ideal to offer different services, for example, for different business purposes or customers, within one 5G carrier.

To support ultra-low latency for URLLC, 5G has specified support for the following techniques in Release 15:

- Short slot length via higher subcarrier spacing as shown in Figure 14.24: This reduces latency due to a shorter slot length, reduces the wait for the next slot, and provides

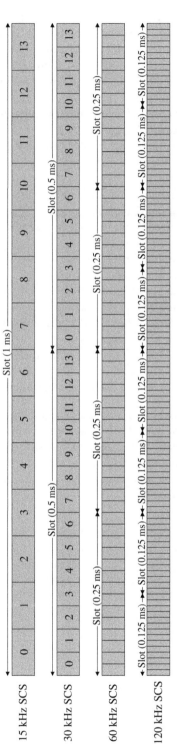

Figure 14.24 5G slot length for different subcarrier spacings (SCSs).

a faster turnaround time. Note that 5G can support multiple slot lengths in the same carrier, enabling different service mixes to be supported simultaneously.

- Data transmission shorter than a slot (i.e. using 2, 4, or 7 symbols in the downlink, and any number of symbols in the uplink). This allows for a packet to be transmitted in a very short time, processed quickly at the receiver, and the acknowledgment transmitted back to the transmitter.
- Flexible slot structure in TDD (i.e. a slot containing symbols for both data and HARQ feedback) to reduce the turnaround time.
- Preemptive scheduling and multiplexing of URLLC data within ongoing eMBB transmission: gNodeB (i.e. 5G base station) can preempt (i.e. puncture) ongoing data transmission with URLLC transmission.
- Preconfigured grant uplink transmission: UEs are assigned a preconfigured grant for uplink transmission and do not have to transmit a scheduling request to the gNodeB. The periodicity of the configured resources can be 2 symbols, 7 symbols, 1 slot, …, up to 5120 slots.
- Short uplink control transmission format (i.e. one or two symbols) to minimize the time to transmit HARQ-ACK feedback.
- Short periodicity for a scheduling request where the UE can be configured with a scheduling request slot as often as every 2 symbols.
- Multi-slot repetition (2, 4, or 8 repetitions) for data channels to allow retransmissions without waiting for HARQ-ACK feedback.
- Shorter UE processing time: 5G design facilitates UE pipelining processing to shorten the UE processing time. This is done by transmitting the DMRS at the beginning of the slot and using frequency-first mapping, which allows symbol-by-symbol processing.

Table 14.9 illustrates an example of the timing delay for transmitting a packet. After packet arrival at the gNodeB, the packet can be transmitted in 8 OFDM symbols for the initial transmission. If a retransmission is needed, the entire process will consume 20 symbols. For a subcarrier spacing of 30 kHz, this corresponds to a total time of 0.7 ms.

To support ultra-reliable packet reception for URLLC, 5G has specified support for the following techniques in Release 15:

Table 14.9 Example timing delay for downlink packet transmission.

Event	Time (OFDM symbols)
Packet arrival	—
gNodeB processing time and scheduling delay	5
Scheduling grant (PDCCH)	1
Data transmission (PDSCH)	2
UE processing time	4
HARQ-ACK (PUCCH)	1
gNodeB processing time and scheduling delay	4
Scheduling grant for retransmission (PDCCH)	1
Data retransmission (PDSCH)	2

- Multi-slot repetitions for the data channels and low code rate for the control channel to improve reliability.
- Enhanced channel state reporting with corresponding modulation and coding level to support 99.999% reliability.
- PDCP layer data duplication which allows a packet to be transmitted from two different carriers.

Further URLLC enhancements are being studied in Release 16. They include downlink control channel enhancements (e.g. compact scheduling grants, additional repetitions, increased monitoring capabilities), uplink control information enhancements (e.g. enhanced HARQ, CSI feedback), uplink data channel enhancements (e.g. mini-slot level hopping), enhancements to scheduling/HARQ/CSI processing timeline, and enhanced uplink configured grant transmission.

In addition, 3GPP is working on supporting time-sensitive networks (TSNs) with 5G. This would replace industrial Ethernet connection with a wireless system that can support time synchronization of machines and deterministic packet delivery. This requires, for example, support of Ethernet over 5G including transporting Ethernet frames over 5G, support of broadcast packets, and automatic address discovery. In addition, QoS requirements for TSNs may be more stringent than for URLLC. For instance, TSNs may require even greater reliability (e.g. 99.9999% reliability instead of 99.999%), and some services may also have a strict jitter requirement. Examples of some TSN services include process automation and electricity distribution.

Evaluations of eMTC and NB-IoT demonstrated that they can satisfy mMTC 5G requirements. Thus, 3GPP agreed not to specify 5G-based technology for mMTC, but to reuse eMTC and NB-IoT technologies to support mMTC services. This also maintains support for legacy 4G eMTC and NB-IoT devices as LTE systems are upgraded to 5G. Many eMTC and NB-IoT devices have a very long lifetime (e.g. water/power meters can have a lifetime of 10–15 years or more), and this allows IoT devices to continue to be supported as part of 5G system migration. Figure 14.25 illustrates eMTC and NB-IoT deployment within an 5G carrier to support mMTC services. 3GPP has evaluated such deployment scenarios and concluded that these scenarios are fully supported by 3GPP.

From a numerology perspective, 5G supports different subcarrier spacing values, while eMTC supports only 15 kHz subcarrier spacing and NB-IoT supports 15 and 3.75 kHz subcarrier spacing. 5G can be deployed with 15 kHz subcarrier spacing, which

Figure 14.25 eMTC and NB-IoT deployment within 5G carrier.

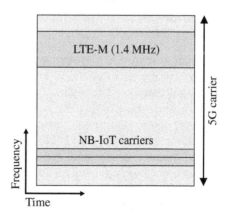

would be identical to eMTC/NB-IoT and allows for subcarrier alignment. When 5G is deployed with a different subcarrier spacing, however, it may not be possible to align subcarriers for both 5G and eMTC/NB-IoT. In this case, 5G supports multiple bandwidth parts within the same carrier, where each bandwidth part can be configured with a different subcarrier spacing. This allows eMTC/NB-IoT to easily coexist with 5G even if they employ different numerology.

In addition to supporting different bandwidth parts, 5G also support a resource reservation scheme. In this case, 5G can reserve time and frequency resources for deployment of eMTC and NB-IoT. This is done using a PRB-level bitmap in the frequency domain and a symbol-level bitmap with repetition pattern in the time domain. 5G UEs will ignore these reserved resources. This allows for efficient coexistence of eMTC/NB-IoT and 5G. For example, 5G resources can be reserved for always-on eMTC and NB-IoT signals and channels such as reference signals and broadcast channels. In addition, when there is no eMTC/NB-IoT data traffic, the gNodeB can schedule unused eMTC/NB-IoT resources to the 5G UE.

Sixteen frequency bands have been specified in 3GPP that can support both 5G and eMTC, for example, Band 8 at 900 MHz. Twelve different frequency bands have been specified that can support both 5G and NB-IoT. So far, only 5G bands below 2.7 GHz support eMTC and NB-IoT deployment as high-frequency bands are not suitable for IoT services due to limited coverage. Almost all common 5G and eMTC/NB-IoT bands share a common 100 kHz raster, so the raster grid for the two systems can be aligned. Band 41, however, supports the 5G channel raster of 15 and 30 kHz, so there are limited locations where eMTC/NB-IoT with 100 kHz channel raster can be deployed.

Although it is possible to align 5G and eMTC on the subcarrier level, it is not possible to align them on the PRB level on the downlink. This is because eMTC utilizes a direct current (DC) subcarrier, whereas 5G does not. This slight mismatch means than 1 additional 5G PRB may need to be reserved for eMTC deployment. NB-IoT, however, does not have a DC subcarrier and therefore does not have this issue.

In the uplink, there is a half-tone shift between the uplink and downlink for eMTC and NB-IoT. To support this in 5G, an optional half-tone shift was introduced. This optional shift, however, is not supported in all 5G bands, and some bands will have a half-tone misalignment. Although this introduces some interference, it can be managed by the network and does not prevent deployment of eMTC and NB-IoT on frequency bands without half-tone shift support.

In Release 16, several enhancements are being studied to further improve the coexistence performance:

- Better coexistence with dynamic TDD operation in 5G: In 5G, each slot can be dynamically allocated for the downlink or uplink. However, eMTC and NB-IoT can only support LTE uplink-downlink TDD configurations which can be changed on a semi-static basis. This would either restrict how 5G TDD can be deployed, or very large guard bands around the eMTC/NB-IoT carrier would be required to minimize uplink-downlink interference. Both approaches will reduce the 5G system performance. One existing eMTC/NB-IoT feature that can be used to provide better coexistence with dynamic TDD in 5G is to use the invalid subframe bitmap. This bitmap can be used to mark a subset of eMTC/NB-IoT subframes as invalid for transmission/reception. These subframes can then be used on a dynamic basis by

5G without any restriction. Further enhancement can be considered in Release 16. For example, dynamic TDD subframe formats can be introduced for Release 16 eMTC/NB-IoT UE. Legacy eMTC/NB-IoT UEs will see this subframe marked as invalid, while Release 16 and beyond NB-IoT UEs can take advantage of the new formats.

- In Release 16, standalone deployment for eMTC is being specified. This will enable the use of the LTE control channel region for downlink control and data transmission. For the control channel, some of the existing symbols will be repeated in the legacy control region. For the data channel, rate matching will be used to extend the data transmission to all OFDM symbols.

- For eMTC and 5G, the PRB cannot be perfectly aligned due to the presence of a DC subcarrier in eMTC. As a result, for 15 kHz subcarrier spacings (SCSs), an additional 5G PRB will be needed to accommodate eMTC. Several potential enhancements can be considered, for example, by puncturing the last subcarrier in eMTC so that only six 5G PRBs would be needed. Note that this may be done via implementation with the eMTC performance impact considered.

- Reduction of Cell Reference Signals (CRSs) on part of the bandwidth, for example, to transmit CRS only on narrowband(s) in which the UE performs measurements. This would reduce the number of PRBs that are marked as reserved and therefore cannot be used by 5G.

As described above, eMTC and NB-IoT are used to support 5G mMTC services. The technologies continue to evolve and improve with new enhancements. In 3GPP Release 15, the following key enhancements were added:

- Reduced UE Power Consumption
  - Wake-up signal for idle mode paging – This is a new physical signal that must be detected prior to monitoring the control channel during idle mode paging. This is illustrated in Figure 14.26. The wake-up signal is an ON/OFF signal that is much shorter than the control channel. It allows the UE to quickly detect whether or not there is an upcoming control channel. If not, it can quickly return to sleep mode. Furthermore, a wake-up receiver can be used where most components of the device stay in sleep mode and therefore requires only a very small amount of power to run. This can significantly reduce UE power consumption.
  - Early data transmission during the random access procedure (only uplink specified in Release 15) – A UE can transmit uplink data (up to 1000 bits) during the random access procedure. This can significantly reduce the latency and overhead associated with small data transmission and also UE power consumption.
  - Uplink HARQ-ACK feedback for early termination (eMTC only) – Early termination supported for (1) uplink data transmission and (2) downlink control channel monitoring. It can be used by the network to quickly release the RRC connection and put the UE in idle mode, which consumes significantly less power.
  - Relaxed monitoring for cell reselection – Relaxed UE monitoring for cell reselection (as low as once every 24 h), for example, for a stationary UE.
- Reduced Latency
  - Reduced system acquisition time using a new resynchronization signal (eMTC only) – This is a new physical Resynchronization Signal (RSS) with dense transmission to reduce the synchronization time.

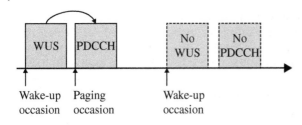

Figure 14.26 Wake-up signal.

- Improved system information demodulation performance – Acquisition time is reduced based on accumulation of multiple transmissions.
- System information update indication – A flag bit in the MIB indicates whether the system information has been updated during the last 24 h. This allows the UE to skip reading system information if it has not been changed.
- Scheduling request support via preconfigured random access channel transmission or via multiplexing of scheduling request onto HARQ-ACK (NB-IoT only).
• Improved Spectral Efficiency
  - Support for 64-QAM for the downlink data channel (eMTC only) – 64-QAM may be used in the downlink to improve system spectral efficiency; however, there is no increase in the peak data rate.
  - Flexible starting PRB resource allocation for a 1.4 MHz UE (eMTC only) – A UE can be allocated PRBs outside of a narrowband to improve efficiency.
  - CRS muting outside bandwidth-limited UE narrowband (eMTC only) – CRS outside of bandwidth UE narrowband may be muted to reduce inter-cell interference.
  - Sub-PRB resource allocation (eMTC only) – Resource can be assigned using three or six subcarriers to improve efficiency and reduce the PAPR; p/2-BPSK modulation was also introduced.
• Improved load control with coverage enhancement level based access class barring – UEs are barred from network access in enhanced coverage per coverage enhancement level.
• Support for TDD in NB-IoT. Note that TDD is supported for eMTC since Release 13.
• NPRACH enhancements (NB-IoT only).
  - New NPRACH format with a longer cyclic prefix of 800 μs to support cell sizes up to 120 km.
  - NPRACH scrambling enhancement to reduce false alarms.

In 3GPP Release 16, further enhancements are being introduced for eMTC and NB-IoT, including group-based wake-up signal, grant-free transmission, support for standalone eMTC deployment, and scheduling multiple transport blocks using a single grant.

It is expected that future 3GPP releases will define new 5G IoT device categories with lower capabilities enabling lower cost implementation. For more information about 5G evolution, see Chapter 15.

14.12 Summary

LTE evolution included major improvements for IoT connectivity with LTE-M and NB-IoT in 3GPP Release 13. The enhancements provided up to 20 dB coverage extension, up to 10 years battery lifetime, and low-cost IoT module implementation. These solutions are focused on low-cost-battery-powered massive IoT with low activity. The target is that a simple upgrade to the LTE network can provide support for a large number of connected devices. Further LTE IoT evolution continued in Release 14 and beyond, which shows the importance of this functionality.

5G IoT focuses initially on industrial IoT with low latency and high reliability in Releases 15 and 16. The target is that low-cost IoT will utilize an LTE-based solution, and therefore, LTE IoT and 5G radios can coexist.

References

1 Kovács, I., Mogensen, P., Lauridsen, M., Jacobsen, T., Bakowski, K., Larsen, P., Mangalvedhe, N., Ratasuk, R. (2017). LTE IoT Link Budget and Coverage Performance in Practical Deployments. IEEE International Symposium on Personal, Indoor and Mobile Radio Communications.

2 Lauridsen, M., Kovács, I., Mogensen, P., Sørensen, M. and Holst, S. (2016). Coverage and Capacity Analysis of LTE-M and NB-IoT in a Rural Area. Vehicular Technology Conference, IEEE 84th.

3 Andres-Maldonado, P., Lauridsen, M., Ameigeiras, P., and Lopez-Soler, J.M. (2019). Analytical Modeling and experimental validation of NB-IoT device energy consumption. *IEEE Internet of Things Journal* 6 (3), 5691–5701.

4 Signals Research Group (2019). It's raining cats and cats – A Performance Benchmark Study of 3GPP-Based IoT Devices in a Lab Environment. *Signals Ahead* 15 (2).

15

LTE-Advanced Evolution

Harri Holma and Timo Lunttila

Nokia, Finland

CHAPTER MENU

15.1 Introduction

Long-Term Evolution (LTE) and LTE-Advanced have turned out to be very successful technologies in terms of practical performance and in terms of subscriber adoption. The first version of LTE specifications was defined in 3GPP Release 8 in March 2009, and the first commercial LTE network was launched in December 2009. The further evolution has continued in the following 3GPP releases. LTE-Advanced in Releases 10–12 considerably enhanced LTE capabilities. The next phase in 3GPP was called LTE-Advanced Pro including Releases 13 and beyond. It is also called 4.9G. The targets of LTE-Advanced Pro are summarized in Figure 15.1. Release 8 provided an excellent starting point with a peak data rate of 150 Mbps and the best case latency of 10 ms. LTE-Advanced and LTE-Advanced Pro improved LTE performance in several areas by a factor of 10: higher data rates, lower latency, larger coverage for the low data rates, longer battery lifetime, and lower-cost devices for Internet of Things (IoT) modems, and higher capacity. These radio enhancements enable a large set of new applications on top of LTE networks. This chapter focuses on the features and the benefits of LTE-Advanced Pro. Section 15.2 briefly illustrates LTE technology timing and evolution. More detailed descriptions of LTE technology can be found from [1–3]. Section 15.3 describes in more detail the technological components of LTE-Advanced Pro. Section 15.4 presents the benchmarking between LTE and 5G technologies. The chapter is summarized in Section 15.5.

5G Technology: 3GPP Evolution to 5G-Advanced, Second Edition.
Edited by Harri Holma, Antti Toskala, and Takehiro Nakamura.
© 2024 John Wiley & Sons Ltd. Published 2024 by John Wiley & Sons Ltd.

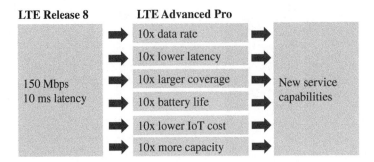

Figure 15.1 Main performance benefits of LTE-Advanced Pro.

15.2 Overview of LTE Evolution

The history of LTE dates back to 2004 when the first 3GPP workshop for LTE was held. The first set of specifications was completed in Release 8 in March 2009. Release 9 added a few additional capabilities on top of Release 8. These two releases enabled LTE to become a successful technology for delivering mobile broadband data and Voice over LTE (VoLTE). The first commercial LTE network was opened by Teliasonera in December 2009 followed by a large number of other networks during 2010–2011. The second phase of LTE work started in Release 10 as LTE-Advanced. Release 10 was finalized in 3GPP 2011, and the first networks with LTE-Advanced features were available in 2013. LTE-Advanced specifications were further enhanced in Releases 11 and 12. Release 11 was completed in 3GPP in December 2012 and Release 12 in March 2015. The third phase of LTE technology was defined in Releases 13–15 under the name LTE-Advanced Pro. Release 13 was completed in June 2016, Release 14 in June 2017, and Release 15 in September 2018. The timing of the 3GPP releases and the corresponding commercial network launches are illustrated in Figure 15.2. We note that it has typically taken 1–2 years from 3GPP specifications to the first commercially available network.

Release 8 was a major new release, including a complete new radio and core network. Release 8 enabled peak data rate of 150 Mbps with 20 MHz bandwidth and 2 × 2 MIMO. Also 4 × 4 MIMO was defined but not included in the first devices. Release 8 supported flat radio network architecture as well. All these capabilities were clear enhancements

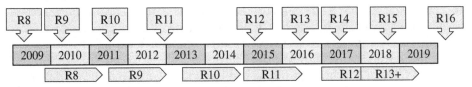

Figure 15.2 LTE schedule in 3GPP and in commercial deployments.

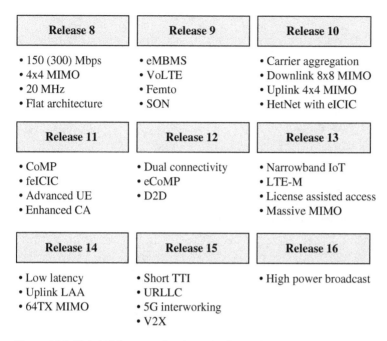

Release 8	Release 9	Release 10
• 150 (300) Mbps • 4x4 MIMO • 20 MHz • Flat architecture	• eMBMS • VoLTE • Femto • SON	• Carrier aggregation • Downlink 8x8 MIMO • Uplink 4x4 MIMO • HetNet with eICIC

Release 11	Release 12	Release 13
• CoMP • feICIC • Advanced UE • Enhanced CA	• Dual connectivity • eCoMP • D2D	• Narrowband IoT • LTE-M • License assisted access • Massive MIMO

Release 14	Release 15	Release 16
• Low latency • Uplink LAA • 64TX MIMO	• Short TTI • URLLC • 5G interworking • V2X	• High power broadcast

Figure 15.3 Main LTE features of each 3GPP release.

compared to the earlier 3G networks. Release 9 brought a few additions including enhanced Multimedia Broadcast Multicast System (eMBMS), VoLTE emergency calls, femtocells, and self-organizing networks (SONs). Release 10 boosted radio capabilities with carrier aggregation and multiantenna MIMO, increasing the peak rate beyond 1 Gbps. Release 10 also included improved Inter-Cell Interference Coordination (eICIC) for Heterogeneous Network (HetNet) optimization, and Release 11 further enhanced ICIC (feICIC). Release 11 enhanced carrier aggregation functionality, brought Coordinated Multipoint (CoMP), and added advanced UE receivers with interference cancelation. The main topics in Release 12 are dual connectivity, enhanced CoMP (eCoMP), IoT optimization, and device-to-device (D2D) connectivity. Release 13 provided LTE downlink operation on the unlicensed bands, also known as License Assisted Access (LAA), enhanced carrier aggregation beyond 100 MHz, three-dimensional (3D) beamforming, and support for Machine Type Communications (MTC) with LTE-M and narrowband IoT (NB-IoT). Release 14 brought very low latency operation as well as uplink LAA, and enhanced 3D beamforming to 64 antenna ports. Release 15 is considered as the last major LTE release, and provides a number of significant new features, including short Transmission Time Interval (TTI) and Ultra-Reliable Low Latency Communication (URLLC), Time Division Duplex (TDD) support for NB-IoT, additional enhancements to MTC, V2X, and many others. Release 16 brought high power broadcast solution to LTE. The main LTE features of each 3GPP release are summarized in Figure 15.3.

The evolution of the LTE peak data rate in commercial devices and networks is illustrated in Figure 15.4. LTE started with 150 Mbps using 20 MHz spectrum and 2×2 MIMO. The main solution for increasing the data rate is carrier aggregation:

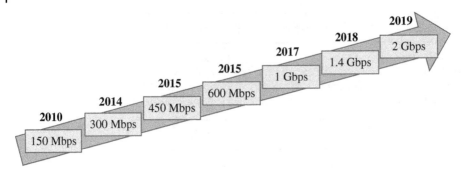

Figure 15.4 LTE peak data rate evolution in commercial networks.

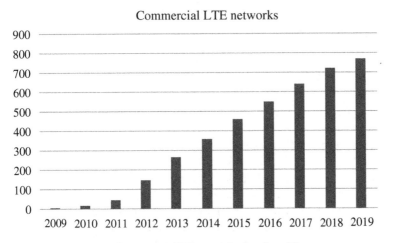

Figure 15.5 Number of commercial LTE networks data from [4].

20 + 20 MHz gives 300 Mbps and 3 × 20 MHz (3CA) gives 450 Mbps. 600 Mbps is obtained with 3 × 20 MHz and 256-QAM. A data rate of 1 Gbps and higher can be obtained with a combination of carrier aggregation and 4 × 4 MIMO. Data rates of 300–600 Mbps became available during 2015 and the peak rate of 1 Gbps during 2017. The latest devices can already support 2 Gbps peak rate.

Since the opening of the first LTE network in 2009, the number of commercial networks has increased rapidly to over 700 in more than 200 countries by 2018. The rapid growth of launches occurred during 2012–2015. The number of launched networks is shown in Figure 15.5.

LTE-Advanced needs a lot of spectrum to deliver high capacity and high data rates. The typical spectrum resources in European or some Asian markets are shown in Figure 15.6. All the spectrum between 700 and 2600 MHz can be utilized together with carrier aggregation. The aggregation of the multiple spectrum blocks together helps in terms of network traffic management and load balancing in addition to providing more capacity and higher data rates. Also, 3.5 GHz band has been used for TD-LTE in some countries. The unlicensed bands at 5 GHz can be utilized in the small cells for more spectrum. In the best case, more than 1 GHz of licensed spectrum is available for the mobile operators and more when considering the unlicensed spectrum.

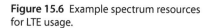

Figure 15.6 Example spectrum resources for LTE usage.

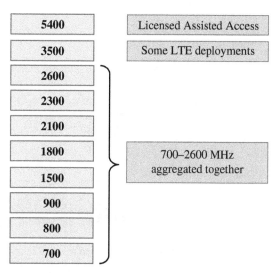

15.3 LTE-Advanced Pro Technologies

LTE technology has become even better with its latest enhancements in 3GPP Releases 13–15. Improved radio capabilities will make mobile broadband services more efficient, providing higher quality and enabling new sets of services on top of LTE networks. These features are called LTE-Advanced Pro and are also referred to as 4.9G. LTE-Advanced Pro enables the Programmable World for billions of connected IoT devices, vehicular communication for Intelligent Traffic Systems (ITS), and public safety/critical communications. LTE-Advanced Pro enhances user data rates to multi-Gbps, reduces latency to a few milliseconds, enables access to unlicensed 5 GHz spectrum, and increases network efficiency, while maintaining backward compatibility with existing LTE networks and devices. LTE-Advanced Pro and 5G use similar technology components in enhancing radio capabilities. 5G is a new non-backward-compatible radio which can operate below and also above 6 GHz frequencies and provide even higher data rates and lower latency. LTE-Advanced Pro operates below 6 GHz and evolves in parallel to the 5G development work.

15.3.1 Multi-Gbps Data Rates with Carrier Aggregation Evolution

LTE started with a 150 Mbps peak rate with 20 MHz bandwidth. In Release 10, the peak data rates were significantly upgraded by the introduction of carrier aggregation and higher-order MIMO. Mainstream carrier aggregation in 2016 delivered up to 450 Mbps on 3 × 20 MHz. 3GPP Release 10 defined a maximum capability up to 5 × 20 MHz with 8 × 8 MIMO, which gives 3 Gbps peak rate with 64-QAM.

The data rate can be increased further with more spectrum and with more antennas. A higher number of antenna elements is feasible with comparatively large base station antennas; however, it is challenging to integrate more antennas into small devices. Therefore, data rates are more easily increased by using more spectrum rather than by using more device antennas. Release 13 enhances the possibilities of carrier aggregation to enable up to 32 component carriers, which makes higher data rates possible through

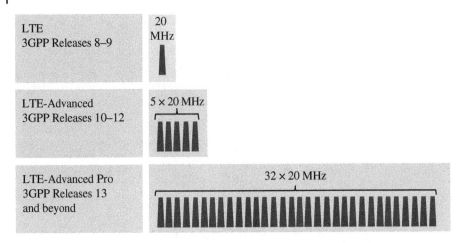

Figure 15.7 Evolution of carrier aggregation up to 32 component carriers.

the use of more spectrum. In practice, the use of unlicensed spectrum enables LTE to benefit from even further carrier aggregation capabilities. Figure 15.7 illustrates carrier aggregation evolution.

15.3.2 Utilization of 5 GHz Unlicensed Band

LTE networks have been deployed using licensed spectrum between 450 and 3600 MHz. The utilization of unlicensed bands in addition to licensed bands will enable capacity and peak data rate improvements for LTE-Advanced Pro. At the 5 GHz unlicensed band, there is plenty of spectrum available, suitable in particular for small cell deployments. This large pool of spectrum allows mobile broadband operators to take advantage of the carrier aggregation evolution provided by LTE-Advanced Pro. The evolution of LTE spectrum usage is shown in Figure 15.8, and the spectrum resources available in the unlicensed 5 GHz band are illustrated in Figure 15.9.

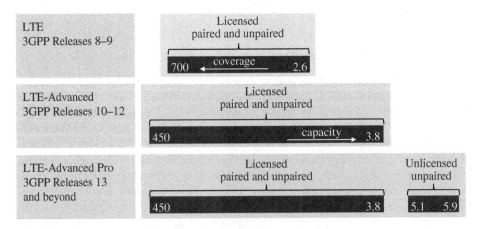

Figure 15.8 Spectrum usage evolution for licensed and unlicensed bands.

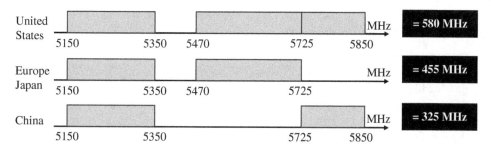

Figure 15.9 Spectrum resources at 5 GHz unlicensed bands.

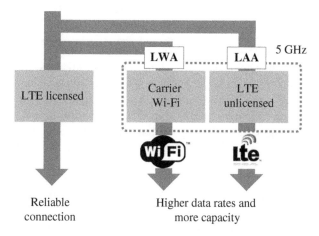

Figure 15.10 Aggregation of licensed and unlicensed bands.

LTE-Advanced Pro can utilize unlicensed band spectrum through two techniques: LAA or by integrating Wi-Fi more closely with the cellular network via LTE-Wi-Fi aggregation (LWA). The two solutions are shown in Figure 15.10. LAA combines the use of licensed and unlicensed spectrum for LTE using carrier aggregation. LAA is an efficient solution for traffic offloading since the data traffic can be split, with millisecond resolution, between licensed and unlicensed frequencies. The primary cell on a licensed bands can provide reliable connectivity, mobility, signaling, and guaranteed data rate services, while the secondary cell(s) on unlicensed band can significantly boost the data rates.

Unlicensed spectrum operation involves various regulatory rules which aim at facilitating fair and equal spectrum usage for different devices and radio access technologies, such as Wi-Fi. Before being permitted to transmit, a user or an access point (such as eNodeB) may need to monitor the given radio frequency for a short period of time to ensure the spectrum is not already occupied by some other transmission. This requirement is referred to as Listen-before-talk (LBT). The requirements for LBT vary depending on the geographic region: for example, in the United States, such requirements do not exist, whereas in, for example, Europe and Japan, the network elements operating on unlicensed bands need to comply with LBT requirements. In case the channel is sensed as free according to LBT rules, the eNodeB or UE may resume transmission for, for example, 6–10 ms. If the channel is sensed as occupied, the eNodeB or UE will need to continue to suspend transmission until the channel is sensed as unoccupied according to LBT rules. In addition to LBT, additional restrictions related to unlicensed band operation include the following:

- Limitations related to the occupied channel bandwidth: According to European Telecommunications Standards Institute (ETSI) regulations, the Occupied Channel Bandwidth, defined as the bandwidth containing 99% of the power of the signal, shall be between 80% and 100% of the declared nominal channel bandwidth.
- Limitations related to the maximum power spectral density (PSD): For most cases, the requirement is stated with a resolution bandwidth of 1 MHz. For example, the ETSI 301893 specifications require 10 dBm/MHz for 5150–5350 MHz. Similar limitations are involved also in the United States, governed by Federal Communications Commission (FCC). The UE's peak PSD for 5.15–5.725 MHz is 11 dBm/MHz in the United States.

These regulatory requirements necessitated a few significant modifications to the LTE air interface. For the downlink, unlike in licensed band operation, all transmissions are omitted unless there is data to transmit for the UE, with the exception of Discovery Reference Signals (DRSs) facilitating, for example, cell search and synchronization. For the uplink, the resource allocation had to be modified to match the requirements related to the occupied channel bandwidth. This was achieved by introducing a block-interlaced resource allocation scheme, where uniformly spaced resources span at least 80% of a carrier.

Figure 15.11 illustrates the supplemental downlink solution using the unlicensed band. The licensed band carries both uplink and downlink traffic including all voice and other traffic requiring reliable mobility and high quality of service (QoS). The unlicensed band provides a lot more downlink capacity and higher data rates. This solution fits very well with the typical traffic patterns where the downlink traffic is 10 times more than the uplink traffic. Therefore, it is beneficial to use the new spectrum blocks mainly for the downlink direction. The second phase of LAA, Enhanced LAA (eLAA) standardized in Release 14, also includes uplink transmission on the unlicensed

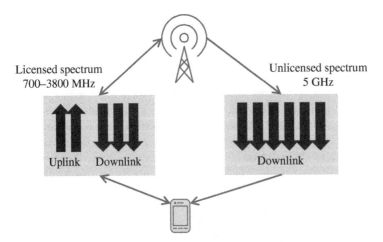

Figure 15.11 Supplemental downlink at unlicensed bands.

Figure 15.12 LTE–Wi-Fi aggregation (LWA).

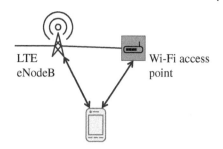

LTE
eNodeB

Wi-Fi access
point

band. LAA was commercially deployed in networks and devices starting 2017 including iPhones starting in 2018.

LTE-Advanced Pro also allows the aggregation of LTE and Wi-Fi transmissions, offering another solution for utilizing unlicensed bands. So far, LTE and Wi-Fi interworking has been implemented on the application layer. 3GPP Release 13 allows dual connectivity of LTE and Wi-Fi on the radio layer, which makes the offloading of data traffic to Wi-Fi more efficient. The LTE eNodeB splits data traffic between LTE and Wi-Fi transmissions. The user device can then receive data simultaneously via both LTE and Wi-Fi, thereby aggregating the spectrum of the LTE eNodeB, and of the Wi-Fi access point. The data transmission can be dynamically split between LTE and Wi-Fi instead of using slow switching between the technologies. Wi-Fi radio can be utilized to the maximum while still maintaining the connection to LTE for reliable mobility and connectivity. The LWA solution is illustrated in Figure 15.12. The commercial interest in LWA has remained low.

15.3.3 Enhanced Spectral Efficiency with 3D Beamforming and Interference Cancelation

LTE has provided high spectral efficiency since its introduction. The downlink efficiency of LTE has been boosted further via MIMO technology, including 4×2 MIMO, 8×2 MIMO, and 4×4 MIMO, and with interference cancelation. LTE-Advanced Pro introduces the next step in spectral efficiency with three-dimensional (3D) beamforming; also known as full dimensional MIMO (FD-MIMO) or massive MIMO. Increasing the number of transceivers at the base station is key to unlocking higher spectral efficiencies. Release 13 specifies MIMO modes for up to 16 transceivers at the base station, while in Release 14 there may be as many 64 transceivers. Figure 15.13 shows the assumptions for the physical antenna structures in 3D beamforming studies. 3D beamforming will bring efficiency gains for the downlink transmissions as shown in Figure 15.14: 16×2 provides a 2.5-fold gain in spectral efficiency compared to 2×2, and 64×2 shows a 3.0-fold gain. The gain available from 64×2 compared to 8×2 is +50%. Note that the 8×2, 16×2, and 64×2 transceiver configurations are each associated with four columns of cross-polarized antenna elements of approximately the same physical dimensions, while the 2×2 transceiver configuration is associated with only one column of cross-polarized antenna elements. In all cases, the total transmission power is the same.

The uplink efficiency can be enhanced with multiantenna reception at the eNodeB, and this has been widely deployed in the field by aggregating the signals received from four, eight, or even more transceivers in the form of the Centralized Radio Access

Figure 15.13 Antenna assumptions in 3D beamforming studies.

Antenna for
4 × 2 simulations

Antenna for 8 × 2, 16 × 2
and 64 × 2 simulations

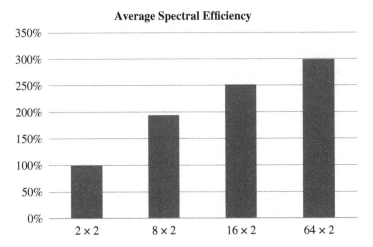

Average Spectral Efficiency

Figure 15.14 Gain of 3D beamforming in spectral efficiency.

Network (CRAN). These uplink improvements are mostly transparent to user devices and can be implemented with Release 8 devices, without requiring any LTE-Advanced feature support.

The introduction of a large number of transceivers is made possible with active antenna arrays, which consist of passive radiating elements, RF amplifiers, filters, and digital processing integrated into the array. The use of such active arrays effectively moves the radio closer to the radiating elements, reducing feeder loss, and this integration reduces the footprint of the tower-top structure.

The spectral efficiency is often limited by the inter-cell interference. Therefore, the efficiency can be improved by using inter-cell interference cancelation at the UE. A relatively simple solution is to cancel the interference caused by Cell-specific Reference Signals (CRSs). When the Resource Utilization (RU) is low, most of the interference is caused by CRSs. Figure 15.15 illustrates the gain of CRS interference cancelation as a function of the network loading. When the RU is low, the gain can be 30–50%. Further improvements are obtained with the cancelation of the Physical Downlink Shared Data Channel (PDSCH). This feature is called Networks Assisted Interference Cancelation and Suppression (NAICS). The network assistance improves the cancelation performance since the network provides assistance information to the UE regarding the source of the interference, including cell identity, number of CRS antenna ports, power offsets, transmission modes, resource allocation, and precoding granularity.

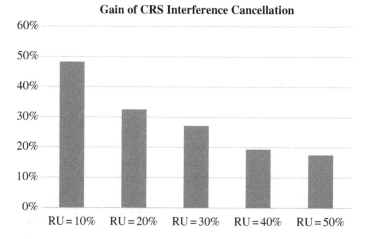

Figure 15.15 Gain of Common Reference Signal Interference Cancelation (RU = Resource Utilization).

15.3.4 Extreme Local Capacity with Ultra-Dense Network

LTE-Advanced radio has demonstrated high efficiency in the macro-cellular layer, but it has not been enough to satisfy the needs of the most extreme traffic hot spots. Small cells are the next step in providing increased local capacity. Such high-traffic hot spots typically appear in dense urban areas, in sports venues, or in music festivals. 3GPP has defined a number of enhancements that make small cell deployment more efficient in coordination with the existing macro layer. Further Enhanced Inter-Cell Interference Coordination (feICIC) was added in Release 11, and dual connectivity in Release 12. Dual connectivity allows the user device to receive data simultaneously from two different sites served by backhaul having non-negligible latency. Dual connectivity is enhanced in Release 13 to also support uplink transmission to two sites. Dual connectivity can be utilized for HetNets consisting of macrocells and many small cells. Dual connectivity brings a number of benefits for small cell deployments by combining reliable mobility in the macro layer with the high capacity and enhanced data rates available from the small cells. Despite these benefits, LTE dual connectivity was not implemented in live networks. The number of small cell deployments was too low to justify the complexity of dual connectivity. However, the functionality was still an important starting point for the dual connectivity between LTE and 5G. The principle of dual connectivity is illustrated in Figure 15.16.

15.3.5 Millisecond Latency with Shorter Transmission Time Interval

Reducing latency is an evergreen quest, similar to increasing bit rates. Low latency is not only driving Transmission Control Protocol (TCP) throughput performance, it is

Figure 15.16 Dual connectivity between macrocells and small cells.

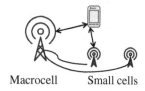

Macrocell   Small cells

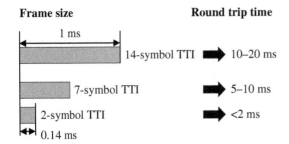

Figure 15.17 Shorter TTI for faster round trip time.

also important in reducing, for example, buffering requirements at high data rates in the radio equipment. Even more importantly, low latency determines user experience in interactive applications, and will drive new delay-sensitive use cases. LTE-Advanced Pro tackles the latency problem by reducing the TTI length and optimizing the physical layer control of the air interface resources. Figure 15.17 illustrates the evolution of the LTE frame structure and TTI. A shorter TTI directly scales down the air interface delay. Shortening the TTI by reducing the number of symbols is the most promising approach when seeking to maintain backward compatibility and usability in existing LTE bands. The regular 1 ms TTI gives in practice a 10–20 ms round trip time, while an LTE-Advanced Pro solution should provide a round trip time of even less than 2 ms and a below 1 ms one-way delay. The evolution of the round trip time from High Speed Packet Access (HSPA) to LTE, LTE-Advanced Pro, and 5G is illustrated in Figure 15.18.

The practical end-to-end latency also depends on the network architecture in addition to the radio capability. The latency measurements in live networks show that the LTE latency can be below 15 ms in some networks while other networks show values in excess of 50 ms. Figure 15.19 shows latency measurements in a live network in Helsinki in Finland where the measurement server is also located in Helsinki. The measurements show that the practical end-to-end latency can be low if the transport network is well designed and if the content is close to the radio. When the radio latency becomes lower, it becomes even more important to optimize the other components including Multi-Access Edge

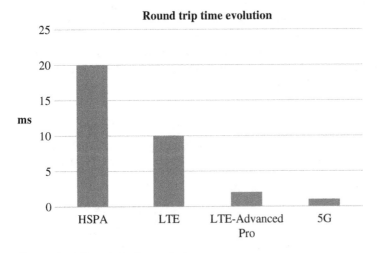

Figure 15.18 Round trip time evolution.

Figure 15.19 Example latency measurements in live LTE network.

Computing (MEC). The MEC concept is being standardized in ETSI. In addition to low latency, MEC allows the exposure of the radio network edge to authorized third parties, which enables them to deploy innovative applications and services for mobile subscribers, enterprises, and vertical segments.

A large part of the physical layer latency today is incurred by a fixed, synchronous timing relation in providing a transmission grant to the device, and acknowledging the received data. Optimizing those timing relations provides another avenue to shortening latency.

Reducing latency was a major theme for LTE Releases 14 and 15, and a few rather comprehensive enhancements were specified. Release 14 provides enhancements to uplink access delay, by allowing for UEs to skip the Physical Uplink Shared Channel (PUSCH) transmission (i.e. an empty buffer status report) unless there is no data in the buffer. Combined with semi-persistent scheduling with 1 ms periodicity, this facilitates a mode of operation where a UE is able to transmit uplink data immediately, without having to first transmit a scheduling request, and wait for an uplink grant from the eNodeB.

Release 15 added a few more advanced features to reduce latency. First, for the regular 1 ms TTI, shorter processing times were specified, reducing the delay between downlink data transmission and the corresponding Hybrid Automatic Repeat Request (HARQ)-acknowledgment, as well as the time between an uplink grant transmission by the eNodeB and the transmission of the corresponding uplink data by the UE, down to 3 ms compared to 4 ms, which has been assumed since the start of LTE.

Second, Release 15 adds support for a shorter TTI (sTTI). The supported sTTI durations are seven symbols for both Frequency Division Duplex (FDD) and TDD, and two or three symbols for FDD. Figure 15.20 shows an example with one-symbol Physical Downlink Control Channel (PDCCH) in the downlink. A similar structure applies to the uplink as well. There can be up to six subslot allocations in a single 1 ms period. Shorter TTIs resulted in major redesign of large part of LTE, especially the physical layer. For the downlink, a new control channel – sPDCCH – was defined, along with new downlink control information formats. Modifications to HARQ-operation and reference signals were also necessary. In the uplink, the control channels and reference signals were also redesigned to match sTTIs having fewer symbols.

An LTE cell can operate with both regular 1 ms TTI and sTTI simultaneously. sTTI-related channels are multiplexed with ordinary LTE channels and signals in the frequency domain, and the eNodeB needs to take their presence into account in scheduling the downlink or uplink resources. For a given UE, dynamic switching between 1 ms TTI and sTTI is also supported, allowing for adaptation to instantaneous traffic needs. Note that sTTI operation is not supported on unlicensed bands.

Another latency enhancement in Release 15 is the introduction of enhanced support to URLLC. The standardized enhancements mainly improve the reliability of data transmissions, by introducing additional repetitions that facilitate higher reliability of reception.

Release 8 with 1 ms TTI

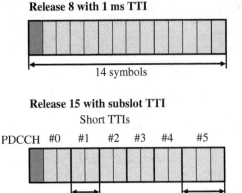

14 symbols

Release 15 with subslot TTI

Short TTIs

PDCCH #0   #1   #2   #3   #4   #5

2 symbols = 0.14 ms   3 symbols

Figure 15.20 Subslot TTI structure within LTE subframe.

15.3.6 IoT Optimization

IoT refers to the interconnection and the autonomous exchange of data between devices which are machines or parts of machines; often involving sensors and actuators. The future connected world will have tens of billions of IoT devices. LTE evolution will enhance LTE's capabilities to serve IoT by improving coverage, device power consumption, cost, and connectivity. IoT optimization is discussed in detail in Chapter 14.

15.3.7 D2D Communications

3GPP has defined direct communication between two devices with the D2D feature in Release 12. Earlier releases supported only communication between the device and the base station. D2D functionality can be utilized for multiple use cases: for public safety, for vehicle-to-vehicle (V2V) communication, for social media, and for advertisements. D2D covers critical use cases including out-of-coverage communications, relaying for extended coverage and communications in case of network failure, for example, due to earthquakes. D2D communication is also known as Proximity Services (ProSe).

The principle of D2D communication is shown in Figure 15.21. The traditional communication has uplink and downlink connections between the device and the base station. The new D2D communication is called the sidelink and is known as the PC5 interface in 3GPP. D2D uses mainly uplink resources, that is, uplink spectrum in the case of FDD and uplink subframe in the case of TDD. D2D uses half-duplex operation even if the cellular network is FDD. Network synchronization is assumed for D2D communication.

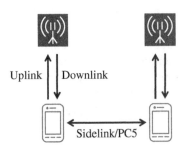

Uplink Downlink

Sidelink/PC5

Figure 15.21 Device-to-device communication.

D2D functionality added new logical channels:

- Sidelink discovery channel (SL-DCH), which is a fixed size, predefined format periodic broadcast transmission. There is support for both UE autonomous resource selection and scheduled resource allocation by the eNodeB.
- Sidelink shared channel (SL-SCH) for broadcast transmission. There is support for both UE autonomous resource selection and scheduled resource allocation by the eNodeB.
- Sidelink broadcast channel (SL-BCH), which is a predefined transport format.

ProSe Direct Discovery is the procedure by the UE to discover other UEs in its proximity, using LTE direct radio signals via PC5. Direct Discovery has two models:

- Model A involves one UE announcing "I am here."
- Model B involves one UE asking "who is there" and/or "are you there."

There are two types of Direct Discovery: open and restricted. Open discovery does not need explicit permission from the UE being discovered. Restricted discovery requires explicit permission from the UE that is being discovered.

D2D communication has two modes:

- Mode 1: The eNodeB schedules the exact resources used by a UE to transmit direct data and direct control information. Downlink Control Information (DCI) format 5, which has the same size as the existing DCI format 0, is used for allocating D2D Data and Physical Sidelink Control CHannel (PSCCH), carrying the Sidelink Control Information (SCI).
- Mode 2: A UE on its own selects resources from resource pools for the transmission of direct data (PSSCH) and direct control channel (PSCCH).

The time-frequency resources where control and data can be transmitted are referred to as resource pools. All communication pools (Mode 1 and Mode 2) are defined in terms of a Sidelink control period (40, 80, 160, 320 ms) and a bitmap defining the subframes used for the transmission of control information. Mode 2 pools have in addition a bitmap defining the subframes for data communication. In addition, there are several other parameters defined for a resource pool like physical resource block (PRB) pairs, offsets, and some transmission parameters.

Present-day communication equipment installed in cars is used for remote car diagnostics, providing in-car entertainment or fleet tracking. Networking vehicles in a more integrated fashion will allow the transformation of the automotive environment in a more fundamental fashion. The target is to introduce public safety systems as, for example, specified in ETSI ITS, to provide the environment for self-driving cars, perform real-time management of traffic in cities, and last but not least to provide car and traffic related services to users.

LTE-Advanced Pro is in a uniquely suitable position to support vehicular communications, as it provides network coverage for vehicle-to-infrastructure or –to-pedestrian communication, enables D2D communications, and will support low latencies as specified in ITS. The main application for V2V communication is enabling safety applications such as collision avoidance.

V2I (infrastructure), V2N (network), and V2P (pedestrian) are other facets of this topic. These use cases are illustrated in Figure 15.22. A network-aiding vehicular communication may be accomplished by routing V2V communication via a base station, or

Vehicle-to-Infrastructure (V2I)   Vehicle-to-Vehicle (V2V)

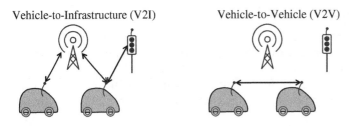

Figure 15.22 LTE for vehicular communications.

by base stations broadcasting or group-casting traffic information to cars. The availability of LTE MEC solutions is very well suited to vehicular communications applications, for example, dynamic map retrieval or localized car traffic management.

15.3.8 Public Safety

Current public safety networks such as TETRA or Project 25 (P25) support mission-critical voice communication, but are limited to narrowband data. Mobile broadband data can significantly help emergency services, for example, with live mobile video, situation-aware dispatching, and remote diagnostics. LTE, and later 5G, will be the solution for public safety data transfer initially, and also for voice later. The LTE standard includes solutions for public safety starting from Releases 12 and 13. The basic technology components, like QoS prioritization and emergency calls, are available already in Releases 8–11. Public safety users need to have the priority over other users in the congested area. Such functionalities include access control with access class barring, admission control with preemption, and prioritization in resource allocation. Proximity services (D2D) are included in Release 12 and enhanced D2D in Release 13. Also, group communications is included in Release 12 and mission-critical push-to-talk in Release 13 as well as standalone base station. The main 3GPP features for public safety are summarized in Figure 15.23. Some of the features are implemented in the mobile networks and some of the features in the application servers and devices.

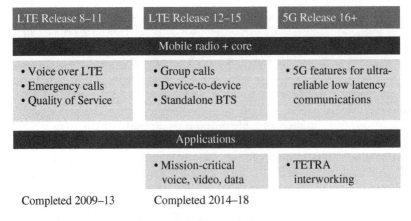

Figure 15.23 Public safety features in 3GPP.

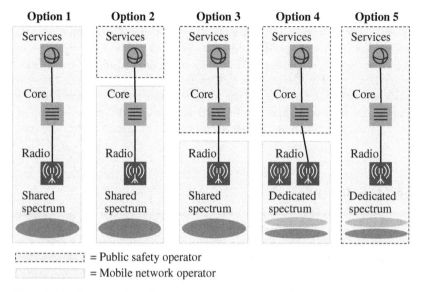

Figure 15.24 Functionality split options for public safety implementation with mobile network operator.

LTE public safety service can be organized in a number of different ways, and the responsibilities and functionalities can be divided between the public safety operator and the mobile network operator. The main five options are shown in Figure 15.24. The target in most options is to take advantage of the existing mobile network coverage and capacity. Options 1–3 use shared spectrum with the mobile operator, while options 4–5 have a dedicated spectrum reserved for the public safety use:

- Option 1: The mobile operator takes the full responsibility for public safety including radio connectivity, core network, and application servers.
- Option 2: The mobile operator takes care of radio and core network, while the public safety operator takes care of the public-safety-related services.
- Option 3: The mobile operator provides access to their radio network, while the public safety operator runs its own packet core network and services. This option can be implemented with Multi-Operator Core Network (MOCN) or roaming architecture.
- Option 4: The public safety operator has dedicated spectrum available. The radio network is built by the mobile operator by reusing its existing cell sites and potentially also existing base stations.
- Option 5: The public safety operator takes care of the complete network rollout including radio, core, and services.

Different options can also be mixed in different parts of the network, for example, using dedicated spectrum in the busy areas while sharing the spectrum in low-traffic areas.

LTE public safety activity started in the United States in 2012 when FirstNet got the responsibility to coordinate the use of Band 14 at 700 MHz, and many other governments are proceeding to use LTE networks for public safety. The United Kingdom selected one of the existing mobile operators to offer LTE network services for the public Emergency Service Network (ESN). Safenet in Korea is provided by the mobile operators on dedicated spectrum at 700 MHz. Nedaa, the Dubai Government secure networks provider, uses LTE for the public safety network. The public safety operator in Finland has its own services and packet core, and uses commercial mobile networks for radio connectivity. LTE-based public safety is underway globally with multiple different deployment models. Public safety is one of the first use cases where a critical service is run on top of mobile networks. Many other critical services will be run on top of 5G networks. LTE public safety can be considered as a pioneering case of the new type of services.

15.4   5G and LTE Benchmarking

This section presents benchmarking of 5G targets from Next Generation Mobile Network (NGMN) [5] and ITU-R [6] with LTE-Advanced Pro capabilities. The reference case in ITU considerations is IMT-Advanced. We use LTE-Advanced Release 12 as the reference case. Table 15.1 shows a summary of LTE-Advanced capability, LTE-Advanced Pro, and 5G targets. We note that LTE-Advanced Pro is able to fulfill in theory most of the 5G targets including data rate, latency, and mobility, but the data rate target requires so much spectrum that it is not available on the current LTE frequency bands. The traffic and device density targets would also be difficult to fulfill with the current LTE spectrum and practical site densities. 5G targets are so high that more than 1 GHz of spectrum and very high site densities are required to fulfill them. That means in practice that higher frequencies with high site densities are needed to fulfill 5G targets.

Table 15.1 Benchmarking of LTE capabilities and 5G targets.

	LTE-Advanced Release 10–12	LTE-Advanced Pro Release 13–15	5G targets	LTE vs. 5G
Peak data rate in downlink	3.7 Gbps FDD 3.55 Gbps TDD	21 Gbps (640 MHz)	10–50 Gbps	OK
Peak data rate in uplink	0.89 Gbps TDD	5.7 Gbps (640 MHz)	10–50 Mbps	Not OK
Cell edge data rate (downlink)	8–40 Mbps (2–10 UEs/cell, 100 MHz)	60–100 Mbps (2–10 UEs/cell, 300 MHz)	50–100 Mbps	OK
One-way latency	5 ms	1 ms	1 ms	OK
Mobility	500 km/h below 1.4 GHz	500 km/h up to 3 GHz	500 km/h	OK
Traffic density	10 Gbps/km^2	20 Gbps/km^2	750 Gbps/km^2	Difficult without mmW
Device density	100–300 k/km^2	100–300 k/km^2	1 M/km^2	Difficult without mmW

15.4.1 Peak Data Rate

The peak data rate can be increased by the following technical solutions:

- More spectrum, which is exploited with carrier aggregation.
- More multistream, which is exploited with higher-order MIMO.
- Higher-order modulation, which is exploited with 256-QAM.

LTE-Advanced Pro includes all those components for increasing the peak data rate. The maximum bandwidth was increased from 20 MHz in Release 8 to 100 MHz in Release 10 and up to 640 MHz in Release 13. The number of multistreams was increased from two to eight in Release 10 and the modulation from 64-QAM to 256-QAM in Release 13. The maximum downlink peak data rate is more than 20 Gbps in Table 15.2, and the uplink rate more than 5 Gbps in Table 15.3. It is straightforward to achieve the high theoretical data rate simply by adding more spectrum. The real challenge is how to find the spectrum.

15.4.2 Cell Edge Data Rate

The 5G target is to provide a minimum data rate of 100 Mbps for the whole coverage area including cell edges. LTE-Advanced simulations [7] show a cell edge spectral efficiency of 0.07 bps/Hz/cell. A cell edge data rate of 100 Mbps would then require 1400 MHz of spectrum for FDD downlink or 1800 MHz for TDD (80:20), which clearly is not practical. The low cell edge efficiency is explained by the simulations having 10 simultaneous users. With a lower number of users, the cell edge data rate would be higher. If the cell density is increased, the number of simultaneous users is lower per cell, and the cell

Table 15.2 LTE-Advanced Pro downlink peak data rate for TDD operation.

	Bandwidth	MIMO streams	Modulation	Peak data rate
LTE Release 8	20 MHz	2	64-QAM	0.13 Gbps
LTE Release 10	100 MHz	8	64-QAM	2.6 Gbps
LTE Release 13	640 MHz	8	256-QAM	22 Gbps

Uplink: Downlink configuration 5 and special subframe 3.

Table 15.3 LTE-Advanced Pro uplink peak data rate for TDD operation.

	Bandwidth	MIMO streams	Modulation	Peak data rate
LTE Release 8	20 MHz	1	64-QAM	0.04 Gbps
LTE Release 10	100 MHz	4	64-QAM	0.89 Gbps
LTE Release 13	640 MHz	4	64-QAM	5.73 Gbps

Uplink: Downlink configuration 0 and special subframe 3.

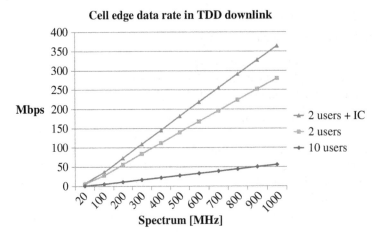

Figure 15.25 Cell edge data rate with TD-LTE.

edge data rate can be increased. Therefore, the practical cell edge data rate is not only dependent on the technology but is also dependent on the deployment scenario. The cell edge data rate is limited by the inter-cell interference. Therefore, it is possible to improve the cell edge performance by inter-cell interference cancelation. Figure 15.25 illustrates the cell edge data rate with 10 and 2 simultaneous users per cell with TD-LTE including the interference cancelation case. Providing a cell edge data rate of 100 Mbps with two simultaneous users and interference cancelation would require 300 MHz of spectrum. The cell edge data rate could be improved further by assuming beamforming.

15.4.3 Spectral Efficiency

The 5G target is to improve spectral efficiency by a factor of 3.0. The main solutions for increasing the efficiency are a higher number of antennas and interference cancelation. We consider the following three solutions in increasing the spectral efficiency:

- Beamforming in the network
- Interference cancelation in the UE
- Four-antenna reception in the UE

The interference cancelation is assumed to increase the efficiency by 20%. 3D beamforming with 64 transmitters is assumed to provide 100% gain and four-antenna UE an additional 50% gain. When all these gains are combined, the theoretical spectral efficiency of LTE-Advanced Pro can be three times more than LTE-Advanced.

15.4.4 Mobility

3GPP Release 12 includes performance requirements for high-speed mobility up to 350 km/h at the 2 GHz band. The requirement is defined for a Doppler shift of 1340 Hz in the uplink and 750 Hz on the downlink. The uplink direction is more critical than the downlink due to the lower reference signal density in time and multiuser interference with any frequency error. The uplink Doppler shift takes account both uplink and downlink Doppler spreads.

If the carrier frequency is lower, the supported speed in terms of km/h is higher with the same Doppler. Therefore, even 800 km/h is supported with Release 12 at frequencies below 1 GHz. On the other hand, at high frequency of 3.6 GHz, the maximum mobile speed is reduced to 200 km/h.

15.4.5 Traffic Density

The 5G target is to provide 0.75 Tbps/km^2 traffic density in urban hot spots. The traffic density is not dependent only on the technology but is also rather dependent on the amount of spectrum and on the deployment scenario. This section illustrates example estimates about the achievable traffic densities. Three scenarios are considered: a typical hot spot case in a live LTE network, maximum practical traffic density with LTE-Advanced Pro, and a 5G case using mmWave. We assume that LTE currently uses 50 MHz of downlink spectrum, LTE-Advanced Pro 200 MHz including usage of 5 GHz unlicensed band, and 5G a total of 3000 MHz spectrum at mmWave. The base station site densities are 30, 50, and 300 per km^2. More sites could be added at LTE bands below 6 GHz, but it makes more sense to use higher bands for dense networks because there is more spectrum available at higher frequencies. The high bands can be efficiently used in dense networks from the propagation point of view. We assume 2.5 sectors on average per site. The spectral efficiency is assumed to grow from 2.0 bps/Hz/cell to 6.0. We also consider the unequal traffic distribution between the cells. The assumption for LTE is that the highest loaded cell carries four times more traffic than the average cell. When the site density increases, we assume that the traffic distribution becomes more unequal. The high site density assumption is 6 for LTE-Advanced Pro and 15 for 5G. The calculations show that the typical LTE network carries approximately 2 Gbps/km^2 traffic density, LTE-Advanced Pro can carry more than 20 Gbps/km^2, and 5G at mmWave can carry even 1 Tbps/km^2. We can conclude that LTE cannot fulfill 5G traffic density requirements at below 6 GHz frequencies (Table 15.4).

15.4.6 Device Density

The 5G target is to enable a device density of 1 M/km^2. We consider the maximum device density from two perspectives: setup capacity and connected UE capacity. We assume a dense deployment of 50 sites/km^2 and 2.5 sectors per site. We further assume 10 cells per sector by fully exploiting the available spectrum. We assume that the maximum

Table 15.4 Maximum traffic density.

	LTE typical 2016	LTE-Pro maximum	5G at mmW
Spectrum	50 MHz	200 MHz	3 GHz
Site density	30/km^2	50/km^2	300/km^2
Sectors per site	2.5	2.5	2.5
Spectral efficiency	2.0 bps/Hz/cell	6.0 bps/Hz/cell	6.0 bps/Hz/cell
Traffic distribution factor	4	6	15
Traffic density	1.9 Gbps/km^2	26 Gbps/km^2	0.9 Tbps/km^2

Table 15.5 Device density from the setup capacity and connected capacity point of view.

	Setup capacity	Connected capacity
Max capacity per cell	50 cell/s	1000 UE/cell
Site density	50/km^2	50/km^2
Sectors per site	2.5	2.5
Cells per sector	10	10
Traffic distribution factor	10	10
Max. capacity per km^2	6250 setups/km^2/s	125 k connected UE/km^2
Max. IoT device density	375 k/km^2 with 1 message/min 1.87 M/km^2 with 1 message/5 min	1.25 M devices/km^2 with 10% connected
5G target	1 M/km^2	1 M/km^2

setup capacity is 50 connection setups per cell per second and the maximum connected capacity is 1000 UE/cell. The calculations show that LTE can support 6250 setups/km^2/s. If we want to transfer one message from every UE once per minute, the maximum UE density is 375 000/km^2, which is below the 5G target. If the transmission frequency is once every five minutes, the UE density is more than 1 M/km^2. The corresponding connected UE capacity is up to 125 000/km^2. If we assume that 10% of the UEs are Radio Resource Control (RRC) connected, it gives a UE density of more than 1 M/km^2. We conclude that LTE can support the required device density from the signaling and connectivity points of view (Table 15.5).

15.5   Summary

LTE-Advanced Pro brings great enhancements in radio performance on top of LTE-Advanced: multi-Gbps data rates, higher spectral efficiency, and below-1-ms one-way latency. LTE-Advanced Pro also enables a number of new application scenarios including IoT optimization for the Programmable World, vehicular connectivity, and public safety. LTE-Advanced Pro is supported by new features in 3GPP Releases 13, 14, and 15. LTE-Advanced Pro is backward compatible with existing LTE networks and can coexist on the same frequencies, which makes a smooth LTE-Advanced Pro rollout possible. The quality and performance of LTE-Advanced Pro is not only important for the short term but also in the long term when 5G radio is deployed. The first phase 5G uses dual connectivity, which allows the data rates from 5G and LTE radios to be combined. A high-quality LTE network is a great starting point for 5G deployment.

References

1 Holma, H. and Toskala, A. (2011). *LTE for UMTS: Evolution to LTE-Advanced*, 2e. Wiley.

2 Holma, H. and Toskala, A. (2012). *LTE Advanced: 3GPP Solution for IMT-Advanced*. Wiley.

3 Holma, H., Toskala, A., and Reunanen, J. (2015). *LTE Small Cell Optimization: 3GPP Evolution to Release*, vol. 13. Wiley.

4 Global Mobile Suppliers Association (2018). www.gsacom.com.

5 NGMN white paper about 5G requirements (2015). https://www.ngmn.org/fileadmin/ngmn/content/downloads/Technical/2015/NGMN_5G_White_Paper_V1_0.pdf

6 ITU-R 5G requirements. (2015). http://www.itu.int/dms_pubrec/itu-r/rec/m/R-REC-M.2083-0-201509-I!!PDF-E.pdf

7 3GPP Technical Report (2010). Feasibility study for Further Advancements for E-UTRA (LTE-Advanced). 3GPP TR 36.912.

16

5G-Advanced Overview

Antti Toskala and Harri Holma

Nokia, Finland

CHAPTER MENU

16.1 Introduction

5G-Advanced refers to the 5G evolution defined in 3GPP Releases 18 and beyond. 5G-Advanced brings many enhancements to 5G radio, boosting its capabilities towards the 6G era beyond 2030. The first part of 5G-Advanced specifications with Release 18 will be completed in 3GPP during the first half of 2024 and is expected to be available commercially during 2025, followed by Release 19, planned to be ready at the end of 2025, with market availability likely in the 2027 timeframe.

5G-Advanced targets utilizing 5G mobile networks for a number of other use cases beyond mobile broadband and ultra-reliable communications. Metaverse future requires extended reality (XR) support, which is one of the 5G-Advanced capabilities. Other focus areas include uplink improvements, mobility enhancements, Internet of Things (IoT) connectivity, super-accurate positioning and timing, artificial intelligence (AI), and machine learning (ML) utilization and energy efficiency. This chapter illustrates 5G-Advanced schedule, the main topics included in the specifications, and the expected benefits and use cases. This chapter concludes with an outlook for 6G.

5G Technology: 3GPP Evolution to 5G-Advanced, Second Edition.
Edited by Harri Holma, Antti Toskala, and Takehiro Nakamura.
© 2024 John Wiley & Sons Ltd. Published 2024 by John Wiley & Sons Ltd.

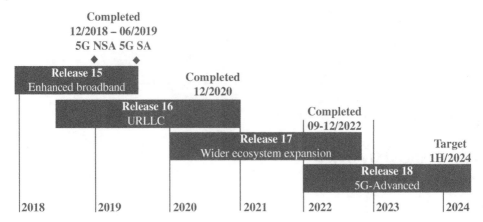

Figure 16.1 3GPP schedule for 5G and 5G-advanced.

16.2   3GPP Schedule

3GPP completed the first set of 5G specifications for non-standalone (NSA) operation as part of Release 15 at the end of 2018, as discussed in Chapter 2, leading to the first commercial 5G network opened in April 2019. Standalone (SA) version of 5G was completed in June 2019. Release 15 provided an excellent starting point for enhanced broadband services – both mobile broadband and fixed wireless access. The next step, Release 16 was completed at the end of 2020, including capabilities for offering ultra reliable low latency communication (URLLC). 3GPP work for 5G evolution continued in Release 17, providing expansion of the 5G ecosystem towards new areas including IoT optimization, non-Terrestrial networks (NTN), public safety, and drones. The next phase of 5G evolution in Release 18 started in 2022, with the target completion date of the first half of 2024. 3GPP decided to label 5G evolution in Release 18 and beyond as 5G-Advanced. This chapter provides a short summary of Releases 16 and 17 and an overview of the main contents of Release 18. The timing of 3GPP release is shown in Figure 16.1.

16.3   5G-Advanced Key Areas

5G-Advanced Release 18 features can be divided into four different categories, as shown in Figure 16.2. The topics can be divided as follows:

- Extension, which enables new use cases, such as IoT, NTN, or unmanned aerial vehicle (UAV) operation. Alternatively, reaching larger areas with better uplink coverage.
- Experience, with enhancements to the user experience with services like XR, enablers including, besides radio improvements and service awareness, edge computing or better capacity with multiple input multiple output (MIMO) improvements.
- Expansion beyond connectivity, which includes new services with more accurate positioning or proving Time as a Service (TaaS).
- Excellence refers to operation excellence, which covers new possibilities for network topology, better energy efficiency, and the use of AI/ML for network performance improvements and optimization.

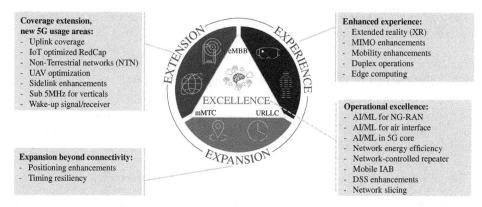

Figure 16.2 Key areas in 5G-Advanced.

5G-Advanced enhances 5G capabilities in multiple areas. The key areas are illustrated in Figure 16.3. It also shows how to bring improvements to 5G-Advanced. The key areas can be highlighted as:

- XR, including virtual reality (VR) and augmented reality (AR) improvements for more efficiency, guaranteed data rates, lower device power consumption, and support for edge computing.
- Radio performance enhancements include extended coverage, MIMO improvements, and mobility enhancements.
- IoT optimization with reduced capability (RedCap) devices for lower cost and lower power consumption.
- Superaccurate positioning and exact timing delivery, TaaS.
- Narrowband sub-5 MHz 5G for replacing GSM-Railway (GSM-R) with 5G.
- Enhanced sidelink for direct device-to-device operation, especially for public safety use cases but also for consumer use cases with unlicensed sidelink.
- Network operation efficiency includes flexible use of the Time Division Duplex (TDD) spectrum, increased utilization of AI and ML, and improved energy efficiency.

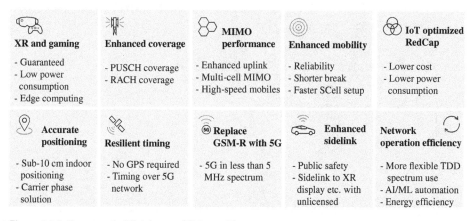

Figure 16.3 Key areas in 5G-Advanced Release 18.

The following sections present these key areas in more detail, including how they are impacting Release 18 standards.

16.4 Extended and Augmented Reality

The use of XR is creating different requirements than the classical mobile broadband type of traffic. The traffic has more stringent requirements for reliability and latency, and data rates also tend to be relatively high, with dependency if we are talking about AR or VR.

The XR service, as also evaluated in [1], has the following characteristics:

- Frame rate, which typically would take values like 60, 90, or 120 frames per second (fps).
- Data rate ranging from 10 Mbps (AR), or from 30 Mbps (VR) upwards, with high dependency on the particular service and equipment being used and resulting capabilities for high-resolution graphics use.
- The reliability in terms of packet success rate has been often assumed to be 99%, and the packet delay budget over the air, depending on the case, has been varying between 7 and 13 ms.
- Additionally, when dealing with services like VR, there is also a trade-off between latency and the required maximum data rate for the connection. When the connection has low enough latency, the XR service (or reaching a good user experience) allows sending accurately only content for covering the field of view, as the service can react fast enough for the change of gaze focus. Would there be larger latency instead, it would need to be compensated by sending those parts of the view with high accuracy, which would not necessarily be in the current field of view otherwise. This also requires that the servers providing the service are not physically located too far from the mobile user to benefit from taking the gaze focus into account, as shown in Figure 16.4.

More information on the different XR services can be found in [2].

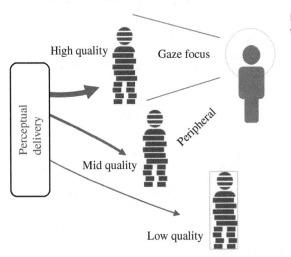

Figure 16.4 Accuracy adaptation in XR for outside gaze focus.

In order to enable RAN to deliver full support for XR operation, one needs first to identify which traffic is XR traffic, which can be achieved by extension of the 5G quality-of-service (QoS) flow. The 3GPP Release 18 XR work item [3] covers the following points for service awareness:

o 5G Core provides RAN semi-static information per QoS flow on the QoS service parameters as well as dynamic info for each PDU set and identification. The end of the data burst indicated is included.
o 5G user equipment (UE) providing RAN assistance information, covering for example periodicity and UL traffic arrival information.
o 5G RAN providing congestion information to the core.

Once the service characteristics are known, the scheduling functionality can be adapted to the timing characteristics of a given XR service, such as the already mentioned frame rate as well as the jitter that can be tolerated.

The enhancements in Release 18 on the RAN side (besides the service awareness) are addressing improvements for capacity and reduction of the UE power consumption when running XR service. The use of the continuous high data rate with low latency requirements causes UE to consume a lot of power, thus enabling power savings with DRX, otherwise reduced processing pipeline is important when possible, with XR service. The Release 15-based network can support XR-type service, but as shown in the Release 17 XR study [4], the number of simultaneous users is limited when there are no enhancements done. The Release 18 improvements, concluded from early Release 18 study [5, 6] for power saving and capacity, are addressing the following areas:

o Power saving, to be able to accommodate the resulting periodicities from the different frame rates since for example, 60 fps does not go even with 10 ms 5G frame structure, thus configuring a DRX/DTX operation with Release 15 principles.
o Capacity improvements to enhance uplink configured grant operation as well as enhancements for Buffer Status Reports (BSR). As part of the Release 17 study on XR, more improvements were identified, expected to be seen in Release 19 to enable a large number of parallel XR per cell.

In general, the XR service can have quite a varying range of requirements, depending on if any is dealing with full XR type of service, which often includes information in the uplink direction about users' actions/movement, which is time critical while limited on the data rate needs.

Besides the items part of the Release 18 work item, the topics in other work items are also important to make the XR service work well. One of the key elements is performance with mobility. Even if the user would not be moving too much, the change of serving cell may take place. In this situation, it is vital to minimize the interruption duration, as otherwise the interruption in the data flow becomes visible in the end-user video stream, or the motion-related control feedback experiences too much delay. Many other improvements foreseen by 5G-Advanced will be of benefit to XR services provision, especially those related to improved uplink coverage and system capacity. In addition to the radio interface aspects, edge computing-related improvements have a direct relationship with XR, as XR application processing in the network needs to be performed sufficiently close to the end user to achieve low enough end-to-end latency. Also, the architecture aspects on the core side need to be considered for full XR support [7].

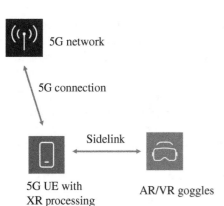

Figure 16.5 Support of XR via sidelink for XR glasses.

One element of XR operation is the connectivity with the XR glasses or other head-mounted display. While the XR glasses could be in SA mode of operation, another possibility is to have smartphones to take care of the XR-related radio reception, and provide the data only needed for rendering the image to the XR glasses. For that connectivity, a new alternative is being developed by enabling support for 5G-based sidelink operation over the unlicensed spectrum, as shown in Figure 16.5. Alternatively, the existing non-3GPP solutions could be used for this link.

If we assume the XR-glasses to have 5G connectivity functionality in them, the topic has been raised if one can expect such a device to have 4 receiver antennas, as is the current standard requirements for example for 3.5 GHz band. The fact has been recognized that such glasses, when not talking about massive VR-helmet, have limitations on how many antennas can be placed on them. Thus, 3GPP will continue to address the boundary conditions for wearable XR devices to allow for reducing the number of antennas it supports. This is to be covered still in Release 18 specification during 2024.

16.5 Superaccurate Positioning

5G-Advanced Release 18 includes a large variety of different improvements for better positioning accuracy. The solution that seems to have the most potential for a real step change in accuracy is the carrier phase positioning (CPP). The approach is known from satellite-based solutions and with the CPP done based on the 5G gNB transmission, accurate positioning becomes possible in indoor environments, underground facilities, and other situations when satellite visibility cannot be ensured.

The resulting centimeter-level accuracy, both indoors and outdoors, using 5G-Advanced is beneficial for industrial use cases like robotics or logistics, and for remote control, automotive, and public safety. 5G-Advanced-based positioning becomes an important complement and backup solution on top of satellite-based options for applications that need to ensure positioning availability in all situations.

The principle of CPP is illustrated in Figure 16.6. The network is also expected to deploy one or more reference devices, which are located in a known position. The use of reference devices allows the elimination of errors that come, for example, from timing errors on the network side, and thus helps to calibrate the network for the best possible

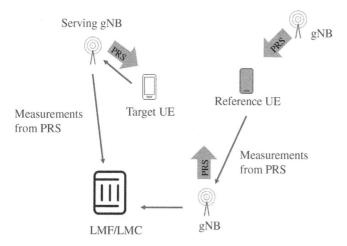

Figure 16.6 Carrier Phase Positioning (CPP) principle.

accuracy. The UE measures the timing from the carrier phase and reports this to the network, and then location management function (LMF) will calculate the position while taking into account the correction factors derived from the reference device reporting.

The CPP is not only about downlink measurements but also uplink measurements as part of Release 18. When downlink measurements are done from Downlink Positioning Reference Signals (DL-PRS), the uplink measurements are done from Uplink Sounding Reference Signals (UL SRS).

The other approach for improved accuracy is to consider aggregating more than a single 100 MHz carrier, but in very few cases there are multiple 100 MHz carriers available for a single operator within a single frequency band. Thus, improvements with this approach are not foreseen to be practical for most deployments, and achievable accuracy improvement is not expected to be that massive as amount of spectrum for intra-band aggregation is often not much more than for a single carrier case, were more than would be possible for a single carrier. If the carrier is 100 MHz and the other carrier, for example, only 30 MHz, then the achievable improvement is not going to be very significant either.

Earlier releases have brought many positioning solutions known from the LTE to the 5G side, as the first 5G release was rather limited in the positioning capabilities, supporting only assistance data provisioning to satellite solutions and the use of cell ID. Naturally, the first phase 5G UEs did not need those solutions on the 5G side as they were using LTE as the anchor carrier.

16.6 Radio Performance Boosters

16.6.1 Enhanced Coverage

The coverage work for 5G-Advanced is built on top of the enhancements done in Release 17, as was covered in Chapter 17, with the following main areas being addressed:

- Power domain enhancements, which basically aims to get more power out form the UE, like with frequency domain spectrum shaping (FDSS).
- Dynamic uplink waveform switching between capacity-optimized OFDM operation and coverage-optimized single carrier operation (DFT-S-OFDMA).
- Random Access Channel (RACH) repetition enhancements.

The power domain enhancements target getting more power out of the UE to boost uplink capacity. The specification has different power classes, with most typical ones being power class 2 (PC2) with 23 dBm, and power class 3 (PC3) with 26 dBm output powers. The coverage improvements are addressed in more detail in Chapter 17.

16.6.2 Multiple Input Multiple Output Performance

The key target in the MIMO area is to enhance uplink data rates and capacity. Higher uplink data rates are required for services like XR. The interest in uplink MIMO operation has also increased with the move towards SA 5G. When not needing to have uplink transmission to LTE anymore, as with the 5G-LTE dual connectivity case, one can use both uplink transmitters to enable uplink MIMO. This allows better performance and typically better coverage as power is no longer shared between LTE and 5G transmissions.

- Higher uplink single-user peak rates with 8-layer MIMO targeting special use cases like fixed wireless access (FWA) or customer premises equipment (CPEs).
- Higher uplink multiuser cell capacity with multiuser MIMO by enabling up to 24 orthogonal DM-RS ports.
- Performance optimization with higher mobile speeds, to have the CSI feedback to better tolerate mobile movements.
- Further enhancements for the multi-TRP framework (coherent joint transmission and asynchronous operation), when sending from multiple sites, when there is timing and phase synchronization available.
- Multi-TRP uplink enhancements.
- mmWave multi-panel UE simultaneous transmission control.

More details on 5G MIMO evolution on top of Release 15 can be found in chapter 17.

16.6.3 Enhanced Mobility

The mobility performance was addressed in Release 16, with the introduction of solutions like conditional handover (CHO) for better reliability or dual-active protocol stack (DAPS) for reduced service interruption time, as discussed in Chapter 17. However, the DAPS resulted in a rather complex solution and was not working together with CHO such that both elements would be available simultaneously.

The Release 18 improvements are addressing the following items:

o L1-triggered intercell mobility (LTM) for reduced mobility latency. The approach for fast cell change is having multiple candidate cells configured and then using the L1/L2 signaling to dynamically switch between the candidate cells. This is limited to intrasite cases only in Release 18. The solution includes enhancements for intercell beam management by defining the L1 measurements and reporting, as well as beam indication.

o CHO enhancements for RAN to optimize data forwarding and in general reduction of signaling.

o Dual Connectivity (NR-DC) with selective activation of cell groups via L3 enhancements. This further allows cell group change after cell group change without reconfiguration and re-initiation of CPA and CPC.

The interruption time with L3 mobility & CHO can be reduced with RACH-less operation, as included in Release 18 specifications. Originally for use with NTN and mobile IAB but can be extended for regular L3 mobility as well.

16.7 New Vertical Use Cases

Release 18 continues to extend 5G towards new vertical use cases with the following directions:

- Enabling support of UAVs via the 5G network, which allows better use of the 5G network for different UAV use cases like surveillance, delivery, or operation as support for public safety operations. The interference created by UAVs is also considered, especially if the UAV is used for providing live video feed from an event.
- Train communication networks with support of smaller than 5 MHz operating bandwidth to enable migration from GSM-R. The train control solution is generally known as the Future Railway Mobile Communication System (FRMCS), where 5G acts as the connectivity layer.
- The same solution is foreseen to be usable for power grid networks in some regions with 3 MHz allocations in place.
- Further steps for enabling support of the public safety ecosystem over 5G, with enhancements for UE relaying operation with sidelink as well as sidelink positioning. Some of the enhancements would also fit for use with automotive.
- NTN enhancements, on top of the basic definitions from Release 17 for transparent payload operation to allow better service continuity and mobility between NTN and terrestrial networks, as well as verification of the UE location and use of bands above 10 GHz for satellite operation. The enhancements for enabling NTN operation could also be used for air-to-ground (ATG) or high-altitude platforms (HAPS) operation as those use cases are not foreseen requiring anything extra.
- V2X has roots in Release 16 for vehicle-to-vehicle (V2V) communications and was extended to enable more power-efficient operation in Release 17 for vehicle-to-pedestrian use case. Further enhancements are included in Release 18 for V2X operation, such as extending operation with carrier aggregation (CA) and extending the operation to unlicensed bands.

The NTN principles in Release 17 and 18, and support for other vertical use cases, are covered in more detail in Chapter 19.

16.8 Resilient Timing

The critical infrastructure of society relies on the timing acquired from satellite systems. That includes telecoms, finance, transportation, power, and utilities. It is estimated that 3% of businesses rely on timing from satellites.

The delivery of TaaS via the 5G network enables the provision of alternative timing source via the 5G network, as shown in Figure 16.7. The timing is today typically obtained either from satellite or via fixed network infrastructure. Both of those solutions have their challenges, either ensuring satellite availability or installing fixed connectivity for accurate timing delivery in a way that high accuracy can be ensured. In some cases, there are also requirements in place for critical infrastructure to have alternative solutions in place, would the satellite-based timing be unavailable for any reason, while in many indoor cases ensuring satellite visibility is not always even feasible.

The timing information would be provided from the core network to the 5G radio network, which would then deliver the timing information as well as information on the clock information quality. Release 17 already introduced some of the key elements as propagation delay compensation (PDC), intended for the needs of time sensitive communications (TSC) for URLLC operation. PDC is also very beneficial feature for timing delivery provision.

The Release 18 solution then allows to sell TaaS, with the RAN network enabling adapting the scheduling based on the RAN feedback, as well as necessary feedback on the clock quality information for the core and for the UE [8].

The critical infrastructure of society relies on timing acquired from satellite systems. That includes telecoms, finance, transportation, power, and utilities. It is estimated that 3% of businesses rely on timing from satellites. 5G-Advanced enables us to provide accurate timing information to the devices via the network. Resilient timing is required for indoor use cases where satellite signal is not available. Another use case would be as a back-up solution when satellite would be unavailable.

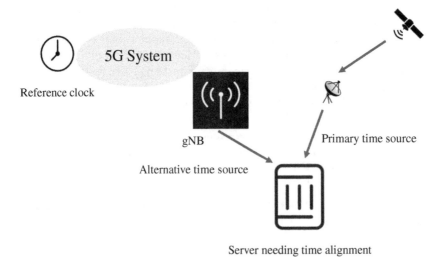

Figure 16.7 Alternative timing via 5G network.

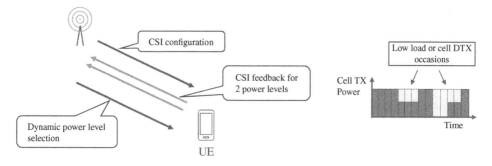

Figure 16.8 Dynamic power adaptation for power saving in connected mode.

16.9 Network Automation and Energy Efficiency

The importance of network energy efficiency has increased in recent years, and while 5G was clearly more efficient than 4G in terms of the overall design, improvements are important considering the increased traffic volumes experienced. Release 18 started work on network energy efficiency and concluded to enhance operation by allowing dynamic power or spatial domain adaptation, as shown in Figure 16.8, as well as reduce when SSB and SIB1 need to be transmitted. The principle of energy saving during active transmission is to take advantage of the variations in the traffic load. When the UE is providing the Channel State Information (CSI) feedback for the reduced power level situation, the gNB scheduler can use the reduced power level when there is not enough data to fill the full carrier. The UE feedback then provides information about what kind of transmission is still expected to work when the power level is reduced, for example, by 9 dBs from the nominal original power level.

Further, the area of network automation is progressing with AI/ML for NG-RAN addressing data collection enhancements and signaling support within existing NG-RAN interfaces and architecture. The AI/ML optimization cases, covered in Release 18, followed by earlier studies in Release 17, are network energy saving, load balancing, and mobility optimization. These are addressed further in Chapter 17.

16.10 RedCap/NR-Light for IoT

Release 17 introduced RedCap operation, known also as NR-Light. RedCap is basically lightweight mobile broadband for 5G. It was identified already as early as 2018 [9] that several solutions, like surveillance cameras, have little need for 2 Gbps reception capability, thus 3GPP conducted the study on the potential cost savings to go for 20 MHz bandwidth and reduced number of antennas. The study found that there can be a 60–70% cost compared to the Release 15 full-blown 5G modem. The cost savings would be naturally more for the lower capability Release 18 modem which is expected to do 10 Mbps, while Release 17 modem definition is around 100 Mbps for a single antenna device with FDD operation as shown in Table 16.1. At the time of writing this first, Release 17-based platforms have been announced already and are expected to enter the market during 2024 timeframe.

Table 16.1 Different steps in RedCap in Release 17 and 18.

	Release 15 baseline	Release 17 RedCap	Release 18 RedCap
Bandwidth	100 MHz	20 MHz	5 MHz (for data)
Peak Rate	2 Gbps	100 Mbps (single RX)	10 Mbps
Total cost	100%	−60%	−70%

16.11 Outlook For 5G Release 19

3GPP has started the work on defining Release 19 content, with the Release 19 package being agreed to in December 2023, and then eventually the specifications to be ready for the end of 2025. Figure 16.9 shows the most popular items based on the Release 19 RAN workshop, which took place in June 2023, with the key items reflected in [10].

The key items for Release 19 are expected to address improvements in areas like mobility, XR, MIMO, and energy efficiency, as shown in Figure 16.10. Also, the outcomes from the studies in Release 18 are going to be reflected in many areas like AI/ML for air interface or duplex evolution.

While some of the topics like XR enhancements or mobility enhancements are a direct continuation of Release 18 5G-Advanced items, some of the topics are clearly bridges towards 6G with less expectations for actual deployment on 5G side. An example of such an item is sensing, which is expected to see preparatory work done in the area of channel modeling, but otherwise, the actual specification work and more detailed studies are foreseen for Release 20 and Release 21 timeframe, when work on 6G is also expected to get started. Other new areas of study in Release 19 include duplex evolution which would not be trivial to use with existing deployments, and new use cases with Ambient IoT which aims to study solutions for low-cost and short-range type of IoT use cases.

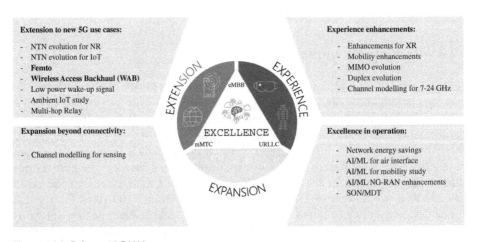

Figure 16.9 Release 19 RAN items.

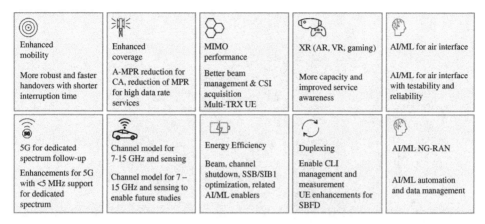

Enhanced mobility	Enhanced coverage	MIMO performance	XR (AR, VR, gaming)	AI/ML for air interface
More robust and faster handovers with shorter interruption time	A-MPR reduction for CA, reduction of MPR for high data rate services	Better beam management & CSI acquisition Multi-TRX UE	More capacity and improved service awareness	AI/ML for air interface with testability and reliability
5G for dedicated spectrum follow-up	Channel model for 7-15 GHz and sensing	Energy Efficiency	Duplexing	AI/ML NG-RAN
Enhancements for 5G with <5 MHz support for dedicated spectrum	Channel model for 7 – 15 GHz and sensing to enable future studies	Beam, channel shutdown, SSB/SIB1 optimization, related AI/ML enablers	Enable CLI management and measurement UE enhancements for SBFD	AI/ML automation and data management

Figure 16.10 5G-Advanced Release 19 key items.

16.12   Outlook For 6G

With 6G there will be new capabilities and technologies taken into use on top of the 5G and 5G-Advanced capabilities, with expected deployments to take place before 2030. Currently, 6G is in the pre-standard phase with different regional research activities looking into candidate technologies to be considered for 6G standards and eventually deployments, as shown in Figure 16.11. The work in 3GPP on 6G radio requirements and technologies is expected to start in 2024, with more detailed studies, and then in 2025 timeframe to allow the creation of specifications for the first phase of 6G for the end of 2028 timeframe. Like with other generations, more releases are then expected to follow the initial 6G release, adding further functionalities to 6G.

What are then the new technology areas envisaged for 6G? The work for 6G is foreseen to cover the key areas shown in Figure 16.12, with the following focus:

- Security, trust, and privacy will be vital for getting the 6G network to be future-proof in resisting threats that networks may face from attackers and possible jammers. For

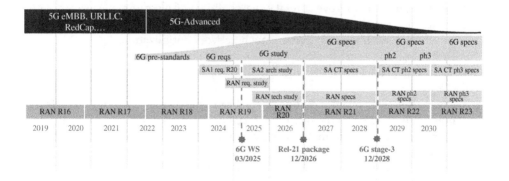

Figure 16.11 Expected 6G schedule in 3GPP.

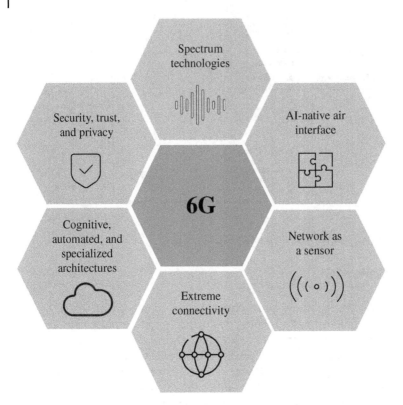

Figure 16.12 6G key technology areas.

networks to take care of critical communications, the design needs to consider the necessary protection mechanisms to address the threats of the future.

- For the spectrum, there needs to be capability to accommodate the bands currently being used for 5G (and likely continued with 5G also in the near term), as well as to cover for possible new mobile bands, like from the 7–24 GHz range or even higher frequencies.
- The increased use of AI and ML is also visible in 6G development. On one hand it is foreseen to be used to improve several functionalities and algorithm performance for the air interface, while on the other hand, AI/ML is to be used to improve operability and automation, enabling zero human touch network optimization.
- Network as a sensor, which basically looks, at on one hand, more accurate positioning, and on the other hand, using network (possible together with UEs) for sensing passive objects.
- Extreme connectivity enables on one side extremely low latency and high-reliability communications, while on the other hand facilitating low energy communications.
- For the architecture evolution, there is expectation to continue with the open interface principle to allow interworking with different networks. The new dimension foreseen is also 6G subnetworks to be considered for specific cases.

Besides these technologies, the theme to be reflected in all areas is sustainability, which on the radio side also covers the importance of energy efficiency, as with the increasing

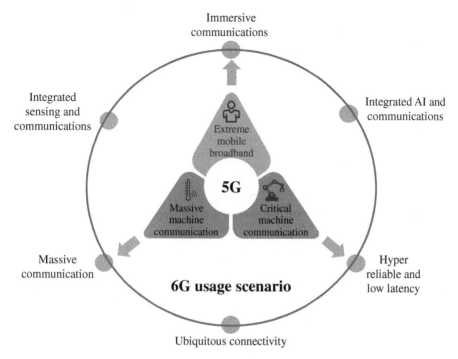

Figure 16.13 6G usage scenarios.

data volumes it will be vital to improve energy efficiency to reduce overall energy consumption of the cellular networks.

The usage scenarios are identified in ITU-R for 6G (IMT-2030) [11], as an expansion of 5G (IMT-2020), are shown in Figure 16.13.

The new capabilities identified in ITU-R reflect the usage scenarios, with capabilities both being enhanced from IMT-2020 as well as new capabilities for IMT-2030, as shown in Figure 16.14. The exact value for new capabilities is still to be identified, except for positioning where the capabilities are foreseen to be centimeter-level positioning accuracy.

To meet these requirements on new capabilities, new technical solutions are needed, thus one can expect many novel areas to be raised as part of the system definition, including considerations of metamaterials for improving indoor coverage [12] with new solutions, in addition to the classical repeater approach.

Spectrum is the key requirement for the wireless system. The target of the 6G design is to utilize any spectrum blocks from sub-1 GHz frequencies to above 100 GHz. 6G can utilize newly available spectrum blocks and support efficient refarming from 5G to 6G on the existing spectrum bands. The key new spectrum blocks for 6G are shown in Figure 16.15 and listed below:

- New mid-band spectrum at 6–15 GHz range for providing high data rates and high capacity using existing base station locations.
- New sub-Teraherz spectrum above 90 GHz for enabling sensing and extreme data rates for short-range communication.
- New low bands below 600 MHz for extreme coverage for low data rates.

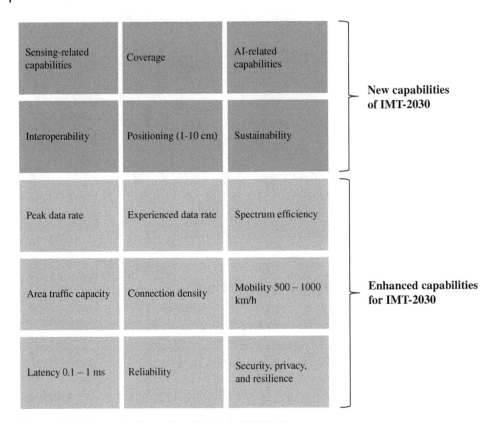

Figure 16.14 New and enhanced capabilities for IMT-2030.

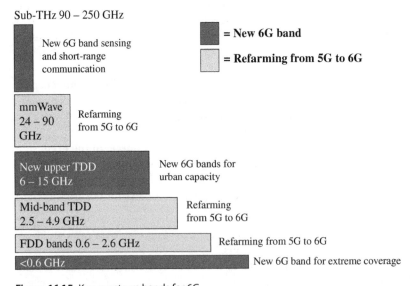

Figure 16.15 Key spectrum bands for 6G.

The new mid-band spectrum deployment should ideally be able to utilize the existing site grid in urban areas and provide similar coverage at least in the downlink direction. The target is to boost radio network capacity and data rates substantially compared to 5G without major network densification. The main solutions are more bandwidth and an extremely massive MIMO antenna. The ideal bandwidth per operator should be a minimum of 400–500 MHz to provide more capacity than typical 5G with 100 MHz. If there are a total of 3–4 operators per country, the amount of new spectrum would ideally be 1–2 GHz. Lots of work will be required globally to make such spectrum blocks available in World Radio Conferences (WRC) in 2023 and 2027.

Sub-THz spectrum refers to frequencies above 90 GHz. The signal propagation is extremely limited on such high-bands, which makes sub-THz suited for very short-range and very high data rate communications. Another use case for sub-THz is sensing which is one of the new 6G capabilities.

Smooth and dynamic refarming of spectrum from 5G to 6G is one of the design targets. The target is to maximally reuse 5G assets, including spectrum, sites, and even RF units, when rolling out the coverage for 6G. If 6G can enable the dynamic of 5G spectrum, it can simply speed up 6G coverage rollout and 6G deployments.

Radio capacity and data rates can be boosted with more bandwidth. The other solution for boosting radio performance on 6–15 GHz bands is to have even more advanced massive MIMO antennas, with more antenna elements and more TRXs. The antenna element spacing is relative to the wavelength. When the transmission frequency gets two times higher, the antenna element spacing gets 50% smaller, which allows four times more antenna elements in the same area. If a 5G antenna at 3.5 GHz has 192 antenna elements, a 6G antenna at 7 GHz can have 768 antenna elements in the same area. Having more antenna elements brings two benefits: (1) Antenna gain increases, which helps to mitigate the path lost, and (2) Higher spectral efficiency can be obtained with more advanced beamforming with more TRXs. The antenna sizes, the number of antenna elements, and theoretical antenna gains are illustrated in Figure 16.16. The illustration

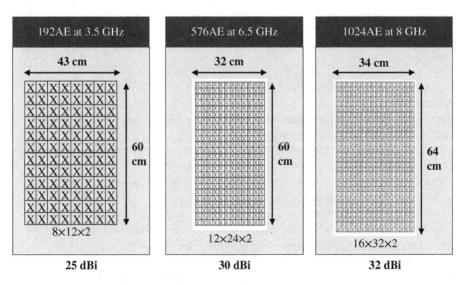

Figure 16.16 Extreme massive MIMO antenna for new 6G band at 6–15 GHz.

shows that it will be feasible to support more than 1000 antenna elements on the new mid-band spectrum. Therefore, massive MIMO evolution with more powerful hardware and more advanced software features is expected to be important for 6G radio networks.

16.13 Summary

Since the start of 5G, we have seen rapid deployment in many markets with high-speed connectivity being enabled to be used for different applications. New services, user segments, and vertical industries are adopting 5G technology.

5G-Advanced is addressing those requirements to achieve more, with lower latencies, fewer interruptions, and less power consumption, both on the UE and network side. Also, the special needs of various verticals are being addressed, as well as the needs from some segments to have lower cost devices for less demanding services, which as of yet has to be future-proof in terms of technology to be supported for long in the future.

While writing this, we see the first Release 17 features coming to the market in early 2024. When we consider the first Release of 5G-Advanced, Release 18, to be stable in June 2024, one can foresee the first 5G-Advanced features to appear in the market in the 2025 timeframe.

The development of new services, which take benefit of 5G capabilities, is ongoing and it remains to be seen whether it is metaverse or some other new use case that will drive the traffic growth towards the next decade.

Some of the technologies being investigated are paving the way for the 6G activity in 3GPP. We expect to start 3GPP in 2025 [13] with detailed 6G studies towards the expected 6G specification availability in 2028, to meet the timelines also for the IMT-2030 submission towards ITU-R. 6G will bring new usage areas and will shape the society for the next decade.

References

1 Petrov, V., Gapeyenko, M., Paris, S. et al. (2021). *Standardization of Extended Reality (XR) over 5G and 5G-Advanced 3GPP New Radio.* Magazine: IEEE Wireless Commun.

2 Shannon, L. (2023). *Interconnected Realities, How the Metaverse Will Transform Our Relationship with Technology Forever.* Wiley.

3 3GPP Tdoc RP-230786 "Updated WID for WID on XR Enhancements for NR", Nokia et al, March 2023.

4 3GPP Technical report TR 38.838, "Study on XR (Extended Reality) evaluations for NR", version 17.0.0, December 2021.

5 3GPP Tdoc RP-220285, "Study on XR Enhancements for NR", Nokia, March 2022.

6 3GPP Technical report TR 38.835, "Study on XR enhancements for NR", April 2023

7 3GPP Tdoc SP-220705, "Architecture enhancement for XR and media services", China Mobile, June 2022.

8 3GPP Tdoc RP-230754, "New WID on NR Timing Resiliency and URLLC enhancements." March 2023.

9 3GPP Tdoc RP-180430, "Considerations for NR-IoT work in Release 16", Nokia, March 2018.

10 3GPP Tdoc RWS-230488. "RAN Chair's Summary of Rel-19 Workshop", June 2023.

11 ITU-R Draft New Recommendation, "ITU-R M.[IMT.FRAMEWORK FOR 2030 AND BEYOND], Framework and overall objectives of the future development of IMT for 2030 and beyond", Document 5/131, June 2023.

12 Tsuboi J. et al., "Evaluation of indoor area improvement in the high frequency band using metasurface lenses, FSS technology and relay stations" In proceedings of IEEE VTC2022 Spring WS, June, 2022.

13 3GPP Tdoc RP-233985 "High-level Consideration for 6G Timeline", December 2023.

17

Radio Enhancements in Release 16–18
Harri Holma and Antti Toskala

Nokia, Finland

CHAPTER MENU

17.1 Introduction

Following the 5G introduction in Release 15, there have been several improvements to boost the 5G radio performance. This chapter will address the improvements for coverage, MIMO operation, UE power saving, as well as improvements with dual connectivity and carrier aggregation. Further, the steps in mobility improvements are addressed along with artificial intelligence (AI)/machine learning (ML) for the air interface and integrated access and backhaul. The XR-Related radio improvements were addressed in Chapter 16.

17.2 Coverage Enhancements

The importance of coverage is continuously recognized in 3GPP, and following the introduction of 5G, especially the 3.5 GHz bands, and in some cases even higher bands, emphasizes the its importance when compared to the long term evolution (LTE) coverage. In the first phase, 5G device coverage was impacted by the 5G-LTE dual

5G Technology: 3GPP Evolution to 5G-Advanced, Second Edition.
Edited by Harri Holma, Antti Toskala, and Takehiro Nakamura.
© 2024 John Wiley & Sons Ltd. Published 2024 by John Wiley & Sons Ltd.

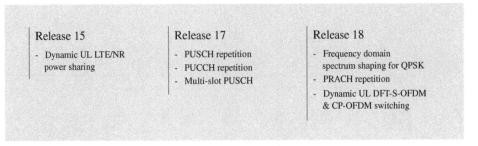

Figure 17.1 Steps to improve uplink coverage in 5G specifications.

connectivity as the LTE anchor carrier took the power it needed and only what was left was available for the 5G uplink. In the early phase, the issue was even more severe as the power was not dynamically shared but had to be configured in slower fashion to ensure that enough power was booked for LTE. LTE had to be prioritized as the control plane was working over the LTE, thus that connection needed to be maintained at all times due to the control plane traffic on the LTE carrier. In some early deployments only downlink data was sent over 5G connection while all uplink user data was sent via LTE.

The first step of improvement came with the introduction of more dynamic power sharing for the UEs and with the start of the standalone (SA) 5G operation. When there was no longer a need for the UE to keep LTE connection running, one could use all the transmit power available for the 5G uplink.

Figure 17.1 indicates the solutions introduced for improving the coverage in Release 17 and 18, with work foreseen to carry on in Release 19 as well [1].

Release 17 improvements for coverage were focusing on physical uplink shared channel (PUSCH) and physical uplink control channel (PUCCH) repletion, and multiple slot PUSCH as follows:

- PUSCH repetition, defining support for up to 32 repetitions.
- PUCCH repetition, which included dynamic repetition and inter-Slot frequency hopping (FH).
- Multiple slot PUSCH, often referred also as transport block processing over multiple slots (TBoMS), which addresses how multiple uplink slots could be processed together even if not sent in consecutive slot uplinks.

The use of repetition has limitations regarding how long the UE transmitter is able to ensure phase and power level continuity between multiple slots. Ideally, the phase and power level continuity would allow the gNB receiver to rely on these properties in detection and make better channel estimation with DMRS bundling, and thus allow reception with smaller signal levels, thus reaching better uplink coverage. The UE side challenge is how to keep the phase and power level continuity if there is gap in transmission operation between the slots.

The work in Release 18 addressed further aspects as:

- Enabling repetition for physical random access channel (PRACH) for the Message 1 (Msg1). Enabling UE to repeat the PRACH message for the same beam allows up to 5 dBs improvement with 4 repetitions for the link budget for PRACH when considering rural deployments [Cov1]. The principle of PRACH repetition is shown in Figure 17.2.

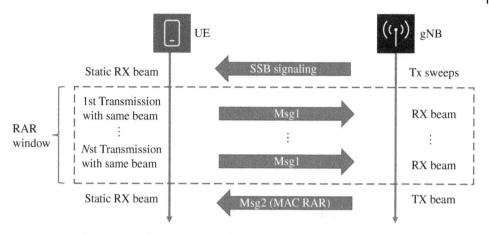

Figure 17.2 PRACH Msg1 repetition with the same beam.

- Dynamic waveform switching. In the Release 15-based operation, the waveform is adjusted with RRC signaling, which adds delay to the process and causes the network to instruct UE to change the waveform early enough. When the waveform switching is more dynamic, one can wait with the waveform change only until the UE hits power limitation with OFDM based on power headroom report (PHR). Once the power limit is reached, the gNB can instruct the UE to switch to a more amplifier-friendly single-carrier FDMA (SC-FDMA or DFT-S-OFDMA as discussed in Chapter 6) operation, which allows higher uplink power thanks to the lower signal peak-to-average ratio (PAR). Again, once UE has more power available (based on PHR), then the network can dynamically change back to OFDM operation.
- UE transmitter PAR reduction, which is done in a transparent way in Release 18, using solutions such as frequency domain spectral shaping (FDSS). When the PAR is reduced, the UE is able to operate with reduced MPR value, and thus boost more power from the device. Release 18 includes performance requirements for a UE which would support such operations. There is no impact on the UE spectrum use or spectral efficiency achieved with such an operation, as shown in Figure 17.3.

Figure 17.3 FDSS principle for UE power boost.

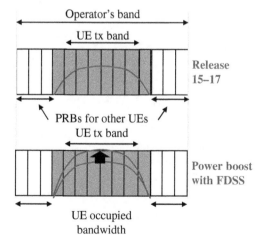

17.3   MIMO Enhancements

The evolution of 5G multiple input multiple output (MIMO) operation has continued intensively after the Release 15 which enabled scalable and flexible MIMO framework, and MIMO support on Sub-6 GHz as well as with mmW frequencies. This included support for beam-based operation, a necessary enabler for mmW deployments.

The MIMO evolution is summarized in Figure 17.4 from Release 16 to 18, and MIMO evolution is foreseen to continue in Release 19 as well. The interesting element is also the increased interest for uplink MIMO, as when more and more networks enable SA 5G, then UEs do not need to use the extra transmitter for LTE anymore, and thus can enable easier use of uplink MIMO on 5G side.

Release 16 addressed the most obvious shortcomings of Release 15, continuing the work to deliver the full potential of 5G MIMO operation. The key enhancements were:

- The Type II CSI codebook operation enabling MU-MIMO was improved by major reduction in the CSI feedback overhead and enabling 4-stream MIMO (rank 4) operation.
- Multi-TRP/panel transmission enhancements for better operation when dealing with different transmission points. Both the cases with ideal and nonideal backhaul were considered such that one would use either single PDCCH reception when the backhaul would be ideal, or multi PDCCH reception when using nonideal backhaul, as shown in Figure 17.5. The work covered both downlink and uplink control signaling for non-coherent joint transmission, while user data was enabled only in downlink operation with multi-TRP. This was considered especially relevant for URLLC operation for improved reliability by replicating transmissions from multiple transmitters. Operation is limited to two transmission points/panels in Release 16.
- Multi-beam operation enhancements were primarily targeting mmW operation, to improve operation for reduced latency and overhead. Also, enhanced measurement and reporting of L1 SINR was included along with UE multi-panel beam indication.
- Other elements covered in Release 16 were full Tx-power UL transmissions with multiple power amplifiers and low PAPR reference signals (RS) for lowering uplink PAPR.

The Release 17 continued to address the topics raised in Release 16, such as:

- Extension of multi-TRP transmission for additional channels to enable uplink data transmission.

Release 16	Release 17	Release 18
- CSI enhancements for MU-MIMO - Multi-TRP panel enhancements - Multi-beam operation enhancements - Full TX-power uplink - Low PARR Reference signals	- Improved mmW multi-beam operation - Extension of multi-TRP transmission - More flexible SRS - Further Enhanced Type II CSI-RS	- Multi-TRP framework for coherent joint tranmission - Multi-TRP uplink enhancements - Multi-panel UE enhancements - CSI improvement for higher speed - 8 Tx uplink

Figure 17.4 MIMO evolution after Release 15.

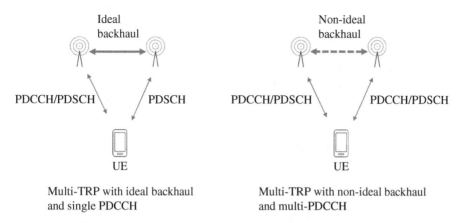

Figure 17.5 Multi-TRP operation with ideal and nonideal backhaul.

- Continuation of the Type II CSI-RS to address additional cases like the non-coherent multi-TRP/panel transmission, and utilization of partial reciprocity with FDD deployments.
- More flexible SRS transmission for better capacity and coverage.
- Improved support for multi-beam operation, especially for mmWave frequencies and devices with multiple antenna panels. For the UEs with multiple panels, there was no means for the network to know, for example, if a UE could communicate with two different gNBs with two different panels as shown in Figure 17.6, or only with a single panel. A typical mmW UE would contain more than just a single antenna panel to allow communication if, for example, the other panel is covered by the user.

As part of Release 18, further steps were taken on MIMO development:

- The CSI reporting as Type-II codebook refinement was improved to better address the performance with high and medium UE velocities, especially with sub-6GHz operation. The improvements identified not only allow good performance gain, especially for cell edge operation, but also improve the cell average spectral efficiency for high/medium UE velocities.
- Support of up to 8 Tx uplink operations to enable 4 or more layers in the UE. This is not a smartphone feature but is intended for fixed wireless access (FWA) and consumer premises equipment (CPE), as well as for vehicle and industrial devices.
- The work on coherent joint transmission proceeded as well, while the achievable benefits are based on very ideal assumptions and thus are not foreseen to be achievable

Figure 17.6 Multi-panel UE.

in the field with practical deployments. Release 19 is expected to discuss if solutions for the realistic use of coherent joint transmission can be found.
- Specification of larger number of orthogonal DMRS ports for MU-MIMO operation in uplink and downlink directions.
- Enabling the simultaneous multi-panel mmW uplink transmission for higher throughput/reliability, for up to 2 TRPs and up to 2 panels, thus also targeting FWA and CPE as well as for vehicle and industrial devices.
- The multi-TRP enhancements include support of two timing advance (TA) values for uplink multi-DCI for multi-TRP operation.

17.4 Mobility

Release 15 included the support of mobility based on the UE layer 3 measurement. There was need to improve operations in Release 16, and again in Release 18, to consider improved reliability as well as reduced interruption time as shown in Figure 17.7. Release 16 concluded with the following solutions:

- **Conditional handover enables the handover** execution phase to be handled separately from the handover preparation phase. As shown in Figure 17.8, the actual

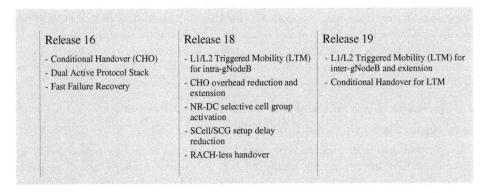

Release 16	Release 18	Release 19
- Conditional Handover (CHO)	- L1/L2 Triggered Mobility (LTM) for intra-gNodeB	- L1/L2 Triggered Mobility (LTM) for inter-gNodeB and extension
- Dual Active Protocol Stack	- CHO overhead reduction and extension	- Conditional Handover for LTM
- Fast Failure Recovery	- NR-DC selective cell group activation	
	- SCell/SCG setup delay reduction	
	- RACH-less handover	

Figure 17.7 Mobility evolution after Release 15.

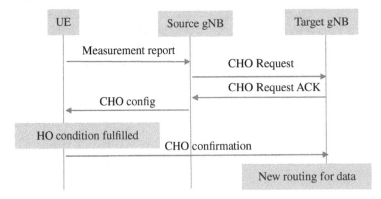

Figure 17.8 Conditional handover principles.

handover is triggered only after a specific handover condition is met. The network configures based on UE measurement or other information (for example expected UE movement), and UE can then execute the handover faster once the actual handover conditions are met. Too many handover candidates should be avoided as well due to the network resource considerations.

- **Dual active protocol stack (DAPS)** for HO interruption time reduction, which reduces the interruption time when a new connection is basically established with other protocol stacks for new gNodeB, before the old one is terminated.
- **Fast failure recovery** (T312 timer based) to speed up recovery from failure situations.

While solutions sounded good at the time when specifications were done for Release 16, it seems that especially DAPS turned out to be too heavy for UE implementation and did not get traction on the implementation side. Thus, after some break in mobility enhancements in Release 17, 3GPP continued the work in Release 18 to address again how the reliability and interruption time could be further improved, as shown in Figure 17.7. Especially URLLC and new emerging services like XR would need shorter handover interruption time than available without DAPS implementation.

Release 18 work on the mobility continues in the following areas:

- L1/L2 triggered mobility (LTM) to allow for reduced interruption time. This is going to be usable only for the intra-gNB case in Release 18, thus further work will be needed in Release 19 to extend this to the inter-gNB cases as well. There are some other elements expected to be finalized in Release 19 for LTM to enable full support, as well as improve reliability with possible combination with the conditional handover (CHO). The L1 measurement is done based on synchronization signal block (SSB).
- The work on CHO improvements aims to reduce the signaling, both on the radio interface as well as on the network internal interfaces, by reusing earlier configurations. The CHO framework is also being expanded to include the target master cell group (MCG) and target secondary cell group (SCG) with NR-dual connectivity (NR-DC) for conditional PSCell addition (CPA) and conditional PSCell change (CPC). The L3 mobility can benefit from the work done for NTN & LTM mobility and use RACH-less handover for reduced interruption time for for "normal" UEs.

17.5 UE Power Saving

The UE power consumption has played a key role since the 3G days, when the use started to move towards always connected applications in the UE, thus resulting in the increased power consumption, together with the introduction of bigger screens in UEs to add to the power consumption. Thus, with the introduction of the larger bandwidths and capabilities with 5G, one had to pay special attention to the power consumption of the UE.

Release 15 already included a set of capabilities like bandwidth part (BWP), as discussed in Chapter 7, as well as discontinuous reception (DRX) in connected mode, SCell dormancy, and UE overheating assistance information, as shown in Figure 17.9.

Release 15	Release 16	Release 17
- Bandwidth part (BWP) for power saving - C-DRX - Scell Dormancy - UE overheating assistance information	- PDCCH-based power saving channel triggering UE adaptation in C-DRX - Cross-slot scheduling - Adaptation of MIMO layers - Indicator to transition out of RRC-Connected. - UE assistance information - RRM measurement relaxation	- Idle mode improvements, incl. paging enhancements - Enhanced DCI-based adaptation for PDCCH monitoring for power saving - Performance impact of Relaxing UE measurements for RLM and/or BFD

Figure 17.9 UE power saving features evolution from Release 15.

Release 16 continued to add more UE power-saving features such as:

- PDCCH-based power saving signal which can control the UE adaptation in RRC-connected mode, as shown in Figure 17.10. The UE will monitor only wake-up signals, and on receiving such a signal move back to full reception with PDCCH detection and PDSCH buffering. The active BWP is naturally taken into account as well.
- Cross slot scheduling basically avoids buffering the PDSCH data when the PDCCH indicated earlier than just before that there is data coming, as shown in Figure 17.11. This allows UE to buffer only the PDCCH for decoding, and then there is enough time to decode PDCCH and start buffering for PDSCH reception at later point in time if data was scheduled for the UE. The use of cross-slot scheduling increases the latency a bit but would be suited for cases when an application can tolerate some increase in latency experienced.

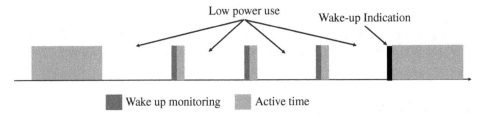

Figure 17.10 Use of the PDCCH-based power saving signal in 5G.

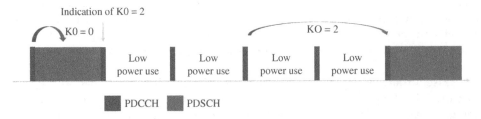

Figure 17.11 Cross slot scheduling.

- Adaptation of MIMO layers could be done when the data rate is small, and UE could save power when using fewer MIMO layers for reception.
- Indicator to transition out of RRC connected is an indication which that UE could send to trigger move to idle/inactive when the application runs out of data.
- The UE assistance information framework was extended from Release 15. The new details added cover information on preferred connected mode DRX (C-DRX) configuration, bandwidth, number of MIMO layers, RRC state preference, and number of component carriers to use.
- Intra- and inter-Frequency RRM measurement relaxation in IDLE and INACTIVE modes.

Release 17 continued the work further on the power savings, addressing the following areas:

- Enhanced PDCCH-based power-saving signal/channel triggering UE adaptation in RRC-Connected, which is intended to address cases when there is a high peak data rate with dense packet arrival.
- Idle mode improvements in the form of paging early indication and provision of additional TRS/CSI-RS occasions. Compared to LTE, which continuously sends common reference signals (CRS) the 5G UE, needs to listen longer to receive enough energy, especially in low-signal situations. The use of PEI means the UE is given PEI configuration to monitor one PEI occasion per DRX cycle, and it then avoids more frequent reception/decoding of paging occasions. The network can also provide additional TRS/CSI-RS signals to help the reception of paging and to shorten the time the UE receiver needs to decode the 5G signal for paging reception.
- Intra- and inter-Frequency RRM measurement relaxation in IDLE and INACTIVE modes targets to reduce the number of SSB occasions the UE needs to measure. A typical case when this could be considered is when UE is in a strong signal area, with limited mobility, then the radio link failure (RLF) detection/beam failure detection (BFD) could consider longer evaluation period.

17.6 AI/ML for Air Interface and NG-RAN

The use of AI/ML was considered for both air interface as well as for NG-RAN operation. The use of AI/ML for NG-RAN was studied first in Release 17 [2], and topics were then specified in Release 18. For the A/ML for air interface, the study was done in Release 18, and the work item was then planned in Release 19 to address the use cases studied during the study phase [3].

Besides the actual use cases studied and specified, both topics represent important enablers, how to bring AI/ML into 3GPP standards, determine what needs to be specified, what is left for individual vendor implementation, and how to ensure the use of AI/ML does not have negative impacts on system performance and reliability. Further elements are how are the models managed, how is related data collected, how are the AI/ML models trained, how is the resulting performance tested, and how to ensure conformance to the regulation and requirements.

The resulting AI/ML framework should be the future, enabling evolution towards 6G.

17.6.1 AI/ML for Air Interface

Part of the overall framework is the AI/ML model life cycle management, which considers how data is collected, model trained, and transferred, as well as actual model inference operation (actual use of the model), and later possible model update.

3GPP has defined different levels of possible collaboration for the AI/ML operation:

- No collaboration.
- Signaling-based collaboration without model transfer.
- Signaling-based collaboration with model transfer.

The operation with the model transfer, as shown in Figure 17.12, is naturally the most complicated to specify as well, as it could cover possible specifications of the model structure and get transferred in possible open format. There are different variations identified in terms of training location and how the model is transferred. It remains still to be determined which level of collaboration would be supported in the specifications in the first phase of Release 19.

The use cases covered for AI/ML for air interface were as follows:

- CSI feedback enhancements
- Beam management
- Positioning accuracy enhancements

The CSI feedback enhancements target reducing the CSI feedback overhead and also improving the CSI feedback accuracy and considering time-domain CSI prediction. The CSI feedback compression was considering the use of a two-sided AI model where there would be models both at the UE side (for encoding) and gNB side (for decoding), while otherwise relying on the existing CSI feedback framework. The time domain CSI prediction was based on the model on the UE side.

Beam management was considering beam prediction in time or spatial domain, based on UE sided model, to achieve a reduction in the overhead and latency, and better beam selection accuracy. The two specific cases were spatial domain downlink beam prediction based on measurement results and temporal downlink beam prediction based on historic measurement results.

The positioning was having either UE location as the output from the AI/ML model based on, for example, fingerprinting from channel measurements. Another approach was AI/ML assisted positioning where AI/ML was used to improve measurements, for example, with the identification if a connection was line-of-sight (LOS) or not.

The conclusions and results captured in Ref. [3], provide the direction for normative work in Release 19, which starts with relatively simple cases for the air interface, and is expected to address functionality/model management as well as ensure robust and improved performance, and predictable UE behavior through UE requirements and test cases. All this is paving the way for more sophisticated AI/ML solutions for additional use cases, like mobility, to be supported in further releases and towards 6G.

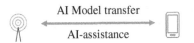

AI Model transfer

AI-assistance

UE with AI capability   UE with AI capability

Figure 17.12 Signaling-based collaboration with model transfer from network to UE.

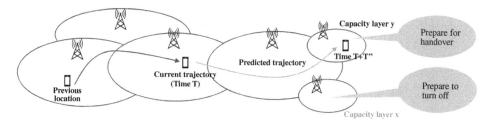

Figure 17.13 Example use of AI/ML for NG-RAN optimization with mobility and energy saving.

17.6.2 AI/ML for NG-RAN

The optimization of the 5G network is challenging with a large number of factors to measure, and consideration of the impact on capacity as well as other performance aspects of the network. The use of automation and AI/ML in general has the potential to help the data collection and analysis of the collected data.

In Release 17, 3GPP studied the use of AI/ML for different use cases and concluded to proceed with normative work in Release 18 for the use of AI/ML in NG-RAN. The following use cases cover the work item:

- Network energy saving, addressing how network energy consumption could be reduced with solutions like traffic offloading, coverage modification, and cell deactivation.
- Load balancing, aiming to optimize the network operation with load prediction in the network among multiple cells, or part of cells over multiple frequency bands and radio technologies. The objective of the operation is to distribute load effectively among cells or areas of cells in a multi-frequency/multi-RAT deployment for improved network performance.
- Mobility optimization, aiming to improve network performance by predicting how UEs would be moving in the future and selecting optimal mobility targets based on this.

The use cases were studied first in Release 17, with findings captured in Ref. [2], and then work progressed to define the necessary support for data collection and other operation in RAN internal interfaces in Release 18.

The different use cases are not mutually exclusive, addressing, for example, mobility optimization, as shown in Figure 17.13, with the intention to have the AI/ML predict the UE movement, and thus be prepared for handovers, accordingly, as shown in Figure 17.13. At the same time, the information obtained can be also used to aid network power saving operations.

17.7 Integrated Access and Backhaul

The integrated access and backhaul (IAB) was first introduced in Release 16 specifications for 5G, and further enhancements were done in Release 17 and 18, as shown in Figure 17.14, to enhance first the duplexing, and then in Release 18 to add IAB mobility to the picture. There is similarity with the work done in Release 18 for network controlled

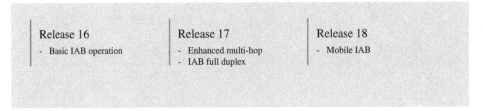

Release 16	Release 17	Release 18
- Basic IAB operation	- Enhanced multi-hop - IAB full duplex	- Mobile IAB

Figure 17.14 IAB development in different 3GPP releases.

repeater (NCR), which is another approach for improving coverage without fiber-based backhaul for a given location.

The IAB principle is to enable wireless backhaul operation using the same bands as being used for access. The IAB transmission would be using the same band as the actual access operation, avoiding the need for fiber connection to every site. The donor gNB would then have fiber connectivity. The IAB operation is being defined for both FR1 and FR2 ranges, but naturally at higher frequency bands, there are more resources available to be used for the backhaul operation.

The IAB architecture model is as follows [4]:

- The interfacing between donor and IAB node is based on the CU/DU split, with an additional adaptation layer with RLC to handle the routing information, since there can be multiple IAB nodes in a chain.
- The interface between donor and IAB nodes is called F1*, with the PDCP and SDAP in the CU side in line with the 3GPP higher layer functional split. The F1* functional split is not visible for UE, but UE assumes being connected to a regular gNB in any case.
- The radio connection ends in the mobile termination (MT) part in the IAB node, with the radio connection being the normal NR Uu interface for the UEs connected to the IAB node(s), and to the IAB donor gNB. This is shown in Figure 17.15.

When there are further IAB nodes in the chain, F1* is established between the IAB donor and each IAB node. The IAB nodes in the middle simply route the traffic to the

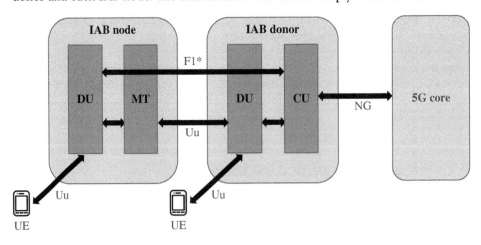

Figure 17.15 IAB architecture and functional split.

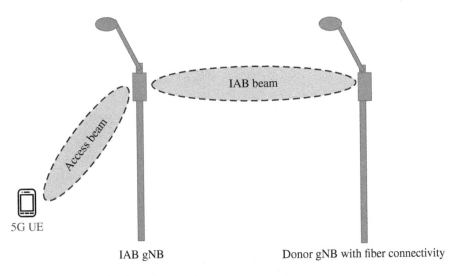

Figure 17.16 Example IAB configuration.

correct IAB node based on the adaptation layer routing information. The UE is not aware of whether it is connected to an IAB node, or a regular base station, nor if there is more than a single hop in place. Multiple hops naturally add to the delay budget extra for the connection, depending on the parameters in use (scheduling interval) and connection quality (level of retransmissions in use), as well as on the load of the connection(s) due to the scheduling delay experienced. An example of IAB configuration is shown in Figure 17.16 with vertical beam used between lamp post installed nodes, and a downward beam for serving UEs. Additional considerations for the IAB link configuration and installation are needed due to the interference considerations between nodes. The presence of other operators using the same band for IAB operation is likely to complicate the interference considerations if using the same or close-by locations for IAB node installation.

The support for network controlled repeater (NCR) was introduced in Release 18, with some elements suggested to continue in Release 19. The key elements include providing information of the TDD configuration for NCR to allow controlling the amplification better, with the scope of Release 18 work covered more in [5]. The IAB work has similarities to the relay operation specified for LTE, but that was not ever deployed widely in the field.

17.8 Dual Connectivity and Carrier Aggregation Enhancements

The 5G–5G dual connectivity (DC) support was included in the so-called "late drop" on Release 15, but that requires the synchronous operation for dual connectivity. The example case of 5G–5G DC is the connection from the UE to both lower band (FR1) and mmW (FR2) 5G gNBs, as shown in Figure 17.17. Support of this requires the support of FR2 which is not available in many markets outside the US.

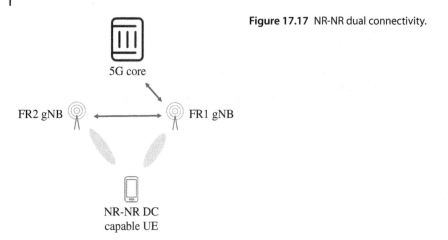

Figure 17.17 NR-NR dual connectivity.

Release 16 enabled the then asynchronous DC and reduced latency for CD and carrier aggregation (CA) activation, with the key solutions here, to consider early measurement reporting as well as minimizing the signaling needed. For CA also the new element to consider was cross-carrier scheduling, which could potentially address PDCCH congestion on one of the carriers.

Release 17 further continued to seek efficiency improvements for secondary cells (SCells) and SCG activation and deactivation. Also, as there was no specific mobility work item in Release 17, some of the conditional handover-related scenarios were covered in Release 17 for conditional PSCell change/addition.

17.9 Small Data Transmission

The small data transmission is intended to improve the situation when the amount of data to be transmitted is very small, thus Release 17 introduced possibility of transmitting small data payloads in an inactive state. This will extend the Release 16 2/4 step RACH operation, as shown in Figure 17.18 by also enabling data besides the control signaling.

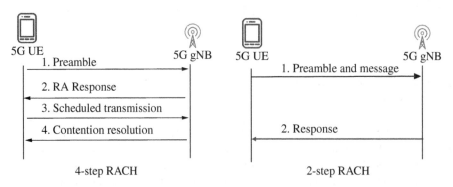

Figure 17.18 Small data transmission with 2/4 step RACH.

17.10   Conclusion

The work on radio improvements has continued steadily from Release 15, addressing the identified improvement needs ranging from mobility interruption time to UE power consumption and many more. Some of the solutions are intended for more near-time use while some of the items foreseen, especially to be included in Release 19, are starting to build the radio capabilities towards 6G use, with the 6G studies expected to start toward the end of Release 19.

References

1 3GPP Technical Report, 3GPP TR 38.830, "Study on NR coverage enhancements", Version 17.0.0, December 2020.
2 3GPP Technical Report TR 37.817, "Study on enhancement for data collection for NR and ENDC", April 2022.
3 3GPP Technical Report TR 38.843, "Study on Artificial Intelligence (AI)/Machine Learning (ML) for NR air interface", June 2023.
4 3GPP Technical Report TR 38.874, "Study on Integrated Access and Backhaul", version 16.0.0, December 2018.
5 3GPP Tdoc RP-230175, "Revised WID on NR network-controlled repeaters", ZTE, March 2023.

18

Industrial Internet of Things

Harri Holma and Antti Toskala

Nokia, Finland

CHAPTER MENU

18.1 Introduction

One of the key targets in 5G network design is to support wireless connectivity for industrial use cases. 5G can help make the industry 4.0 vision a reality, where digitalization and automation are utilized for resource sourcing, manufacturing, and transportation. 3GPP Release 15 provides a baseline for high-performance radio in terms of high data rates, extreme capacity, and low latency. Further, 3GPP evolution in Releases 16–18 provides the complete set of features required to support different industrial Internet of Things (IoT) applications. Figure 18.1 shows a summary of the key industrial IoT features in 3GPP Releases 16–18. Those features are discussed in more detail in this chapter.

5G Technology: 3GPP Evolution to 5G-Advanced, Second Edition.
Edited by Harri Holma, Antti Toskala, and Takehiro Nakamura.
© 2024 John Wiley & Sons Ltd. Published 2024 by John Wiley & Sons Ltd.

	Release 16	Release 17	Release 18
Reduced capability (RedCap) IoT devices		• 100-200 Mbps capability	• 10 Mbps lower cost capability
New spectrum options	• Unlicensed 5 GHz and 6 GHz bands	• New bands up to 71 GHz	
Ultra reliable low latency communications (URLLC)	• URLLC features	• URLLC enhancements	
Time sensitive networking (TSN)	• TSN	• TSC	• TSN / TSC enhancements
Positioning	• Positioning (sub-1 meter)	• Enhanced positioning	• Superaccurate (sub-10 cm)
Accurate timing			• Timing information from 5G network
Non-public networks (NPN)	• Standalone NPN • Integrated NPN	• Enhanced NPN	• Further enhanced NPN

Figure 18.1 Summary of industrial IoT features in 3GPP releases 16–18.

- Reduced Capability (RedCap) devices for lower cost and lower power consumption IoT applications.
- New spectrum options beyond the licensed band, including unlicensed 5 and 6 GHz bands in Release 16, and higher frequencies up to 71 GHz in Release 17.
- Ultra-reliable low latency communication (URLLC) features in Releases 16 and 17.
- Time Sensitive Network (TSN) provides guaranteed time-sensitive and mission-critical traffic communication.
- Accurate positioning and timing information is important for many industrial use cases. The basic positioning solution is in Release 16, while precise centimeter-level positioning and accurate timing information is in Release 18.
- Low-cost IoT device optimization under the title RedCap in Releases 17 and 18.
- Non-Public Network (NPN) feature in Release 16 enables more efficient support of local private networks which can be isolated from the public networks or integrated with the public network.

This chapter is organized as follows: Section 18.2 illustrates RedCap IoT categories, Section 18.3 shows the new spectrum options, Section 18.4 presents URLLC capabilities, and Section 18.5 describes the solutions for TSN and accurate timing service. Positioning is illustrated in Section 18.6, followed by NPNs in Section 18.7. The chapter is summarized in Section 18.8.

18.2 Reduced Capability (RedCap) Devices

The availability of low-cost device chipsets is expected to boost the adoption of 5G technology for Internet of Things (IoT) applications. 3GPP Releases 17 and 18 bring

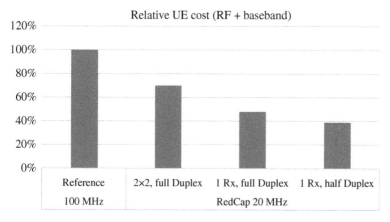

Figure 18.2 Relative cost of RedCap devices compared to the reference 5G device. Source: Adapted from [1].

RedCap devices, which enable lower cost 5G device implementation and lower power consumption, bringing longer battery lifetime. RedCap is also known as a new radio light (NR-Light). RedCap's overall performance analysis is shown in [2]. The relative cost of RedCap devices compared to reference 5G device is shown in Figure 18.2.

18.3 RedCap Device Complexity

The solution is to minimize the device requirements both in the Radio Frequency (RF) and baseband sections. The RF requirements in RedCap devices are minimized by using less RF bandwidth, fewer antennas, and MIMO layers, and half duplex operation. 5G reference broadband User Equipment (UE) must support a minimum 100 MHz bandwidth at sub-6 GHz frequencies, while RedCap can be implemented with only 20 MHz. For the mmWave spectrum, the bandwidth is reduced from 200 to 100 MHz. These bandwidths in RedCap still allow all the physical channels and signals specified for initial acquisition to be readily reusable for RedCap devices, therefore minimizing the impact on network and device deployment when introducing RedCap to support the new use cases.

Reference UEs must support two or four receiver antennas, while RedCap UEs can live with just one receive antenna. There is no Multiple Input Multiple Output (MIMO) required for a single antenna case.

The RF section for Frequency Division Duplex (FDD) bands is simplified by using half duplex (HD-FDD) where transmission and reception do not happen at the same time. No duplex filter is required in HD-FDD operation, which helps to minimize device RF complexity. Instead, a switch can be used to select the transmitter or receiver to connect to the antenna. As a switch is less expensive than multiple duplexers, cost savings are achieved. The support of FD-FDD is optional for RedCap UEs. In practice, 5G devices need to support multiple FDD bands, therefore, multiple duplex filters may be needed to support the FD-FDD operation. The multi-band requirement increases the benefit of HD-FDD operation.

The baseband requirements are minimized by supporting only lower peak data rate, and lower order modulation. The peak rate for 5G reference UE is more than 1 Gbps,

Table 18.1 RedCap requirements for minimized complexity.

	Sub-6 GHz FR1 FDD		Sub-6 GHz FR1 TDD		mmWave FR2 TDD	
	Reference UE	RedCap UE	Reference UE	RedCap UE	Reference UE	RedCap UE
Device bandwidth	100 MHz	20 MHz	100 MHz	20 MHz	200 MHz	100 MHz
Device antennas	2RX	1RX, optional 2RX	4TX	1RX, optional 2RX	2RX	1RX, optional 2RX
Downlink MIMO	Yes	Yes for 2RX	Yes	Yes for 2RX	Yes	Yes for 2RX
Duplex operation	FD-FDD	FD-FDD, HD-FDD	TDD	TDD	TDD	TDD
Downlink modulation	256QAM	64QAM, optional 256QAM	256QAM	64QAM, optional 256QAM	64QAM	64QAM
Peak data rates	>1 Gbps	FD-FDD, 1RX: 85 Mbps DL and 91 Mbps UL. Release 18: 10 Mbps	>1 Gbps	50:50 DL/UL 1RX: 42 Mbps DL, 45 Mbps UL	>1 Gbps	50:50 DL/UL 1RX: 213 Mbps DL, 228 Mbps UL
Carrier aggregation	Yes	No	Yes	No	Yes	No
Dual connectivity	Yes	No	Yes	No	Yes	No

while RedCap UE data rate is less than 100 Mbps with full duplex and less than 50 Mbps with half duplex with one receiver (1RX) devices in sub-6 GHz bands. The data rate will be higher with 2RX devices. Release 18 limits the maximum data rate to 10 Mbps regardless of the number of antennas. The highest modulation order is reduced from 256QAM to 64QAM in RedCap in sub-6 GHz bands.

The overall RedCap UE complexity is minimized also by supporting only standalone (SA) connection, but no Non-StandAlone (NSA). Also, RedCap UE does not support carrier aggregation (CA). RedCap must transmit and receive only one technology and one frequency band at a time. RedCap requirements are summarized in Table 18.1.

3GPP studies showed that there is a major cost reduction in RedCap UE compared to the reference UE. The expected cost and complexity reduction is 65–70% for sub-6 GHz and 50% for mmWave compared to the reference UE. This helps establish RedCap become a major solution for low-end mobile broadband or time-critical communication device segment.

RedCap UE can be identified by the network during random access procedure via MSG3/MSGA from a RedCap-specific logical channel identity, or optionally via MSG1/MSGA preamble. Operators may decide to support RedCap UEs only on specific frequencies, for example on low bands only.

18.4 RedCap Device Power Consumption

Another major target for RedCap is to minimize the device power consumption and maximize the battery lifetime. The main solutions for minimizing power consumption are: (1) Extended discontinuous reception (DRX) in radio resource control (RRC) idle and inactive states, (2) Radio RRM relaxation for stationary devices, and (3) Reduced PDCCH monitoring. The maximum value of the eDRX period for RedCap is extended to 10.24 seconds for inactive in Release 17, and 2.9 hours for idle. RedCap UE must wake up less seldom to receive the downlink paging. The drawback is that there can be substantial delays in accessing the device from the network. Shorter DRX periods should be used if RedCap UEs must support the reception of emergency broadcast services within a pre-defined delay of 4 seconds.

The second power saving solution is the relaxation of Radio Resource Management (RRM) measurements for neighbor cells for low-mobility UEs that are not the cell edge. Two important characteristics for many IoT use cases are: first, the device is stationary and second, the device is mostly in an idle or inactive state with infrequent data transmissions. In idle mode, power consumption is dominated by RRM measurements. Therefore, 3GPP solutions are designed to allow the UE to relax RRM measurements for neighboring cells. There are new criteria defined for UEs being not-at-the-cell edge and for being stationary. RedCap devices can then extend the time between neighbor cell measurements. All this RRM relaxation is under network control and can be configured on a device basis.

The relative RedCap device battery lifetime is shown in Figure 18.3 as a function of the eDRX period and packet inter-arrival time. The reference point is eDRX of 2.56 seconds. When the eDRX period is extended to 1 minute, the battery lifetime grows by a factor of 10–50, depending on the packet inter-arrival time. There is some further gain in extending the eDRX period to 15 minutes if the inter-arrival time is large. The battery lifetime extension can be 50–70 times if the eDRX period is at least 15 minutes,

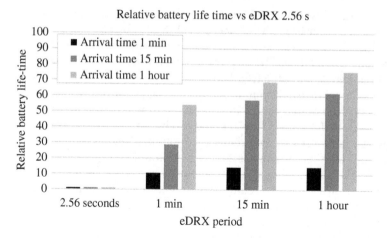

Figure 18.3 Relative battery lifetime for RedCap device compared to eDRX period of 2.56 seconds. Source: Adapted from [4].

and if the inter-arrival time is at least 15 minutes. Assuming the reference case provides 1 week of battery lifetime, RedCap can extend the battery life beyond 1 year.

18.5 RedCap Benchmarking with LTE-Based IoT

IoT optimization for lower cost and extended battery life was included in LTE in 3GPP Release 13 with two solutions – NarrowBand IoT (NB-IoT) and Category-M (Cat-M), known also as LTE-M and enhanced machine type communication (eMTC). RedCap has a few main differences compared to NB-IoT and Cat-M:

- RedCap has a substantially higher capability which enables a number of new use cases including video, extended reality, wearables, and larger file transfers. NB-IoT has a very limited data rate designed mainly for transmitting limited data volumes, for example from electricity meters.
- RedCap is designed backward compatible with minimal requirements for the control channels and for the network features. NB-IoT was more complex to deploy in practice since it was designed in a such narrowband that the LTE control channels had to be redesigned for NB-IoT devices.

Table 18.2 illustrates high-level benchmarking of LTE and 5G IoT solutions. NB-IoT has been designed for extremely low-cost and for extreme coverage with 0.2 MHz bandwidth and a peak data rate of approx. 0.1 Mbps. The low data rate in practice implies that the transmission latency is relatively high. NB-IoT is not designed for latency-critical applications. Cat-M supports a higher data rate than NB-IoT by using 1.4 MHz bandwidth. Cat-M practical deployment is simpler than NB-IoT because a normal LTE control channel can be utilized. Cat-M also supports voice over LTE (VoLTE). RedCap provides substantially higher capabilities than Cat-M with data rates in the order of 100 Mbps, and also practical latencies lower than in NB-IoT or Cat-M. We could say that RedCap capability is similar to the first generation LTE Release 8 devices with 20 MHz bandwidth and 100 Mbps data rate.

Table 18.2 LTE and 5G IoT benchmarking.

	LTE		5G	
	NB-IoT	**Cat-M**	**Broadband**	**RedCap**
3GPP release	Release 13	Release 13	Release 15	Release 17
Bandwidth	0.2 MHz	1.4 MHz	100 MHz	20 MHz
Peak rate (approx.)	0.1 Mbps	1 Mbps	2000 Mbps	100 Mbps
Battery life	Years	Years	Days	Months
Latency (approx.)	10 s	1 s	10 ms	100 ms

18.6 New Spectrum Options

Industrial wireless connectivity can be provided by the public-wide area public network with slicing, or by the dedicated private network. The public network is suited especially for wide area use cases like public safety, or remote control of machinery where nation-wide connectivity is required. The dedicated private network is preferred for cases where great local coverage is needed, full control of the network is required, and the data has to be kept inside the local premises. Such cases can be, for example, factories, airports, harbors, or mines. These two options are shown in Figure 18.4. The practical industrial cases can include a combination of these two options depending on the location of the device.

The dedicated private network operator can make an agreement with the public network operator to lease the spectrum from the public operator for the local network, but that may not be feasible in all cases. There are also situations where the private network needs to use a different spectrum than the public network. There are a number of spectrum options globally for the private network cases:

– Shared spectrum, for example, Citizens Broadband Radio Service (CBRS) at 3550–3700 MHz in the USA.
– Part of the licensed sub-6 GHz band reserved for local use, for example, 3700–3800 MHz in Germany.

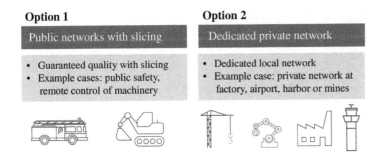

Option 1

Public networks with slicing
• Guaranteed quality with slicing
• Example cases: public safety, remote control of machinery

Option 2

Dedicated private network
• Dedicated local network
• Example case: private network at factory, airport, harbor or mines

Figure 18.4 Deployment options for industrial IoT connectivity.

- Some country regulators have defined obligations for the national licensees that they need to lease local spectrum to private wireless use at predefined conditions.
- European Commission (EC) is considering using the spectrum at 3.8–4.2 GHz for local area connectivity.
- Part of licensed mmWave band reserved for the local use, for example, 24.25–25.1 GHz in Finland.
- Unlicensed band at 5 GHz covering 5150–5925 MHz and 6 GHz covering 5925–7126 MHz are defined in Release 16.
- Unlicensed band at 60 GHz covering 57–71 GHz defined in Release 17 [5].

These new unlicensed band options provide high flexibility for the local 5G networks because there is no need for any specific spectrum license. So far, Wi-Fi has been the main radio option for the unlicensed bands, while 3GPP Release 16 brings for the first time another real alternative for Wi-Fi. The other benefit of the unlicensed band is that it is nicely harmonized globally. The same 5 GHz band, or part of it, can be utilized in most countries globally.

18.7   Ultra-reliable Low Latency Communication

URLLC is one of the key targets for 5G mobile networks. URLLC is required to support new critical communication applications on top of the 5G network. Providing high reliability and low latency requires a number of steps in terms of features, network deployment aspects, and network architecture. This section focuses on the URLLC features provided by 5G evolution in 3GPP.

Figure 18.5 illustrates a general list of techniques that can improve reliability and latency. High reliability can be provided by strong error correction coding, using robust modulation and coding schemes, repetition in time, frequency, and spatial domain, and using frequency diversity. Low latency can be provided by using a very short transmission time, minimizing the processing times in gNodeB and in UE, using a short control channel format, or using transmission without a control channel, using higher priority

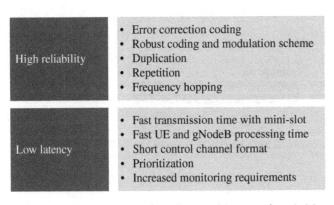

Figure 18.5 General solutions for minimizing latency and maximizing reliability.

Ultra reliable features

Low latency features

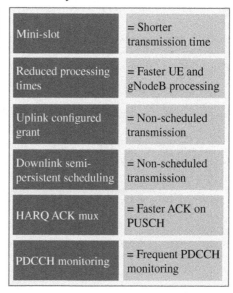

PDSCH/PUSCH repetition	= Repetition of data transmission
Ultra reliable CQI/MCS	= Use more reliable coding scheme
PDCP duplication	= Transmit data via multiple cells

Mini-slot	= Shorter transmission time
Reduced processing times	= Faster UE and gNodeB processing
Uplink configured grant	= Non-scheduled transmission
Downlink semi-persistent scheduling	= Non-scheduled transmission
HARQ ACK mux	= Faster ACK on PUSCH
PDCCH monitoring	= Frequent PDCCH monitoring

Figure 18.6 URLLC features in Release 15.

during the congestion, and increasing UE monitoring requirements. All these solutions are utilized in 3GPP Releases and explained in detail in this section.

Release 15 includes a toolbox for providing high reliability and low latency as shown in Figure 18.6. High reliability can be provided by repetition, more robust coding scheme, and data duplication. Low latency capability is enabled by short transmission time, reduced processing time, non-scheduled transmission, faster acknowledgments, and more frequency control channel monitoring. Release 16, with new features, enables even higher capabilities up to 10^{-6} and 0.5 ms one-way latency.

The service requirements are indicated from the core network to the radio using 5G QoS Identifiers (5QI). Standardized 5QI values are specified for services that are assumed to be frequently used. Dynamically assigned 5QI values can be used for services for which standardized 5QI values are not defined. The mapping of standardized 5QI values to 5G QoS characteristics is specified in Table 18.3 for delay critical Guaranteed Bit Rate (GBR) 5QIs.

Priority value indicates the relative priority in the scheduling between QoS flows. Lower priority value corresponds to higher priority in the scheduling. Delay budget defines an upper bound for the time that a packet may be delayed between the UE and the core network user plane function. That is one-way delay, and the same value is used for downlink and for uplink. For GBR flows, the packet is considered lost if the packet is not delivered within the delay budget. The error rate target devices the limit of non-congestion related packet losses. This target allows configuration of link layer protocols. The same error rate target is used both for downlink and for uplink.

Table 18.3 5QI categories for delay critical GBR.

5QI value	Priority	Delay budget	Error rate	Example services
82	19	10 ms	10^{-4}	Discrete automation
83	22	10 ms	10^{-4}	Discrete automation, V2X messages, platooning, advanced driving
84	24	30 ms	10^{-5}	Intelligent transport systems
85	21	5 ms	10^{-5}	Electricity distribution with high voltage, V2X messages for remote driving.
86	18	5 ms	10^{-4}	V2X messages (Collision avoidance, platooning with a high level of automation
87	25	5 ms	10^{-3}	Interactive service – motion tracking data
88	25	10 ms	10^{-3}	Interactive service – motion tracking data
89	25	15 ms	10^{-4}	Visual content for cloud/edge/split rendering
90	25	20 ms	10^{-4}	Visual content for cloud/edge/split rendering

Source: Adapted from [3].

18.8   Low Latency Communication

18.8.1   Low Latency Solutions

The latency caused by the radio network and the radio transmission has multiple components that need to be addressed. The main components in downlink are illustrated in Figure 18.7 and in Table 18.4 from the data packet arriving to gNodeB buffer until the data is available in UE.

gNodeB processing time: gNodeB has an internal processing time which is dependent on the base station's internal architectures, processor design, and potentially on the resource sharing and loading. 3GPP has not specified any processing time requirements for the base stations.

Waiting time for control resources (frame alignment): The data transmission must be preceded by the control channel transmission. The control channel transmission cannot be started immediately, but we need to wait for the next possible transmission slow which is impacted by time division duplex (TDD) frame structure and slot boundaries. In the case of TDD, only uplink or downlink transmission happens at a certain point in time, but both transmission directions are not running simultaneously. Therefore, there is a waiting time needed until the corresponding transmission direction starts. We need to wait for the beginning of the next slot in case of FDD. There are a few options to minimize the waiting time – a shorter TDD frame and usage of shorter slots. In practice, it is difficult to change the TDD frame structure because all operators in the same area must use the same TDD frame to avoid interference. Therefore, FDD has an inherent benefit in terms of latency since both uplink and downlink can run continuously.

Downlink control information (DCI) transmission: The data transmission in 5G is typically scheduled, which requires control channel transmission before data transmission.

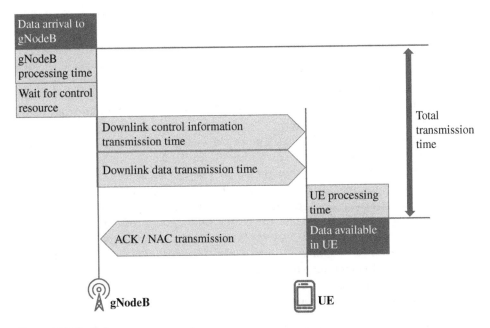

Figure 18.7 Radio latency components.

Table 18.4 Low latency solutions.

Delay component	Solution	Description
gNodeB processing time	Short processing time	gNodeB architecture and processor design for latency minimization.
Waiting time	FDD usage instead of TDD	FDD has simultaneous transmission and reception which minimizes the waiting time.
	TDD slot format optimization	TDD frame length and uplink and downlink locations impact latency. Self-contained slot including both downlink and uplink.
	Enhanced scheduling	Reduce delay by eliminating the handshake required by dynamic scheduling between UE and gNodeB for uplink traffic. Semi-persistent scheduling. Uplink configured grant.
	QoS differentiation	Prioritize latency sensitive traffic in loaded networks. Intra-UE prioritization.
		Scheduling deadlines according to the 5QI latency budget.
Transmission time	Flexible subcarrier spacing	Higher subcarrier spacing reduces symbol duration and transmission time.
	Mini-slot	Mini-slot can use 2, 4, or 7 symbols instead of 14 symbols which reduces transmission time.
UE processing time	UE capability 2	UE requirements for low latency processing.

The control channel transmission can be avoided by using semi-persistent scheduling and uplink grant-free transmission.

Downlink data transmission: The data transmission latency can be minimized by using a shorter transmission time with a mini-slot. Also, the subcarrier spacing has an impact on the latency – larger subcarrier spacing leads to shorter symbol length and shorter latency. Subcarrier spacing cannot be changed in practice. It is a predefined value for each spectrum block – 15 kHz for the FDD bands, 30 kHz for mid-band TDD, and 120 kHz for mmWaves typically.

UE processing time: 3GPP defines maximum processing times for UEs. There are two categories defined – processing time capability 1 is targeted for mobile broadband services, while capability 2 is tighter requirements for delay-critical services. The tightest processing time requirement in the reception for 15 kHz subcarrier spacing with capability 2 is 3 symbols, which equals 0.2 ms processing time. The requirement for capability 1 is 8 symbols, which equals 0.53 ms. The corresponding requirements for the uplink transmission processing are 5 and 10 symbols which equals 0.33 ms and 0.67 ms.

Acknowledgement and potential retransmission: If the packet is received correctly, it can be forwarded to higher layers. If there happens to be an error in the reception, there is an additional delay caused by the negative acknowledgment and retransmission.

Examples of low latency solutions are illustrated in Figure 18.8.

(1) More granular CQI reporting + URLLC CQI/MCS table which allows faster and more reliable scheduling.
(2) Shorter gNodeB processing time.
(3) No control channel transmission is needed for semi-persistent scheduling.

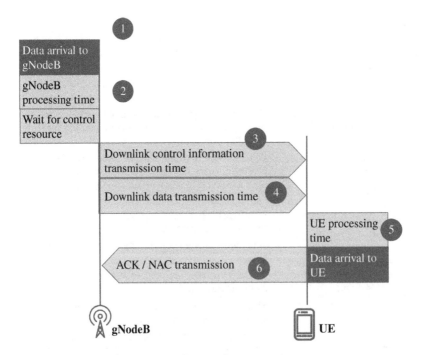

Figure 18.8 Latency improvements in case of no congestion.

(4) Short transmission time with mini-slot, and with short symbol duration, with large subcarrier spacing.

(5) Short UE processing time.

(6) Short PUCCH format for fast ACK/NACK.

The scheduling procedure, including scheduling requests and resource allocation, takes some time. Latency can be minimized by uplink grant-free transmission. It utilizes multi-channel slotted contention-based transmissions in time-frequency resources. This eliminates signaling related to scheduling requests and resource allocation. Uplink grant-free transmission is illustrated in Figure 18.9.

Grant-free configuration can be activated and deactivated by the following schemes in 3GPP Release 15:

1. *RRC signaling*: This solution is similar to LTE semi-persistent scheduling (SPS), where uplink data transmission is based on RRC reconfiguration without any layer 1 signaling. RRC provides the grant configuration to UE through higher-layer parameters. SPS scheduling can provide the suitability for deterministic URLLC traffic patterns because the traffic properties can be well-matched by appropriate resource configuration.

2. *PDCCH grant:* This solution uses additional layer 1 signaling (downlink control indication), where the uplink transmission is semi-persistently scheduled by an uplink grant via activation DCI. The grant is activated and deactivated through DCI scrambled with CS-RNTI. The DCI signaling can enable fast modification of semi-persistently allocated resources. It enables the flexibility of uplink grant-free transmission in terms of URLLC traffic properties, for example packet arrival rate, number of UEs sharing the same resource pool, and/or packet size.

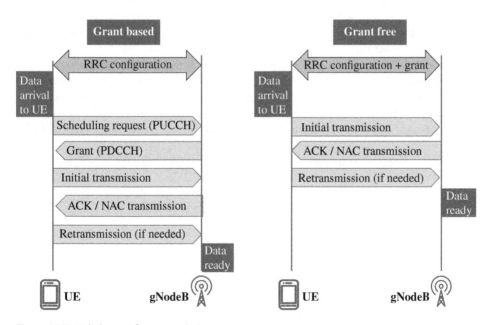

Figure 18.9 Uplink grant-free transmission.

18.8.2 Low Latency Simulations

Dynamic system-level simulations have been conducted to illustrate the achievable latency and reliability in FDD and TDD networks. The simulations include 3GPP features from layer 1 to layer 3, as well as three-dimensional radio propagation channel modeling. We consider the urban macro cellular environment with three-sector sites operating at 3.5 GHz. The subcarrier spacing is 30 kHz, and the default Transmission Time Interval (TTI) size is a 4-symbol mini-slot. We assume only URLLC type of traffic, where payloads are generated according to homogenous Poisson point process for the UEs. UEs are uniformly distributed in the network. Figure 18.10 shows the experienced packet latency at the 1e-5 outage probability. The results are displayed for two different average offered traffic loads per macro cell area: 0.25 and 1 Mbps/cell. The results show that FDD has the advantage of always having simultaneous uplink and downlink transmission opportunities in each cell. FDD can provide 1 ms latency for both loading cases, while TDD shows latency of 2.6–4.2 ms. One option to improve TDD latency is to adjust TDD configuration dynamically between downlink and uplink. If the adjacent cells use different TDD configurations, there is a potential interference between uplink and downlink. These interference cases must be controlled by network RRM algorithms aided by Cross Link Interference (CLI) measurements. For more details, see Ref. [6].

18.8.3 Low Latency Measurements

Latency measurements in the laboratory conditions at 3.5 GHz band are illustrated in Figure 18.11. The left side shows commercial 2-way round trip time measurements with 0.5 ms slot and 2.5 ms TDD frame. The median latency is 9.2 ms. The right side shows trial setup with short 2-symbol mini-slot and self-contained subframe where both downlink and uplink are included in every slot. The median 2-way latency was 1.95 ms with the low latency features which indicates that 5G radio can deliver one way latency even below 1 ms.

Additionally, measurements were done at 28 GHz with 120 kHz subcarrier spacing and compared to 3.5 GHz with 30 kHz subcarrier spacing. Figure 18.12 summarizes the roundtrip time measurements with the commercial system at 3.5 and 28 GHz bands, and with the trial system at 3.5 GHz. The measured latency is slightly lower at 28 GHz than at 3.5 GHz because of the shorter frame size – 3.5 GHz uses a 2.5 ms frame, while 28 GHz

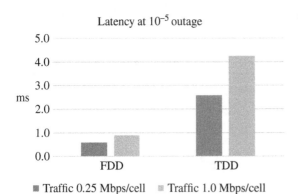

Figure 18.10 System simulations for latency at 10-5 outage level.

Latency at 10^{-5} outage

■ Traffic 0.25 Mbps/cell   ■ Traffic 1.0 Mbps/cell

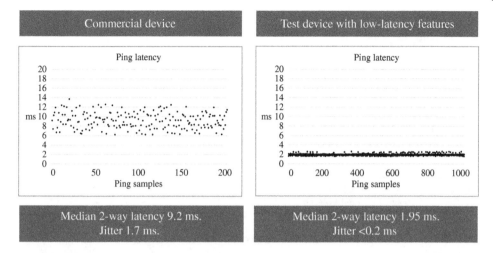

Figure 18.11 Round trip time measurements at 3.5 GHz.

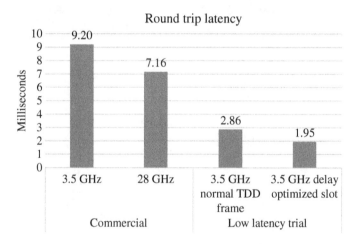

Figure 18.12 Round trip time measurements at 3.5 GHz and at 28 GHz.

uses a 0.625 ms frame. The trial results show even lower latency with the mini-slot and with the delay optimized slot, with both uplink and downlink transmissions.

18.8.4 Low Latency Architecture

It is not enough from the application's point of view to provide low latency only in the air interface, but low latency is required end-to-end including all components beyond radio network. The signal propagation creates some latency – 100 km round trip time in direct fiber adds 1 ms latency and 1000 km adds 10 ms. In practice, the round trip time is even longer because the fiber connection is not straight line, and because transport network elements add some delays. Therefore, the application server and the core network User Plane Function (UPF) must be close to the radio network to provide very low end-to-end latency. The practical radio round trip time in the 5G networks without

Mobile broadband with centralized core

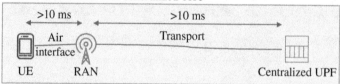

Low latency with local core

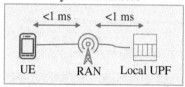

Figure 18.13 Local user plane function (UPF) required for low latency.

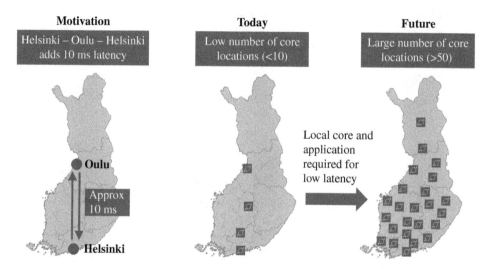

Figure 18.14 Latency minimization requires larger number of core UPFs.

low latency features is approximately 10–15 ms. If the transport latency adds another 10–15 ms, the combined end-to-end latency is 20–30 ms, and it is typically still fine for general mobile broadband use cases. If low latency features are activated in the radio network and the air interface latency is reduced to 1 ms, the end-to-end latency would be dominated by the transport latency and most of the low latency benefits would be lost. Local UPF and local application servers are important to minimize the transport latency as illustrated in Figure 18.13. Figure 18.14 shows a practical case in Finland in which the number of UPF locations is approximately five currently (2023), which can easily add up to 10 ms transport latency. Larger number of UPF locations will be required to provide lower latency in the wide area networks. The private local networks, like factories, mines, or harbors, can use local core and then provide very low latency for the local applications. It is expected that the low latency deployments will first take place in the local 5G networks.

18.9   Ultra-Reliable Communication

Radio connection has inherent challenges in providing reliable connection because of fading, low signal levels, interference, and congestion. There are multiple options for providing ultra-reliable connections in 5G. Table 18.5 shows an overview of the options for boosting the reliability. Reliability can be improved by using stronger error correction coding with low order modulation, which will make the signal more robust against interference and fading. Repetition is a variant of channel coding by sending multiple copies of the same data in the time domain. Data duplication in the frequency, or in the spatial domain, enhances reliability by exploiting two frequencies or two cells for the data transmission. Higher transmission power is a clear option for improving reliability. Frequency diversity can be utilized to avoid frequency-dependent fading and improving reliability. Channel State Information (CSI) is utilized for the link adaptation. If CSI quality can be improved, it can turn into higher reliability. Many of these reliability improvement options have a trade-off by reducing data rate, or by using more resources, or by increasing interference levels. Therefore, the reliability solutions should be applied only when the ultra-reliability is required by the application.

We will look at some of the 3GPP reliability solutions in more detail starting with packet duplication. The channel impairments like frequency-dependent fading, shadowing, interference, and path loss make it challenging to obtain ultra-reliable connectivity. The target of packet duplication is to transmit the same data from two different cells, or using two different frequencies, in order to improve the reliability,

Table 18.5 High reliability solutions.

General reliability solution	5G solution	Description
Channel coding	Robust modulation and coding scheme	Low code rates with BPSK or QPSK.
Repetition	Time domain duplication	Multiple copies of the same data transmission in time domain. Also, HARQ and RLC retransmissions and slot aggregation.
Duplication	Frequency and space domain duplication	Parallel transmission of multiple copies of the same data in frequency/spatial domain including multi transmission point (TRP) and packet data control protocol (PDCP) duplication.
Power	Higher power level	Power boosting or power allocation.
Frequency diversity	Mini-slot frequency hopping	Avoiding frequency-dependent fading with frequency hopping.
Channel state information (CSI), modulation, and coding scheme (MCS) selection	CSI feedback based on lower Block Error Rate (BLER) target.	BLER target lowered from 10% to 1% to improve reliability.
Prioritization	Intra-UE priority	Parameters and rules for handling different priorities within UE.

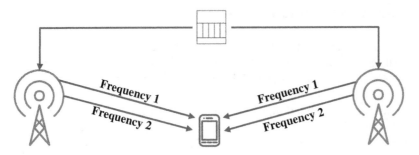

Figure 18.15 Packet duplication in spatial and frequency domain.

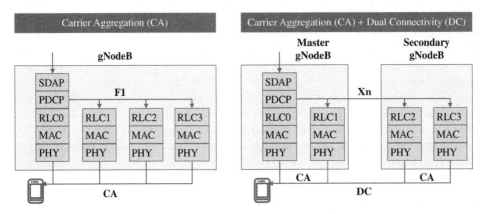

Figure 18.16 Packet duplication with carrier aggregation and dual connectivity.

but naturally at the expense of increased resource usage. UE keeps the first successfully detected packet and discards all others. It is enough if one of the links can provide a reliable connection, while the other links fail. Packet duplication also helps minimize latency by avoiding retransmissions. The packet duplication concept is shown in Figure 18.15 and the performance aspects are studied in more detail in Ref. [7].

3GPP supports a combination of Carrier Aggregation (CA) in the frequency domain and Dual Connectivity (DC) in the spatial domain. Up to four Radio Link Control (RLC) legs can be configured on different frequencies, and up to two base stations are allowed. Figure 18.16 shows CA option on the left and CA combined with DC on the right. The data is split at Packet Data Protocol Convergence (PDPC) layer.

Network can activate and deactivate PDCP duplication in uplink using a MAC signaling to save resources when duplication is not needed. When duplication is deactivated for a radio bearer, all secondary RLC entities associated to this bearer are deactivated. The concept is shown in Figure 18.17.

We take a look at the reliability benefits of packet duplication in the system simulations by considering three cases – traditional single connection transmission, blind duplication, and selective duplication. Maximum two copies of the same data were transmitted in these simulations. Blind duplication is used for the cell edge UEs. Selective duplication refers to the case where the duplicate transmission only takes place if the first

Figure 18.17 Secondary path activation and deactivation in uplink.

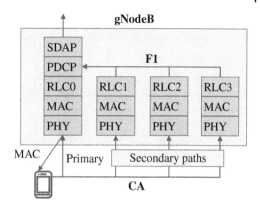

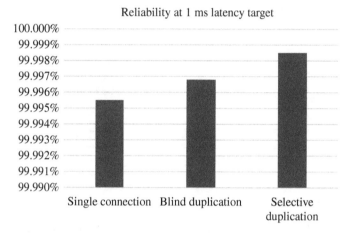

Figure 18.18 Reliability of achieving 1 ms latency target.

transmission fails, and if the delay budget allows an additional transmission. The simulation results for the reliability are shown in Figure 18.18 and the resource consumption in Figure 18.19 [8]. The reliability describes the probability of receiving the packet correctly within a 1-ms window. The results show that packet reliability improves with blind duplication compared to a single connection, while the physical resource block usage increases as expected. Even higher reliability can be obtained with selective duplication. The explanation is that selective duplication reduces the resource usage and the queuing delay of the packets.

Another solution added in Release 16 was an intra-UE priority, which refers to the multiplexing of URLLC traffic together with mobile broadband inside a UE where UE can stop transmitting low priority traffic in favor of a higher priority transmission within the same UE. The latency and reliability requirements were not considered in the treatment of the overlapping channels in Release 15. New parameters were added in Release 16 to indicate priority and a new rule for handling overlapping physical channels.

Reliability was also enhanced with modified CSI reporting and MCS tables. CSI reporting for mobile broadband services is designed for a BLER target of 10%. The CSI feedback reporting can be modified for URLLC services to allow link adaptation

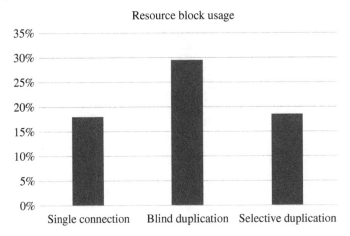

Figure 18.19 Physical resource block usage.

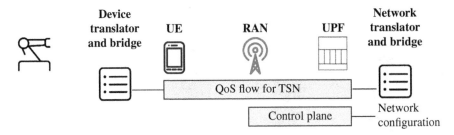

Figure 18.20 TSN system.

decisions targeting lower BLER in the air interface. A lower BLER target is reflected similarly in MCS selection.

3GPP brought improvements to the reliability of the control channel transmission. The payload size matters because it affects the required signal quality. Compact Downlink Control Information (DCI) with a small payload size was defined to improve the reliability for Physical Downlink Control Channel (PDCCH). Similarly, a smaller payload size is also helpful in the uplink. In addition, higher aggregation levels can be supported for the downlink control channel to improve reliability.

18.10 Time Sensitive Network

Time sensitive networking (TSN) is designed to manage latency between devices in Ethernet networks. TSN is important for industrial applications such as robotics and process control, where low latency and jitter are critical for the control requirements. TSN is a set of layer 2 Ethernet protocol standards defined in IEEE802.1. 5G networks must support TSN in order to make 5G accepted as the connectivity solution in the industrial cases. 3GPP Release 16 added TSN to offer layer 2 deterministic network capability for industrial communications. 5G system, modeled as a local or virtual bridge, provides TSN ports and control plane connectivity at the user plane. 5G system can receive TSN traffic QoS information from the Central Network Controller (CNC) via the interface standardized in IEEE 802.1Q, and then map TSN traffic to 5G QoS flow in a corresponding PDU session, together with the appropriate QoS configuration. TSN architecture is shown in Figure 18.20.

There are two concurrent synchronization processes in an integrated 5G-TSN system – 5G system synchronization and TSN synchronization. 5G systems provide an internal system clock for 5G internal synchronization for RAN, UE network translator, and device translator. TSN synchronization provides the synchronization service to the devices in the TSN network. The network translator generates an ingress timestamp based on the 5G reference time for every generic Precision Time Protocol (gPTP) message entering the 5G system at the UPF and embeds the timestamp within that gPTP message. UPF forwards the gPTP message to the UE via the user plane. Once a UE receives the gPTP message, the UE forwards it to the device translator which then creates an egress timestamp for the gPTP message of the external gPTP working domain. This timestamp is also based on the 5G system reference time which was provided to the UE by the 5G internal synchronization process. The difference between ingress and egress time stamps is the time spent in the 5G system. The device translator modifies the TSN timing information accordingly and sends it to the TSN system connected to the device translator. The 5G system is considered a TSN bridge. The translator functions hide 5G system procedures from the TSN bridged network. For more details about the system performance of integrated TSN and 5G, see Ref. [9].

Many services with precise time synchronization requirements currently rely on GNSS (Global Navigation Satellite System). The satellite-based solutions can provide accurate time information, but there are some concerns including indoor availability, the impact of solar storms, and potential interference issues. 5G-based time synchronization can be used as an alternative or as a backup for satellite-based solutions.

3GPP Release 17 introduced enhanced support of deterministic applications by exposing network capabilities to support time sensitive communication services and optimizations for UE-UE time sensitive communication. Release 17 capabilities include reservations between 5G ingress and egress points at UPF and UE, TSC capability exposure by any Application Function (AF), and UE-UE traffic with local switching by UPF.

18.11 LAN Service

5G Local Area Network (LAN)-type service offers private communication using IP and/or Ethernet type communications over 5G systems. Devices typically access the enterprise network via Wi-Fi and fixed network, while the mobile network is used outside offices. 5G LAN allows integration of the different access options. The main drive for 5G LAN is to be able to replace, or to extend, Ethernet and Wi-Fi with 5G. 5G systems can support the routing of non-IP packets (Ethernet frame) efficiently for private communication between UEs within a 5G LAN-type service. 5G systems can also support 5G LAN virtual networks (VNs). The technology allows a specific group of users to communicate with each other, or a specific user to communicate with existing private network users. This enables flexible group management, direct communication,

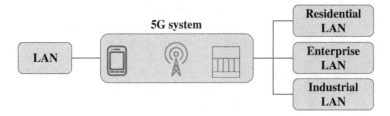

Figure 18.21 5G LAN service.

and access to the enterprise cloud anytime and anywhere. 5G LAN targets to address the enterprise requirements for convenient management, flexible interoperability, and reliable communication. 5G LAN provides services with similar functionalities to LAN, but improved with 5G capabilities, including mobility and security. The 5G LAN type service enables the management of 5G VN group identification, membership, and group data. 5G LAN overview is illustrated in Figure 18.21.

18.12 Positioning Solutions

Accurate positioning information is critical for many enterprises and consumer applications. Satellite-based solutions can be used for positioning as well as for timing information, but there are a number of benefits that 5G-based positioning can provide in terms of indoor performance, accuracy, and resilience. 3GPP has defined multiple position solutions for 2G, 3G, and 4G technologies, and further options are added for 5G. An overview of positioning solutions is shown in Figure 18.22.

– Cell identity (Cell-ID) and enhanced CID (E-CID) use cell information and the location of the base station for the positioning, which is fast and simple, but the accuracy is low, especially with large outdoor cells. Downlink and uplink cell-ID solution was added to 5G in Release 16. Cell-ID-based solution does not require any support from the device.

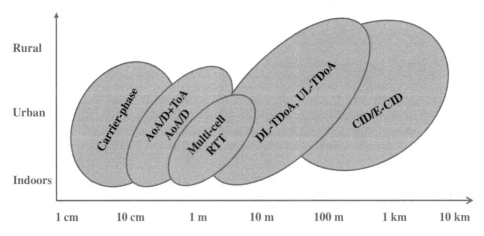

Figure 18.22 Positioning solutions and their accuracies.

Table 18.6 3GPP 5G positioning roadmap.

Release 16	Release 17	Release 18
Positioning methods	Positioning enhancements	Positioning enhancements
– Downlink E-CID	– Improved accuracy for timing	– Carrier phase positioning
– Uplink E-CID	and angle	– RedCap positioning
– Downlink AoD an TDoA,	– LOS indicator and multipath	– Low Power High Accuracy
– Uplink AoA and TDoA,	mitigation	Positioning (LPHAP)
– Multi-RTT	– Positioning in RRC Inactive	– Sidelink positioning
	– Latency reduction	

- Time difference based solutions can provide better accuracy including downlink and uplink Time Difference of Arrival (TDoA). The accuracy can be sub-1 meter level. Downlink and uplink TDoA were added in Release 17 as well as multi-Round Trip Time (RTT) measurement.
- Uplink Angle of Arrival (AoA) and downlink angle of departure (AoD) solutions are defined in Release 16 and can be combined with time difference-based solutions to further improve the accuracy.
- Carrier phase positioning enhances the accuracy to sub-10 cm level in Release 18. The technology is explained in more detail in Ref. [10]. The carrier phase solution was not available in 4G radio networks.

3GPP defined similar position options in Release 16 that have already been defined for 4G earlier. Release 17 added multiple options for improving positioning performance including Line-of-Sight (LOS) indicator, enhanced multipath reporting, and higher accuracy for timing and angle information. Release 17 also added positioning in RRC inactive state and faster measurements with less than 100 ms latency. Release 18 brings a number of enhancements including carrier phase positioning, IoT positioning for RedCap, and Low Power High Accuracy Positioning (LPHAP) and sidelink positioning. Carrier phase solution can become attractive for the indoor industrial use cases because the accuracy is better than with satellite-based solutions, and it works indoors. Carrier phase positioning can provide ten times higher accuracy than time-based positioning. RedCap positioning evaluates the accuracy achievable with reduced bandwidth and considers enhancements where possible. Low Power High Accuracy Positioning (LPHAP) investigates functionalities for minimizing the power consumption for the positioning which is relevant for low power IoT devices. Sidelink positioning is targeted specifically for automotive use cases. 3GPP positioning roadmap is shown in Table 18.6.

18.13 Non-Public Networks

The target of Non-Public Networks (NPN) is to support private wireless 5G networks in order to provide better coverage and radio performance or offer increased security with dedicated credentials or isolate operations from the public networks for resiliency reasons. Private 5G networks can be utilized in industries, in campus areas, in harbor and airports, in mines, and in wind farms. NPN can be deployed as StandAlone (SA)

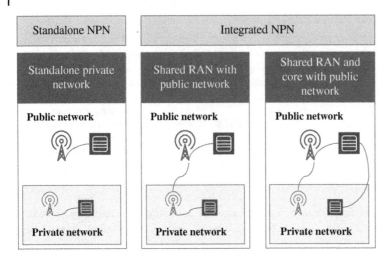

Figure 18.23 Non-public network options.

NPN which does not rely on network functions provided by a Public Land Mobile Network (PLMN), or as public network integrated NPN which relies on the support of a PLMN. SA NPN is operated independently from the public PLMN network, and no business level cooperation is required between NPN and public PLMN. The combination of PLMN ID and network ID identifies SA NPN. Public network integrated NPN uses PLMN while providing Closed Access Groups (CAG) for NPN. PLMN ID identifies the network and CAG ID identifies the CAG cells. Integrated NPN can share RAN or both RAN and core with the public network. Figure 18.23 illustrates the different architecture options.

3GPP Release 16 covers the ability to identify, discover, and select access control for NPNs as well as interworking between NPN and PLMN. Release 17 adds more NPN architecture features, including credentials owned by another entity from SA NPN, support for service continuity, simultaneous service from NPN, and PLMN and support for IMS voice and emergency services with SA NPN.

NPN can be deployed on licensed, unlicensed, or shared spectrum. NPN is mainly a core network feature, and only minor RAN enhancements are needed to enable it.

18.14 Summary

LTE and 5G Release 15 have already been widely utilized for supporting industrial use cases in public and in local networks. 5G evolution in Releases 16–18 brings major enhancements for supporting further industrial IoT use cases. Low-cost RedCap IoT devices enable to the expansion of 5G connectivity to more use cases. New spectrum options at 5–6 and 60 GHz unlicensed bands allow more flexible deployment options. URLLC capability can open new use cases over 5G radio where the latency and reliability are extremely important. Superaccuracy timing and location information help in many industrial use cases, and private 5G networks can be deployed to support local industrial use cases.

References

1 3GPP TR 38.875 Study on support of reduced capability NR devices, 2021.
2 Ratasuk, R., Mangalvedhe, N., and Lee, G. David Bhatoolaul "Reduced Capability Devices for 5G IoT", 2021 IEEE 32nd Annual International Symposium on Personal. *Indoor and Mobile Radio Communications (PIMRC)*.
3 3GPP TS 23.501 3rd Generation Partnership Project; Technical Specification Group Services and System Aspects; System architecture for the 5G System (5GS).
4 Veedu, S.N.K., Mozaffari, M., Hoglund, A. et al. Toward Smaller and Lower-Cost 5G Devices with Longer Battery Life: An Overview of 3GPP Release 17 RedCap. *IEEE Communications Standards Magazine* 6 (3): [6]September 2022.
5 3GPP TR 38.808 Study on supporting NR from 52.6 GHz to 71 GHz, 2021.
6 Pedersen, K., Esswie, A., Lei, D. et al. (August 2021). Advancements in 5G New Radio TDD Cross Link Interference Mitigation. *IEEE Wireless Communications* 28 (4).
7 Centenaro, M., Steiner, J., Pedersen, K., and Mogense, P. (December 2019). System-Level Study of Data Duplication Enhancements for 5G Downlink URLLC. *IEEE Access* 8.
8 Centenaro, M., Laselva, D., Steiner, J. et al. Resource-efficient dual connectivity for ultra-reliable low-latency communication. In: *2020 IEEE 91st Vehicular Technology Conference (VTC2020-Spring)*.
9 Peter M. Rost and Troels Kolding "Performance of Integrated 3GPP 5G and IEEE TSN Networks", *IEEE Communications Standards Magazine* (Volume: 6, Issue: 2, June 2022)
10 Fouda, A., Keating, R., and Cha, H.-S. Toward cm-Level Accuracy: Carrier Phase Positioning for IIoT in 5G-Advanced NR Networks. In: *2022 IEEE 33rd Annual International Symposium on Personal, Indoor and Mobile Radio Communications (PIMRC)*.

19

Verticals

Antti Toskala and Harri Holma

Nokia, Finland

CHAPTER MENU

19.1 Introduction

5G and 5G-Advanced radio with high capabilities can be applied to many use cases beyond mobile broadband. Verticals refers to the use of 5G for these new application areas. This chapter presents vertical use cases from satellites and drones to public safety, train networks, and automotive. 5G started to provide support for some verticals and 5G-Advanced added wider support for different verticals. In many cases, the solutions have commonalities and are suitable for more than just a single vertical use case. Also, unlicensed use is covered.

19.2 Non-Terrestrial Networks (NTN)

3GPP studied support for NTN for 5G in Releases 15 and 16 [1], and specified in Release 17 the support for the transparent NTN architecture with the gNB on Earth and the satellite payload acting as a repeater for providing coverage on Earth's surface. This

5G Technology: 3GPP Evolution to 5G-Advanced, Second Edition.
Edited by Harri Holma, Antti Toskala, and Takehiro Nakamura.
© 2024 John Wiley & Sons Ltd. Published 2024 by John Wiley & Sons Ltd.

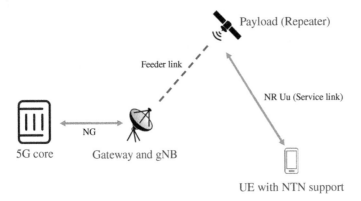

Payload (Repeater)

Feeder link

NR Uu (Service link)

NG

5G core   Gateway and gNB

UE with NTN support

Figure 19.1 NTN in 3GPP architecture.

has the advantage that the changes and maintenance needed in the gNb are naturally easier to implement compared to the case when the gNB functionality would be part of the satellite payload. This architecture is shown in Figure 19.1. In Release 19, 3GPP is expected to introduce support for the nontransparent architecture with the gNB functionality located in the satellite.

For the radio connection, the challenges with NTN are related to the following factors:

- Distance to the satellite, which causes extra delay simply due to the propagation distance. Depending on whether the satellite is low Earth orbit (LEO) or geostationary orbit (GEO), the extra latency ranges from 8 ms to nearly 600 ms.
- Satellites, other than GEO, also move very fast, which causes additional issues with the resulting doppler, for both uplink and downlink directions, as well as variable round trip time causing drift in the timing between UE and the satellite.
- The cell size covered by a satellite can be massive, as satellites at high altitudes will have visibility to a large area. Depending on the beam width of the satellite beams, the cell diameter may range from 75 to 1000 km.
- A satellite at 600 km altitude will have a velocity of approximately 7500 m/s, and hence there will be a need for frequent handovers due to the rapid movement of the LEO satellites, especially for small cells.

3GPP has specified a set of solutions in Release 17 to address these challenges. As some recent satellite systems will end up using unmodified 3GPP UEs [2], the corresponding solutions are needed on the satellite network side.

The use of pre-compensation by the UE has been specified to address the time and frequency shift observed with satellites. This is based on the NTN network providing SIB information about feeder link properties, satellite location, and the direction of travel for the satellite. Such information is carried in the so-called "serving satellite ephemeris information." As part of the specifications, it is assumed that the UE is aware of its location (through implementing satellite positioning like GPS, Galileo, Baidu, or Glonass), and thereby the UE can calculate the delay components of the system, and autonomously compensate for both feeder link delays (via the common delay calculations) and service link delays (through knowledge of propagation path distance between the UE position and the satellite position) when accessing the system. Also, when transmitting in the uplink, the UE is expected to compensate for the Doppler shift such that signals from different UEs are aligned when received at the satellite antennas.

For addressing the impacts of the long delay, depending on the type of satellite deployment, there are different possibilities included in Release 17, mainly to accommodate continuous or near-continuous scheduling for a single UE, which are:

- Turning off hybrid automatic repeat request (HARQ) for a selected set of HARQ processes, thus, operating without L1 retransmissions.
- Increase number of HARQ processes from 16 to 32, which allows the use of the L1 HARQ with LEO satellites in most cases while being able to continuously serve a single UE.
- Slot aggregation, which effectively creates longer HARQ process duration and improves coverage.

Also, several timings have extra value compared to the terrestrial operations, for example, the random access response (RAR) timing starts to run only after the configured and calculated additional delay has elapsed, and then another timing value is used for downlink/uplink frame alignment at the so-called uplink synchronization reference point.

Dealing with large cells has been made easier by supporting multiple tracking areas in an NTN cell. UE is then selecting one of the tracking areas supported within a coverage area. Further, an NTN cell can transmit multiple public land mobile network (PLMN) codes as a large cell may provide coverage for multiple countries simultaneously.

Mobility is especially important in the case of LEO satellites. A UE which is not moving is still subject to possible handover due to the movement of the satellite. The fact that satellite movement along the orbital plane into the future is well-known and predictable, based on the provided serving satellite ephemeris information, solutions such as conditional handover (CHO) from Release 16 could be used with NTN operation as well, including the NTN specific trigger conditions based on timing and location to improve the reliability. Two possible approaches can be used with mobility with LEO satellites:

- Earth-Moving cells (EMC), where the radio beams of the satellite are fixed and thus the cell or cells provided by a single satellite will move corresponding to the Earth-projected path, corresponding to the satellite movement.
- Quasi Earth-Fixed cell (EFC) aims to steer the satellite beams such that each cell remains in a specific location when the satellite is moving, thus making the radio coverage stationary on Earth, as shown in Figure 19.2. Handover occurs then when another satellite approaches and is able to provide the coverage, instead of the satellite that is moving out of the provided coverage area.

In general, the satellite coverage cannot always be assumed to be continuous, as constellations may be sparse and thus there may be periods when no satellite can be detected with a UE. To avoid a UE having to search a long time for satellites, and thereby waste battery power, 3GPP has also defined the means for providing assistance information to UEs on satellite movement and coverage. In this case, the UE has prior knowledge of when and where to start searching for a satellite. The use of satellites for providing coverage to remote areas is foreseen to be an energy efficient solution once the satellite is up in the orbit, as satellites are powered by solar energy.

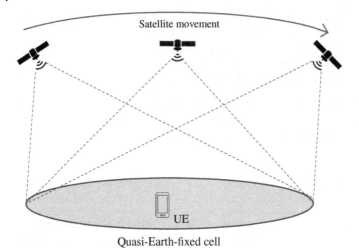

Figure 19.2 Quasi-Earth-Fixed cell for mobility management simplification.

Releases 18 and 19 cover further improvements to address the areas such as:

- Improved coverage, which will limit data rates with satellites, and especially with handheld devices without specific external antennas optimized for satellite communications.
- Mobility between NTN and terrestrial networks, as the interworking between the two is seen as important due to the complementing role of NTN for situations when terrestrial network availability is limited.
- Regenerative payload is expected to be addressed in Release 19, with the gNB functionality included in the satellite payload.

19.3 High Altitude Platform Stations (HAPS)

The HAPS are considered part of the NTN framework in 3GPP work. The use of HAPS does not need changes in the devices as the resulting extra delay or doppler due to the HAPS movement can be accommodated within the existing requirements. The extra delay depends on the altitude and distance to the UE on the ground but could be estimated to be in the order of 1 ms.

The idea of HAPS is to have an airplane or balloon platform that stays up in the stratosphere for extended periods of time and will provide coverage to large areas while maintaining a smaller impact on the latency compared to satellite-based NTN. The HAPS expected altitude would be between 20 and 50 km. Depending on the desired data rate, the service area could have a radius in the order of 100 km, depending on the shape of the terrain as well [3].

The HAPS is served by feeder link (or links) from the ground, providing the necessary backhaul as shown in Figure 19.3. Key design questions would then be also how the airplane is powered to enable long intervals between change/refueling of the airplane providing the HAPS service, and how much power is then available for the radio equipment on board.

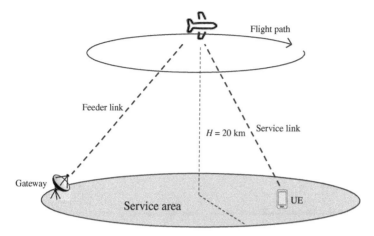

Feeder link

$H = 20$ km   Service link

Gateway

Service area

UE

Figure 19.3 HAPS principle.

19.4 Drones

5G-Advanced Release 18 also addresses the support of unmanned aerial vehicle (UAV) with 5G [4], while some support for LTE was done already earlier [5]. The work in Release 18 reaches anyway further than the limited LTE support available. The key aspects supported in Release 18 are:

- UAVs are supported to report to network information such as flight path or height for proper operation.
- Another important element, with also regulatory impacts, is UAV identification. This has become important to help control where the UAVs are allowed to fly and where not. There is interest to be able to identify UAVs that could cause problems (e.g. flying in areas not permitted for UAV use) to ensure UAVs to fly only in authorized air space. Another aspect is also that UAVs should be operated only using such aerial subscriptions and special hardware implementation that meets possible extra regulatory requirements for the protection of specific frequency bands when such tighter requirements have been defined in different regions.
- Part of the identification is also to have support for UAV identification broadcast to have UAVs broadcasting their IDs. This is not happening on any random frequency band but would be defined by the local regulation to ensure that such an ID can be also decoded easily. This sidelink-based UAV ID broadcast is an aspect that is also to be extended for LTE side.
- Beyond the pure UAV identification broadcast, considerations have also been included to support detect and avoid (DAA) operation, as also other UAVs could identify based on the sidelink approaching UAV, and then course could be altered accordingly. The further details of such protocol are handled outside 3GPP.
- It is also critical to consider the impact of UAVs on the network, as UAVs associated with heavy traffic (e.g. video feed) may easily create lots of interference in the network. This is due to the fact that a UAV up in the air may have LOS connections with many cells. With the use of directive/beamforming antennas in the UAV, one can replace the omnidirectional transmission, making the data flow from a UAV to be more directive,

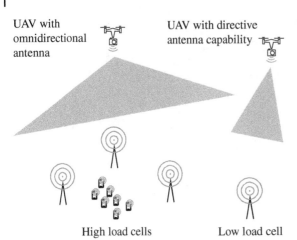

UAV with omnidirectional antenna

UAV with directive antenna capability

Figure 19.4 Reduced interference from UAVs with directive antennas.

High load cells

Low load cell

and thus causing less interference. This is also shown in Figure 19.4, where a UAV with an omnidirectional antenna causes interference to all cells visible, while a UAV with a directive antenna can reduce drastically the interference caused and also connect to less loaded cells in the network. This has been studied in Release 18 and remains to be seen if the special UAV capability for directivity support would be included to support this operation and help network understand level of interference caused by a UAV with certain service characteristics.

Further work on UAV remains open if there is more work in Release 19, but the basic aspects necessary have been covered in Release 18 for 5G and some extensions were done for LTE side as well.

19.5 Vehicle Connectivity

3GPP 5G V2X covered cellular vehicle-to-vehicle (V2V) and vehicles and infrastructure (V2I) in Release 16. 3GPP had earlier developed the basic solution to be used on top of LTE, but now the same capability and additional extended elements have been added for the 5G side as well.

The basic elements include functionality like collision avoidance for V2V cases (for example breaking information) and traffic light information in V2I cases. The key motivation to enable V2X for the 5G side has been the possibilities for more advanced use cases in addition to the basic traffic safety use cases supported with LTE. An example of such a use case is sensor data sharing between two different vehicles (with the vehicle in front sharing the data with the vehicle(s) following behind), as shown in Figure 19.5.

Release 17 extended the operation to include vehicle-to-pedestrian (V2P) operation, which also takes into account the constraints on the battery power when the other end of the V2P communication is operating on a smartphone type of device and is not powered by a car battery.

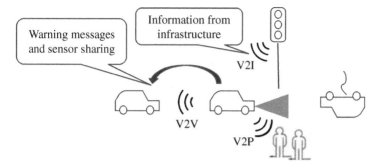

Figure 19.5 The use of 5G V2X with V2V, V2I, and V2P.

Outside 3GPP, especially the 5G automotive alliance (5GAA) [6], is working on the requirements and use cases for V2X.

The V2V operation is possible also without network support/coverage, with the sidelink (the communication link between two vehicles) being either network-controlled (sidelink mode 1) or autonomous between UEs (sidelink mode 2).

19.6 Public Safety

The use of 5G provides a highly capable platform with plenty of new capabilities compared to the capabilities of TETRA or P25, enabling, for example, high quality video transmission. The needs of the public safety community have been also taken into account when progressing the work on 5G-Advanced. Already in Release 15 there were solutions to be used for public safety, including end-to-end network slicing and the new QoS solutions enabled with standalone (SA) 5G operation.

The later releases have included many features well-suited, or in some cases absolutely required, for public safety use, such as:

- The device-to-device (D2D) has been covered with sidelink communications, which includes important aspects including sidelink relay, to enable communications between terminals directly. The automotive industry is going to be the sidelink technology as well, which will lower the bar for implementation as the solution is not for the public safety sector only.
- The support for UAV operation, introduction of mobile IAB nodes, and NTN operation can be considered as well in connection with public safety network deployments.

When implementing a public safety network, there are different alternatives, as shown in Figure 19.6, ranging from a fully separate network to a fully hosted network by an operator. A middle solution would be to have only shared RAN, but then a dedicated core network for public safety. In the fully hosted network, one could envisage a slicing solution being used for the public safety operations to ensure sufficient capacity and service quality.

More combinations can be created if considering, for example, use of NTN as additional solution alongside the terrestrial network.

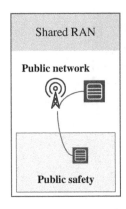

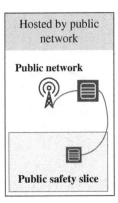

Figure 19.6 Different public safety deployment options.

19.7 Dedicated Networks with less than 5 MHz of Spectrum

There has been a clear need identified to support 5G for dedicated networks when spectrum available is less than 5 MHz. The following three use cases have been identified:

- The networks supporting railway communications today are based on GSM-R technology, and the need to migrate to 5G has been identified. The train control solution is generally known as future railway mobile communication system (FRMCS) where 5G acts as the connectivity layer [7]. The amount of spectrum available is 5.6 MHz typically, but the upgrade needs to be done such that for while both systems need to be operated simultaneously, as shown in Figure 19.7. Thus, 5G needs to be operated on smaller allocation until GSM-R can be turned off. During this transition period operation, around 3 MHz bandwidth will be needed.
- For smart grids there exists in the US 2 × 3 MHz FDD channels in bands n26 and n8 in the 800–900 MHz frequency range, and similar 2 × 3 MHz allocations in 400 MHz range is planned in Europe as well.
- For public safety, Europe has allocated 2 × 3 MHz FDD channels for public protection and disaster relief (PPDR) communications, in the 700 MHz frequency in band n28.

In order to facilitate the use of 5G with around 3 MHz bandwidth, one needs to use puncturing, especially on the physical broadcast channel (PBCH), while primary and secondary synchronization channels (PSS/SSS) are not impacted, thus allowing to keep the L1 implementations in the UE unchanged and use only some PRB level puncturing

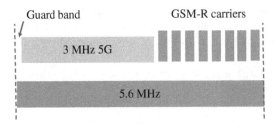

Figure 19.7 Example of coexistence between GSM-R and 5G operation.

operation. Further, to optimize the spectrum use, one needs to address the channel raster possible with 3 MHz bandwidth. It has been important to keep the changes reasonable in order to encourage chipset implementation to support the dedicated network use cases when the overall volumes are clearly smaller than, for example, with smartphone use cases. With radical redesign, the implementation threshold would have easily been too big.

The use of puncturing creates some loss in the link level performance. The smallest loss can be assumed when UE is aware of which PRBs are punctured from the PBCH. In such a case, the loss has been estimated to be in the order of 1.4 dB based on the studies presented in [8].

Release 19 is foreseen to address enhancements like carrier aggregation for the consideration of additional spectrum being combined especially with the train networks.

19.8 Unlicensed

The 5G unlicensed, which is called in 3GPP as new radio unlicensed (NR-U), covers the following deployment scenarios in Release 16:

- Carrier aggregation between licensed band 5G and 5G unlicensed.
- Dual connectivity (DC) between licensed band LTE and 5G unlicensed.
- SA 5G unlicensed, which would be operated without any licensed carrier necessary.
- 5G unlicensed downlink with licensed band uplink (this is a special scenario intended for cases where the listen-before-talk (LBT) operation would be done only at the gNB side.
- DC between licensed band 5G and 5G unlicensed.

Out of these scenarios shown in Figure 19.8, the enabling of both DC as well as SA 5G unlicensed operation offers much more deployment flexibility compared to the

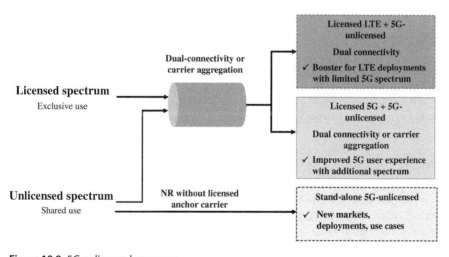

Figure 19.8 5G unlicensed use cases.

LTE licensed assisted access (LAA) operation. The DC solution facilitates deployment when 5G (or LTE) inside a building has been done with distributed antenna system. In that case, connecting unlicensed operation to the distributed antenna system in carrier aggregation scenario is not possible, as lower power would be distributed for the whole antenna system, and one would need to listen to the whole area covered by the antenna system for LBT operation. The SA allows to use 5G unlicensed for coverage extension in locations without licensed band coverage as well as to roll out, for example, in-building networks provided by the building owner.

The 5G unlicensed supports wideband operation, with bandwidths exceeding 100 MHz (with integer multiples of 20 MHz). The coexistence with LTE-based systems and Wi-Fi will be ensured with the energy detection, and possible other methods to be considered as well, especially for cases when there is no other system used on the same channel than 5G unlicensed (5G-U).

The key motivation for the use of 5G-U instead of Wi-Fi can be summarized as follows:

- Better capacity.
- Better tolerance for interference.
- Better coverage.
- Seamless integration in the licensed 5G or LTE operation, including mobility and security.
- Flexible deployment both with aggregation with the licensed band operation, as well as with DC approach or with SA 5G unlicensed operation.

While Release 16 included the support for both 5 and 6 GHz spectrum, the work on 5G-unlicensed was continued in Release 17, with the introduction of support for the operation on 60 GHz spectrum as well as making the unlicensed spectrum operation more suited for ultra reliable, low latency communications (URLLC) with the introduction of UE initiated channel occupancy time for frame based equipment with unlicensed operation. This allows for reduced latency compared to Release 16 NR-U when operating with the unlicensed band.

19.9 Summary

5G and 5G-Advanced are expanding the supported vertical use cases for every release. With 5G-Advanced, new vertical cases were added and already started ones were made more complete. Further enhancements identified will be done in Release 19 for many areas, including NTN operation and multi-hop relay nodes for public safety.

References

1 3GPP Technical Report TR 38.821, "Solutions for NR to support non-terrestrial networks (NTN)," Release 16, May 2021.
2 https://ast-science.com/ (last visited July 14th, 2023).
3 Xing, Y., Hsieh, F., Ghosh, A., & Rappaport, T.S., "High altitude platform stations (HAPS): Architecture and system performance", in 2021 IEEE 93rd VTC2021-Spring.

4 3GPP Tdoc RP-213600, "New WID on NR support for UAV (Uncrewed Aerial Vehicles)," 3GPP Technical Document RP-213600, Dec. 2021.

5 Lin, X. et al. (2018). The sky is not the limit: LTE for unmanned aerial vehicles. *IEEE Communications Magazine* 56 (4): 204–210.

6 www.5GAA.org (last visited July 14th, 2023).

7 https://uic.org/rail-system/frmcs/

8 Hooli, K. et al. (2023). Extending 5G to narrow spectrum allocations. *IEEE Journal on Selected Areas in Communications* 41 (6) June.

20

Open RAN and Virtualized RAN

Harri Holma and Antti Toskala

Nokia, Finland

20.1   Introduction

Open RAN and virtualized RAN (vRAN) can lead to the transformation of how operators and large enterprise customers procure, deploy, manage, and optimize cellular equipment. The primary focus of Open RAN and vRAN are: (1) decoupling radio unit (RU) and baseband distributed unit (DU) and centralized unit (CU), and (2) decoupling hardware (HW) from software (SW). Two additional targets are: (3) to enable RAN automation using artificial intelligence (AI) and machine learning (ML), and (4) using centralized location for baseband processing. The architecture targets and challenges are shown in Figure 20.1.

5G Technology: 3GPP Evolution to 5G-Advanced, Second Edition.
Edited by Harri Holma, Antti Toskala, and Takehiro Nakamura.
© 2024 John Wiley & Sons Ltd. Published 2024 by John Wiley & Sons Ltd.

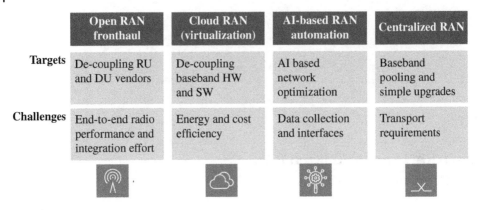

Open RAN fronthaul	Cloud RAN (virtualization)	AI-based RAN automation	Centralized RAN
Targets De-coupling RU and DU vendors	De-coupling baseband HW and SW	AI based network optimization	Baseband pooling and simple upgrades
Challenges End-to-end radio performance and integration effort	Energy and cost efficiency	Data collection and interfaces	Transport requirements

Figure 20.1 Open RAN and vRAN targets and challenges.

Open RAN open fronthaul enables more flexible RF and baseband deployment options with different vendors. The ideal target is to have plug-and-play multi-vendor deployment. The practical deployment has turned out to be more challenging in the areas of radio performance optimization in wide area macro networks, and in the required multi-vendor integration efforts. The term "O-RAN" refers to the interfaces and architecture elements specified by the O-RAN Alliance on top of 3GPP specifications. The term "Open RAN" is more generic and is not only limited to O-RAN Alliance definitions.

Cloud RAN or vRAN enables attractive evolution for the radio networks in terms of deployment flexibility and in terms of decoupling SW and HW. The ideal target is to have independent HW and SW vendors for the baseband. The practical challenges are related to the HW efficiency for the radio-specific processing, latency requirements, and integration cloud RAN with legacy networks.

AI-/ML-based network automation can bring improvements to the network configuration, optimization, and anomaly detection. The challenge is to collect the required information from RAN using standardized interfaces.

Centralized RAN can ideally enable pooling of baseband resources and make the baseband upgrades simpler. The challenges with centralized RAN are related to the tight latency requirement between RU and DU. The targets and challenges are shown in Figure 20.1.

This chapter discusses Open RAN and vRAN technologies and presents solutions for the optimized implementations. This chapter starts by presenting the ongoing trends in the mobile network architectures in Section 20.2. Open RAN interfaces and specifications are illustrated in Section 20.3, and baseband centralization and virtualization are described in Section 20.4. RAN Intelligent Controller (RIC) is discussed in Section 20.5, O-RAN Alliance in Section 6, and the chapter is summarized in Section 20.7.

20.2 Radio Network Architecture Trends

The typical mobile network is currently based on the setup where the same vendor provides both RU and baseband unit, including DU and CU. The baseband unit is located

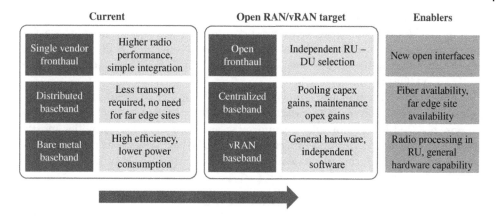

Current	Open RAN/vRAN target	Enablers

Figure 20.2 Radio network architecture trends.

at the base station site close to the antenna and RU. The baseband processing runs on purpose-built bare metal platforms. The target architecture with O-RAN interfaces and vRAN virtualization is to bring more flexibility to the deployment – multiple vendors can be used in the radio network, baseband processing can be centralized, and baseband processing can run on the generic HW. The ultimate target is to enable independent RU and DU/CU vendors, independent HW and SW for DU and CU, and centralized DU/CU location for simpler operability. The architecture trend is illustrated in Figure 20.2. There are a few technology enablers that could make these changes feasible in the future architecture:

- New open RAN-internal interfaces have been defined in O-RAN Alliance and in 3GPP.
- There is more fiber capacity and far-edge sites available, which allows baseband centralization.
- A 5G massive MIMO antenna has more processing in RU than the traditional RF heads, making DU virtualization simpler.
- Generic HW has more capabilities, which allows running DU and CU functionalities.

The chapter will illustrate how to take benefit of these trends and enablers in 5G radio network evolution.

20.3 Open RAN Fronthaul

20.3.1 Fronthaul Functionality Split

The success of mobile networks has been based on open interfaces since GSM times in 1990s. The open interface between the device (user equipment, UE) and the network, was the starting point for the global success – the same device works in different networks in different countries, provided by different vendors. Another important open interface is between the radio network and core network. Both interfaces are well-defined in 3GPP and widely implemented in commercial multivendor networks. 5G brings further potential to add open interfaces within the radio access network

(RAN) to allow for network virtualization and flexible deployment. The target is to allow optimal location of the different network functions and to allow mixing between different equipment providers. The interface between DU and CU is defined in 3GPP, and the interface between RU and DU is defined in O-RAN Alliance. The open interfaces are illustrated in Figure 20.3.

A more detailed functionality split within the RAN is shown in Figure 20.4. The most important unit in 5G radio network is the RU including antenna, radio frequency (RF) components, and layer 1 low-like beamforming and receiver. RU capabilities mostly define the network coverage and capacity which are the key radio performance indicators. RU also defines most of the network power consumption as well as the site solution

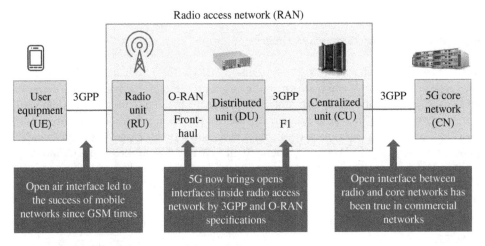

Figure 20.3 Open interfaces in 5G mobile network.

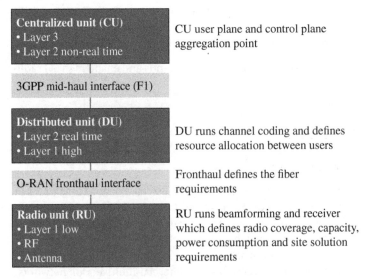

Figure 20.4 Functionality split within radio access network.

requirements. RU makes typically a large part of the costs of the wide area radio network because of high power and high-performance silicon technology and advanced algorithms.

DU runs layer 1 high functions including channel coding and delay critical layer 2 including packet scheduling. Therefore, DU is an important element for network slicing between different applications, use cases, and customers. CU runs less delay critical layer 2 and 3 functions and can utilize centralized aggregation points.

There are a few options on how to exactly split the functionality between DU and RU. Figure 20.5 and Table 20.1 illustrate more detailed fronthaul split options. The differences can be summarized as follows:

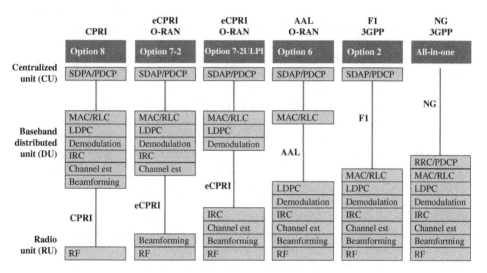

Figure 20.5 Fronthaul, midhaul, and backhaul functionality split options.

Table 20.1 Benchmarking of RAN architecture options.

	Option 8	Option 7-2	Option 6	Option 2	All-in-one
Interface throughput requirement	>100 Mbps	15–20 Gbps	10–15 Gbps	10–15 Gbps	<10 Gbps
Interface latency requirement	<0.1 ms	<0.1 ms	<0.1 ms	1 ms	>1 ms
HW acceleration in baseband	Required	Required	Not required	Not required	Not required
Inter-frequency and inter-sector combining in baseband	Yes possible	Yes possible	Requires extra interface	Requires extra interface	Requires extra interface
Interface standardization	CPRI	O-RAN but not truly open	O-RAN but not properly defined	3GPP F1	3GPP RAN – core interface

Option 8 is the Common Public Radio Interface (CPRI) -based interface where all baseband processing is in DU. The challenge with CPRI is that the fronthaul data rate requirement is too high for practical deployment, especially with a wideband massive MIMO antenna. Also, the latency requirement is tough with CPRI with less than 0.1 ms latency required.

Option 7-2 interface was defined in O-RAN Alliance in 2021. Option 7-2 has beamforming in RU, which reduces the fronthaul throughput requirements substantially compared to CPRI. Option 7-2 evolution includes more functionalities in RU than Option 7-2 – channel estimation and the receiver interference rejection combining (IRC) are implemented in RU. This option is known as 7-2 Uplink Performance Improvement (ULPI). Option 7-2ULPI has a benefit in the uplink capacity with large number of multiuser MIMO (MU-MIMO) layers and advanced receiver. Option 7-2ULPI was defined in the O-RAN Alliance in 2023 and 2024. Both 7-2 options need low latency between RU and DU. These options also need HW acceleration in DU for processing the channel decoding. Both options are well suited for multi-sector distributed MIMO, and for carrier aggregation, because the signal combination and MAC layers are located in the common DU.

Option 6 divides the functionality between layer 1 and layer 2. All layer 1 functions are located in RU, and all layer 2 functions in DU. The interface is called Acceleration Abstraction Layer (AAL) and it refers to the fact that RU contains all functions that require HW acceleration in this split. AAL interface has been considered in the O-RAN Alliance, but it is not properly standardized. Option 6 has slightly lower fronthaul throughput requirements compared to Option 7-2, but it still requires low latency because MAC processing is in DU.

Option 2 is based on the 3GPP F1 interface between DU and CU. The full DU is integrated with RU in this functionality split. All delay critical components in layer 1 and layer 2 are included in RU + DU, which is known as radio access point (RAP). Option 2 split can tolerate more transport latency than the earlier options since all the delay critical processing is integrated into the radio. This split is also known as Higher Layer Split (HLS). Option 2 results in a higher cost RU, but lower transport costs, and simplifies the system through the elimination of the DU. It is well suited when connected to Cloud RAN where CU is virtualized in the cloud, and RU and DU are integrated together. Option 2 has a few major limitations – carrier aggregation between frequencies is difficult and signal combining between sectors, also known as distributed MIMO, is difficult because there is no common DU between frequencies and sectors. Therefore, Option 2 is not expected to be practical for sub-6 GHz bands. Option 2 has been utilized for mmWave deployments because mmWave integration with sub-6 GHz bands can use dual connectivity instead of carrier aggregation.

An example of the practical implementation of RAN functionality split with eCPRI fronthaul is shown in Figure 20.6. Massive MIMO antenna in the tower includes a lot more functions than the traditional passive antenna. Massive MIMO includes large number of RF units, digital front ends, beamforming, and layer 1 low processing. Baseband DU + CU includes the rest of the digital processing of layer 1, layer 2, and layer 3. Massive MIMO antenna makes a large part of RAN capital expenditure (CAPEX) because so many functionalities are included in the antenna. Baseband DU + CU makes just smaller parts of CAPEX. RU and DU are connected with optical fiber with an eCPRI interface using typically 25 Gbps data rate.

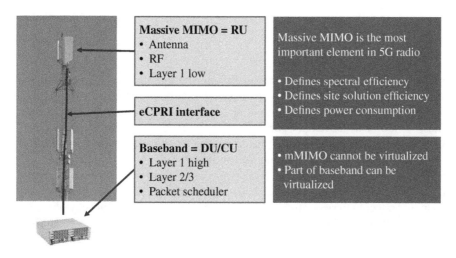

Figure 20.6 Radio network functionality and capex split.

20.4 Uplink Capacity Optimization

The uplink spectral efficiency can be enhanced with Option 7-2ULPI compared to Option 7-2 when the receiver processing is located within RU. Figure 20.7 illustrates simulation results with 8-IRC, 16-IRC, and 64-IRC receivers with 4-layer and 8-layer MU-MIMO assuming 25G fronthaul capacity. The results show that the uplink spectral efficiency with 7-2 and 7-2ULPI are similar if we limit to a maximum of 4 layers. When the number of receiver branches and MU-MIMO layers is increased, Option 7-2ULPI can enhance uplink capacity by 35–45%. These results illustrate that both options can be both deployed for the low–medium loaded uplink cases, while Option 7-2ULPI is preferred for the extreme uplink loaded cases. Note that the functionality split does not have any impact on the downlink performance.

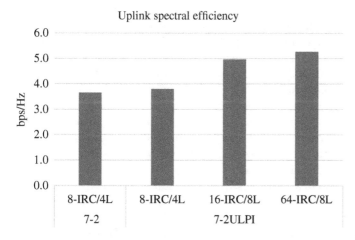

Figure 20.7 Uplink spectral efficiency with different features and fronthaul split.

20.5 O-RAN Alliance

20.5.1 O-RAN Alliance Background

O-RAN Alliance was founded in February 2018 with the intention of promoting open and intelligent RAN. It was formed by a merger of two different organizations, namely the C-RAN Alliance and the XRAN Forum. C-RAN Alliance consisted of China Mobile and Chinese vendors, while XRAN Forum consisted of US, European, Japanese, and South Korean vendors and operators. The fronthaul interface is defined in working group WG4 including also a large number of fronthaul profiles [1].

Another related forum is the Telecom Infra Project (TIP) which was formed by Facebook in 2016 as an engineering-focused, collaborative methodology for building and deploying global telecom network infrastructure, with the goal of enabling global access for all. TIP is jointly steered by its group of founding telecom companies, which form its board of directors. TIP has announced partnership with O-RAN. They have active projects on RU, DU, DU, outdoor macro, and indoor small cell [2].

20.6 O-RAN Fronthaul

Multi-vendor integration is an important consideration when connecting RU and DU from two different vendors. O-RAN has specified and published the O-RAN fronthaul interface specifications for user plane, control plane, and management plane. O-RAN profiles facilitate multi-vendor interoperability testing by defining test parameters, including duplexing type (FDD/TDD), TDD configuration, subcarrier spacing, bandwidth, number of antenna streams, fronthaul delay, beamforming type, and data compression (number of bits, floating/fixed point). The specifications and the profiles can be found under working group WG4. The main profile categorization is illustrated in Figure 20.8.

The profile defines the basic radio- and transport-related parameters. The practical challenge is that the O-RAN profiles and test cases are not designed to validate and optimize the radio performance, especially in the case of massive MIMO beamforming. The profiles are not expected to provide plug-and-play multi-vendor functionality with optimized performance. Multi-vendor integration with a basic RF head is relatively simple and has been done in bilateral exchanges between baseband vendors and radio vendors. O-RAN has specified a 7-2 fronthaul interface for the mMIMO case, but the integration gets more complex because the beam calculation and beamforming are done in different units. That likely implies that vendor-specific inter-operability testing will be needed with the beamforming antennas in order to achieve adequate radio performance, and to allow the evolution of beamforming algorithms over time from simple grid of beams to more advanced UE beamforming and MU-MIMO. The trade-off between multi-vendor integration and radio performance is illustrated in Figure 20.9.

Figure 20.10 shows the practical learnings from the multi-vendor RU – DU integrations. In case of RF head with 4T4R (4 transmit 4 receiver) functionality, the multi-vendor deployments are feasible and has been utilized in the live networks. mMIMO beamforming adds complexity to the integration – careful vendor-specific testing is required for Option 7-2 interface integration to achieve reasonable radio

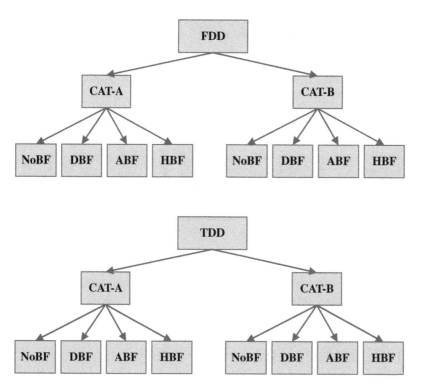

Figure 20.8 O-RAN profile main categorization (BF = Beamforming, DBF = Digital BF, ABF = Analog BF, HBF = Hybrid BF).

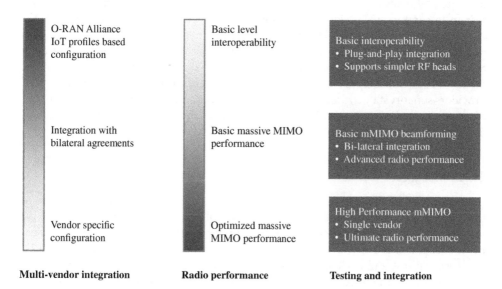

Figure 20.9 Multi-vendor fronthaul integration effort vs. radio performance.

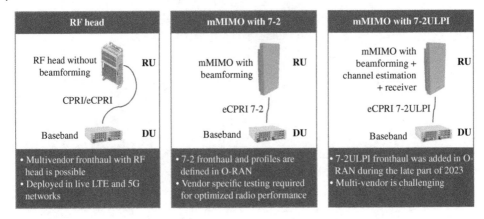

Figure 20.10 Multi-vendor fronthaul integration effort with different RF and fronthaul options.

performance. Option 7-2ULPI tends to make the multi-vendor cases more challenging. It is not trivial to achieve simple multi-vendor integration and optimized radio performance at the same time. Practical O-RAN learnings are illustrated here [3].

20.7 Open Test and Integration Center and PlugFests

20.7.1 Open Test and Integration Center (OTIC)

Open Test and Integration Center (OTIC) was formed to enable O-RAN multi-vendor interoperability testing. OTIC participants include operators, equipment vendors, and system integrators. OTIC initiative was launched in September 2020. As the O-RAN Alliance focuses on open RAN interfaces and the use of open-source solutions, OTIC is providing an environment for validating these solutions. By creating a common platform and set of processes, OTIC will speed up multi-vendor testing and practical deployments. OTIC's initial focus is to ensure RAN components from participating equipment providers support standardized open interfaces and can successfully interoperate based on the test specifications published by O-RAN Alliance. O-RAN Alliance and TIP have established OTIC with the TIP community lab in Berlin.

20.7.2 PlugFest

The purpose of the PlugFest Project Group is to define and accelerate the development of test materials, test plans, or other documents that will support a TIP sponsored PlugFest for interoperability among products. Effectively, the PlugFest Project Group will quantify the readiness of TIP-sponsored technologies for integration into operator networks. PlugFest will test and evaluate multiple solutions in one common environment as a representative of the operator's environment. PlugFest brings partners and operators together to agree on a common test plan and environment to later facilitate the integration of each solution into the operator's networks. PlugFests have been running at multiple venues across the globe with more than 100 participating companies.

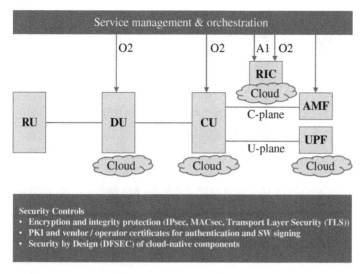

Figure 20.11 O-RAN interfaces and security impacts.

20.8   O-RAN Security and Orchestration

Security is a key capability of the wireless networks, providing critical services for the connected societies. Opening interfaces, and the creation of new interfaces via network disaggregation, or via the decoupling of HW to SW or exposing new "open" interfaces does increase threat exposure to operating networks as well as risks to subscriber and device access to networks. When comparing the new network architecture, it is clear that the disaggregation of the network has a consequence of not only moving HW from a site with trusted HW enabled via a trusted connection. O-RAN security builds on 3GPP 5G security methods plus additional security controls to support disaggregated networks. O-RAN security framework is shown in Figure 20.11. The target is that O-RAN-based solutions with new interfaces are able to provide the same level of security as the networks without those interfaces. O-RAN security aspects are discussed in [4].

20.9   Baseband Virtualization and Cloud Ran

The ideal target of vRAN configuration is to take full benefit of virtualization and centralization in terms of flexibility and upgrade capability. There are number of underlying technology requirements to make these targets feasible. These requirements are listed in Figure 20.12.

1. Efficient mMIMO RU for maximizing the radio performance.
2. Fiber availability for connecting RU to centralized DU.
3. 100G optics capability for supporting 5G bandwidths with massive MIMO for three sectors, together with other bands with passive antennas.
4. Low latency connection between RU to vDU. The latency should typically be less than 100 μs (0.1 ms) which limits the practical distance to less than 20 km.

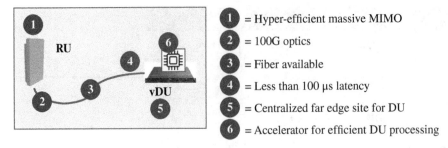

Figure 20.12 Ideal vRAN target configuration.

5. Far edge sites that are close enough to RU sites to allow the low latency connection between RU and DU.
6. HW accelerator for processing layer 1 high in DU in far edge sites.

Such an ideal solution is not generally available today. The following sections discuss the challenges and the solution.

20.10 Baseband Virtualization and Centralization

Virtualization and centralization of baseband do not need to come at the same shot. Figure 20.13 illustrates the four main options:

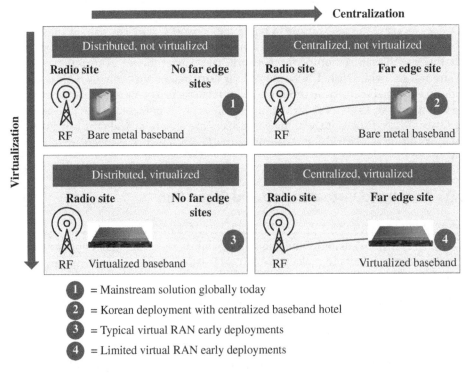

Figure 20.13 Virtualization, centralization, and open fronthaul are independent.

Case 1: Most typical architecture today has bare metal baseband located next to the RF.

Case 2: Centralized bare metal baseband has been used, for example in Korean networks. Such a deployment is enabled by the wide availability of fiber networks. Centralized baseband brings some benefits in terms of baseband upgrades and capacity expansions.

Case 3: Distributed vRAN is utilized today in vRAN early deployments if there are no suitable far edge sites available. Such a deployment is relatively inefficient because the cloud server environment at every site adds to the costs.

Case 4: Centralized vRAN is used in limited scale in some LTE and 5G networks.

Open O-RAN fronthaul is independent of virtualization and centralization. The open fronthaul interface is more common with vRAN deployments just because the new and small vendors providing baseband functionality are not capable of developing efficient baseband HW and must use generic HW in vRAN.

20.11 Far Edge Availability and Network Topology

Figure 20.14 illustrates today's typical network architecture and the future one. The current architecture has fully distributed radio network processing and highly centralized core network processing. DU and CU are located close to the RF at base station sites. It is expected that the radio architecture will become more centralized gradually, and core architecture will become more distributed. Ideally, there would be harmonization in the radio and core locations in the long term.

The largest benefit from DU virtualization can be obtained by placing several DUs into a centralized location. Such a solution allows easier HW upgrade and enables resource pooling. Virtualized DUs must be placed relatively close to RU because of the low latency requirement between RU and DU. The distance should typically be less than 20 km. Such locally centralized sites are called far edge sites. There are a few challenges related to the far edge sites:

Traditional

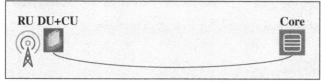

Future

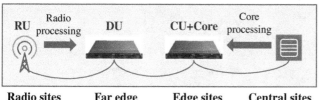

RU = Radio unit
DU = Distributed unit (baseband)

Radio sites	Far edge	Edge sites	Central sites
10k	1k	100	<10

Figure 20.14 Future network architecture with DUs located in far edge sites.

- Web scalers' edge sites are too far away from the radio. For example, Google has thousands of global edge nodes, but those sites are too far away from the radio and cannot be used for virtualized DU.
- There are no synergies between vRAN and core network deployments from the architecture point of view. vRAN must use far edge sites that are close to the radio sites, while the distributed core will likely not be deployed close enough to the radio. vRAN requires typically at least 10× more far edge sites than distributed core.
- There is very low synergy in resource pooling across the vRAN radio function and virtualized core function even when those are co-deployed at the same far edge site.

Figure 20.15 illustrates the locations of mobile operator's far-edge sites and web scale edge sites. Each far-edge site with centralized vRAN processing can typically connect to a few tens of RF sites. Therefore, large mobile networks need hundreds of far-edge sites for hosting vRAN processing. The number of webscale edge sites tend to be substantially lower and cannot be utilized for vRAN DU processing. The simple reason is that vRAN requires a latency of less than 0.1 ms, while webscale edge sites are designed for application performance and typically a few milliseconds of latency is perfectly fine.

Typical mobile network transmission topology is illustrated in Figure 20.16. Star configuration is used for connecting base station sites to the first aggregation point.

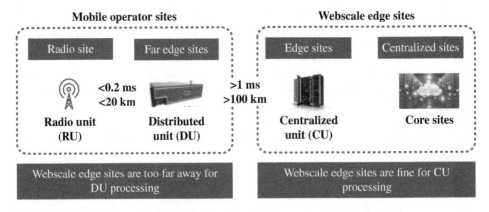

Figure 20.15 Mobile operator and webscale edge sites. Source: Scanrail/Adobe Stock.

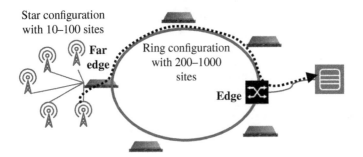

Figure 20.16 Typical mobile network transmission topology.

The first aggregation point connects to a few tens of base stations. This point could be used for placing vDU processing. The next level is ring configuration with up to 1000 sites connecting to the edge node. Ring brings resilience but adds also potential delay variance if the path changes. The edge nodes are generally too far away from the radio sites for vDU locations.

20.12 Fiber and Optics Availability

Centralized DU requires low latency fiber connection between RU and DU. Some networks have high fiber availability, which makes it simpler to utilize centralized DUs. Japan, Korea, and China have, in general, fiber widely available. Also, many incumbent mobile operators with fixed networks have access to the fiber network, but there are also many operators in advanced countries where fiber is not available. An example case for a European operator with networks in three countries is shown in Table 20.2. The simple conclusion is that centralized baseband is not an all-around solution for all sites and operators. Baseband DU products should be designed both for centralized as well as for distributed installations depending on the operator case and deployment area.

Fronthaul also requires high optics capability in addition to the fiber availability. The fronthaul eCPRI transport requirement for a massive MIMO site, with 100 MHz bandwidth, is in practice 25 Gbps per sector and 3x25 Gbps for 3-sector case. When also the other bands are also considered, the total transport requirement is approximately 100 Gbps. The backhaul requirement is substantially lower because pooling gains can be utilized in DU/CU, typically just 10 Gbps. The backhaul requirements can be fulfilled with fiber but also with E-band microwave radio links.

Table 20.2 Fiber availability in three example European networks.

	Operator 1	Operator 2	Operator 3
Fiber availability in urban	All urban sites are fiber-connected	70% of urban sites are fiber-connected	Most urban sites are fiber-connected
Fiber availability in rural	Many rural sites are fiber-connected	Only a few rural sites have fiber	Some rural sites have fiber
Fiber network ownership	Operator owns the fiber network	Operators do not own the fiber network	Operators do not own the fiber network
Readiness for centralized baseband	Well-prepared	Centralization would result in clearly higher opex	Centralization would result in clearly higher opex

20.13   Baseband Hardware Efficiency

RU implementation requires optimized HW including analog components, and front-end circuits, like linearization, beamforming, and layer 1 low. DU runs layer 1 high processing which is also computationally complex, in particular channel decoding also known as Forward Error Correction (FEC). Common of the Shelf (COTS) HW can run channel encoding and decoding as well as other physical layer processing, but it is not as efficient as the purpose-built HW developed and deployed in the bare metal products. The higher layers 2 and 3 can be implemented on COTS HW. The high-level functionality split and the required HW solutions are shown in Figure 20.17. We will focus on the baseband HW acceleration solutions and efficiency in this section.

There are multiple options for layer 1 high HW acceleration:

- Central Processing Unit (CPU) combined with in-built Field-Programmable Gate Array (FPGA) or Application Specific Integrated Circuit (ASIC) based HW accelerator.
- Graphics Processing Unit (GPU) or RAN-NIC card with System on Chip (SoC) solution for acceleration.

The first cloud RAN implementations were CPU-based solutions using FPGA accelerators for FEC. These first-generation cloud RAN deployments, however, turned out to be inefficient in terms of required HW, and in terms of power consumption for high capacity 5G networks. Therefore, the later trend is to utilize RAN-NIC (network interface card) with SoC accelerator for running layer 1 high in an efficient way. The same SoC solution can be utilized in the classic DU which allows to the reuse of the same layer 1 SW in virtualized and in classic DU. Layer 2/3 can run on advanced risc machine (ARM) or x86 processors. Figure 20.18 illustrates the typical implementation of cloud DU and classic DU. We can note that layer 1 SW is tightly coupled with layer 1 HW. It is not possible to use the same SW on different layer 1 HW solutions. The HW and SW decoupling is not feasible in layer 1, while the decoupling is practical for layers 2 and 3. Therefore, it is suggested to utilize independent HW solutions for layer 1 rather than for layer 2 and 3. The optimized layers 2 and 3 may utilize ARM-based HW in one implementation and x86-based HW in another implementation. More details about the HW acceleration options can be found in [5].

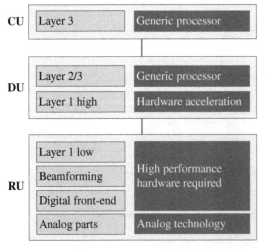

Figure 20.17 Radio network functionalities and the required hardware.

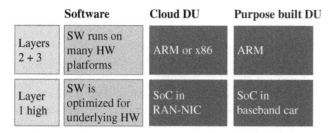

Figure 20.18 Typical cloud RAN hardware and software design.

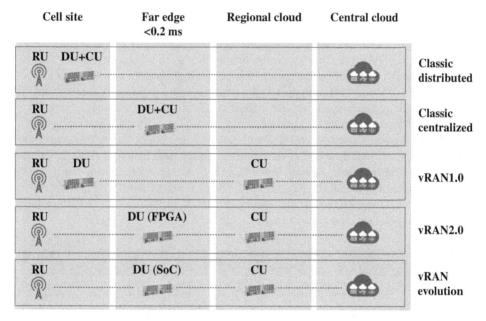

Figure 20.19 Virtual RAN evolution steps.

20.14   Virtual RAN Evolution

As RAN virtualization is complex and takes time, we illustrate here an example of vRAN evolution steps: vRAN1.0 is the first step in radio virtualization where DU is distributed, and CU is centralized and virtualized. CU virtualization is simple from the technology point of view because CU processing is not delay-sensitive nor computationally complex. vRAN1.0 was deployed commercially in mmWave networks in 2019. vRAN2.0 has virtualized DU running on top of Intel FlexRAN. FPGA-based HW acceleration is used for channel coding. vRAN2.0 has been deployed since 2021. Further evolution can utilize optimized silicon for DU processing. The evolution steps are illustrated in Figure 20.19.

20.15   RAN Intelligent Controller

5G networks are expected to grow increasingly complex due to various new requirements and technologies including low latency, control and user plane separation

(CUPS), functional RAN splits, network slicing, and RAN virtualization. Autonomous network operations using AI and ML are essential to optimize 5G network. RIC is designed to solve these challenges and bring new innovations into RAN optimization. Applications like mobility management, admission control, and interference management are available as apps on RIC, which enforces network policies via a southbound interface towards the radios. RIC provides advanced control functionality, which delivers increased efficiency and better radio resource management. These control functionalities leverage analytics and data-driven approaches, including advanced ML/AI tools, to improve resource management capabilities. RAN functionalities are summarized in Figure 20.20.

O-RAN has defined overall RIC architecture and functional SW elements. They all are deployed as virtual network functions or containers to distribute capacity across multiple network elements with security isolation and scalable resource allocation. They interact with RU HW to make it run more efficiently and to be optimized in real-time as a part of the RAN cluster to deliver a better network experience to end users.

RIC consists of a non-real time part and a near real time part. Non-real time RIC is located with an orchestration layer and includes service and policy management, RAN analytics, and model training for the near real time RIC. A1 interface is used between the orchestration layer and non-real RIC and near real time RIC. Network management applications in non-real time RIC receive and act on the data from the DU and CU via a standardized A1 interface. AI-enabled policies and ML-based models generate messages in non-real time RIC and are conveyed to the near real time RIC. E2 interface runs between near real time RIC and RAN. Fast optimization loops run via the E2 interface. O1 interface is used for data collection. RIC interfaces are shown in Figure 20.21.

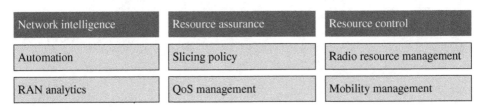

Figure 20.20 RIC functionalities.

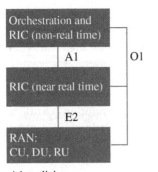

Figure 20.21 RIC interfaces.

A1: policies
E2: optimization
O1: data collection

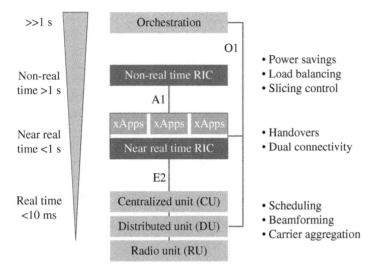

Figure 20.22 RIC operation time scales.

Near real time RIC is SW platform for hosting applications called xApps. Near real time RIC platform allows xApps to control RAN network elements via the E2 interface. xApps use an E2 interface to collect near real time information on a UE basis or a cell basis. That is, the so-called southbound interface. Operators can control near real time RIC via A1 northbound interface. Near real time RIC runs on the time scale from approximately 10 ms to 1 second.

Non-real time RIC functionality includes policy management, slicing management, load balancing actions, power savings, configuration management, device management, and lifecycle management for all network elements. It is similar to the combined element manager and reporting functionalities in legacy networks. Non-real time RIC can be applied to get better understanding of the network performance and consequently, optimize the network by the policy parameters. Non-real time RIC can use data analytics and AI and ML training and inference to determine the RAN optimization actions. Non-real time RIC runs on the time scale of 1 second and higher (Figure 20.22).

5G operators want the optimization applications to be separate from the RAN equipment vendor, allowing them to select the application providers and RAN equipment providers, and interwork them through standardized interfaces. Also, RIC apps may be developed by any entity – by the RAN vendor, by the operator, or by any third party. RIC is intended to be multi-vendor in order to support end-to-end network slicing across different RAN infrastructures, however, this requires support from all vendors. RIC hosted apps can be also implemented directly in the RAN if the apps are provided by the RAN vendor.

20.16 Summary

O-RAN new interface and RAN virtualization will change the way radio networks are procured, deployed, and optimized. The evolution is towards open multi-vendor interfaces, centralized baseband, virtualized baseband, and to bring AI into the network optimization. The target is to make the deployment faster and more cost-efficient by separating radio and baseband vendors, and by separating baseband HW and SW. O-RAN Alliance has defined new interfaces inside RAN, in particular, for the fronthaul connecting different radio and baseband units together. Multi-vendor integration will still need substantial amount of testing to achieve the optimized radio performance. Open interfaces to RIC are targeted to bring new level of AI-based optimization into the radio networks.

RAN virtualization is another major trend in the network where the target is separate baseband HW and SW. Baseband processing can be run on generic HW platforms, while purpose-built HW acceleration can improve the efficiency considerably. Baseband centralization can be used together with virtualization if the transport network can provide high bandwidth and low latency connection to far-edge sites.

The design of future radio networks must consider these different deployment scenarios to provide a high level of flexibility.

References

1 O-RAN Alliance. https://www.o-ran.org/
2 Telecom Infra Project (TIP). https://telecominfraproject.com/
3 Deutsche Telekom O-RAN learnings blog and white paper. https://www.telekom.com/en/company/details/bundled-in-a-white-book-learnings-from-o-ran-town-1026846
4 Liyanage, M., Braeken, A., and Shahabuddin, S. (2023). Pasika Ranaweera "Open RAN security: Challenges and opportunities". *Journal of Network and Computer Applications.* 214, May: 103621.
5 Nokia Cloud RAN blog and white paper. https://www.nokia.com/blog/in-line-architecture-bringing-efficiency-and-performance-to-cloud-ran/

21

Machine Learning for 5G System Optimization

Riku Luostari[1], Petteri Kela[1], Mikko Honkala[2], Dani Korpi[2], Janne Huttunen[2], and Harri Holma[1]

[1] Nokia, Finland
[2] Nokia Bell Labs, Finland

5G Technology: 3GPP Evolution to 5G-Advanced, Second Edition.
Edited by Harri Holma, Antti Toskala, and Takehiro Nakamura.
© 2024 John Wiley & Sons Ltd. Published 2024 by John Wiley & Sons Ltd.

21.1 Introduction

This chapter explores the application of machine learning (ML) in wireless systems, primarily focusing on opportunities in 5G-Advanced, although not exclusively so. The early sections deliver an introduction, highlighting the role of ML in the wireless technology domain, and an overview of training and inference of (ML) models in wireless systems is discussed. Various categories of ML and selected key algorithmic techniques are introduced.

The subsequent sections are dedicated to ML for wireless systems. A broad spectrum of use case examples are explored, such as signal linearization, deep neural network (DNN)-based receivers, and radio resource management (RRM) and scheduling. While some topics are explored in depth to provide a detailed understanding of the technical aspects, others are only briefly introduced, with references to published papers for further reading, ensuring broad coverage of the subject while offering avenues for more profound exploration where readers are most interested.

Although the example list is not exhaustive, and new use cases are frequently emerging, it offers a snapshot of the current research direction and acknowledges the continuously evolving landscape.

An AI-based air interface is expected to be part of next-generation 6G radio access targeting for higher flexibility and increased efficiency. The first steps towards AI integration took place in 5G-Advanced with a study item in Release 18. See Chapter 11 for more details about AI-specific topics in 3GPP evolution.

21.2 Motivation

Wireless networks are transitioning from traditional paradigms that rely on physical and heuristic mathematical modeling to those based on big data and machine learning (ML). This shift is primarily driven by wireless networks' increasing complexity and scale, and the need for more adaptive and efficient systems.

The wireless radio environment in cellular networks is inherently complex, with various factors affecting performance, such as interference, user mobility, propagation, and traffic patterns. Traditional mathematical models often simplify assumptions that may not accurately capture this complexity. ML techniques can better handle the complexity by learning from real-world data and adapting to changing conditions in real-time.

The rapid increase in wireless devices and network sizes has led to large volumes of data, which can be harnessed to train ML models. This data can be used to extract valuable insights and develop more accurate and robust models for various aspects of wireless networks, such as channel estimation, user behavior prediction, and traffic management. As wireless networks grow and the number of connected devices increases, traditional mathematical methods for computing, for example the resource allocations, or processing of the physical layer signals, may struggle to scale efficiently. ML techniques can adapt to these growing demands by parallelizing computations, leveraging distributed computing resources, and employing advanced optimization algorithms.

ML can be a good solution for solving problems where the physical principles impacting the system's behavior are understood, but the relation between inputs and outputs is too complex to be captured by a heuristic or physics-based model. ML models can learn from data and adapt to changes in the environment, making them suitable for applications such as control systems, where the system's behavior may be affected by variables that can change over time.

21.3 Model Training and Inference in Wireless Systems

ML can be broadly divided into two phases: training and inference. The training phase involves feeding a large amount of data to a model and adjusting its parameters so that it can accurately make predictions on new, unseen data. The training phase can be computationally costly. This phase is also referred to as model fitting or model training. The inference phase involves using the trained model to make predictions on new data. The model uses the parameters learned during the training phase and generally is computationally less expensive than the model training.

21.3.1 Training

ML model training pipeline typically involves: (1) data collection and preprocessing in a suitable format for ML models, (2) selecting the ML model architecture and algorithm suitable for solving the problem, (3) training the model by using the training data iteratively until it fits the training data well, (4) model evaluation using testing data to assess performance, (5) model hyperparameter tuning and retraining the model in case the performance is not satisfactory, and (6) deployment of the trained model in the real-world setting (Figure 21.1).

The above steps can be iterated until the performance is satisfactory. Once deployed in a production network, the deployed model performance should be monitored over time for any model drift. Model drift may occur, for example, when the radio environment changes due to new cells added to the network, or end users slowly changing their behavior. When the model performance is no longer optimal, the above process may need to be repeated to update the model.

Some of the problems in mobile wireless systems are suitable for solving with ML models trained in a factory environment. For such models, retraining or fine-tuning

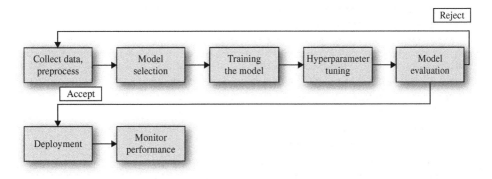

Figure 21.1 Model training process.

the model in the target environment is unnecessary. The benefit is that the data may be collected at the factory with more consistent and controlled procedures, making it easier to train an ML model accurately and easily deploy it in the field through software updates.

In contrast, other models may benefit from a detailed understanding of the live cell's radio environment or traffic profile; training the model at the cell is beneficial in such cases. However, model training on the field may be difficult without negatively impacting the network performance before the optimal solution is derived. Training the models in a live network also requires model management systems capable of storing, documenting, and versioning models, and monitoring the model performance. This area is also actively researched, and methods, such as transfer learning, can help in the initial stages of the learning process.

Some models can learn and update the model weights online. Unlike traditional batch learning, where a model is trained on a large dataset and then deployed, online learning models continuously learn from new data, adapting their weights and parameters as they encounter further information.

21.3.2 Inference

ML inference refers to using the trained model to make predictions based on new, unseen data. For example, the prediction could be a traffic forecast in a cell for switching some BTS functions off early to save energy. It could also be a radio channel prediction to improve beamforming (BF) accuracy in the coming scheduling intervals.

Depending on the latency requirements, this inference function can be performed on various platforms. For example, processing radio channel estimations may have a delay budget of just tens of microseconds, making it imperative to have the model embedded in the BTS hardware. In contrast, tuning radio network parameters based on infrequently updated performance data may be performed only once a day, allowing the model placement in, for example, Self-Organizing Network (SON) platform or network management systems.

21.4 Machine Learning Categories

Various types of ML systems can be classified based on what kind of supervision is applied during the model training process.

- In *supervised learning*, the model training data includes input examples and labeled output data, and the goal is to learn the input-output relation of a function. After training, the model can make predictions on previously unseen input examples. Supervised learning can be used, for example for regression, prediction, and classification tasks.
- In *unsupervised learning*, the model is trained to find patterns or relationships from unlabeled training data. The training data does not contain the desired output values. The goal is to learn functions that describe the training data. Example use cases are dimensionality reduction, clustering, density estimation, and anomaly detection.

- *Semi-supervised* learning falls in between the last two categories. The combined capability is useful when only a limited amount of labeled data is available. The model learns structures from the labeled data and can use the unlabeled data to improve the model further.
- *Reinforcement learning* (RL) is a sequential decision-making process where an agent learns to interact with the environment to maximize the rewards. Here, the training data is collected during the interaction and is thus optimized for improving the RL agent's performance. RL is helpful in dynamic environments where defining clear rules is difficult, e.g. to play games. The training is performed by taking actions by trial and error and receiving rewards or penalties as feedback. Typically, RL is modeled using Markov Decision Process (MDP) framework. Essential and widely used algorithms include Q-Learning, Deep Q-learning, SARSA, and Multiarmed Bandits (MAB).
- In *transfer learning*, a model is pre-trained with a similar task using a large dataset, but not explicitly made to solve the problem. The model is afterward fine-tuned for the specific task with relatively small training data.

21.5 Key Algorithm Techniques

A few key algorithm families are shortly introduced. The list is not comprehensive.

- *Classic* ML techniques have been used for many years; they are based on mathematical probability and statistics and are easy to implement and understand. Techniques include logistic regression [1], linear regression [2], support vector machines [3], and random forests [4].
- *Deep Neural Network (DNN)* [5] is a popular family of ML models, often also called Deep Learning (DL) models. They consist of several layers of neurons and can learn complex nonlinear patterns in data using a back propagation technique during the training process. DL has been successful in a wide range of applications, and due to active research, it is evolving quickly; new innovations and model architectures are popping up daily.
- *Deep RL* [6] combines DL and 'classic' RL. A DNN is used to process and analyze the data from the environment and make decisions. It is useful when the number of observable states of the system is too large for conventional RL. Deep Q-Networs (DQN) are popular and widely used architectures.
- *Autoencoders* are used for dimensionality reduction, compression, and feature learning. They consist of two neural networks, an encoder, and a decoder, connected with a 'noisy information bottleneck' layer. The model is typically trained unsupervised, where the decoder tries to recover the original data as accurately as possible [7].
- *Bayesian Optimization* [8] is a classic ML technique that uses a probabilistic model, typically a Gaussian process, to efficiently find the optimum values of a black box function that is expensive to evaluate. It can converge quickly with relatively small training data and provides estimates and the associated uncertainty.

21.6 Machine Learning for 5G Wireless Systems

This section introduces selected examples of use cases in a broad area of wireless systems. While some examples go into detail with simulation results, others are only described at a high level, with reference for further reading where available.

21.6.1 Linearization of the Signal

The impairments of analog components in the transmitter and the receiver result in distortions to the linearity and increase the Error Vector Magnitude (EVM). Higher modulation schemes, such as 256QAM and above, are more sensitive to nonlinear distortions and require high signal phase and amplitude accuracy. Minimizing the distortions caused by analog components, for example, in the power amplifier (PA) and the preamplifier in the receiver, is essential.

21.6.2 Signal Linearization at the Transmitter

The digital pre-distortion (DPD) technique works by applying a pre-distortion to the input signal of a power amplifier (PA), to compensate for the nonlinear distortion introduced at the PA operating at high power levels. Applying DPD is typical for linearizing PA, and the process is depicted in Figure 21.2.

ML models can be trained to apply learned nonlinear pre-distortion to the signal, allowing the system to adapt to the PA characteristics based on feedback from PA output to the ML algorithm, which estimates the DPD function. Learning the linearity, instead of using a fixed function, is helpful for various reasons; the nonlinear behavior may change over time, for example because of retuning the cell to a different frequency, temperature variations, or component aging, and an ML model can learn the changes [9]. Figure 21.3 further illustrates the idea. ML-based DPD has learned the nonlinear characteristics of the PA and applies inverse pre-distortion to the input signal, resulting in linear output.

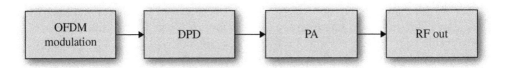

Figure 21.2 DPD is typically located before the PA in the RF chain.

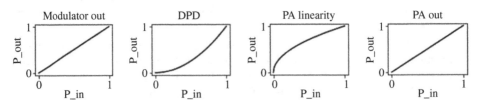

Figure 21.3 The DPD aims to inverse the nonlinear behavior of the PA.

ML-based DPD can be applied to various kinds of radio technologies, and it does not have direct requirements or dependencies from the 3GPP specifications point of view. DPD with ML can be used to improve the performance of any radio communication system, regardless of the specific standards or specifications that it adheres to.

21.6.3   Signal Linearization at the Receiver

Neural network-based approaches can also be used to model and compensate for non-linear distortion at the receiver side. For instance, Pihlajasalo et al. in [10] proposed an ML-based solution for demodulating Orthogonal Frequency Division Multiplexing (OFDM) signals under severe PA-induced nonlinear distortion. The main motivation for this work is to facilitate an accurate detection of signals even under high error vector magnitude (EVM) levels, which results in less stringent signal quality requirements at the transmitter side. This, on the other hand, can be used for extending the coverage of the system, as mobile devices can push their transmitter PAs closer to saturation and thereby increase their transmit power. This will also increase the transmitter power efficiency.

The solution proposed in [10, 11] involves developing a DL-based convolutional neural network (CNN) receiver with layers in both time and frequency domains. It was observed that having ML processing capabilities, both in time and frequency domains, is crucial for the accurate detection of nonlinearly distorted signals. The link distance of such an ML-based receiver is evaluated in Figure 21.4 using link-level performance results and an urban micro (UMi) non-line-of-sight (NLOS) path loss model. The circle denotes the highest coverage for each considered receiver. It can be observed that

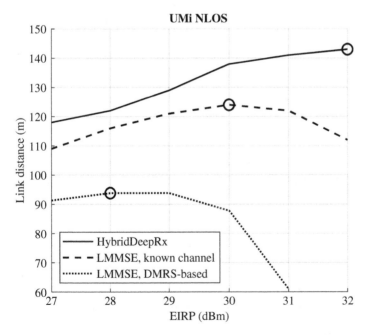

Figure 21.4 Link coverage analysis using an ML-based HybridDeepRx receiver, compared to conventional linear receivers.

the proposed ML-based HybridDeepRx receiver outperforms classical linear receivers in terms of the achievable link distance, owing to the fact that it can still detect the signal when the EVM is high compared to the modulation order. Therefore, such an ML receiver can facilitate pushing terminal PA systems deeper into saturation, improving terminal power efficiency, radiated power, and network coverage.

21.7 Channel State Information (CSI) Improvement and Channel Prediction

Radio channel estimation is used for determining the characteristics of a wireless communication channel. Channel estimation is a critical task; it enables the receiver to estimate the transmitted signal and compensate for the effects of the radio channel. OFDM receiver performance highly depends on the quality of the radio channel estimate. The estimate is typically calculated from known OFDM pilot signals, such as demodulation reference signals (DMRS) and channel sounding reference signals (SRS).

21.7.1 SRS-based Channel Estimation Improvements

A key characteristic of TDD systems is channel reciprocity; the radio channel response from the BTS to the UE (downlink) is the same as from the UE to the BTS (uplink), given the two transmissions occur within a short period of time. This reciprocity allows the BTS to estimate radio channels based on received uplink SRS and DMRS transmissions. However, as the UE moves, the radio channel changes rapidly, and the channel estimate may not be optimal for longer than just some milliseconds. This phenomenon, called channel aging, reduces practically achievable BF and massive multiple input multiple output (mMIMO) performance.

The fundamental concept in radio channel prediction is to measure a subset of signal attributes based on, for example, SRS or DMRS values and forecast the evolving radio channel. The objective is to improve the channel estimation quality or reduce the required pilot signals overhead, without negative performance impact.

21.7.2 CSI Feedback

In most cases, the SRS is the preferred method for providing the base station with wideband radio channel information. However, this reciprocity does not hold for FDD systems because the uplink and downlink transmissions occur on different frequency bands, and hence SRS cannot be used in FDD-based systems. Also, in the case of a TDD system, SRS resources might be exhausted in a cell, for example, due to too many RRC-connected UEs, and other methods for delivering the radio channel must be used. In one such method, the UE measures the channel based on downlink pilot signals and reports it to the base station on control channels. To reduce the overhead, the UE matches the measured and vectorized radio channels of OFDM subcarriers as closely as possible with an entry from a carefully designed codebook and reports the codebook indexes to the base station.

There is a delay from when the UE measures the channel to when the next opportunity is to send the CSI to the base station. Once the base station receives the CSI, it needs to

Figure 21.5 Autoencoder for CSI compression.

use it until a new CSI report from the UE arrives. Due to radio channel aging, codebook quantization, and the relatively large amount of data required for the CSI information transmission, this process is not optimal. It can be improved with ML.

One option considers using ML on the user equipment (UE) side for a more accurate and predictive selection of codebook indexes based on historical channel measurements. Such a method can improve CSI quality at the base station and reduce reporting overhead. The second option is to use ML on the base station side to predict the channel's frequency and time variations until the following report becomes available. Both options can be implemented without changes to the 3GPP specifications.

The third option considers using an autoencoder for CSI-reporting overhead reduction and overall channel state quality improvements [12]. This approach trains the encoder neural network to compress the measured channel state into lower-dimensional latent weights transmitted over the control channels. The decoder neural network reconstructs the original CSI data from the received weights. This auto-encoder can be trained separately or together. The autoencoder can learn a good radio channel representation with much-reduced data compared to heuristic approaches without sacrificing performance. This approach impacts the 3GPP specifications as the weights need to be transmitted over the radio interface (Figure 21.5).

An autoencoder-type solution implemented across the air interface poses various challenges to the implementation, standardization work, and system performance monitoring. In the above example: (1) a new signaling interface is required for transferring the model weights, (2) BTS would need to be trained in such a way that it can efficiently use the received weights from UE, (3) to monitor the performance, a specific data collection method could be required, (4) signaling for switching a model in case of failure could be required, (5) interoperability between different BTS and UE vendor equipment needs to be ensured, (6) management and delivery of new encoder and decoder versions should be considered.

21.8 Deep Neural Network-Based Receivers and DeepRx

The motivation for improving OFDM receiver performance is to increase spectral efficiency and reduce the overhead of pilot symbols. OFDM receivers use the pilot symbols to produce a channel estimate of the time- and frequency-variant signal, which in mMIMO systems is used not only for signal detection but also for BF and spatial multiplexing of UEs. The quality of the channel estimate is arguably the most crucial factor for generating accurate beams and a functional and efficient mMIMO system.

Despite the well-optimized heuristic algorithms in the current OFDM systems, ML- and, in particular, DNN-based replacement functionalities have shown promising results. For example, channel estimation based on a neural network has been demonstrated to provide gains [10, 13] by learning to compensate for distortions in the received signal and better tracking of the time- and frequency-variant signal.

In another example, one Fully Convolutional Neural Network (FCNN) [14] replaced the channel estimation, equalizer, and demodulation functionalities [15], thus jointly learning the entire OFDM receiver. This solution can perform better than a non-ML state-of-the-art solution, especially at higher UE speeds where rapid temporal changes in the radio channel can become challenging for non-ML solutions.

Traditionally, a receiver includes many separate processing blocks, such as raw channel estimation from the reference signals, smoothing and interpolation, equalization, demapping, and decoding. Therefore, the traditional receiver splits the processing of the channel estimate using reference signals from the processing of the unknown data signal. It only combines these two in the equalization step, relying on an accurate channel estimate. Therefore, the receiver performance is bound by the channel estimation accuracy, which is inherently limited, e.g. for high mobility UEs. We can overcome this limitation by training the receiver algorithm all the way from the antenna signal into the uncoded bits, as is done in the DeepRx receiver, depicted in Figure 21.6. By also using the unknown received data, such an ML receiver implicitly learns a data-aided algorithm (e.g. an iterative one) that can track the changes in the channel response in time and frequency, even in the case of sparse reference signals, as shown in Figure 21.7. The improvement in the physical layer performance leads to higher system-level throughput.

In theory, a DL-based receiver can be implemented today as it does not have 3GPP specifications-related dependencies. However, it also provides excellent flexibility for future innovation. Practical implementation of ML-based receivers in multi-layer MIMO systems can still be challenging [16].

For example, a DeepRx-type receiver can also be extended to support MIMO detection. Conventional MIMO detection algorithms contain an equalization block that relies on input-input multiplications. To train a neural network to replace such algorithms efficiently, we must equip neural networks with additional components to represent input-input multiplications; otherwise, the neural network needs to learn to approximate them, leading to higher complexity and approximation error. The DeepRx MIMO

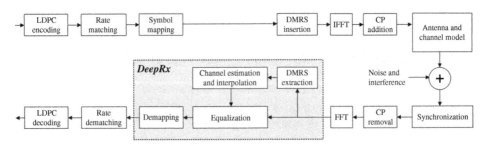

Figure 21.6 DeepRx replaces most of the processing blocks in the frequency domain receiver processing in the physical layer. The blocks inside the DeepRx box describe the traditional receiver processing blocks that DeepRx replaces with a trained neural network.

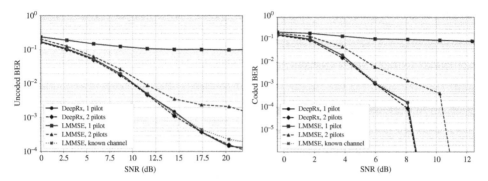

Figure 21.7 DeepRx SIMO reaches LMMSE with known channel performance, even for highly mobile UEs of 0–130 km/h, as in these figures.

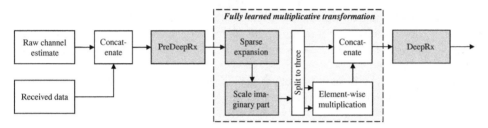

Figure 21.8 DeepRx can be extended to MIMO by adding the capability of input-input multiplication to the neural network, which allows more efficient learned equalization.

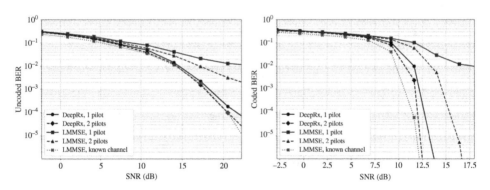

Figure 21.9 DeepRx MIMO can surpass the LMMSE receiver in TDL-E channels and reach near LMMSE with known channel performance.

has multiple options on how to represent the multiplications. One of them is shown in Figure 21.8, in which the network learns which inputs should be conjugate multiplied, and which should be multiplied. The performance results of the DeepRx MIMO network are shown in Figure 21.9. DeepRx with one pilot symbol outperforms an LMMSE receiver with two pilot symbols. DeepRX, with two pilot symbols, gets very close to the performance of an LMMSE receiver with a known channel.

DeepRx learns a state-of-the-art receiver algorithm based on data. Since the learned method has very few assumptions about the actual waveforms or channels, it is possible to train this kind of a receiver for other types of waveforms, e.g. ones that utilize novel constellation shapes, as described in the next section.

21.9 Pilotless OFDM

A particularly lucrative approach for taking full advantage of DeepRx-type receivers' data-aided capabilities is to train it to operate without any channel estimation pilots. This is possible by learning the mapping from bits to symbols jointly with DeepRx, which results in a novel constellation shape [17].

An example of such a learned constellation is shown in Figure 21.10 for a case of four bits per symbol (16-QAM). A particularly evident feature of the learned constellation is its apparent asymmetry. Unlike conventional QAM constellations, which appear identical with every 90-degree phase shift, this is not the case with the learned constellation. Indeed, it has no ambiguity in terms of the correct orientation, which means that the receiver can detect the correct phase shift blindly as long as a sufficient number of samples are observed.

This property is the basis for pilotless detection by DeepRx. When trained jointly with the constellation shape, it will learn to implicitly expect a certain constellation and

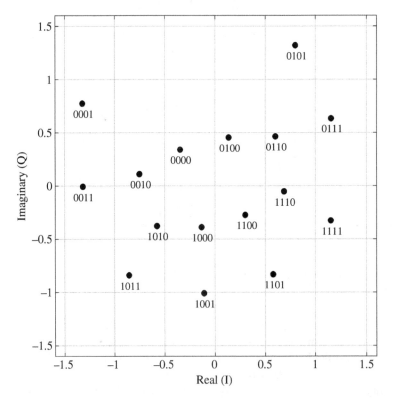

Figure 21.10 Example of a learned constellation.

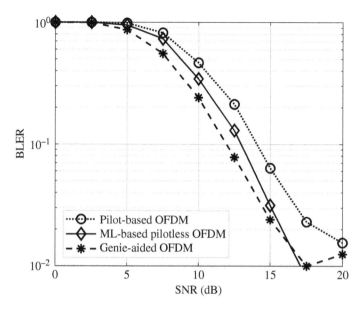

Figure 21.11 Block error rate results of a pilotless OFDM system, compared to 5G baseline.

correct for the amplitude and phase mismatches caused by the channel. The main benefit is the increased spectral efficiency, as all the resource elements (REs) can be used for data transmission.

To demonstrate the effectiveness of this approach, Figure 21.11 shows the block error rates (BLERs) of the pilotless OFDM scheme, compared to a pilot-based approach and a genie-aided approach with perfect channel knowledge. It can be observed that despite not having any pilots to aid detection, the learned scheme with DeepRx achieves very competitive BLER. Since no overhead is reserved for pilots, this translates to increased spectral efficiency. In these example results, the baseline approach utilized 2 DMRS symbols per slot of 14 OFDM symbols, which means that the efficiency of data allocation increased by nearly 17% with the pilotless approach. Moreover, the pilotless DeepRx approach can reach the BLER of 10% with approximately 1 dB lower SNR.

This concept can also be extended for learning more advanced waveforms with additional desired properties, such as better resilience against PA-induced nonlinear distortion. For instance, as shown in [18], it is possible to learn a waveform that produces considerably less out-of-band emissions under a nonlinear PA, in addition to communicating without pilots. By introducing a neural network-based time-domain transform to the transmit signal, and training it under a randomized PA model, the adjacent channel leakage ratio (ACLR) was increased by [17] as much as 10 dB.

21.10 Massive MIMO, Beamforming, and DeepTx

In mMIMO systems, a large number of antennas can be used for transmitting and receiving signals. The amplitudes and phases of the individual antenna elements can

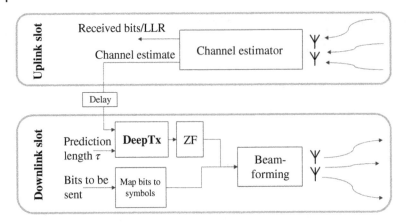

Figure 21.12 A convolutional neural network (DeepTx) can process SRS-based channel estimates to provide better input for ZF. DeepTx takes additional input specifying how long in the future (with respect to UL slot) beamforming should be predicted.

be controlled, enabling BF and multiuser MIMO (MU-MIMO) spatial multiplexing technologies. These technologies can significantly increase spectral efficiency, coverage, and radio interface capacity, but also introduce signal processing and intra-cell interference challenges. The massive MIMO technology (mMIMO) [19] was introduced in 3GPP-based mobile networks in the late stages of 4G and further improved in 5G. mMIMO is a very active research topic.

The mMIMO performance of the base station, including BF and multiuser spatial multiplexing functionalities, relies heavily on the accurate channel estimate. In particular, when technologies such as zero forcing (ZF) BF and MU-MIMO are combined, the CSI needs to be as precise, fresh, and highly granular as possible. Furthermore, a significant issue in BF is channel aging, which can be significant for mobile UEs, especially with high speed. As the radio channel estimate can only be made periodically during UL transmissions, the channel conditions of a moving UE can significantly change between UL and DL slots and result in non-optimal BF weights in DL slots that are farther from the most recent UL transmission.

ML can bring several benefits to mMIMO and BF. First, the quality of the channel estimate in the receiver can be improved in various ways using, for example, DNN or RL, for channel estimation, prediction, and compression, as was described in 1.1. In addition, ML can be applied to channel prediction and BF to tackle channel aging. For example, Huttunen et al. in [20] propose to use a CNN referred to as DeepTx, which processes the channel estimate before feeding it to ZF BF (see Figure 21.12). The CNN can modify the channel estimate to update it for predicted channel evolution and include possible other augmentations to improve ZF further. This solution significantly improves the BF quality, as seen in Figure 21.13.

In principle, as a DL transmitter a neural network beamformer does not have 3GPP specifications-related dependencies. It could also be implemented for today's transmitter but the practical implementation in real-time hardware involves several challenges.

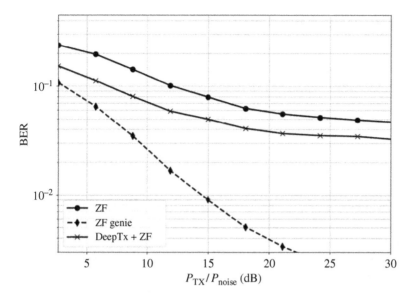

Figure 21.13 DeepTx results compared to ZF beamformer and ZF with a known channel (genie). The results are for the DL slot, which is nine slots ahead of the SRS channel estimate. Results are for 4×2 MU-MIMO.

21.11 Beam Tracking for mmWaves

Millimeter-wave (mmWave) frequencies are attractive due to their large bandwidth. However, mmWaves tend to have high penetration loss generating challenges in providing radio coverage. This challenge can be mitigated using large antenna arrays with hybrid BF architecture. In mmWave band 5G systems, beam management is often based on the analog beam domain and designed to be compatible with a hierarchical beam searching structure.

The base station periodically sweeps a predefined set of beams, UEs report back the received signal power, and BTS selects the best beam for each UE based on the measurement reports. Hence this type of BF does not require a channel estimate. As the antenna array size grows, the beams become increasingly narrow; consistently selecting the best SSB or CSI-RS beams for a mobile user can become problematic.

Maggi et al. in [21] discuss how Bayesian optimization can be used to perform effective beam tracking and help maintain the best beam for mobile users.

21.12 Channel Coding

Channel coding is a method that adds redundant information to a digital signal before it is passed onto the modulator and transmitted over the radio channel. The added redundancy allows the detection and correction of errors due to noise, interference, and non-

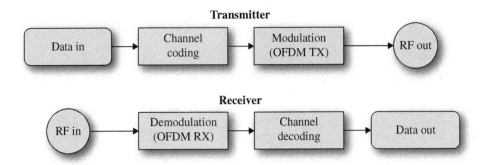

Figure 21.14 Channel coding in OFDM-system.

linearity in the radio channel. 5G uses two algorithms for coding: Polar coding [22] and Low-Density Parity-Check (LDPC) [23]. Channel coding occurs before OFDM modulation in the transmitting end and decoding after the OFDM receiver in the receiving end, as shown in Figure 21.14.

The channel coding can be optimized by using ML. One option is to apply it simply on the transmitting end, where the algorithm can learn the characteristics of the radio channel and then select the optimal coding scheme accordingly.

The second option is to implement it in the receiving end, where it can independently learn to improve the decoding performance based on, for example, the radio channel estimate. As far as the writer found, the use of DL for channel coding was first proposed in [24].

Another option is to use autoencoders and Variational Auto Encoders (VAE) for designing channel codes, e.g. [25]. This approach trains the encoder neural network to compress the input data into a lower-dimensional representation. The lower-dimensional representation is modulated and transmitted over the radio channel. The decoder neural network in the receiver is used to reconstruct the original input from the received representation. By training the encoder and decoder networks together, the autoencoder can learn a representation of the input data that is robust to noise, and errors introduced by the impairments in the channel.

While the models applied only in the transmitting or receiving end can be implemented directly on the existing systems for improved performance, the end-to-end autoencoder or VAE would require changes in 3GPP specifications and is unlikely in the scope of 5G-Advanced.

21.13 MAC Scheduler and Radio Resource Management

In a wireless 3GPP-based mobile network, the Medium Access Control (MAC) scheduler allocates radio resources in the cell within a carrier and between any aggregated carriers. The scheduler makes highly complex decisions; its decision-making process is critical in ensuring that network radio resources are used efficiently and that each data stream's Quality and Service (QoS) requirements can be met. The scheduler relies on heuristically optimized algorithms that consume significant computational resources to complete the tasks for each UE, with data awaiting scheduling in the uplink or downlink buffers. An ML-based approach can reduce the computational requirements, provide better spectral efficiency, and significantly reduce the heuristic algorithms' development requirements.

A massive MIMO scheduler functionality involves making decisions as follows:

- Time Domain (TD) decisions which shortlist mobiles to schedule candidates for which resources are allocated in the next time slot.
- Spatial Domain (SD) decisions about which of these scheduling candidates can be spatially multiplexed, and in some cases, how to most efficiently select beams between those UEs.
- Frequency Domain (FD) decisions about which resource block groups (RBGs) to allocate.
- In the case of Carrier Aggregation (CA), the combinations of different frequency bands in each time slot for each UE.

The decision-making inputs include QoS requirements of each data stream, current load in the cell, radio channel conditions between the base station, and transmit power reserve. Various Radio Resource Management (RRM) algorithms are closely linked with the scheduler, such as Power Control (PC) and Link Adaptation (LA), impacting the scheduler's decisions. To further complicate the decision-making process, 3GPP specifications contain strict rules on when specific resources can be scheduled, e.g. discontinued reception (DRX) and retransmission of the Hybrid Automatic Repeat Request (HARQ) processes.

Many approaches can be used for optimizing the allocated resources with learned models, either by applying ML algorithms to individual processes, or by combining several functions under one model. For example, a single DL model could combine and calculate the PC and LA algorithms for more efficient power and modulation, and coding scheme selections.

Other examples include using reinforcement learning (RL) algorithms for utilizing historical data. Such a system can learn patterns in user behavior or cell load and predict when CA cells or individual PAs can be switched off to save energy without impacting the end-user experience in the cell.

The knowledge of the application can help the network evaluate the resource requirements and optimize the resource allocation for better end-user experiences and increased spectral efficiency. Observing the end user behavior at the milli-second level, for example by monitoring the arrival of the data packet's volume and periodicity into the scheduling buffers, can also help estimate the application. Such a model can learn to identify application types and allocate low scheduling priority to applications that aren't interactive, e.g. firmware upgrade or downloading a new email, while providing high priority to voice calls and video.

21.13.1 Deep Scheduler

Since the most exhaustive schedulers are becoming rather complex, one major potential aspect of ML is the complexity reduction while still keeping performance at the level of the most exhaustive heuristic schedulers. This has paramount importance in real-time computing because scheduling decisions must be made with very strict time windows determined by transmission time intervals (TTIs). To demonstrate this, a dynamic packet scheduling framework (originating from [26]) was used in a realistic 5G system level simulator configured for an urban dense macro environment consisting of 21 cells with 200 intersite distance and 40 MHz bandwidth on a 3.5 GHz center frequency. The packet scheduling framework's frequency domain scheduler was replaced with a deep scheduler, i.e. a neural network trained with RL principles, as depicted in Figure 21.13. The basic principle is that a real-time scheduler makes scheduling decisions with neural network forward passes. To keep real-time computing as minimal as possible, a non-real-time trainer is collecting state, scheduling decisions, and rewarding information from the physical layer and real-time scheduler to train the actor neural network for real-time scheduler apart from real-time computing resources (Figure 21.15).

To obtain an ML-based frequency selective downlink deep scheduler, the expert knowledge of proportional fair scheduler was used to train Double Deep Q Network (DDQN) [27]. Input values for the model are similar and are used to calculate scheduling metrics and optimizations with heuristic schedulers. In this example, normalized values of sub-band CQI, wideband CQI, number of already allocated RBs, past average throughput, and PDU size in the transmission buffer were used. Once input data is propagated through the neural network, the output layer provides likelihoods for each scheduling candidate for being the best selection for the RBG being scheduled.

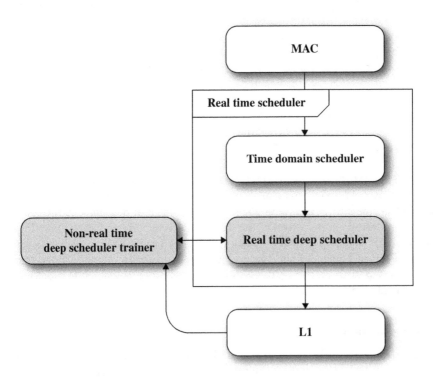

Figure 21.15 Deep scheduler framework. Real-time scheduling decisions are made with neural network forward passes. Online RL-based training can be performed from replay memory without stressing real-time computing resources.

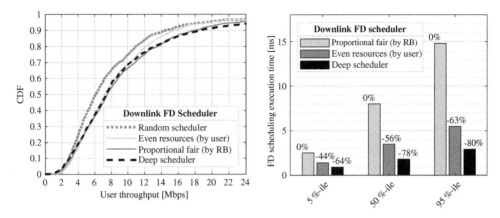

Figure 21.16 Downlink user throughput distribution and scheduler execution times in system-level simulation environment. The deep scheduler can be trained to perform similarly to the most exhaustive frequency selective scheduler. The most significant benefit comes from the real-time computing complexity reduction.

As shown in Figure 21.16, once trained, deep scheduler can provide ~21% gain compared to the random scheduler and has a similar performance to its rather exhaustive expert trainer. Scheduler complexities are compared by measuring execution times during the simulations. This gives a relatively good idea of the complexity because all schedulers within this example, including deep scheduler, are implemented with C++ only and executed with the same CPU resources. Because every scheduling decision boils down to a single actor neural network forward pass (i.e. a couple of matrix operations per neural network layer), deep scheduler can outperform heuristic frequency selective schedulers with a clear margin. The difference in execution time comes from the vast amount of optimization loops and physics-based calculations required for throughput estimations used for scheduling metric calculations, and sorting scheduling candidates either by the user or even additionally for each RB separately. Hence, sub-band CSI from each scheduling candidate is converted into throughput estimations with modulation and coding scheme (MCS) reselections and physics-based calculations for each available resource.

21.13.2 Deep Scheduler for 5G Uplink Waveform Options

The heterogeneity of transmitter devices, relatively limited transmit power budgets requiring power spectral density optimization, and a more diverse interference environment make uplink scheduling a rather complex problem, even for the most exhaustive and complex man-made heuristic schedulers. Hence, it is an appealing problem to be solved by ML algorithms. Moreover, ideally, an ML-based scheduler should be trainable for both 5G uplink waveform options. While downlink 5G uses CP-OFDM, for the uplink, there is also another alternative, DFT-s-OFDM, which is beneficial for scenarios with limited link budgets. Since DFT-s-OFDM is a single carrier waveform, the deep scheduler has to be able to learn to make continuous RB allocations as well. Since the deep scheduler does the scheduling of RB by RB for each MIMO layer, one way to train

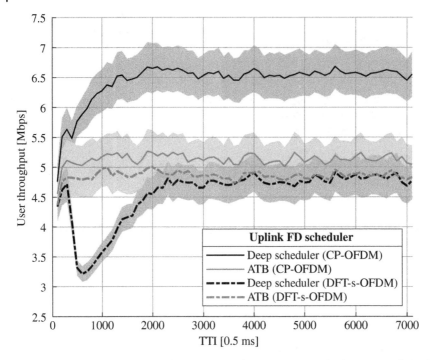

Figure 21.17 Windowed mean uplink user throughput evolution at the beginning of simulations (100 OFDM symbol averaging window). Background fills depict a 95% confidence interval.

it to keep allocations continuous in the frequency domain is to mask neural network inputs and outputs for UE selections for which RB allocation continuity has become broken. Because of random exploration and the fact that it takes some time for the deep scheduler to learn to keep allocations, continuous portion of RBs will be unnecessarily unscheduled at the beginning of the training. Thus, performance dips for a short period once training kicks in at the beginning of the simulations, as can be observed for DFT-s-OFDM from Figure 21.17, where the deep scheduler is benchmarked against the adaptive transmission bandwidth (ATB) uplink scheduler [28]. Once trained with samples collected to replay memory during the first 50 000 simulation steps, equal to ~3570 subframes with 30 kHz subcarrier spacing, deep scheduler can match the performance of the heuristic scheduler generating continuous allocations that can be allocated with 5G NR Type 1 allocation, i.e. start RB and number of RBs within DCI. With CP-OFDM uplink, on the other hand, the deep scheduler can truly shine and immediately (during the first throughput averaging windows) get on with learning more spectral efficient 5G NR Type 0 allocations, where allocations are provided with a bitmap in DCI. To achieve this, deep scheduler's decisions are rewarded based on the number of bytes per allocated RB it can deliver through its scheduling decisions. It should be noted that a small "learning curve" visible also for the ATB scheduler during the very first simulation steps is caused by the convergence of various heuristic algorithms, such as link adaptation, modeled in the simulator.

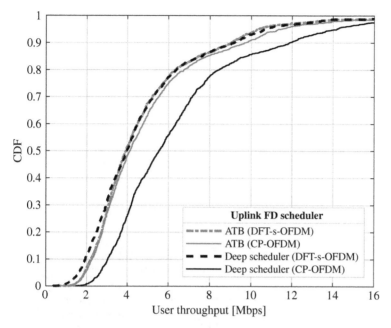

Figure 21.18 Uplink user throughput distribution. Due to more flexible frequency allocations, CP-OFDM can outperform DFT-s-OFDM in this simulation scenario. Especially, the deep scheduler is able to learn to exploit this freedom in scheduling decisions.

The deep scheduler's capability of optimizing frequency selective scheduling for uplink with CP-OFDM can be seen more clearly in the uplink user throughput distribution given in Figure 21.18. The power of ML in this scenario lies in the capability of learning scheduling decision patterns and their consequences that especially help cell edge UEs. This leads to reduced retransmissions and more efficient spectrum usage, leaving even more resources for UEs in better radio conditions. Therefore, improvement can be seen throughout the user throughput distribution.

21.13.3   Deep Scheduler for MU-MIMO

When a spatial domain is added on top of frequency selective uplink scheduling, the complexity rises to another level, especially for heuristic schedulers. In simulations, we can afford throughput estimations and exhaustive throughput estimation calculations and checks to ensure that every scheduling decision for the new MIMO layer increases throughput. For example, the Greedy spatial domain scheduler in Figure 21.19 ensures that each addition in the spatial domain will increase the sum throughput for scheduled resources. However, such exhaustiveness cannot be afforded with real-time computing when frequency domain scheduling and MU-MIMO user pairing are done dynamically for each resource in every TTI. Especially for this issue, ML can provide an answer. As illustrated in Figure 21.19, the deep scheduler can be trained within ~3570 TTI

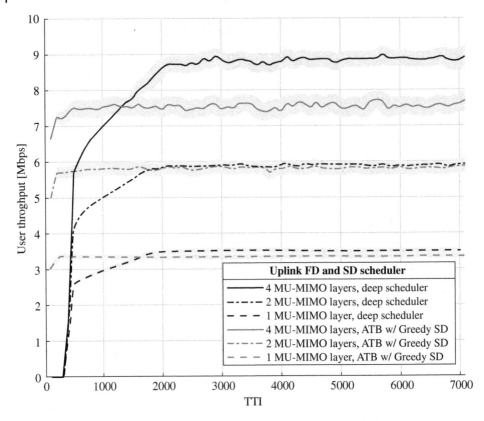

Figure 21.19 Windowed mean uplink user throughput evolution at the beginning of simulations (100 OFDM symbol averaging window). Background fills depict a 95% confidence interval.

simulation warm-up period for additional MU-MIMO layers by providing additional input parameters hinting at the spatial correlation of scheduling candidates and candidates already scheduled for previous MU-MIMO layers. Simulations were carried out with 10 MHz bandwidth on a 4 GHz center frequency. For receiving BF, ZF was used to provide up to four MU-MIMO layers.

Figure 21.20 illustrates user throughput distribution comparisons between the heuristic uplink scheduler combination and deep scheduler. Due to receiving BF's natural capability to reduce interference, ML cannot find much more optimal frequency domain scheduling patterns without UE pairing. Hence, ML provides only a marginal gain with a single beam. However, when MU-MIMO layers are added, deep scheduler can find a more optimal frequency domain and, more importantly, better spatial domain UE pairing strategies compared to heuristic scheduling algorithms basing their decisions on instant throughput estimations.

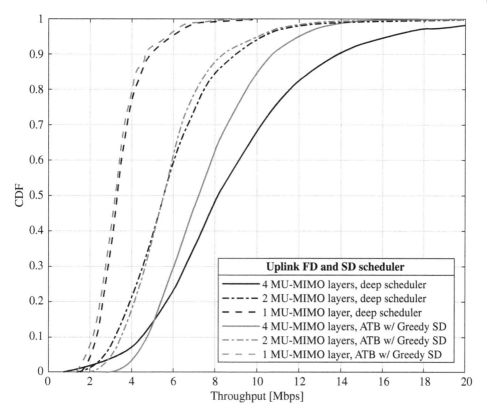

Figure 21.20 Uplink user throughput distributions of different frequency and spatial domains scheduling alternatives for ZF-based MU-MIMO. The deep scheduler can select users for each frequency domain resource and pair UEs on spatial MU-MIMO layers.

21.14   Learned Communication Protocols

The telecommunications industry has traditionally relied on protocol standards to develop complex signaling schemes and procedures for wireless networks. However, this process is costly, time-consuming, and can result in ambiguous technical specifications. As a result, there is growing interest in automating this process using ML techniques.

One promising approach is to use deep multiagent RL to train wireless devices to learn communication protocols [29]. Protocol implementations trained this way may outperform expert systems due to customized signaling and channel access policy. This approach could replace protocol interpretation, implementation, and testing efforts, resulting in significant cost savings and reduced time-to-market for 5G-Advanced and 6G network development.

Moreover, training protocols through supervised learning and self-play, rather than being limited to human-designed ones, may lead to harder-to-interpret protocols but could also result in highly tailored radio systems, improving 5G-Advanced and 6G capacities. However, it is essential to develop metrics that measure the difference between the two protocols, as this can improve intelligibility and facilitate fault detection and performance monitoring.

AI-driven protocol learning could revolutionize the telecommunications industry by reducing future signaling overheads and development efforts for complex radio technologies. In a 5G-Advanced and 6G future, parts of the radio stack development cycle could be replaced by the click of a button, allowing for highly customized radio systems tailored to their deployment environment and niche markets.

21.15 Network Planning and Optimization

21.15.1 Radio Network Planning

Radio network planning involves designing the network and optimizing the placement of new base stations and wireless access points. An ML-based system can use traffic patterns, geological locations of traffic generation, radio channel conditions, and QoS requirements to determine the optimal cell locations for coverage, capacity, and energy efficiency.

21.15.2 Network Optimization

Radio network optimization [30–32] involves adjusting various Radio Resource Management (RRM) parameters and physical cell configurations to achieve better performance or improve radio coverage. Physical configuration here refers to properties such as antenna azimuths, downtilts, and cell radiation power.

Network optimization is a complex and challenging task, as it is essential to understand practical radio propagation in various environments, interference generation, and the algorithms implemented in a BTS. Secondly, the cells can interact strongly with each other. Parameter modifications can impact the performance of the optimized cell and often also the performance of nearby neighboring cells. Downtilting an antenna of a cell provides an example of strong interaction; the action may boost the cell's spectral efficiency, but it may either improve or degrade the performance in the neighboring cells. Reduced interference towards neighboring cells would improve the overall spectral efficiency as the coverage overlap between the neighboring cells would reduce. On the other hand, if the antenna downtilt is too large, mobility issues and coverage holes between the cells could be introduced. Thus, changing parameters in one cell can significantly affect the surrounding network, making the network optimization task a distributed decision-making problem requiring a specific kind of ML approach.

An ML-based model could operate independently in each cell with the objective of improving network performance, rather than focusing solely on the performance of the cell it runs on. A distributed approach based on reinforcement learning (RL) can be suitable for solving network optimization problems and automating the process. In this approach, each cell requires an RL agent to operate with its objective function to improve the overall network. A possible solution for implementing such a system would be to run an RL agent locally but co-operatively in each cell. Such a distributed solution would require a new interface for providing observations from other cells to the RL agent.

In an alternative solution, a powerful server could be used for near real-time data collection and running the individual RL agents, or after reformulating the problem, running a single agent with a much larger number of inputs. Open Radio Access Network (O-RAN) includes specifications for RAN Intelligent Controller (RIC) [33], which would be a suitable centralized platform for many network's optimization-related algorithms, data collection, and model training. O-RAN architecture is discussed in more detail in Chapter 7.

21.15.3 Capacity Management

The traffic trends are usually smooth at the network level, so making capacity predictions using simple linear regression is straightforward. However, at the cell level, the traffic variations are significant and often seemingly random; the projection of capacity upgrade requirements at the microscopic level can be complex. Capacity management and optimization are discussed extensively in Ref. [31].

When a hardware upgrade is needed for additional radio capacity, a site redesign, larger transport capacity, or even a more substantial tower may be required to accommodate new antennas and radio units. The process can be lengthy and can even last for several months. Hence, the upgrade process must be started well before the congestion in a cell reaches to maintain committed service levels. ML algorithms can help predict when a certain predefined service level in the cell cannot be guaranteed any longer, helping the operator trigger timely upgrades.

21.15.4 Mass Event Management

Effective network management during mass events aims to ensure access to network services and prioritize emergency traffic, often at the cost of spectral efficiency. Significant events such as music festivals, sporting events, traffic incidents, cyclones, political rallies, or earthquakes, can lead to heavy but highly localized network congestion, resulting in dropped calls, network accessibility issues, and other service degradation. While the demand greatly exceeds the network capacity, it is crucial to adjust the network settings so that as many users as possible can access the network and have some level of service while prioritizing any emergency traffic.

In some cases, the events are predictable, making it easier to prepare for an event, set parameters, and perhaps provide additional capacity to the venue well in advance. In other cases, the event may be very unpredictable and possibly in a location difficult to access.

An ML-based model can be trained with data collected during earlier mass events for early detection. The detection can automatically apply mass event-specific parameter sets and alert network operations. It can reduce the need for the operator to prepare and plan. This is particularly useful with unpredictable events like earthquakes, which may knock parts of the network down and amplify network congestion. Triggering additional actions is possible, for example, for sending additional capacity by flying drones with cells to the location.

21.16   Network Operations

ML can enhance the efficiency and cost-effectiveness of network operations by automating tasks. For instance, predictive maintenance can leverage historical data and patterns to predict equipment failures and reduce downtime and maintenance costs. Autonomous cell configuration can use ML algorithms to plan and configure cell parameters automatically when a new cell is introduced in the network, reducing the need for manual configuration and potential errors. Self-healing systems can detect and troubleshoot network problems automatically, improving network uptime and reducing the need for manual intervention. Lastly, ML-based alarm monitoring can identify anomalies and root causes of issues, enabling faster and more accurate responses.

21.17   Network Security

While several ML techniques can be deployed to build effective intrusion detection systems, outlier detection, also called anomaly detection, is an ML technique often used to identify patterns in data that do not conform to expected behavior. Anomalies can indicate unusual events, errors, faults, or malicious activities in various domains. Typically unsupervised ML methods, such as clustering, are used for grouping similar traffic, system log events, and user behavior. Outliers or anomalies in these groups can indicate the presence of malicious or compromised network elements or suspicious end-user behavior. In this section, a few network security-related examples are discussed.

Anomaly detection can identify unusual patterns in network traffic, such as unauthorized attempts to access the transport network. By detecting these anomalies, intrusion prevention systems can be implemented to safeguard the network and maintain its security and integrity.

A malicious or misconfigured UE can cause accessibility issues for legitimate users. Examples include bombarding a base station with an excessive number of preambles or radio resource control (RRC) setup requests. Anomaly detection can help identify such abnormal behaviors, enabling network operators to take corrective actions to maintain network accessibility.

Wireless networks can be vulnerable to Distributed Denial of Service (DDoS) attacks, where multiple devices flood the network with traffic to overwhelm it and disrupt service for legitimate users. Anomaly detection can help identify sudden surges in network traffic that may indicate a DDoS attack in progress, allowing for timely mitigation.

Rogue base stations or International Mobile Subscriber Identity (IMSI) catchers could be used to intercept or manipulate wireless communications. Anomaly detection can identify unusual signaling patterns or unregistered base stations, indicating the presence of malicious equipment in the network.

Anomaly detection can help identify unusual data transfer patterns, such as large volumes of data being sent to IP addresses, which may indicate data exfiltration or other forms of information leakage, or simply unfair use of the subscription.

21.18   Positioning

Positioning is used in various applications such as indoor navigation, asset tracking, and emergency response. While multiple methods can be used for estimating the UE location, Channel Impulse Response (CIR)-based fingerprinting is a promising method for indoor and urban environments, where GNSS signals may be obstructed or are less accurate. ML methods can improve CIR-based fingerprinting techniques' performance by creating and recognizing location-specific fingerprints. This section discusses the steps of implementing an ML-based CIR fingerprinting solution.

Creating a location estimation system based on CIR fingerprinting involves several steps. The first step is to collect CIR measurements at various reference locations throughout the environment. These measurements can be taken using specialized equipment, such as a network analyzer or a software-defined radio.

Once the CIR measurements have been collected, relevant features such as path delays, amplitudes, and phases are extracted from them. These features can be further processed to reduce noise, improve the fingerprint's uniqueness, and facilitate pattern recognition. For example, dimensionality reduction techniques such as Principal Component Analysis (PCA) can be applied to reduce a number of features while retaining essential information.

The extracted features and the corresponding location information (e.g. coordinates or rooms in a building) are then stored in a fingerprint database. This database serves as the training dataset for the Machine Learning (ML) model. A suitable ML algorithm, such as k-Nearest Neighbors (k-NN), Support Vector Machine (SVM), or a neural network, is trained on the fingerprint database. The model learns to map the CIR features to their respective locations, allowing for location estimation based on previously unseen CIR measurements.

The trained model is evaluated using a separate dataset of CIR measurements to determine its accuracy and performance. This evaluation can be performed using various metrics such as the mean error distance or the percentage of correct location estimations.

Once the ML model has been trained and evaluated, it can be deployed in the target environment to estimate the location of devices based on their CIR measurements. When a device's CIR is captured, the ML model processes the data, matches it to the most similar fingerprints in the database, and predicts the device's location accordingly.

However, over time, the environment may change due to the movement of objects, and alterations in building structures, leading to changes in radio signal propagation characteristics and radio channel fingerprints. The ML model should be periodically updated with new CIR measurements to maintain accuracy and adapt to these changes. This is known as fingerprinting maintenance.

The CIR-based fingerprinting method can be combined with other localization techniques to further improve the location estimation accuracy. Combining a CIR fingerprint with a Received Signal Strength Indicator (RSSI) or Time of Arrival (ToA) can further enhance the robustness and precision of a positioning system.

21.19   Challenges

Even though AI and ML have emerged as powerful tools to address wireless systems' complexities and challenges, integration into wireless systems is not without challenges. This chapter will discuss ML difficulties in wireless systems and explore potential solutions.

21.20   Scalability

As the size of networks and data for ML algorithms continue to grow, traditional computing architectures can struggle to scale effectively. The challenges of large models, time-critical inference, memory limitations, power consumption, and the dynamic number of users sharing the capacity require scalable solutions.

Although scaling can present challenges, it is worth noting that often ML models can be pruned and quantized to accommodate limited RAM and lower computational power at the cost of model accuracy. The hardware typically sets the limit for computational resources, which must be shared among the users utilizing the network element in question. One exciting solution is to allow dynamically reduced accuracy during peak load or while waiting for a hardware upgrade.

21.21   Uncertainty

In the wireless communications industry, uncertainty is a common challenge that ML algorithms must overcome to make well-informed decisions. This uncertainty can arise from various sources, including insufficient data, unexpected measurement samples, interference, and dynamic environmental conditions.

Approaches like Bayesian optimization [8] could be used to address the challenges. Bayesian method provides a probabilistic framework for integrating prior knowledge and updating beliefs based on incoming data, enabling more robust and accurate decision-making in uncertain environments. Additionally, incorporating domain-specific constraints into the learning process can be implemented. These constraints can come from various sources, such as the physical properties of the wireless medium, network protocols, or insights gained through empirical testing. By incorporating these constraints into the learning process, algorithms can make more informed decisions, even in the face of significant uncertainty. This can lead to more efficient and effective wireless systems, which can help address the growing demand for high-speed, reliable wireless communication in various applications and industries.

21.22   Time Criticality and Computational Requirements

Specialized hardware may be required for time-sensitive ML applications, especially for DNNs. In OFDM-based wireless systems, numerous functional blocks or entities are subject to stringent time constraints for inference, with some use cases demanding time budgets as low as microseconds. Although not universally applicable, ML algorithms can be more computationally intensive than heuristic models, posing practical challenges. Furthermore, DL may necessitate substantial memory to store model weights.

Several DL architecture design techniques can help decrease computational and memory demands. Methods such as model pruning, quantization [34], and distillation [35] are examples that contribute to creating smaller, faster, and more memory-efficient models.

21.23 Standardization and Specifications Impact

Implementing ML into wireless systems is not straightforward. It involves a tremendous amount of consideration from the system architecture point of view, so that required functions for various use cases can be implemented. Aspects such as data collection, model development and training, model management in the network, and model performance monitoring are still largely open from the specifications point of view when writing this book.

21.23.1 Data Collection

The foundation of AI/ML applications is data collection. While synthetic data created, e.g. by simulations, can be helpful for evaluation purposes, real-world data collected from live networks can be essential for ensuring the proper performance and robustness of ML models for many use cases in practical deployments.

This data can be used to train the AI/ML model to fit the real-world wireless environment, model performance monitoring, and decision-making in the case of model failure. The BTS or UE could collect this data upon a request initiated by any network element. Depending on the use case and implementation, the data needs to be transferred to the location of storage, which could be in the BTS or some other network elements, such as the RAN Intelligent Controller (RIC) or network management system.

Given that data collection requests may be transmitted across multiple interfaces, standardization researchers and specification bodies must consider the need for signaling enhancements. Inter-vendor interfaces, such as the radio interface, can be particularly challenging.

21.23.2 Model Development

Developing and training ML models involves several complex tasks, such as finding suitable model architecture and training and optimizing the model. Further work is often required to reduce the size to fit the hardware memory and minimize the latency by quantizing and pruning the model. Finally, the model needs to be compiled for the target hardware. When writing this book, these development processes required offline engineering. However, the research in this area is very active, and these steps may be automated in the future, and the process may operate online.

The specification work needs attention to support the offline model development process. Still, the online development process also requires consideration for use cases where model training is performed outside the target network element.

Particularly complex are the cases where auto-encoder-type models are used. For example, the coder could be physically located in the UE and the encoder in the BTS; hence, inter-vendor scenarios in auto-encoder implementation should be considered.

21.23.3 Model Performance Monitoring

Performance monitoring is crucial for managing the life cycle of AI/ML models. Monitoring is required to assess the need for model retraining, switching to another model, or identifying model failures.

Several key performance indicators (KPIs) can be used to assess the model inference, such as various error metrics, BLER, or system throughput. The model monitoring can be conducted at the element where the model is operating or outside, e.g. at RIC or SON. Monitoring could occur on either side for use cases where an auto-encoder-type solution operates across an open interface.

One particularly complex scenario worth highlighting is the monitoring and observation of what is referred to as model drift. This situation can arise, for instance, when there are changes in end-user behavior since the model was last trained, resulting in the model no longer being optimal. A performance comparison between the learned model and baseline performance would be necessary to identify such a situation.

21.23.4 Model Transfer

There can be several methods for delivering AI/ML models to network elements and choosing an appropriate strategy would depend on the specific use case. The delivery of an AI/ML model may involve providing either the parameters of a model structure already known at the receiving end, or a completely new model with parameters. For instance, a new model could be delivered through a scheduled firmware update to the base station (BTS), transmitted over the user plane to the user equipment (UE), or sent over a 3GPP standardized interface with dedicated signaling.

As with previous scenarios, standardization considerations are essential to delivering the model weights or entire models to support different use cases and implementations. New signaling procedures may be needed to trigger model retraining or switching actions.

21.24 Summary

In this chapter, an in-depth exploration of ML's role in wireless networks, emphasizing 5G-Advanced, was carried out. Firstly, the reasons underpinning the integration of ML in these systems were outlined. A discussion on training and inference of ML models within wireless systems followed this.

A broad range of ML applications for 5G wireless systems was discussed, including signal linearization, improvement of CSI, and channel prediction. The significance of DNN-based receivers, notably DeepRx, was highlighted, as were innovative approaches such as pilotless OFDM and applications in massive MIMO, BF, beam tracking for mmWaves, and channel coding and RRM. The effects on network planning and optimization were assessed, encompassing areas such as radio network planning, network optimization, capacity management, and mass event management. Additional consideration was given to the potential influence of ML on network operations, network security, and positioning.

While some topics were explored in detail, providing an in-depth understanding of technical aspects, others were briefly introduced, directing readers to published papers for more in-depth exploration. The chapter acknowledged the dynamic nature of the field, with new use cases emerging regularly.

The chapter concluded by addressing the challenges tied to these advancements. Issues such as scalability, uncertainty, time criticality, computational requirements, and the impact of standardization and specifications were introduced, providing an understanding of the obstacles to be overcome.

References

1 Cox, D.R. (1958). The regression analysis of binary sequences. *Journal of the Royal Statistical Society: Series B: Methodological* 20 (2): 215–232. https://doi.org/10.1111/j .2517-6161.1958.tb00292.x.

2 Galton, F. (1894). *Natural Inheritance, by Francis Galton.* London: Macmillan and co. https://doi.org/10.5962/bhl.title.46339.

3 Cortes, C. and Vapnik, V. (1995). Support-vector networks. *Machine Learning* 20 (3): 273–297. https://doi.org/10.1007/BF00994018.

4 Ho, T.K. (1995). 'Random decision forests'. In: *Proceedings of 3rd International Conference on Document Analysis and Recognition,* 278–282. Montreal, Que., Canada: IEEE Comput. Soc. Press https://doi.org/10.1109/ICDAR.1995.598994.

5 LeCun, Y. et al. (1989). Backpropagation applied to handwritten zip code recognition. *Neural Computation* 1 (4): 541–551. https://doi.org/10.1162/neco.1989.1 .4.541.

6 Sutton, R.S. and Barto, A.G. (1998). *Reinforcement learning: an introduction.* In: *Adaptive Computation and Machine Learning.* Cambridge, Mass: MIT Press.

7 Hinton, G.E. and Salakhutdinov, R.R. (2006). Reducing the dimensionality of data with neural networks. *Science* 313 (5786): 504–507. https://doi.org/10.1126/science .1127647.

8 Mockus, J. (1989). *Bayesian approach to global optimization: theory and applications.* In: *Mathematics and Its Applications.* Soviet series. Dordrecht ; Boston: Kluwer Academic.

9 Mkadem, F. and Boumaiza, S. (2011). Physically inspired neural network model for RF power amplifier behavioral modeling and digital predistortion. *IEEE Transactions on Microwave Theory and Techniques* 59 (4): 913–923. https://doi.org/10.1109/ TMTT.2010.2098041.

10 Pihlajasalo, J. et al. (Sep. 2021). HybridDeepRx: Deep Learning Receiver for High-EVM Signals. In: *2021 IEEE 32nd Annual International Symposium on Personal, Indoor and Mobile Radio Communications (PIMRC),* 622–627. Helsinki, Finland: IEEE https://doi.org/10.1109/PIMRC50174.2021.9569393.

11 Pihlajasalo, J. et al. (2023). Deep Learning OFDM Receivers for Improved Power Efficiency and Coverage. *IEEE Transactions on Wireless Communications* 1–1. https://doi.org/10.1109/TWC.2023.3235059.

12 RP-213599, 'New SI: Study on Artificial Intelligence (AI)/Machine Learning (ML) for NR Air Interface'. 3GPP RAN Plenary #94e, Dec 2021.

13 Ye, H., Li, G.Y., and Juang, B.-H. (2018). Power of deep learning for channel estimation and signal detection in OFDM systems. *IEEE Wireless Communications Letters* 7 (1): 114–117. https://doi.org/10.1109/LWC.2017.2757490.

14 J. Long, E. Shelhamer, and T. Darrell, 'Fully convolutional networks for semantic segmentation', 2014, https://doi.org/10.48550/ARXIV.1411.4038.

15 Honkala, M., Korpi, D., and Huttunen, J.M.J. (2021). DeepRx: fully convolutional deep learning receiver. *IEEE Transactions on Wireless Communications* 20 (6): 3925–3940. https://doi.org/10.1109/TWC.2021.3054520.

16 Korpi, D., Honkala, M., Huttunen, J.M.J., and Starck, V. (Jun. 2021). DeepRx MIMO: convolutional MIMO detection with learned multiplicative transformations. In: *ICC 2021 – IEEE International Conference on Communications*, 1–7. Montreal, QC, Canada: IEEE https://doi.org/10.1109/ICC42927.2021.9500518.

17 Aoudia, F.A. and Hoydis, J. (2021). Trimming the fat from OFDM: pilot- and CP-less communication with end-to-end learning. In: *2021 IEEE International Conference on Communications Workshops (ICC Workshops)*, 1–6. Montreal, QC, Canada: IEEE https://doi.org/10.1109/ICCWorkshops50388.2021.9473605.

18 Farhadi, H. et al. (2023). Deliverable D4.3: AI-driven communication & computation co-design: final solutions. *Hexa-X, Apr.* 30: [Online]. Available: https://hexa-x.eu/wp-content/uploads/2023/05/Hexa-X_D4.3_v1.0.pdf.

19 Marzetta, T.L. (2015). Massive MIMO: an introduction. *Bell Labs Technical Journal* 20: 11–22. https://doi.org/10.15325/BLTJ.2015.2407793.

20 J. M. J. Huttunen, D. Korpi, and M. Honkala, '*DeepTx: Deep Learning Beamforming with Channel Prediction*', 2022, https://doi.org/10.48550/ARXIV.2202.07998.

21 L. Maggi, R. Koblitz, Q. Zhu, and M. Andrews, '*Tracking the Best Beam for a Mobile User via Bayesian Optimization*', 2023, https://doi.org/10.48550/ARXIV.2303.17301.

22 Arikan, E. (2009). Channel polarization: a method for constructing capacity-achieving codes for symmetric binary-input memoryless channels. *IEEE Transactions on Information Theory* 55 (7): 3051–3073. https://doi.org/10.1109/TIT.2009.2021379.

23 Gallager, R. (1962). Low-density parity-check codes. *IRE Transactions on Information Theory* 8 (1): 21–28.

24 E. Nachmani, Y. Beery, and D. Burshtein, '*Learning to Decode Linear Codes Using Deep Learning*', 2016, https://doi.org/10.48550/ARXIV.1607.04793.

25 O'Shea, T.J., Karra, K., and Clancy, T.C. (Dec. 2016). Learning to communicate: channel auto-encoders, domain specific regularizers, and attention. In: *2016 IEEE International Symposium on Signal Processing and Information Technology (ISSPIT)*, 223–228. Limassol, Cyprus: IEEE https://doi.org/10.1109/ISSPIT.2016.7886039.

26 Kela, P., Puttonen, J., Kolehmainen, N. et al. (May 2008). Dynamic packet scheduling performance in UTRA Long Term Evolution downlink. In: *in 2008 3rd International Symposium on Wireless Pervasive Computing*, 308–313. Santorini, Greece: IEEE https://doi.org/10.1109/ISWPC.2008.4556220.

27 Van Hasselt, H., Guez, A., and Silver, D. (2016). Deep Reinforcement Learning with Double Q-Learning. *Proceedings of the AAAI Conference on Artificial Intelligence* 30 (1): https://doi.org/10.1609/aaai.v30i1.10295.

28 Calabrese, F.D., Rosa, C., Anas, M. et al. (Sep. 2008). Adaptive Transmission Bandwidth Based Packet Scheduling for LTE Uplink. In: *in 2008 IEEE 68th Vehicular Technology Conference*, 1–5. Calgary, Canada: IEEE https://doi.org/10.1109/VETECF .2008.316.

29 J. Hoydis, F. A. Aoudia, A. Valcarce, and H. Viswanathan, 'Toward a 6G AI-Native Air Interface', *ArXiv201208285 Cs Math*, Apr. 2021, Accessed: May 07, 2021. [Online]. Available: http://arxiv.org/abs/2012.08285

30 Reunanen, J., Salo, J., and Luostari, R. (2015). 'LTE Key Performance Indicator Optimization'. In: *LTE Small Cell Optimization* (ed. H. Holma, A. Toskala, and J. Reunanen), 195–248. Chichester, UK: John Wiley & Sons Ltd https://doi.org/10 .1002/9781118912560.ch12.

31 Reunanen, J., Luostari, R., and Holma, H. (2015). Capacity Optimization. In: *LTE Small Cell Optimization* (ed. H. Holma, A. Toskala, and J. Reunanen), 249–292. Chichester, UK: John Wiley & Sons Ltd https://doi.org/10.1002/9781118912560.ch13.

32 Luostari, R., Salo, J., Reunanen, J., and Holma, H. (2015). VoLTE Optimization. In: *LTE Small Cell Optimization* (ed. H. Holma, A. Toskala, and J. Reunanen), 293–331. Chichester, UK: John Wiley & Sons Ltd https://doi.org/10.1002/9781118912560.ch14.

33 Alliance, O.-R.A.N. (Mar. 2021). *O-RAN Near-RT RAN Intelligent Controller Near-RT RIC Architecture 2.00*. O-RAN Alliance [Online]. Available: https://www .o-ran.org/specifications.

34 Han, S., Mao, H., and Dally, W.J. (2015). Deep compression: compressing deep neural networks with pruning. *Trained Quantization and Huffman Coding* https://doi.org/10.48550/ARXIV.1510.00149.

35 G. Hinton, O. Vinyals, and J. Dean, '*Distilling the Knowledge in a Neural Network*', 2015, https://doi.org/10.48550/ARXIV.1503.02531.

Index

5G Technology: 3GPP Evolution to 5G-Advanced, Second Edition.
Edited by Harri Holma, Antti Toskala, and Takehiro Nakamura.
© 2024 John Wiley & Sons Ltd. Published 2024 by John Wiley & Sons Ltd.